Latin-American Seeds

Over the last few years, Latin-American seeds have gained increased importance (in part due to the increased demand for gluten-free foods). Worldwide demand for Latin-American seeds and grains has risen in high proportion. In parallel, research into seeds and grains from this region in all relevant fields has intensified. *Latin-American Seeds: Agronomic, Processing and Health Aspects* summarizes recent research into Latin-American crops regarding their agronomic and botanical characteristics, composition, structure, use, production, technology, and impact on human health.

Latin-American cultivars studied here are included in the groups of cereals, pseudo-cereals, oilseeds, and legumes that are used in a great variety of innovative and traditional foods. The main crops that are covered in this book are Latin-American maize (*Zea mays* L.), amaranth (*Amaranthus* spp), quinoa (*Chenopodium quinoa* Willd), kañiwa (*Chenopodium pallidicaule* Aellen), chia (*Salvia hispanica* L.), sacha inchi (*Plukenetia volubilis*), and legumes such as black turtle and common beans (*Phaseolus vulgaris*) and tarwi (*Lupinus mutabilis*).

Key Features:

- Contains updated information about recent research work on Latin-American crops.
- Includes a variety of Latin-American plant species that are used in a great variety of innovative and traditional foods.
- Addresses a wide range of topics related to agronomy, plant physiology, and nutritional and technological properties, processing, fractionation, and development of new products for human health.

The book also provides information about milling, fractioning, and oil extraction, and on the elaboration of various derived traditional and novel food products. The nutritional and health implications of these Latin-American crops are also widely addressed.

Food Biotechnology and Engineering

Series Editor:
Octavio Paredes-López

Volatile Compounds Formation in Specialty Beverages
Edited by Felipe Richter Reis and Caroline Mongruel Eleutério dos Santos

Native Crops in Latin America: Biochemical, Processing, and Nutraceutical Aspects
Edited by Ritva Repo-Carrasco-Valencia and Mabel C. Tomás

Starch and Starchy Food Products: Improving Human Health
Edited by Luis Arturo Bello-Perez, Jose Alvarez-Ramirez, and Sushil Dhital

Bioenhancement and Fortification of Foods for a Healthy Diet
Edited by Octavio Paredes-López, Oleksandr Shevchenko, Viktor Stabnikov, and Volodymyr Ivanov

Latin-American Seeds: Agronomic, Processing and Health Aspects
Edited by Claudia Monika Haros, María Reguera, Norma Sammán, and Octavio Paredes-López

For more information please visit https://www.routledge.com/Food-Biotechnology-and-Engineering/book-series/CRCFOOBIOENG

Latin-American Seeds
Agronomic, Processing and
Health Aspects

Edited by
Claudia M. Haros, María Reguera,
Norma Sammán, and Octavio Paredes-López

CRC Press
Taylor & Francis Group
Boca Raton London New York

CRC Press is an imprint of the
Taylor & Francis Group, an **informa** business

First edition published 2023
by CRC Press
6000 Broken Sound Parkway NW, Suite 300, Boca Raton, FL 33487-2742

and by CRC Press
4 Park Square, Milton Park, Abingdon, Oxon, OX14 4RN

CRC Press is an imprint of Taylor & Francis Group, LLC

Library of Congress Cataloging-in-Publication Data
Names: Haros, Claudia Monika, 1966– editor.
Title: Latin-American seeds : agronomic, processing and health aspects /
Claudia Monika Haros, María Reguera, Norma Sammán, and Octavio Paredes-López.
Description: First edition. | Boca Raton, FL : CRC Press, 2023. |
Series: Food biotechnology and engineering |
Includes bibliographical references and index.
Identifiers: LCCN 2022040886 (print) | LCCN 2022040887 (ebook) |
ISBN 9780367531454 (hbk) | ISBN 9781032367392 (pbk) |
ISBN 9781003088424 (ebk)
Subjects: LCSH: Crops–Latin America. | Crops–Health aspects–Latin America.
Classification: LCC SB99.L29 L38 2023 (print) |
LCC SB99.L29 (ebook) | DDC 338.1098–dc23/eng/20221118
LC record available at https://lccn.loc.gov/2022040886
LC ebook record available at https://lccn.loc.gov/2022040887

ISBN: 9780367531454 (hbk)
ISBN: 9781032367392 (pbk)
ISBN: 9781003088424 (ebk)

DOI: 10.1201/9781003088424

Typeset in Kepler Std
by Newgen Publishing UK

Contents

Series Preface

BIOTECHNOLOGY – OUTSTANDING FACTS

The beginning of agriculture started about 12,000 years ago and ever since has played a key role in food production. We look to farmers to provide the food we need but at the same time, now more than ever, to farm in a manner compatible with the preservation of the essential natural resources of the earth. Additionally, besides the remarkably positive aspects that farming has had throughout history, several unintended consequences have been generated. The diversity of plants and animal species that inhabit the earth is decreasing. Intensified crop production has had undesirable effects on the environment (e.g., chemical contamination of groundwater, soil erosion, exhaustion of water reserves). If we do not improve the efficiency of crop production in the short term, we are likely to destroy the very resource base on which this production relies. Thus, the role of so-called sustainable agriculture in the developed and underdeveloped world, where farming practices are to be modified so that food production takes place in stable ecosystems, is expected to be of strategic importance in the future; but the future has already arrived.

Biotechnology of plants is a key player in these scenarios of the 21st century. Nowadays, molecular biotechnology in particular is receiving increasing attention because it has the tools of innovation for agriculture, food, chemical, and pharmaceutical industries. It provides the means to translate an understanding of, and ability to modify, plant development and reproduction into enhanced productivity of traditional and new products. Plant products from seeds, fruits, and plant components and extracts are being produced with better functional properties and longer shelf life; and they need to be assimilated into commercial agriculture to offer new options to small, and more than small, industries and finally to consumers. Within these strategies it is imperative to select crops with larger proportions of edible parts as well, thus generating less waste; it is also imperative to consider the selection and development of more environmentally friendly agriculture.

The development of research innovation for products is progressing; however, the constraints of relatively lengthy delivery times to reach market, intellectual property rights, uncertain profitability of products, consumer acceptance, and even caution and fear with which the public may view biotechnology are tempering the momentum of all but the most determined efforts. Nevertheless, it appears uncontestable that the food biotechnology of plants and microbials is and will emerge as a strategic component for providing the food crops and other products required for human well-being.

FOOD BIOTECHNOLOGY AND ENGINEERING SERIES / OCTAVIO PAREDES-LÓPEZ, PH.D, SERIES EDITOR

The Food Biotechnology and Engineering Series aims to address a range of topics around the edge of food biotechnology and the food microbial world. In the case of foods, it includes agrigenomics, molecular biology, genetic engineering, metabolic aspects, omic sciences, chemistry, nutrition, medical foods and health ingredients, and processing and engineering with traditional- and innovative-based approaches. Environmental aspects and strategies for producing green foods cannot be left aside. At the world level there are foods and beverages produced by different types of microbial technologies to which this series will give attention. It will also consider genetic modifications of microbial cells to produce nutraceutical ingredients, and advances of investigation into the human microbiome as related to diets.

LATIN-AMERICAN SEEDS: AGRONOMIC, PROCESSING AND HEALTH ASPECTS. EDITORS: CLAUDIA M. HAROS, MARÍA REGUERA, NORMA SAMMÁN, AND OCTAVIO PAREDES-LÓPEZ

This book is divided into eleven chapters dealing with the agronomic characteristics of crops from the Latin-American region; seed quality from crops as influenced by genotype and environmental factors; challenges and opportunities facing Latin-American seeds; characteristics and composition of crops and their seeds; gluten-free applications in products from these crops; main components of their grains such as types of flour, oil, proteins, fibers, and starch; food uses of ancient grains; bioactive compounds, peptides, and anti-nutritional factors present in these crops; potential and outstanding contributions from these regional crops in nutrition and health at world level; components and ingredients from these crops for nutrition and nutraceutical purposes; the last chapter reports on current legislation for the rational use of Latin-American seeds, and their updating as compared to some advanced experiences in other countries from other areas of all continents.

This multi-author publication has been produced by 52 scientists and technical experts who come from many academic institutions and organizations located in Latin America: Argentina, Chile, Cuba, Ecuador, Guatemala, Mexico, and Peru; in Europe: Denmark, France, Portugal, and Spain; and in the Arab world: the United Arab Emirates.

This book is of interest to academicians, food scientists, food technologists, biotechnologists, agricultural specialists at universities, technical people in private and public organizations, and baccalaureate and graduate students.

Thanks are due to the editors, professors Norma Sammán, María Reguera, and Claudia M. Haros, and to all of the authors for their excellent contributions to this book and to Sofia Blanco Haros for her contribution towards the writing of the Alt Text. Acknowledgments are also due to the editorial staff of CRC Press, especially to Mr. Stephen Zollo and to Ms. Laura Piedrahita.

Preface

The authors and editors of this book belong to the CYTED Ia ValSe-Food Group (Iberoamerican Valuable Seeds, Ref. 119RT0567) and Chia-Link Network. This book gathers recent investigation and experience of Ia ValSe-Food Group on Latin-American valuable seeds and other crops, and provides comprehensive and up-to-date knowledge within several relevant fields of food science and agriculture.

The main crops included in this book are Latin-American maize (*Zea mays* L.), amaranth (*Amaranthus* spp), quinoa (*Chenopodium quinoa* Willd), kañiwa (*Chenopodium pallidicaule* Aellen), chia (*Salvia hispanica* L.), sacha inchi (*Plukenetia volubilis*), and legumes such as black turtle (*Phaseolus vulgaris*) and tarwi (*Lupinus mutabilis*).

These seeds are still underutilized and their cultivation is reduced but, in recent years, worldwide demand for them has been rising to a remarkable level, resulting in an increasing trend in their production. They have therefore become widely recognized for their excellent nutritional and nutraceutical values by food scientists and producers. They contain high-quality proteins, peptides with outstanding nutraceutical and medicinal properties, some polysaccharides with useful functional performance, and important quantities of minerals, vitamins, and several bioactive compounds; additionally, they are all gluten free, which makes them suitable for people suffering from various gluten intolerances or with a low preference for the consumption of this component. For these reasons, interest in Latin-American crops has increased substantially which has resulted in intensification of research projects in different geographical regions of the world.

This book summarizes the large number of recent investigations performed on these seeds and provides knowledge within several of the relevant fields of food science and agronomy, and their impact on the prevention of various chronic diseases. Along with information on their origin and on their centers of distribution, botanical characteristics, production and utilization, structure, and chemical composition, a good level of attention is also directed toward their interesting components as indicated above. This publication also includes information about milling, fractioning, and oil extraction, and on the elaboration of various derived traditional and novel food products. The nutritional and health implications of these Latin-American crops are widely addressed.

We hope that this book can contribute to raising the scientific importance of these cultivars and at the same time increasing their potential uses in high-quality human nutrition; the latter is fundamental to corresponding health and strengthening the immunological system of all consumers. In view of most of these food treasures' role as natural vaccines, their sustainable use, as classified by the United Nations, is a key element for the preservation of their genetic bio-diversity, a basic tool for the present and future of food availability of world-wide populations.

The Editors

About the Editors

Claudia M. Haros, PhD, graduated Bachelor of Chemistry from the School of Exact and Natural Sciences, University of Buenos Aires (UBA), Argentina, in 1990. She earned an MSc in Bromatology and Food Technology (1992); and an MSc in Biology Analysis (1997) from UBA. She earned her PhD in Chemistry (UBA,1999). From 1991–2003, she worked as university professor in the Organic Chemistry Department, Food Science and Technology Area of UBA. From 1991–1999 she was Research Assistant in the Cereals and Oilseeds Group, Department of Industrial Chemistry, UBA. Later, from 2000–2002 she worked in Spain as a visiting professor in the Cereal Group of the Institute of Agrochemistry and Food Technology (IATA) in Valencia. In 2003, she was a postdoc fellow at the Department of Food Microbiology, Institute of Animal Reproduction and Food Research (CENEXFOOD-EU), Polish Academy of Science, Olsztyn, Poland. From 2003–2004 she received an award for working with Prof. Sandberg of the Department of Chemical and Biological Engineering, Life Science Division, University of Chalmers, Gothenburg, Sweden. In 2005 she became an Associate Researcher for the Spanish Council for Scientific Research (CSIC) in the framework of a Ramón y Cajal Programme. Since 2008 she has been a Senior Scientist at CSIC and continues her investigation into the Cereal Group, Department of Food Science of IATA. Since the early stages of her career, she has mainly been engaged in research in the cereal science and technology field. The major theme in Dr. Haros's research is the utilization of different strategies to improve the nutritional and/or functional value of cereals, their ingredients, and products, with the emphasis on starch, minerals, and proteins. These strategies include use of different physical, biochemical or biological treatments during the cereal milling process; development of new cereal by-products by including novel ingredients; use of new starter phytase producers for regulating content and composition of lower *myo*-inositol phosphates in food with clear nutritional and health benefits. The isolation of chemical components of cereals/pseudo-cereals such as starches, proteins, fibers, or oils by different processes such as dry milling, wet milling, and cold germ pressing, among others, is one of the main topics of Dr. Haros's career. Her investigations include determination of bioaccessibility/bioavailability of minerals, glycaemic index and nutrient inputs according to Dietary Reference Intakes/Adequate Intakes of nutrients. The ultimate objective is to identify dietetic solutions and innovation to prevent diseases and to improve consumers' well-being and health. She was Leader of the International Chia-Link Network (2015–2021) (www.chialink.es/) and from 2018 has been the Leader of la ValSe-Food Group (Iberoamerican Valuable Seeds-Food, with participation of research groups/companies from 12 countries from Iberoamerica (www.cyted.org/es/valse_food). She has been involved as participant or leader in numerous national (Argentina, Chile, Spain, Poland, and Sweden), European, and international research projects related to cereals. She developed extensive activity in transferring knowledge through close cooperation with stakeholders mainly dealing with the inclusion of cereals/pseudo-cereals/ancients crops in the human diet through their characterization and utilization to develop innovative and sensory accepted cereal-based products. Today she belongs to the International Board of the Latam Food Innovation Hub in pursuing diffusion of her research activities in a business forum (www.latamfo odinnovationhub.com/).

María Reguera, PhD, graduated Bachelor of Environmental Sciences and completed her PhD in Plant Biology at the *Universidad Autónoma de Madrid* (UAM, Madrid, Spain), earning the Doctoral Excellence Award in 2009. This work contributed to the understanding of the role of boron in symbiotic nitrogen fixation in legumes. Later, her postdocs at the Department of Plant Sciences at the University of California Davis (UCD, US) (2010–2014) and as Juan de la Cierva Fellow (UAM, Spain) (2015–2017) at the Department of Biology at UAM focused on analyzing the impact of abiotic stress on crops. She also participated as Lecturer of undergraduate courses and taught different postgraduate courses. As Juan de la Cierva postdoctoral fellow at UAM (2015–2017) led different research lines and was the PI of a competitively funded international project working on quinoa studying the impact of agroecological conditions on the nutritional properties of seeds. Currently, she holds a non-permanent assistant professor position as Ramon y Cajal researcher at UAM leading, as PI, a research line studying the biological mechanisms underlying plant responses to changing environmental conditions and their impact on seed quality in quinoa and other emerging crops while she continues her work on plant boron nutrition. She maintains active collaboration with different international and national research groups and with entities that belong to the private sector and contributes actively to scientific conferences. She has organized two scientific meetings on quinoa, participates in I+ D evaluation activities (as a referee of High-impact factor journals and as a member of several international research committees, including the European Commission), in outreach activities, and has also combined her research work with intense teaching and mentoring activity. Altogether, her scientific career gives evidence of her knowledge working on a diversity of crops and plant model systems, her achievements in understanding the regulation of key agronomic traits (including grain yield or seed quality under various abiotic stresses), and has proven results extending basic research findings on applied agronomic research.

Dr. Norma Sammán is Chemistry Engineer, Doctor in Food Technology, and Full Professor at the National Universities of Jujuy and Tucumán. She has extensive experience in the chemical, nutritional, and functional characterization of food and the development of new food products. She gained important achievements in the academic, research, and transfer fields. Since 1999 she has been a member of the Regional Academic Committee of the Doctorate in Network in Food Science and Technology and has contributed to the training of human resources through the development of national and international courses and directing more than 20 doctoral theses. She was Director of the Jujuy Research and Transfer Center (CIT Jujuy) UNJu-CONICET (2012–2018) and President (2003–2009, 2015–2018) of LATINFOODS (Latin American Food Composition Network), and since 2018 she has been the Geographical Representative of the network for South America.

She has contributed to the capacity and standard development in food composition. Her research activities are currently oriented to the study of Andean crops in Northwest Argentina with the main objective of promoting sustainable productive activities. (i) The research group she directs relieved agricultural producers in relation to their productive and economic–social characteristics to detect potentialities. Organizational proposals accepted by the communities involved and compatible with their worldview, values, culture, and history were made. (ii) The biodiversity of Andean crops was preserved through the reinsertion of genetic material and characterization of the different varieties. This material was distributed to agricultural cooperatives to encourage their cultivation and conservation. (iii) Processed foods are developed based on Andean products, using low-cost technologies applicable to the region, which allow high retention of nutrients, bio-functional compounds, and their sensory characteristics. She has published numerous scientific articles, book chapters, and technical or policy-guidance documents and has received awards for her work in the field of food and nutrition.

Octavio Paredes-López, PhD, earned his Bachelor's degree in Biochemical Engineering, and a Master's in Food Science and Technology at the National Polytechnic Institute in Mexico City; a Master's in Biochemical Engineering at the Czech Academy of Sciences; and a PhD at the University of Manitoba, Department of Plant Sciences, Winnipeg, Manitoba, Canada. He has competed postdoctoral and research stays in several universities and institutes in countries such as the USA, Canada, France, UK, Germany, Switzerland, and Brazil. He has published more than 245 scientific papers in refereed international journals, 55 reviews and book chapters, is author/editor/coeditor of 10 books, and 125 articles on the role of science in popular newspapers mainly in Mexico and some in the USA and France; in total more than 435 works. His scientific publications have been on basic and applied aspects of foods (i.e, fruits, cereals, plant foods, molecular biology, genetically modified organisms (GMOs); on microbial topics (i.e., genetic modifications, fermentation technology, overexpression of secondary compounds); and on biotechnology. Students: he has graduated over 100 students doing research in his laboratory; some of them from abroad. He has received all Mexican awards of his area of research including the highest recognition: Mexican Award in Science. Emeritus Professor and Emeritus National Researcher. International recognition: Institute of Food Technologists Fellow USA; WK Kellogg International Food Security Award; Nestlé Award; Academy of Sciences of the Developing World – Agriculture Award, Trieste, Italy. Ordre National du Mérite, Republique Française, Paris, France. Vice President and President of the Mexican Academy of Sciences. Doctor Honoris causa of three Mexican universities. Doctor Honoris causa of the University of Manitoba, Canada, where he earned his PhD.

Contributors

Gabriel Barros
Universidad de Concepción
Concepción, Chile

Luisa Bascuñán-Godoy
Universidad de Concepción
Concepción, Chile

Natalia Bassett
Universidad Nacional de Jujuy
and Consejo Nacional de
Investigaciones Científicas y
Técnicas (CONICET)
San Salvador de Jujuy, Argentina

Alain P. Bonjean
Bonjean & Associates SAS
Orcines, France

Barbara Borczak
University of Agriculture
Krakow, Poland

Edgardo Calandri
Instituto de Ciencias y Tecnología de
los Alimentos Córdoba - ICyTAC
(CONICET)
Cordoba, Argentina

Jehannara Calle-Domínguez
Instituto de Investigaciones para la
Industria Alimenticia (IIIA)
La Habana, Cuba

Sonia Calliope
Universidad Nacional de Jujuy
and Consejo Nacional de
Investigaciones Científicas y
Técnicas (CONICET)
San Salvador de Jujuy, Argentina

Marianela Capitani
Argentina–CONICET, CCT Tandil
Buenos Aires, Argentina

Isabel Castanheira
National Institute of Health Doutor
Ricardo Jorge
Lisbon, Portugal

Catalina Castro
Universidad de Concepción
Concepción, Chile

Doris Chalampuente-Flores
Universidad Santiago de Compostela,
Spain / Universidad
Técnica del Norte, Ecuador

Nancy Chasquibol Silva
Universidad de Lima,
Lima, Peru

Teodoro Coba de la Peña
Centro de Estudios Avanzados en
Zonas Áridas (CEAZA)
La Serena, Chile

José María Coll Marqués
Instituto de Agroquímica y
Tecnología de Alimentos
(IATA-CSIC)
Paterna, Valencia, Spain

José Delatorre-Herrera
Universidad Arturo Prat
Iquique, Chile

Víctor Delgado-Soriano
Universidad Nacional Agraria La
Molina (UNALM)
Lima, Peru

Néstor Fernández del Saz
Universidad de Concepción
Concepción, Chile

Nieves Fernández-García
Centro de Edafología y Biologia
Aplicada del Segura - Consejo
Superior de Investigaciones
Científicas (CEBAS-CSIC).
Murcia, Spain

Susana Fischer
Universidad de Concepción
Concepción, Chile

María Alejandra Giménez
Universidad Nacional de Jujuy
and Consejo Nacional de
Investigaciones Científicas y
Técnicas (CONICET)
San Salvador de Jujuy, Argentina

Claudia M. Haros
Instituto de Agroquímica y Tecnología
de Alimentos (IATA)
Paterna, Valencia, Spain

Vanesa Ixtaína
Universidad Nacional de La Plata
(UNLP)
Buenos Aires, Argentina

Sven-Erik Jacobsen
Quinoa Quality ApS
Regstrup, Denmark

María Dolores Jiménez
Universidad Nacional de Jujuy
and Consejo Nacional de
Investigaciones Científicas y
Técnicas (CONICET)
San Salvador de Jujuy, Argentina

Nete Kodahl
University of Copenhagen
Frederiksberg C, Denmark

Manuel Oscar Lobo
Universidad Nacional de Jujuy and
Consejo Nacional de Investigaciones
Científicas y Técnicas (CONICET)
San Salvador de Jujuy, Argentina

Enrique Martínez
Centro de Estudios Avanzados en
Zonas Áridas (CEAZA)
La Serena, Chile

Marcela Lilian Martínez
Instituto Multidisciplinario
de Biología Vegetal (IMBIV,
CONICET-UNC)
Córdoba, Argentina

Silvia V. Melgarejo-Cabello
Universidad Nacional Agraria La
Molina
Lima, Peru

Ángela Miranda
Instituto de Ciencia y Tecnología
    Agrícolas (ICTA)
Ciudad de Guatemala, Guatemala

Luis Rodolfo Montes Osorio
Universidad de San Carlos de
    Guatemala
Ciudad de Guatemala, Guatemala

Ana Laura Mosso
Universidad Nacional de Jujuy
    and Consejo Nacional de
    Investigaciones Científicas y
    Técnicas (CONICET)
San Salvador de Jujuy, Argentina

Carla Motta
National Institute of Health Doutor
    Ricardo Jorge
Lisbon, Portugal

Ángel Mujica
Universidad Nacional Altiplano
Puno, Peru

Loreto A. Muñoz
Universidad Central de Chile
Santiago, Chile

Enrique Olmos
Centro de Edafología y
    Biología Aplicada del
    Segura - Consejo Superior
    de Investigaciones
    Científicas (CEBAS-CSIC).
Murcia, Spain

José Ortiz
Universidad de Concepción
Concepción, Chile

Enrique Ostria
Centro de Estudios Avanzados en
    Zonas Áridas (CEAZA)
La Serena, Chile

Octavio Paredes-López
Centro de Investigación y de Estudios
    Avanzados (CINVESTAV)
Guanajuato, Mexico

M. Carmen Pérez Camino
Instituto de la Grasa (IG-CSIC)
Seville, Spain

María Reguera
Universidad Autónoma de Madrid
Madrid, Spain

**Ritva Ann-Mari
Repo-Carrasco-Valencia**
Universidad Nacional Agraria La Molina
Lima, Peru

Juan Pablo Rodríguez
International Center for Biosaline
    Agriculture
Dubai, UAE

Inmaculada Roman-García
Centro de Edafología y Biología
    Aplicada del Segura – Consejo
    Superior de Investigaciones
    Científicas (CEBAS-CSIC).
Murcia, Spain

María Constanza Rossi
Universidad Nacional de Jujuy
    and Consejo Nacional de
    Investigaciones Científicas y
    Técnicas (CONICET)
San Salvador de Jujuy, Argentina

Karina B. Ruiz
Universidad Arturo Prat
Iquique, Chile

Norma Sammán
Universidad Nacional de Jujuy
and Consejo Nacional de
Investigaciones Científicas y
Técnicas (CONICET)
San Salvador de Jujuy, Argentina

Carolina Sanhueza
Universidad de Concepción
Concepción, Chile

Adriana Scilingo
Universidad Nacional de La Plata
(UNLP)
Buenos Aires, Argentina

Marten Sørensen
University of Copenhagen
Frederiksberg C, Denmark

César Tapia Bastidas
INIAP, Estación Experimental Santa
Catalina, Quito, Ecuador

Mabel Tomás
Universidad Nacional de La Plata
(UNLP)
Buenos Aires, Argentina

Bernabé Vázquez Agostini
Universidad Nacional de Jujuy
and Consejo Nacional de
Investigaciones Científicas y
Técnicas (CONICET)
San Salvador de Jujuy, Argentina

Chapter 1

Agronomical Characterization of Latin-American Crops

Sven-Erik Jacobsen[1], Angela Miranda[2],
Doris Chalampuente-Flores[3], Juan Pablo Rodríguez[4],
Luis Rodolfo Montes Osorio[5], Alain P. Bonjean[6],
Nete Kodahl[7], César Tapia Bastidas[8] and Marten Sørensen[7]
[1]Quinoa Quality ApS, Regstrup, Denmark.
[2]Instituto de Ciencia y Tecnología Agrícolas
(ICTA), Ciudad de Guatemala, Guatemala
[3]Universidad Santiago de Compostela, Spain / Universidad
Técnica del Norte, Ecuador
[4]International Center for Biosaline Agriculture, Dubai, UAE,
[5]Universidad de San Carlos de Guatemala, Ciudad de Guatemala,
Guatemala
[6]Bonjean & Associates SAS, Orcines, France
[7]University of Copenhagen, Thorvaldsensvej
Frederiksberg C, Denmark
[8]INIAP, Estación Experimental Santa Catalina,
Panamericana sur km 1, Quito, Ecuador

CONTENTS

DOI: 10.1201/9781003088424-1

1.1   THE IMPORTANCE OF AGRO-BIODIVERSITY

Dependence on a few crops has negative consequences for ecosystems, food diversity, and health. Food monotony increases the risk of micronutrient deficiency. A strategy based on utilizing existing plant agro-biodiversity offers much promise when the objective is to feed the world's growing population within the coming decades (Jacobsen et al. 2013). The discussion is which selection criteria should be implemented to ensure the highest level of sustainability and reliability for the biodiversity-based strategy of crop development (Jacobsen et al. 2015).

When considering current global challenges, a reformation of our food systems is imperative to ensure food security, mitigate climate change, and alleviate malnutrition. In this regard, underutilized crops may be essential tools that can provide agricultural hardiness and reduced need for external inputs, climate resilience, diet diversification, and improved income opportunities for smallholders.

Few neglected and underutilized species (NUS) have made their way to export markets worldwide (Akinnifesi et al. 2008). They are interesting for increasing agro-biodiversity, and they often have high nutritional value and can be an essential source of micronutrients, protein, energy, and fibre, contributing to food and nutrition security. Many of these crops require relatively low inputs, and they can be grown on marginal lands and easily intercropped or rotated with staple crops and fit easily into agroecological practices. As frequently adapted to marginal conditions, they have the unique ability to tolerate or withstand stresses, making production systems more sustainable and climate-resilient (Li and Siddique 2018).

There are examples of NUS that have become important commercial crops. For example, quinoa and lentil are two NUS crops that attracted global interest despite being important mainly in subsistence agriculture, either local or regionally (Li and Siddique 2018). The Latin-American crops described in this chapter represent some of the most promising cultivars for the future. Their highlights are listed in Table 1.1.

1.2   THE ANDEAN REGION OF LATIN AMERICA

Reassessing neglected and underutilized crops for maintaining regional food security and improving human nutrition worldwide might be an excellent opportunity to recover forgotten crops at risk of extinction (Leidi et al. 2018; Jacobsen et al. 2013, 2015). For example, global initiatives like the declaration of 2013 as the international year of quinoa (Bazile et al. 2016; FAO 2013) boosted public knowledge of an important seed crop already adopted by vegetarian, vegan, and gluten intolerant consumers, and in general, consumers interested in sustainable food and reducing meat consumption. However, many other crops whose valuable diversity is maintained by local producers and consumers lay far behind in world awareness (Gahukar 2014; Hernández Bermejo and Leon 1994).

The Andean region is an important part of Latin America, and the most fragile. The challenge of increasing Andean crop production while protecting the fragile Andean environment and its endemic biodiversity lies mainly in developing organic conservation agriculture, considering various techniques. It should include the use of reduced or no tillage for increasing soil fertility and reducing purchased inputs, maintain soil cover for minimizing erosion,

TABLE 1.1   HIGHLIGHTS OF SELECTED LATIN-AMERICAN CROPS

Crop species – origin	Origin	Highlights
Quinoa (*Chenopodium quinoa*)	Andean region	High nutritional quality High content of micronutrients such as iron, calcium, and zinc High protein quality Climate proof, tolerant to drought and saline soils Ideal for plant-based food Increasing market demand
Kañawa (*Chenopodium pallidicaule*)	Andean region	High level of tolerance to cold and frost High content of iron High protein quality
Andean lupine (*Lupinus mutabilis*)	Andean region	Highest protein content among pulses, similar to soybean Dual-purpose (protein and oil) Potential for alkaloid use
Common bean (*Phaseolus vulgaris*)	S. Mexico, N Central America	Part of the traditional diet Well-known product with high market demand
Maize, Andean maize (*Zea mays*)	S. Mexico, N Central America, Andean region	Well-known product High global demand Many uses
Chia (*Salvia hispanica*)	S. Mexico, N Central America	High nutritional quality Increasing world market
Sacha inchi (*Plukenetia volubilis*)	Tropical S. America	High nutritive value Exceptional oil composition Good sensory acceptability Well suited for cultivation Numerous potential applications in gastronomy, medicine, and cosmetics Broad adaptation

Source: Authors' own work.

and adopt improved crop rotations, including the use of neglected and underutilized crops.

The cold and rural highland of the Peruvian-Bolivian-Chilean altiplano, and specifically the productive agricultural land close to Lake Titicaca, on the border between Peru and Bolivia, has played an essential role in the domestication of crops of the pre-Hispanic Andean world, as well as *in situ* conservation of genetic variability.

1.3   ANCIENT TECHNIQUES FOR IMPROVING FOOD SECURITY

Tahuantinsuyo, the Inca territory that covered the current area of Peru, Bolivia, Chile, Ecuador, and Argentina (approx. AD 1100–1542), was a highly developed country until the invasion and Spanish conquest in the 16th century. The different pre-Hispanic Andean cultures like Wari, Chimus, Colleagues, Pukara, Tiwanaku, Lupaka, and finally the Incas, answered with technological strategies and wisdom to the challenges of agricultural production in the altiplano, with altitudes up to 4500 m above sea level (a.s.l.), characterized by frost, drought, floods, hailstorms, and high solar radiation. They developed a social and economic structure adapted to the extreme climatic and edaphic conditions, creating sound systems of managing soil and water linked to a cosmovision of the Andean world (Paz Silva 1993).

A range of pre-Hispanic modifications of the climate have been developed. They were constructed and used to offset the adverse climate to enable cultivation of quinoa and other Andean crops. Crop production in these extreme conditions is only possible because of these ancient structures. Therefore, crop production at this high altitude is purely organic, utilizing animal manure, genetic diversity, and diverse technologies to secure food production.

The ethnic diversity of pre-Inca cultures is characterized by various agricultural skills and knowledge invented by these cultures (Erickson 1985). The technologies, such as terraces, high beds, artificial lakes, boundary layers, and stonewalls, allowed them to modify the environment, especially on the altiplano (Figure 1.1).

The Pukara developed the qochas (artificial lakes) to produce food in cold and plane conditions with an excess of humidity. The Lupaka, situated at the shores of Lake Titicaca, constructed waru warus (high beds) to produce food, mainly quinoa and kañawa. These crop species are highly resistant to adverse climatic conditions prevalent in the Andean highlands. Andenes (*terrazas*) were constructed in mountain areas, cultivating Andean mountain slopes to reduce frost risk. Andenes, waru warus, and qochas are still in use by indigenous communities of the altiplano.

A unique production system developed for specific conditions in the southern altiplano of the salt desert of Bolivia enabled quinoa to be grown under harsh conditions, where no other crops can be grown.

The Aynoka is a collection *in situ* of germplasm of the crop, for instance, quinoa, and related species, between which crossings may occur, maintaining

Figure 1.1   (A) Quinoa on the altiplano of Bolivia (left, Sven-Erik Jacobsen, right, Freddy Chipana) **(B)** Waru waru (high beds) in Puno, Peru. **Source: Authors' own work.**
Notes: A) The image on the left shows two men standing in a quinoa crop plantation with a snow-topped mountain in the background.
B) The image on the right shows the traditional high beds of the altiplano with a lake behind them.

genetic variation. Aynokas are systems of farmers' organization of the altiplano for multiple purposes, such as food security, rational soil management, pest control, conservation, and use of genetic diversity, and management of crop production in different altitudes (Ichuta and Artiaga 1986). *In situ* conservation is related to both the cultivated crops as well as wild relatives. The systems are widely distributed in the Andean region under different names such as mandas, laymes, etc. (Mujica and Jacobsen 2000).

In situ conservation is practiced only in a few places, due to lack of technological progress, such as tractors, modern varieties for the market, and lack of farmers' support and stimuli, accentuated by the economic crisis and extreme poverty of many of the rural communities who carry out these systems of cultivation and preservation of genetic resources. When systems are abandoned, it may lead to losing Andean culture and identity.

The wild relatives have particular importance, often seen on fields' borders and in areas regarded as sacred where conservationist farmers conserve them. In addition, the diversity of wild relatives is used for food, medicine, and rituals, especially in times of drought and other climatic adversities in the altiplano.

The ancient technologies are all in danger, and in many places, structures are abandoned. It is important that institutions in charge of preserving genetic resources succeed in changing this development; otherwise, they could disappear in the near future, together with genetic diversity (Mujica and Jacobsen 1998).

1.4 LIMITATIONS OF ANDEAN AND OTHER LATIN AMERICAN CROPS

Since colonial times, there has been a lack of native Andean crops due to devaluation and rejection of crops, and some were even prohibited from cultivation. It led to a reduction in their consumption, later exacerbated by considering it "Indian food" or "poor man's food" (Horton 2014). The challenge is recognizing that Andean grain and pulse crops contribute to food sovereignty (Mercado et al. 2018).

Selection and breeding efforts have been sparse of Andean crops in general. For this reason, they have remained at a marginal level with regard to production, market, and consumption. They have often served as essential subsistence crops but have had difficulties reaching domestic and international markets to create income for the rural population, agro-industries, and the country.

Borlaug and his colleagues (Borlaug 2000) initiated the first green revolution, successfully increasing yields under rising use of fertilisers and pesticides. However, this agronomical revolution created several environmental problems, which should now be corrected, focusing on sustainable production, soil fertility, and increasing agro-biodiversity, using the vast range of underutilized crops with defined potential.

A sustainable green revolution no. 2 should utilize a wider diversity of crops so that food production can benefit from a broader set of species, each adapted for specific marginal conditions. Breeding techniques may be classical or molecular, including CRISPR-Cas-based systems (i.e. Clustered Regularly Interspaced Short Palindromic Repeats), and TILLING (Targeting Induced Local Lesions IN Genomes), and eventually, Genetically Modified Organisms (GMO) (López-Marqués et al. 2020).

1.5 SELECTED CROPS WITH MASSIVE POTENTIAL FOR THE FUTURE

The selected crops are different types of crops with different virtues. They are all highly valuable crops, potentially contributing to an improved agro-biodiversity worldwide and improved food diversity and food security.

1.5.1 Quinoa (*Chenopodium quinoa* Willd.)

1.5.1.1 Introduction

One of the oldest existing crops

It is not known when quinoa was brought to Europe, so the crop literally remained unknown outside Andean countries until the late 1970s, first introduced to the USA. Later studies of adaptation were initiated in England, Denmark, and the Netherlands, followed by a further spread with trials in Europe and areas with a temperate to subtropical climate (Jacobsen 2017).

Until recently, no commercial production of quinoa was found outside the Andes. Nowadays, quinoa is cultivated, or at least tested, in many countries (>120) and the potential for further expansion of global production is high (Alandia et al. 2020). However, there is still little scientific information available regarding the factors limiting quinoa cultivation. Sellami et al. (2021) reviewed scientific papers dealing with quinoa from 2000 to 2020, finding only 33 publications. The main topics addressed were variety, deficit irrigation, water quality, fertilization, and sowing date, with very few focusing on tillage and weed control.

Origin of quinoa

Historically, quinoa has been continuously selected for new environments in the Andean region, as it was spread gradually from its centre of origin around the Titicaca Lake between Peru and Bolivia. Distribution from the lake went both north, to Ecuador, Colombia, and Venezuela, and south, to Chile and Argentina, east, from the highlands down to valleys, and west, to coastal regions of the Andean countries. The process, however, was slow due to environmental

Figure 1.2   **(A)** Quinoa (Vikinga cv.) in the field **(B)** Titicaca plants showing the panicles (end stages of seed development). **Source: Authors' own work.**
Notes: A) The image on the left shows an extensive quinoa field with trees and houses at the back of the field.
B) The image on the right shows multiple quinoa panicles in which seeds are developed.

variability and the unstable climatic conditions in the Andean region (Bertero et al. 2004; FAO 2013).

Lately, tropical species of quinoa have been adapted to temperate regions of the USA and Europe, and even in northern Europe it is now possible to grow quinoa successfully (Figure 1.2).

1.5.1.1.1   *Chenopodium species*
The *Chenopodium* genus includes around 250 species from all over the world. It is considered one of the most nutritious genera due to its protein and dietary fibre content and healthy fat, ash, and minerals (Repo-Carrasco et al. 2003).

The oldest domesticated *Chenopodium* species, identified to date, is the South American quinoa developed in the Andes mountains *c.*7500 years ago (Pearsall 1992). It reached North America *c.*AD 1200.

In Northern Europe, *Chenopodium album* L. with common names fat-hen and lamb's quarters, which is a global weed species, was a secondary crop in Denmark during the Iron Age (1200 BC–AD 400) (Stokes and Rowley-Conwy 2002), where both the seeds and the leaves were consumed. *Chenopodium album* was also used as a pasture for milking cows in Denmark under World War II (1940–1945), as farmers found that it secured good milk production in a situation where there was a lack of quality foodstuffs.

1.5.1.2   Limitations – Breeding
To our knowledge, modern molecular tools such as GMO and CRISP/Cas9 are not used in quinoa. Much research work has been performed using these molecular tools in many different crops, however, they have hardly been accepted by consumers in Europe so far, and, in addition, they are subjected to strict regulation by governments (López-Marqués et al. 2020). Our suggestion is to combine molecular and genetic techniques, such as TILLING, with traditional breeding together with other molecular techniques, not considered GMOs, in order to breed improved quinoa cultivars that can be commercialized without further regulation limitation or consumption acceptance problems.

1.5.1.3   Future Potential
Quinoa, a new world staple
Quinoa is a new crop, presently being tested in various parts of the world. According to the Food and Agriculture Organization (FAO), quinoa is regarded as a new world staple and predicted to spread fast across the globe (FAO 2013). New study sites and commercial productions, apart from South and North America and Europe, are seen in central Asia, i.e. the Aral Sea basin (Khaitov et al. 2020) and Africa, i.e. Zimbabwe (Muziri et al. 2020).

In order to adapt the tropical crop species quinoa to European conditions, day-length neutral varieties have been developed. These varieties have been tested and grown successfully in various countries and climate zones worldwide. Quinoa is being established in various countries globally (Bazile et al. 2016), such as China with 12,000 ha, being the third-largest producer of quinoa in the world (Xiu-shi et al. 2019). Quinoa was tested with success in several harsh environments, with results mainly from 2–4 t ha[-1] in Iran, Turkey, and UAE (Razzaghi et al. 2020). Good results were seen also in Africa, e.g. Morocco (Nanduri et al. 2019), Burkina Faso (Alvar-Beltran et al. 2019), and Egypt (Adel 2020). In Asia aside from in China, quinoa has also been tested in Bhutan (Katwal and Bazile 2020).

1.5.1.3.1 Global demand

Due to the increasing global demand for quinoa, both as an Andean export commodity and for agricultural development purposes, there is enormous interest in testing quinoa for growing under various environmental and geographical conditions.

Quinoa is mainly grown in the Andean countries Bolivia and Peru, with an area of 112,000 ha in Bolivia and 65,000 ha in Peru. The estimated area in Europe is 8000 ha, with the largest producers being Spain and France, with production also in the rest of Europe from Sweden in the north to Italy in the south. China has introduced quinoa and developed different varieties, which are grown around the country on 13,000 ha. In Africa there is mainly small-scale production in several countries, such as Morocco, Egypt, Zambia, and Kenya. Total global quinoa production is 200,000 ha, of which around 88% is produced in the two cited Andean countries (Jacobsen, personal communication, December 2022).

As the market widens, there is great potential for production of quinoa also outside the Andes. The estimated potential area in Europe is 2 million ha. For example, yield levels in Northern Europe are normally 1–2 t ha,[1] but with potentially higher yields as demonstrated in plot trials (Jacobsen 1997, 2017; Jacobsen and Christiansen 2016; Jacobsen et al. 1994, 2010).

1.5.2   Kañawa (*Chenopodium pallidicaule* Aellen)

1.5.2.1   Introduction

Cañahua, cañihua, kañawa, or kañiwa (*C. pallidicaule*) is an Andean species belonging to the Amaranthaceae family. Kañawa, as well as quinoa, is a gluten-free achene type grain (Gorinstein et al. 2015). Saponin content is low in Cañahua. Protein content is 15–18%, with high protein quality like quinoa (Moscoso-Mujica et al. 2017; Rodriguez et al. 2020). *C. pallidicaule* plants produce several dry, single-seeded fruits from small inflorescences distributed in many branches (Mamani 2016a). Before harvest, seed losses are 15–35% due to

Figure 1.3 Kañawa plants in the field. **Source: Authors' own work.**
Note: The image shows different flowered kañawa plant varieties, planted in rows showing the color diversity (purple, green, and yellow).

shattering occurring between flowering and physiological maturity (Rodriguez et al. 2017). Diminutive flowers are distributed in lateral inflorescence branches that are a feature of both cultivated and wild Amaranthaceae species (Spehar and Santos 2002). In Bolivia and Peru, three growth habit forms are known, 'Saihua (Chuqhu)', 'Lasta (Thasa)', and 'Pampalasta (mamakañawa or illamanku)' (see Table 1) (Quispe 2004; Mamani 2016a). The cultivated Saihua and Lasta kañawa growth habit forms produce a fair amount of grains and are widespread in the highland regions (Maydana 2010; Mamani 2016b; Rodriguez et al. 2017, 2020). The third growth habit form, Pampalasta, is a wild ecotype (Estaña and Muñoz 2012; Serrano 2012; Mamani 2016a). This wild kañawa is known for its early ripening grains (Mamani 2016). The glycaemic index is low (Mangelson et al. 2019). A problem is the irregular grain maturation on the branches, which leads to seed shattering, indicating its status as partially domesticated (Rodriguez et al. 2017, 2020) (Figure 1.3).

1.5.2.1.1 *Domestication/cultivation history, including present cultivation*

Kañawa and quinoa cultivation are related, and both belong to the genus *Chenopodium* with *c.*150 species. Kañawa originated in the Andean highlands of Peru and Bolivia. The Tiwanaku culture accomplished its semi-domestication in Bolivia's Collao highland. For millennia, chenopod species were cultivated as food crops in North, Central, and South America. For example, the pit goosefoot (*C. berlandieri* Moq. subsp. *jonesianum* Bruce Smith) has been being cultivated by the native population in the eastern part of North America since before the domestication of maize (*Zea mays* L.) (Bruno 2006, 2014; Fritz et al. 2017).

Kañawa is reported as having been cultivated for 7000 years as a staple crop in pre-Incan societies (Gade 1970). Communities around the Titicaca Lake basin between Peru and Bolivia are the origin of kañawa as observed from collections of landraces (Estaña and Muñoz 2012; Serrano 2012; Mamani 2016b). Bolivia holds an extensive collection of kañawa accessions with 801 entries and Peru with 341 (Mamani 2013, 2016a; Mamani et al. 2018; Mangelson et al. 2019).

Quinoa is well known and scaled-up worldwide (Alandia et al. 2020), while kañawa is only traded in small quantities from Peru and Bolivia (Rodriguez et al. 2020). In Bolivia, kañawa is grown in the high-altitude regions or altiplano ecozone of departments of La Paz, Oruro, Cochabamba, Potosi, Tarija; and in Peru throughout the Sierra in Cusco, Puno, Ayacucho, Junín, and Huaraz (Bravo et al. 2010; Rojas et al. 2010; Mamani et al. 2018). Peru is the first producer. However, breeding is limited, and the few available varieties for scaled-up production derive from selection efforts (Rodriguez and Mamani 2016; Mamani 2016b).

Export of kañawa remains marginal compared to quinoa (Carrasco and Soto 2010; Rodriguez et al. 2010; Rodriguez et al. 2020). Producers of this crop have consolidated into clusters, around La Paz, Oruro, and Cochabamba in Bolivia, supplying urban markets (Mamani 2016b; Rodriguez et al. 2020; INEI 2020). In Peru, producers in Puno, Cusco, and Juliaca commercialize kañawa (Giuliani et al. 2012; INEI 2020).

Until 2019, Peru had higher yields than Bolivia, but in general, it is low, *c.*500 kg ha.[1] Cultivation of kañawa is still practiced with traditional methods (Rodriguez et al. 2020). Barley (*Hordeum vulgare* L.) and oats (*Avena sativa* L.) are grown in preference to kañawa because they are attractive as alternative feeds for dairy cattle (Mamani 2016b), albeit that pilot research demonstrated that kañawa does constitute a good quality alternative feed for minor livestock (Calle 2004; Cortez Quispe 2016).

1.5.2.1.2 *Geographical distribution*
Currently, Bolivia and Peru are the main producers and exporters of kañawa (Gade 1970; Estaña and Muñoz 2012; Rodriguez et al. 2020). Kañawa is not cultivated outside the high-altitude regions of Bolivia (north of Oruro and near Cochabamba) and Peru (Ayaviri, Puno, Cusco, Ayacucho, and Junin). Geographically, the distribution of kañawa is not as extensive as for quinoa (Estaña and Muñoz 2012; Mamani 2016b). Kañawa collections are small and

maintained in germplasm banks (Galluzzi and Lopez Noriega 2014: Bonifacio 2019). Kaňawa is a partially domesticated *Chenopodium* species, which provides nutritional food security to many Andean farmers' households in the highland region (3500–4200 m a.s.l.) of both countries. In the Andean region kaňawa is grown by subsistence farmers due to its tolerance to frost, drought, and salinity (Rodriguez et al. 2017, 2020).

1.5.2.1.3 Uses
In rural households, farmers toast the grains, and they are then milled in a flat stone mill. The flour is known in Peru as '*caňihuacu*' and as '*pito*' in Bolivia. Its local popularity as grain is due to its high nutritional profile (Repo-Carrasco-Valencia 2011). In addition, kaňawa flour demonstrated a high capacity to mix with cereal flours improving the dough quality for bakery products (Pisfil 2017; Zegarra 2018; Chambi 2019).

1.5.2.2 Limitations – Breeding Needs
Genetic erosion is reported for some traditional varieties in Bolivia (Serrano 2012; Mamani 2016a). As a result, farmers have replaced kaňawa with forages to feed dairy cattle. Furthermore, kaňawa has received little or no commercial breeding attention resulting in registered cutivars, and only native landraces exist.

Using Cobalto-60 drums and Gamma radiation, the induced mutation was used in seeds to provoke changes in plant morphology and, if successful, reduce seed shattering. Therefore, this needs to be studied and tested with a broad range of genotypes.

Genetic improvement of kaňawa should be addressed by obtaining varieties according to current climate change while maintaining product quality requirements for local and export customers.

1.5.2.3 Future Potential
Recently the genome of kaňawa was sequenced (Mangelson et al. 2019), meaning that it can be used as a genetic model to understand the evolution of *Chenopodium* spp., particularly the cultivated *C. quinoa*. In addition, development of the genome sequence can help and support the improvement of kaňawa through breeding.

The grain has excellent potential as a nutritive food source with a low-glycaemic index and high in iron and zinc. Studies in Peru and Bolivia have demonstrated that flour mixed with wheat or amaranth can be transformed into nutritive food products such as cookies (Pisfil 2017; Zegarra 2018; Chambi 2019; Coila 2019; Šver et al. 2019). Kaňawa has been introduced as an integrated element in the Complementary School Food Program, but its availability needs to be scaled up and supported to improve nutritional security at the rural and urban household levels. Kañawa may also be used for biodegradable films or packages (Ramirez 2016; Salas 2017; Yanapa Velasquez 2018).

Soil salinity is expanding in many parts of the world, especially in high-altitude regions, and kañawa has demonstrated that it can tolerate high soil salinity (Rodriguez et al. 2020). Therefore, the crop should be promoted for intro-duction as an alternative for alleviating soil salinity.

Further development of the crop ought preferably to be accomplished by util-izing native varieties, where, for instance, early maturity is present.

Kañawa can now be found in non-domestic markets, e.g., in the USA, where it is marketed as baby quinoa. Use of such a denomination is not, however, recommended, as kañawa should instead be marketed on its own virtues.

1.5.3   Andean lupine (*Lupinus mutabilis* Sweet)

1.5.3.1   Introduction
The history of this species as an Andean subsistence crop demonstrates its potential as a crop for low-input agriculture in temperate climates (Cowling et al. 1998). The selection activities of Andean farmers have represented the only means of domestication, giving rise to semi-domesticated forms characterized by indehiscent pods, large seeds, multi-colored flowers, and highly branched architecture (Clements et al. 2008).

It is a robust crop that can be grown in poor soils and dry climates (Williams 1993), and which stands out for its great potential in soil recovery (De Ron et al. 2017). Furthermore, in addition to its high alkaloid content (4.5 g 100g^{-1} DW), the crop has high resistance to microbial infections and insect attacks (Gueguen and Cerletti 1994; Ciesiolka et al. 2005). However, the presence of alkaloids in the seeds and the relatively low yields (800–1300 kg ha^{-1}) have strongly limited the expansion of the crop (Tapia 2015; Galek et al. 2017). Therefore, efforts have been made to re-establish lupine as a crop in South America and adapt it to Europe's conditions (Caligari et al. 2000).

The Andean lupine is characterized by the highest grain quality of all cultivated lupines, presenting an oil content similar to that of soybeans (*Glycine max* (L.) Merr.) (Gulisano et al. 2019). Numerous studies investigating the nutri-tional profile and potential applications of this pulse have found a wide range of possible products ranging from proteins, oils, and food additives to cosmetics, medicines, and bio-pesticides. In addition, the nutritional advantages make it an ideal product for the transition from meat-intensive diets to diets based on vegetable proteins (Gulisano et al. 2019).

1.5.3.1.1   Origin, Diversification, and Domestication
The oldest evidence of cultivated Andean lupine is related to the seeds found in tombs of the Nazca culture and representations in Tiahuanaco ceramics in

Peru (FAO 1992; Tapia and Fries 2007). The oldest archaeological evidence of domesticated *L. mutabilis* seeds has been found in the Mantaro Valley in central Peru and dates to about AD 200. The use of Restriction site Associated DNA Sequencing (RAD-Seq) in the analysis of this archaeological material confirms that *L. mutabilis* was first domesticated in the Cajamarca region (northern Peru) from the wild progenitor *L. piurensis* C.P.Sm. Demographic analysis suggests that *L. mutabilis* separated from its parent around 2600 BC and suffered a bottleneck in domestication, with subsequent rapid population expansion as it was cultivated in the Andes (Atchison et al. 2016). *L. mutabilis* is reported in eastern South America, from Colombia to northern Argentina, and with a wide altitudinal range from 1500 to 3800 m a.s.l. (Jacobsen and Mujica 2008).

In the Andean region, 83 species have been identified; the wild relatives that show diversity and variability found in the Andean lupine are the following species: *Lupinus ananeanus* Ulbr., *L. aridulus* C.P.Sm., *L. ballianaus* C.P.Sm., *L. chlorolepis* C.P.Sm., *L. condensiflorus* C.P.Sm., *L. cuzcensis* C.P.Sm., *L. dorae* C.P.Sm., *L. eriocladus* Ulbr., *L. gibertianus* C.P.Sm., *L. macbrideianus* C.P.Sm., *L. microphyllus* Desr., *L. paniculatus* Desr., *L. sufferuginous* Rusby, *L. tarapacencis* C.P.Sm. and *L. tomentosus* DC. (Jacobsen and Mujica 2006).

Figure 1.4   Andean lupine **(A)** plant and **(B)** seeds and pods. **Source: Authors' own work.**
Notes: A) The image on the left shows a long stem lupine flower with small petals colored in blue, white, and yellow, and palmately lobed leaves within a lupin field. In the background, some trees are visible.
B) The image on the right shows many lupine seeds on graph paper which, together with the rulers in the image, indicate size of approximately 1 cm. These seeds are labeled as DCH-094. Also, two lupine dark-colored pods appear below the seeds and a bunch of pods (dark color) to the right of the seeds.

1.5.3.1.2 Botanical description

Three geographically separated morpho-types of the Andean lupine have been suggested based on the considerable genetic and morphological variability and wide ecological adaptation in the Andean zone: (a) *L. mutabilis*, lupino (northern Peru and Ecuador), of more prolific branching, very late, greater hairiness of leaves and stems, some ecotypes behave as biennials, tolerant to anthracnose; (b) the Andean lupine, tarwi (central and southern Peru), scarcely branched, moderately late, somewhat tolerant to anthracnose; and (c) the Andean lupine, Tauri (highlands of Peru and Bolivia), smaller (1–1.40 m) with a developed main stem, very early, susceptible to anthracnose (Tapia et al. 1980; Gross 1982; Tapia 2015).

High diversity seed characteristics include shape (lenticulate to spherical), primary and secondary seed (seed coat, i.e., testa) color, as well as distribution patterns, can range from pearly white to solid black and includes beige, yellow, brown, dark brown, and those in between such as brownish green and greyish green. Most of the seeds have a secondary color distribution in darker shades of the primary one; the secondary color distribution also varies among a wide range of patterns, such as brown-shaped, crescent, mottled, or spotted, which can be expressed either as solitary or in combination (Falconí 2012; Tapia 2015) colors.

The presence of considerable variation in germplasm is shown by different phenotypic traits, such as a wide range of growth periods, branching patterns, color and shape of grains and flowers, as well as flowering times. Both the inter-simple sequence repeat (ISSR) and single-sequenced repeat (SSR) markers have revealed broad genetic diversity among *L. mutabilis* lines (Galek et al. 2007; Chirinos-Arias et al. 2015), which could be related to the mixed pollination system that the species has, and which could explain the presence of the testa color (Chirinos-Arias et al. 2015).

1.5.3.1.3 Environments where lupine is grown

The requirement for lupine is variable, depending on soil, temperature, and wind. It grows well in temperatures from 20 to 25°C; grain development is optimal below 9.5°C (night temperature), a condition that occurs in the high Andean region (Gross 1982); the early ecotypes of Puno-Peru need 450 mm of precipitation per crop cycle, while the late ecotypes require 600–700 mm (Tapia 2015). Lupine prefers sandy loam soils, with a thick, deep texture, a balance of nutrients, and good drainage with a pH of 5–7 (Meneses 1996; FAO 2000; Jacobsen and Mujica 2006). The seedling stage is susceptible to frost (-4°C); the higher the temperature, the greater the growth and development; at less than 0°C, development and evapotranspiration are inhibited (Tapia and Fries 2007).

1.5.3.1.4 Uses

The Andean lupine can be consumed directly as a snack (Villacrés et al. 2003) and as an ingredient in different products such as fresh salads, soups, cakes, cookies, bread, hamburgers, and baby food (Cremer 1983; Ruales et al. 1988; Villacrés et al. 2003; Güémes-Vera et al. 2008). New uses of lupine are related to the extraction of oil and production of vegetable milk, yoghurt, and obtaining flours and by-products for animal feed (Tapia 1982; Peralta and Villacrés 2015).

Alkaloids such as lupunine and sparteine present in the leaves, stem and seeds of lupine plants were traditionally used, in combination with paico (*Dysphania ambrosioides* (L.) Mosyakin & Clemants, syn. *Chenopodium ambrosioides* L., a wild relative of quinoa) to repel pests on potato crops such as the Andean weevil (*Premnotrypes* spp.), and on quinoa the Kona Kona (*Eurysacca quinoae* P.), the primary pest of quinoa. In livestock, it is used to control internal and external parasites, practices that are disappearing due to the promotion of industrial agrochemicals (Canahua and Román 2016).

Lupine seeds are used for consumption after debittering (Cremer 1983; Villacres et al. 2000), reducing the alkaloid content to 0.02% for people and 0.4 to 0.6% for pigs, ruminants, and poultry (Tapia and Fries 2007; Caicedo and Peralta 2000). To eliminate antinutritive substances (alkaloids), a hydro-thermal process was carried out, which consists of hydrating the dry grain and soaking it for 12 to 14 h, after which it is cooked for 30 to 40 min. Finally, the seed is left in a stream of continuous drinking water for three or four days, or in circulating water from the river or streams between seven to ten days (Caicedo et al. 2001).

1.5.3.2 Limitations

1.5.3.2.1 *From an agronomic point of view*
Limitations are the lack of early maturity and high-yielding genotypes, lack of locally adapted genotypes (Caligari et al. 2000), and lack of good quality seed (Mercado et al. 2018).

There has been limited plant breeding taking place to increase yield and seed quality and decrease susceptibility to diseases such as anthracnose (*Colletotrichum gloeosporioides*). Lack of advanced biotechnological methods in genetics, molecular cytogenetics, and tissue culture have limited the possibility of exploiting natural variability and performing distant crosses and haploidization of material from reproduction (Gulisano et al. 2019).

Domestication and history of reproduction are fragmented in time and space, and participatory approaches with farmers to select local ecotypes are missing.

1.5.3.2.2 *From a nutritional point of view*
The Andean lupine has a bitter taste due to the high content of alkaloids, limiting direct consumption for both humans and animals (Guerrero 1987). Hence, it is necessary to improve the debittering processes (Mercado et al. 2018).

Specific data about the consumption of the Andean lupine or tarwi are needed, which will help make decisions and take actions that allow the promotion of greater consumption (Mercado et al. 2018).

1.5.3.3 Future Potential

The Andean lupine, known locally as 'tarwi' or 'chocho' (*L. mutabilis* Sweet), has been cultivated, processed, and consumed for at least 1500 years, and whose genetic variability has adapted to many microclimates (Williams 1993). Even before the Spanish conquest, this crop played an essential role in high Andean production systems and fed the indigenous population (FAO 2016). Among legumes, lupine is characterized by its high protein content of good quality, suitability for environmentally robust production, and potential health benefits (Lucas et al. 2015).

In the Andean region, the annual *per capita* consumption varies. In Ecuador, it is 4–8 kg person^{-1}, much higher than in Bolivia (0.2 kg person^{-1}) and Peru (0.5 kg person^{-1}). In 2017, production did not meet domestic demand in Ecuador, reporting a deficit of around 6000 tons (Mazón 2018).

The gastronomic versatility and nutritional qualities of this legume crop, combined with the work carried out for more than 20 years by both public and private entities in technological innovation, post-harvest, added value, quality seed, and improved varieties, have renewed interest in this crop (Horton 2014; Nájera 2015; Márquez 2016).

The combined use of germplasm and modern approaches to broaden the genetic base could help the introgression of desirable adaptive traits for environments adapted to certain latitudes (Gulisano et al. 2019). Future work should aim to develop bitter lines, with a sufficient level of alkaloids in vegetative tissues to decrease the presence of pathogens (Wink 1991); and sweet lines, which will be easier to process for consumption. Also, integrated pest and disease management programs are required to improve farmers' production systems (Mina et al. 2018).

Converting production into a dual-purpose alternative (protein and oil) similar to soybean could be an economical alternative for the productive and competitive development of the Andean region (Lucas et al. 2015).

Andean lupine alkaloids could be of commercial importance due to their pharmacological activity (Ciesiolka et al. 2005; Jiménez-Martínez et al. 2003). Furthermore, specific protein isolates and concentrates could be of commercial importance due to their functional properties for the chemical and food industry (Sathe et al. 1981; Gueguen and Cerletti 1994; Doxastakis 2000; Moure et al., 2006). In addition, the presence of ferritin (a protein rich in Fe) in the protein profile (Strozycki et al. 2007) increases the nutritional value of this culture by offering a safe way to increase the intake of iron in the diet (Zielinska-Dawidziak 2015).

Quinolizidine alkaloids (QAs) also have an essential role in the medical field due to multiple properties such as antiarrhythmic, anti-inflammatory, diuretic, and hypotensive effects, among others (Bunsupa et al. 2012). Besides, QAs can also find application in agriculture as a bio-stimulant increasing the growth and yield of other crops (Przybylak et al. 2005) as antibacterial agents (Romeo et al. 2018) or as biocidal agents that replace synthetic toxins (Bermúdez-Torres et al. 2009).

1.5.4   Common Bean (*Phaseolus vulgaris* L.)

1.5.4.1   Introduction

The traditional production system of common beans is in monoculture with bush-type varieties. However, the traditional system known as milpa, which consists of an intercropping arrangement (maize–bean) and other vegetables and tubers, is prevalent in the Highlands but not in all regions of Latin America; irrigation is generally not used, and the crop depends exclusively on rainfall. The dry bean seed constitutes one of the primary protein sources for human consumption, providing essential micronutrients such as iron and zinc (Rodríguez-Castillo and Fernández-Rojas 2003; MAGA 2017).

1.5.4.1.1   Origin, diversification, and domestication

The Mesoamerican region, and especially Mexico, is considered the centre of origin and diversification of beans (*Phaseolus* spp.), cultivated and wild (Hernández-López et al. 2013; Bitocchi et al. 2012, 2017). Common beans are currently grown in various regions and conditions, from 0–3000 m a.s.l., mainly by small producers in Central America, Africa, and Asia, representing 77% of world production. Central America, and Guatemala and Nicaragua are the largest producers (550 and 390 million kg per year, respectively) (Secretaría de Economía 2012; OECD 2021).

Latin America is an important centre of diversity for the common bean (Singh et al. 1991; Beebe et al. 2000; Blair et al. 2013). This diversity should be studied for the generation of improved cultivars with better seed yield potential. The common bean is part of Latin-American people's main diet, mainly because it is an essential protein source (Broughton et al. 2003). The improvement of nutritional aspects through breeding, such as bio-fortification with minerals (iron and zinc), can create a positive impact on food security (Espinoza-Moreno et al. 2016; Morais Dias et al. 2018; Orona-Tamayo et al. 2018; Contin Gomes et al. 2021; OECD 2021).

1.5.4.1.2   Germplasm collections

The National Plant Germplasm System of the United States of America (NPGS) identifies 81 dry bean species accepted in their online Germplasm Resources Information Network (GRIN). This number increases to 117 taxa when including subspecies and varieties. A comprehensive analysis of North and Central American species recognizes no less than 36 species, many with one or more subspecies, five of them of economic importance (FAO 2018). The Mesoamerican region is considered the main centre of origin and genetic diversity of domesticated and wild species of the genus *Phaseolus* (Beyra et al. 2004; Kwak and Gepts 2009; Rossi et al. 2009; Mamidi et al. 2011; Bitocchi et al. 2012).

The germplasm bank at Centro Internacional de Agricultura Tropical (CIAT), located in Cali, Colombia, one of the international germplasm banks of the Consultative Group on International Agricultural Research (CGIAR), operates within the International Treaty Framework on Plant Genetic Resources for Food and Agriculture (TRFGAA). The TRFGAA is primarily funded by the Platform of Germplasm Banks coordinated by Global Crop Trust. The germplasm bank at CIAT maintains the most important collections worldwide of essential crops that provide carbohydrates and protein in the tropical food systems: Beans (*Phaseolus* species, 37,938 accessions), manioc, cassava, tapioca (*Manihot* species, 6155 accessions), and forage crops (22,694 accessions). The CIAT germplasm bank also stores duplicates (security copies) of seeds of world crops in collaboration with the Nordic Genebank, who are using their underground facilities on the arctic island of Svalbard, midway between Norway and the North Pole, and in the International Maize and Wheat Improvement Center (CIMMYT) germplasm bank at Texcoco, Mexico (CIAT 2020). Moreover, the Institute of Agricultural Science and Technology (ICTA) has national germplasm collections from different pulse crop species. Additionally, Guatemalan beans are also in the major world collection, stored in the germplasm bank at CIAT (FAO 2018).

1.5.4.1.3 Botanical description
Botanically, the genus *Phaseolus* – including the common bean species (*P. vulgaris* L.) – belongs to the Leguminosae (Fabaceae) family, Faboideae (Papilionoideae) subfamily, Phaseoleae tribe and Phaseolinae subtribe. It is an annual herbaceous plant, and depending on growth habit, it can reach heights up to 2 m (Fernández et al. 1984).

Common beans can present four growth habits: type I determined (bush type), type II indeterminate (bush type), type III indeterminate (prostrate), and type IV indeterminate (climber). Those of determined growth can reach heights between 30 cm and 90 cm, while the plants with indeterminate habit reach heights from 50 cm to 3 m (Rosas 2003).

The bean has a primary root and many secondary roots with nodules developed from an association with the nitrogen-fixing bacterium *Rhizobium*. The leaves are trifoliate. The flowers have a tubular calyx of five sepals, a papilionoid corolla of unequally sized petals, ten stamens, and a receptive stigma. The flower color can be white, lilac, purple, or bicolored. The fruits are legumes, also called pods, and the seeds have two cotyledons. The seeds inside the pods are rich in protein (Clavijo 1980).

1.5.4.1.4   Production and geographical distribution

Common beans are produced in diverse systems, regions, and environments, e.g., Latin America, Africa, the Middle East, China, Europe, USA, and Canada. In Latin America, it is an essential and in the daily-diet food, especially in Brazil, Mexico, Central America, and the Caribbean.

In Latin America, bean producers and smallholder farmers, on average, have 2 ha of land to produce this pulse (Fischer and Victor 2014). Common bean production in developing countries is considered low input and small-scale agriculture (Miklas et al. 2006).

Estimated data from FAOSTAT (2021) report an average yield for Mexico, Guatemala, Honduras, and El Salvador of 862.58 kg ha^{-1}. Guatemala has the highest yield with 999.1 kg ha^{-1}, and Mexico has the lowest yield with 728.3 kg ha^{-1}. Despite its importance in some countries' diets, on the world stage, the volume of bean production compared to other grains such as maize, wheat, and rice represents only 1%.

1.5.4.2   Limitations

1.5.4.2.1   Agronomy

The highest production area (8 mill. ha) is found in the Latin-American region (CGIAR-Beanfocus 2019). However, bean production has not expanded along with increasing demands from a growing population.

The average yield is relatively low, ranging from 600–1000 kg ha^{-1} (Rathna Priya and Manickavasagan 2020). Yield is affected by heat and drought (Villarino 2018). Besides, diseases and pests may cause up to complete loss. Most varieties will take c.65–110 days to mature, but some landraces and cultivars may take more than 200 days according to their growth habit (Katungi et al. 2009).

1.5.4.2.2   Nutrition

The common bean is a good source of protein, but beans also present anti-nutritional factors that negatively affect human and animal metabolism; fortunately, these factors are thermolabile and their content may be partially or totally destroyed during cooking (Cravioto et al. 1945; Bressani 1990; Martín-Cabrejas et al. 2004). In addition, following consumption, the flatulence factor creates rejection of this pulse (Geil and Anderson 1994; Haileslassie et al. 2016). However, nutrition properties for this pulse can be improved by using different processing methods, including soaking, de-hulling, germination, fermentation,

Figure 1.5   Bean (**A**) plant and (**B**) seeds. **Source: Authors' own work.**
Notes: A) The image on the left shows a small bean plant with a deltoid shape and with small white flowers in the center of the plant, which seems to be in a bean field.
B) The image on the right shows oval, bean-shaped, whitish beans.

and the use of thermal processing methods (Cravioto et al. 1945; Bressani 1990; Tiwari and Singh 2012).

Storage is critical to maintaining quality. The lack of humidity and temperature control during storage results in a hard seed coat (testa) and a hard-to-cook seed, which requires a longer cooking time (Coelho et al. 2007). Selected essential amino acids are lost because of the hard-to-cook phenomenon.

1.5.4.2.3 Breeding
Common bean breeding programs have to focus on improved cultivars adapted to the different crop systems in Latin America. Their studies should focus on disease resistance and abiotic stress tolerances (Villarino 2018).

Future research should include a strategy that encompasses agronomic features, post-harvest management, and marketing since producers in Latin America have limited access to field supplies and facilities for seed storage. These issues directly affect the quality of the product for consumption. In addition to agronomic traits, investigations using traditional and new molecular techniques are needed to generate common bean seeds with improved nutritional and nutraceutical messages.

1.5.5   Chia (*Salvia hispanica* L.)

1.5.5.1   Introduction

Historically, chia nutlets were the principal food of the Nahuatl (Aztec) culture of central Mexico. The Jesuit chroniclers considered chia as the third most important crop of the Aztecs, after maize (*Zea mays* L.) and beans (*Phaseolus* spp.), and before amaranth (*Amaranthus cruentus* L. and *A. hypochondriacus* L.). Moreover, tributes and taxes to the Aztec clergy and nobility were frequently paid in chia nutlet (Miranda-Colin 1978; Cahill 2004; Muñoz et al. 2013). In pre-Columbian times, chia was an essential crop for the civilizations of that time; it was cultivated in great quantity by the Aztecs, including nutlets in their diet, in medicine and art, and even as offerings in pagan religious rituals; it also functioned as currency and as a tax on the people subjected to this empire.

Mayan culture called it chihaan, which means strong or fortress (Cahill 2003). Since then, it has remained part of the culinary culture of the ethnic groups of the region. One popular way was through chia fresco, which, following the introduction of the citrus fruits, is generally an orange or lemon soda to which chia nutlets are added. In recent years, due to its properties as a superfood, being a source of Omega-3, antioxidants, and fiber for human nutrition, it has been in great demand. As a result, 'La Chia' has become established as an emerging crop in different parts of the world (Pascual-Villalobos et al. 1997; Beltrán and Romero 2003; Peiretti and Gai 2009; Jamboonsri et al. 2012; Norlaily et al. 2012; Muñoz et al. 2013; Bochicchio et al. 2015; FAO/WHO 2016; López et al. 2017; Rivera-Cabrera et al. 2017; Peláez et al. 2019; EFSA 2020).

Figure 1.6   Chia crop **(A)** plant and **(B)** seeds. **Source: Authors' own work.**
Notes: A) The image on the left shows green chia plants with many leaves and flowers (with blue petals) in the field, and, at the back of the chia field appear other types of green plants with different sizes and shapes.
B) The image on the right shows a bulk chia stall accompanied by other plant-based products such as onions, tomatoes, purple beans in bags, and chickpeas in bags.

1.5.5.1.1 Botanical description

Chia (*Salvia hispanica* L.) is part of the Lamiaceae family, in which are found other species such as mint (*Mentha* L. sp.), rosemary (*Rosmarinus officinalis* L. syn. *Salvia rosmarinus* Schleid.), and oregano (*Origanum vulgare* L.). They are annual herbs up to 1 m, robust, erect, sparsely branched, often with a woody base; green stems, sometimes with purple pigmentation. Leaves are petiolate, ovate, or lanceolate, membranous, with a green upper surface, serrated margins, and acuminate apex; inflorescence is 4–10 cm, terminal, in dense racemes; fruiting calyx acrid, markedly urceolate. The corolla is 8–9 mm, blue or purple; stamens appear with exerted anthers; exerted style. Nutlets/mericarps are 1.75 × 1.2 mm, flattened, greyish-brown, with dark brown mottling. In the wild state, they flower and fructify from May to October and can be found in the forests of tree genera like *Ulmus* L., *Pinus* L., *Quercus* L.) and in open grasslands as well as in bean and maize fields or cultivated (Bentham and Oersted 1854; Cahill 2003; Hernández-Gómez *et al.* 2008; Lobo Zavala et al. 2011).

1.5.5.1.2 Origin and distribution

Chia is an annual species native to the mountainous areas of western and central Mexico, and Guatemala. It is found naturally in oak or pine-oak forests and is distributed in semi-warm and temperate environments of the Transversal Neovolcanic Axis of the western Mother Sierras and southern Chiapas altitudes 1400 and 2200 m a.s.l. (Cahill 2001, 2003; Ayerza and Coates 2005; Norlaily et al. 2012; Sosa-Baldivia et al. 2018; Peláez et al. 2019).

1.5.5.1.3 Uses

The most common way to use chia is by soaking the nutlets in water; they form a mucilaginous-gelatinous mass used as a flavouring in fruit juices which are consumed as refreshing drinks. The gelatinous mass of the nutlets can also be prepared as a pudding. In addition, the germinated seedlings are frequently eaten in salads, sandwiches, soups, and stews. Due to their mucilaginous properties, they are commonly sown on clay or other porous material, which is kept moist. From ground nutlets, flour is obtained that can be used to make bread, stews, and cakes, generally mixed with cereal flours, and it is an excellent source of easily digestible proteins and fats. Some other uses were as a revitalizer for combatants who went to war, and for women preparing for childbirth (Cahill 2003, 2005; Bresson et al. 2009; Capitani et al. 2013; Cabrera and Cerna 2014).

1.5.5.2 Limitations

Chia has great potential as a superfood. The ancient Aztec and Mayan people knew its beneficial characteristics. Several studies have shown that it can be a rich source of Omega-3, easy to use. Unfortunately, not much is known about its cultivation and agronomic practices when cultivated as a monoculture. There are few experiences in this regard. In addition, there are no varieties or hybrids that can be recommended for different conditions. Currently, creole material is used with a mass selection process, as it is likely unknown how it will behave in different conditions worldwide. In view of its great demand, developing breeding programs and agronomic practices are necessary to boost this potential superfood; fortunately, there are some actual efforts in this direction in a few countries of Latin America.

1.5.5.3 Future Potential

Chia has great potential as a functional food for modern life. Several studies have focused on the nutritional potential as a supplement for preventing diseases such as cancer, diabetes, cardiovascular diseases, inflammatory disorders, and nerves (Muñoz et al. 2013; Hernández-Pérez et al. 2021). In addition, its soluble dietary fiber helps to counteract problems of constipation, diverticula, and colon cancer (Alvarado Rupflin 2011). To do this, ingest 15 to 25 g of nutlets soaked in water for 15 min. for 20 days (Bernal et al. 2015). This property caused commercialization of the nutlets to begin in the 1990s.

It is grown in Argentina, Mexico, Bolivia, Paraguay, and Australia. In 2011–2012 Argentina had 35% of the production, with Australia, Mexico, Bolivia, and Paraguay producing 15% and 3000 ha each (Busilacchi et al. 2015). Nicaragua and Southeast Asian countries recently joined as producers (Jamboonsri et al. 2012). As a result, world production has proliferated; an example is Nicaragua, where chia production went from 5000 quintals (= 227 metric tonnes) in 2013 to 180,000 quintals (= 8165 metric tonnes) in 2014 (Miranda 2014). Mexico, which registered 15 ha harvested in 2006, in 2014 this increased to 16,550 ha. Due to this, chia has a high potential to become an essential crop for human consumption worldwide (Valdivia-López and Tecante 2015).

1.5.6 Andean maize (*Zea mays* ssp. *mays* L.)

1.5.6.1 Introduction

Maize is one of the world's top three crops, the two others being rice and wheat, with a small production under adverse conditions in the highland Andes. The Andean countries currently cultivate a traditional maize population, frequently amylaceous, white, or colored, for human consumption, and yellow hybrids primarily for feeding animals. However, they are far from becoming self-sufficient and usually import additional quantities of maize grain from other American countries.

Origin and distribution

The center of origin of maize is Mesoamerica. Most researchers now consider that maize was domesticated 9000 years ago from a lowland tropical teosinte of the Balsas valley of Mexico, *Zea mays* ssp. *parviglumis* Iltis & Doebley. This single domestication occurred in a forest located close to the tropic of Cancer between 700 and 1800 m of elevation with 1500 mm of summer rainfall. In a second step, this proto-maize has been crossed above 1500 m through gene-fluxes from another teosinte, *Zea mays* ssp. *mexicana* (Schrad.) Iltis thus adapted to highlands (Bonjean 2020).

From 7000 years ago, human migration dispersed this semi-domesticated crop to South America. It seems that *Tripsacum* introgression happened in this subcontinent, where teosinte is not present (Grobman Tversqui et al. 1961; Mangelsdorf 1961). Archaeological discoveries proved that maize was present in Ecuador from 4400 BC, in Peru between 2800 and 2600 BC, in Colombia between 2745 and 2380 BC, and in Uruguay between 2800 and 2500 BC (Bonjean 2020).

Therefore, South America, with its wide variety of agroclimatic environments, became an expansive secondary stratified domestication centre where numerous semi-domesticated subgroups emerged, in parallel with similar processes in Mesoamerica (Kistler et al. 2018). Recent studies (Kistler et al. 2020) demonstrated that as the second wave of Mesoamerican cultivars reached South America and intercrossed with Andean maize resulting in more productive landraces between 2300 and 500 BC before their progenies were introduced back into Central America around 300 BC.

1.5.6.1.1 Genetic Diversification and Cultivation

Three phenomena, each in its own way, have strongly influenced the cultivation, early spread, and genetic diversification of maize in Mexico, Mesoamerica, South America, and the Andean area:

- Around 1500–1200 BC, *nixtamalization* was developed in Mesoamerica; a hot-treatment and aqueous process, which uses lime, and allows a safe daily consumption of maize grain, making it at the same time more nutritious for the human population (Cravioto et al. 1945; Staller 2010).
- Since at least 3000 years BC, Mesoamerican farmers were primarily vegetarians. However, they then invented the *milpa* system, an intercropping of maize, squash (*Cucurbita pepo* L.), and bean (*Phaseolus* spp.), which limited the use of water, pesticides, and fertilizers, offering a better nutritional diet than maize alone (Gianoli et al. 2006; Almaguer González et al. 2015).
- Simultaneously, native populations associated maize with several gods in their native religions and numerous linked rituals in their culture, the most common being the production and share of chicha, a low-alcoholic drink with great religious and social importance (Pérez-Suárez 1997; Kulas 2015) in all of Latin America.

During the pre-Colombian period, maize was a primary staple food for the Andean populations together with potato (*Solanum tuberosum* L.), quinoa (*Chenopodium quinoa*), amaranth (*Amaranthus cruentus* L.), tarwi (*Lupinus mutabilis*), and other crops, and also a part of their cosmovision.

During the 16th century, when the *Conquistadores* invaded the Andean regions, they forced populations to cultivate wheat and barley and adopt the Christian religion; they pronounced a ban on some Inca crops such as quinoa and amaranth involved in previous cults. However, due to its high productivity and large grain, the European invaders adopted maize. Thus, they introduced it into Europe, where it progressively replaced millet crops (both *Panicum miliaceum* L. and *Setaria italic* (L.) P. Beauv.). Moreover, due to the better adaptation of maize than wheat to tropical environments, the Europeans also introduced it in Africa and Asia during their successive waves of colonial expansion.

Later, at the beginning of the 20th century, maize became hybrid in the USA (Crown 1998) and genetically modified in the 1990s (Lundmark 2007).

1.5.6.2   Limitations

According to CIMMYT, the average yield in Andean countries is 3.6 t ha^{-1} compared to a world average yield of 5.4 t ha^{-1}, mostly because Andean farmers do not systematically use improved or hybrid seeds, and they lack financial funds and technical data. Moreover, mechanization is not always easy to implement in their mountainous zone. Each year Andean nations import part of the maize grain they consume.

Andean countries' increasing food security from maize production would need better quality, improved genetics, and durable field practices, including irrigation and better post-harvest storage facilities (González 2018) (see Figure 1.7).

1.5.6.3   Future Potential

All Andean countries have maintained a large diversity of traditional maize populations adapted to various Andean region environments, both *in situ* with local farmers and ex-situ within national and CIMMYT genebanks (Roberts et al. 1957).

From the beginning of the 1990s, CIMMYT and Andean institutions have developed a breeding program dedicated to subtropical, mid-altitude, and highland maize (Bjarnason 1994; Galeano et al. 2019). In 2008 CIMMYT, CIAT, and Harvest Plus also launched a zinc-enriched maize program to address zinc deficiency, which is usually characterized by growth retardation, loss of appetite, and impaired immune function, and affects part of the Andean population (Michail 2019).

The local Andean maize varieties own wide phenotypic diversity (Staller 2016) and include a floury and soft grain with varied colors, shapes, and sizes, which constitute an attractive breeding reservoir for future selection of staple

Figure 1.7   Andean maize (**A**) Plants in Bolivia, and (**B**) maize ears. **Source: Authors'
own work.**
Notes: A) The image on the left shows an Andean maize plantation at the back of a sparse
bean field. In the backgound, some tall trees are visible.
B) The image on the right shows many different Andean maize ear types in yellow, black,
and white in which the grains also vary in color within an ear.

food that needs to be protected from genetically modified (GM) maize pollen
fluxes. In addition, their natural pigments and antioxidants are related to health
benefits.

Climate change affects the Andean region (higher temperatures, reduction in
glacier sizes, instability of rainfall and water flow, etc.) as with other parts of the
planet. Second are damage from disease, insects, and other pests increase, and
risks of soil erosion and loss of plant and animal biodiversity. Finally, the warmer
climate also changes cropping systems: maize and potato can now be grown at
higher altitudes than previously, frequently accompanied by the development of
terracing (Skarbø and van der Molen 2016).

In addition, drought is becoming a more common issue. It requires improvement in water management and selection of maize material more tolerant to
abiotic stresses. Technologies such as tilling and gene editing offer hope in this
direction. Agroforestry with native trees may be another solution, which allows
stabilizing mountain slopes and riverbanks.

The challenge of increasing Andean maize production while protecting the
fragile Andean environment and its endemic biodiversity lies mainly in the
development of conservation agriculture.

1.5.7   Sacha Inchi (*Plukenetia volubilis* L.)

1.5.7.1   Introduction

Plukenetia volubilis is an underutilized oilseed crop native to the Amazon
basin, where humans have utilized it since Incan times. The large seeds contain

Figure 1.8 Sacha inchi plant with fruits. **Source: Authors' own work.**
Note: The image shows a Sacha inchi plant with fruit capsules of four lobes and condiform shaped leaves in the field. At the back of the Sacha inchi field, some short palm trees are visible in a dark green color.

45–50% lipid, of which 35.2–50.8% is α-linolenic acid (C18:3 n-3, ω-3) and 33.4–41.0% is linoleic acid (C18:2 n-6, ω-6), the two essential fatty acids required by humans. The seeds also contain 22–30% protein and have antioxidant properties (Hamaker et al. 1992; Follegatti-Romero et al. 2009; Gutiérrez et al. 2011; Maurer et al. 2012; Chirinos et al. 2013; Cisneros et al. 2014; Triana-Maldonado et al. 2017). As a result, its excellent nutritional composition and good agronomic properties have attracted increasing attention in recent years, and cultivation is expanding.

1.5.7.1.1 *Morphology, phylogeny, and distribution*

Plukenetia volubilis is a perennial liana with large, oleaginous seeds. The plants are monecious; the leaves are triangular to ovate with a truncate to cordate base, palmate venation and basilaminar glands, usually with a small knob between them. The racemose inflorescence is axillary or terminal with one to two pistillate flowers situated basally and numerous small, inconspicuous, staminate flowers in condensed cymes situated above (Kodahl and Sørensen 2021). The winged ovary has four carpels, and the style column is elongate and cylindrical, four-lobed at the apex. During fruit maturation, the ovary develops from green and fleshy to brown, woody, and dehiscent. The seeds are lenticular, around 1.8 × 0.8 × 1.6 cm in size, and the testa is hard and brown, with dark brown markings. In cultivation, the fruit is often larger sized and is five- or six-carpellate (Gillespie 1993; Gillespie and Armbruster 1997). The genus *Plukenetia* L. (Euphorbiaceae) comprises 25 species, several of which have only recently been described. Circumscription of the genus has undergone several changes during the last four centuries, but Cardinal-McTeague and Gillespie (2020) recently revised the classification. *Plukenetia* belongs to tribe Plukenetieae, subtribe Plukenetiinae, and is distinguished by four-carpellate ovaries and two extrafloral nectaries situated basally on the adaxial surface of the leaf blade. The genus is divided into two major clades: the pinnately veined clade, one primary vein, and the palmately veined clade, including *P. volubilis*, with three to five primary veins (Cardinal-McTeague and Gillespie 2020).

Plukenetia volubilis is distributed widely in South America and the Lesser Antilles; it is found in the northern and western parts of the Amazon Basin in Surinam, Venezuela, Colombia, Ecuador, Peru, Bolivia, and Brazil (Gillespie and Armbruster 1997). The most common ecological niche for *P. volubilis* is a moist to wet lowland forest, but the species complex comprises two morphologically differing groups: an open savannah and a mid-elevation species group. The open savannah group generally has thicker leaf blades and smaller seeds and fruits, while the mid-elevation group has narrower leaf blades and differing leaf base morphology compared with typical *P. volubilis*. However, further studies are needed to define species group boundaries better and assess whether the definition of a new, additional species from within the complex is warranted (Cardinal-McTeague and Gillespie 2020).

1.5.7.1.2 Uses

Plukenetia volubilis has traditionally been consumed in Peru and other countries of Latin America, and has been associated with humans since pre-Hispanic times. Artifacts depicting *P. volubilis* fruits and vines have been found in Incan burial sites along the coast of Peru, indicating that the plant may have been cultivated by pre-Incan cultures 3000–5000 years ago (Brack Egg 1999).

The vernacular name 'Sacha inchi' is Quechuan and is the most commonly used name for *P. volubilis* and other large-seeded species in the genus. However, 'Sacha inchik' or 'Sacha inchic' is also used depending on the dialect, and the meaning is the same; 'sacha' can be translated to 'mountain' or sometimes to 'false' or 'resembling', and 'inchi' means groundnut/peanut. Less common names for *P. volubilis* include Sacha Yachi, Sacha Yuchi, Sacha yuchiqui, Yuchi, sampannankii, suwaa, Correa, amauebe, amui-o, maní de Arbol, maní del monte, and maní Estrella, several of which hint at the nut-like texture of the seed (Brack Egg 1999; D. Cacique pers. comm.). Accordingly, some of the most often mentioned culinary uses are similar to the uses of groundnuts; *P. volubilis* seeds are most commonly consumed roasted and salted as a snack but are also used in confectionery, e.g. dipped in chocolate, or ground to a butter-like substance, milled to flour, or used in a large variety of traditional dishes. These include 'inchi cucho' (a spicy, savoury sauce or dip), 'lechona api' (plantain porridge), and 'inchi capi' (chicken or beef soup). Likewise, the young leaves are occasionally eaten in salads or in tea (Flores 2010; Flores and Lock 2013).

However, while *P. volubilis* has many culinary applications, the most reported use was shown by an ethnobotanical study performed in San Martín, Peru, which was related to health (67% of answers) (Rodríguez del-Castillo et al. 2019). Accordingly, several Peruvian ethnic groups, including the Mayorunas, Chayuhitas, Shipibas, and Boras, have traditionally used a mixture of *P. volubilis* ground seeds and seed oil as a skin cream to rejuvenate and revitalize the skin. Similarly, the Secoyas, Candoshis, Amueshas, and Cashibos, among others, have rubbed *P. volubilis* oil on the skin to relieve muscle pain and rheumatism (Flores and Lock 2013). In addition, the oil and roasted seeds have been consumed for cholesterol control, and for cardiovascular and gastrointestinal problems (Rodríguez del-Castillo et al., 2019).

With the increasing awareness and popularity of *P. volubilis* in international markets in recent years, several new or differently branded products have also become available, e.g. gourmet oil, protein powder, and encapsulated oil marketed as a dietary supplement. In addition, roasted and salted, or candied, seeds are marketed.

1.5.7.2   Limitations and Breeding Opportunities

Plukenetia volubilis is an up-and-coming crop, primarily due to the nutritional composition of the seeds; however, further study of the plant and development of breeding strategies will probably prove beneficial. Importantly, although there is a focus on sustainability in the plants' native range, exploration of more sustainable management practices will be advantageous, both regarding biodiversity and climate, but might also improve product quality and provide opportunities for product branding and marketing. Moreover, even though the seed oil has been approved for consumption in the European Union, the seeds of *P. volubilis* are not yet approved due to a lack of knowledge concerning alkaloid content and composition in the seeds (EFSA 2020). Studies indicate that thermal processing significantly reduces the amount of alkaloid compounds in the seeds (Srichamnong et al. 2018). However, the recent decision from the European Food Safety Authority (EFSA 2020) nevertheless underlines the need for further documentation of the safety of consumption of the seeds, including details on potentially allergenic or toxic compounds.

Very little breeding of *P. volubilis* has been carried out, so the plant is not considered to be fully domesticated (Vašek et al. 2017). Breeding should be aimed at improving the agronomy of the crop, higher yield and improved sensory qualities. Similarly, further exploration of the domestication potential of other large-seeded species in the genus would be of interest, both as crops in their own right but possibly also as material for the development of hybrids with *P. volubilis*.

1.5.7.3   Future Potential

Current global challenges include climate change, degradation of land and environment, population growth, and lack of food security. Accordingly, our food systems need to be optimized to ensure food security while avoiding global ecosystem collapse and the subsequent loss of ecosystem services. It is becoming ever more apparent that upscaling of current agricultural systems, particularly monocultures, is not the best procedure for this purpose, and alternative strategies are needed (Jacobsen et al. 2015; Funabashi 2018). Furthermore, neglected and underutilized crops may prove necessary resources for the reformation of our food production systems by improving, e.g., climate change resilience, genetic diversity, and the nutritional value of agricultural products.

Plukenetia volubilis has considerable potential for contributing to these goals, as the plants can thrive in a broad range of environmental conditions, have an exceptional nutritional composition, and possibly be an economically beneficial alternative crop for small-scale farmers. Further, a wide variety of cultivars and a high genetic diversity of germplasm is available, providing outstanding opportunities for further domestication and breeding, ultimately to establish the integration of *P. volubilis* in sustainable cultivation systems. Thus, on a local scale, *P. volubilis* may aid in food security, alleviate malnutrition, and provide economic benefits, while simoultaneously being part of the solution to our global challenges. Consequently, *P. volubilis* has considerable potential for further domestication.

1.6   GENERAL DISCUSSION

There is great potential for increasing production of Latin-American crops, such as those presented in this chapter. However, it must be done in a sustainable and wise way. In Ecuador, an increase in demand for Andean lupine intensified the production system of the crop by using improved varieties and a larger cultivated area. It caused an escalation in pests and the indiscriminate use of insecticides (Mina et al. 2018). It is obvious that increased production of new Andean or Latin-American crops may lead to new problems, which we must be prepared for in order to choose the best possible strategies.

The failure of agricultural development implemented in the Andean highland and Latin America among other factors has perpetuated poverty and malnutrition in the population. Ancient technologies – in danger of abandonment – seem to be one of the options for increased food production and improvement of the actual standard of living.

Production of healthy, natural, and nutraceutical food, like the crops analyzed in this report, associated with environmental modifications, is the best way to obtain economically acceptable organic production under challenging conditions of climate, soil, and slope. The ancient knowledge of cultivation suggests rational use of soil, water, and nutrients; integral control of pests and diseases; and processing and use of foods. We must solve the nutritional problems that rural communities face throughout the Andes mountains as in other regions of the Latin-American and Caribbean subcontinent.

The altiplano of Peru and Bolivia and traditional Latin-American agricultural systems can be considered under-utilized agroecosystems, capable of reverting the current food deficiency by applying a rational and suitable use of existing environmental modifications and using the great diversity and variability of highly nutritious native Andean and Latin American crops.

Specific data about the consumption of Andean and Latin-American crops are needed, which will be useful for making decisions and taking actions that allow their promotion for more consumption. Several crops offer an option for future food demand, and some of them have been included in this chapter. They are examples of crops with a great potential for increased use for food locally and regionally, and also serving as export crops thus creating income for farmers and in general for the Latin-American region. The species mentioned here have high nutritional values, are healthy, climate-resilient, and tasty, and they may be important components for more plant-based foods, replacing meat. These crops are considered superfoods because of their outstanding nutritional and nutraceutical characteristics; therefore, they should be key components in the diet of people in Latin America and in the rest of the world.

REFERENCES

Adel, H. 2020. Towards expanding quinoa cultivation in Egypt: The effect of compost and vermicompost on quinoa pests, natural enemies and yield under field conditions. *Agricultural Sciences* 11:191–209. https://doi.org/10.4236/as.2020.112012.

Akinnifesi, F. K., R. R. B. Leakey, O. C. Ajayi, et al. 2008. *Indigenous Fruit Trees in the Tropics: Domestication, Utilization and Commercialization.* CAB International Publishing, Wallingford, UK.

Alandia, G., J. P. Rodriguez, S.-E. Jacobsen, D. Bazile, and B. Condori. 2020. Global expansion of quinoa and challenges for the Andean region. *Global Food Security* 26: 100429. https://doi.org/10.1016/j.gfs.2020.100429.

Almaguer González, J. A., H. J. García Ramírez, M. Padilla Mirazo, and M. González Ferral. 2015. *La dieta de la milpa: modelo de alimentacion mesoamericana biocompatible.* Secretaria de Salud, Mexico. https://alianzanahuaca.org/2018/07/10/la-dieta-de-la-milpa-modelo-de-alimentacion-mesoamericana/.

Alvarado Rupflin, D. I. 2011. Caracterización de la semilla de chan (*Salvia hipanica* L.) y diseño de un producto funcional que la contiene como ingrediente. *Revista de la Universidad del Valle Guatemala* 23: 43–49. https://xdoc.mx/preview/caracterizac ion-de-la-semilla-del-chan-salvia-hispanica-l-y-diseo-602f482d946c8.

Alvar-Beltran, J., A. Dao, A. D. Marta, et al. 2019. Effect of drought, nitrogen fertilization, temperature and photoperiodicity on quinoa plant growth and development in the Sahel. *Agronomy* 9: 607. https://doi.org/10.3390/agronomy9100607.

Atchison, G. W., B. Nevado, R. J. Eastwood, N. Contreras-Ortiz, et al. 2016. Lost crops of the Incas: Origins of domestication of the Andean pulse crop Tarwi, *Lupinus mutabilis. American Journal of Botany* 103(9): 1592–1606. https://doi.org/10.3732/ajb.1600171.

Ayerza, R., and W. Coates. 2005. *Chia: Rediscovering a Forgotten Crop of the Aztecs.* University of Arizona Press, Tucson, AZ.

Bazile, D., S.-E. Jacobsen, and A. Verniau. 2016. The global expansion of Quinoa: Trends and limits. *Frontiers in Plant Science* 7: 622. https://doi.org/10.3389/fpls.2016.00622.

Beebe, S., P. W. Skroch, J. Tohme, M. C. Duque, F. Pedraza, and J. Nienhuis. 2000. Structure of genetic diversity among common bean landraces of Middle American origin based on correspondence analysis of RAPD. *Crop Science* 40(1): 264–273. https://doi.org/10.2135/cropsci2000.401264x.

Beltrán, O. M. C., and M. R. Romero. 2003. *La chía, alimento milenario. Departamento de Graduados e Investigación en Alimentos.* ENCB. Instituto Politécnico Nacional (IPN), México.

Bentham, G., and A. S. Oersted. 1854. Labiatae centroamericanae. *Videnskabelige Meddelelser fra Dansk Naturhistorisk Forening i Kjøbenhavn* 1853(1–2): 32–42.

Bermúdez-Torres, K., J. M. Herrera, R. F. Brito, M. Wink, and L. Legal. 2009. Activity of quinolizidine alkaloids from three Mexican *Lupinus* against the lepidopteran crop pest *Spodoptera frugiperda. BioControl* 54: 459–466. https://doi.org/10.1007/s10 526-008-9180-y.

Bernal, A. E., J. J. Iñaguazo, and B. Chanducas. 2015. Efecto del consumo de chía (*Salvia hispanica*) sobre los síntomas de estreñimiento que presentan los estudiantes de una universidad particular de Lima Este. *Revista Científica de Ciencias de la Salud* 8(2): 8–24. https://doi.org/10.17162/rccs.v8i2.468.

Bertero, H. D., A. J. de la Vega, G. Correa, S.-E. Jacobsen, and A. Mujica. 2004. Genotype and genotype-by-environment interaction effects for grain yield and grain size of quinoa (*Chenopodium quinoa* Willd.) as revealed by pattern analysis of multi-environment trials. *Field Crops Research* 89: 299–318. https://doi.org/10.1016/j.fcr.2004.02.006.

Beyra, A., and G. Reyes Artiles. 2004. Revisión taxonómica de los géneros *Phaseolus* y *Vigna* (Leguminosea-Papilionoideae) en Cuba. *Anales del Jardín Botánico de Madrid* (1979) 61(2): 135–154. https://doi.org/10.3989/ajbm.2004.v61.i2.41.

Bitocchi, E., L. Nanni, E. Bellucci, et al. 2012. Mesoamerican origin of the common bean (*Phaseolus vulgaris* L.) is revealed by sequence data. *PNAS* 109: E788–E796. https://doi.org/10.1073/pnas.1108973109.

Bitocchi, E., D. Rau, E. Belucci, et al. 2017. Beans (*Phaseolus* ssp.) as a model for understanding crop evolution. *Frontiers in Plant Science* 8(122): 1–21. https://doi.org/10.3389/fpls.2017.00722.

Blair, M. W., A. J. Cortés, R. Varma Penmetsa, A. Farmer, N. Carrasquilla-Garcia, and D. R. Cook. 2013. A high-throughput SNP marker system for parental polymorphism screening, and diversity analysis in common bean (*Phaseolus vulgaris* L.). *Theoretical and Applied Genetics* 126(2): 535–548. https://doi.org/10.1007/s00122-012-1999-z.

Bochicchio, R., T. D. Philips, S. Lovelli, et al. 2015. Innovative crop productions for healthy food: The case of chia (*Salvia hispanica* L.). In *The Sustainability of Agro-Food and Natural Resource Systems in the Mediterranean Basin*, ed. A. Vastola, 29–45. Springer Nature AG, Cham, Switzerland. https://doi.org/10.1007/978-3-319-16357-4_3.

Bonifacio, A. 2019. Improvement of Quinoa (*Chenopodium quinoa* Willd.) and Qañawa (*Chenopodium pallidicaule* Aellen) in the context of climate change in the high Andes. *International Journal of Agriculture and Natural Resources* 46(2): 113–124. http://dx.doi.org/10.7764/rcia.v46i2.2146.

Bonjean, A. P. 2020. Le maïs, rouleau compresseur américain. *Paysans & société* 383: 38–46. https://doi.org/10.3917/pes.383.0038.

Borlaug, Norman E., 2000. Ending world hunger. The promise of biotechnology and the threat of antiscience zealotry. In *Plant Physiology*. American Society of Plant Biologists (OUP).124(2): 487–490. doi:10.1104/pp.124.2.487. ISSN 1532-2548.

Brack Egg, A. 1999. *Diccionario Enciclopedico de Plantas Utiles del Peru*, p. 400. PNUD, Cuzco, Peru.

Bravo, R., R. Valdivia, K. Andrade, S. Padulosi, and M. Jager. 2010. *Granos andinos: avances, logros y experiencias desarrolladas en quinua, cañihua y kiwicha en Perú*. Bioversity International, Roma. www.bioversityinternational.org/e-library/publications/detail/granos-andinos-avances-logros-y-experiencias-desarrolladas-en-quinua-canihua-y-kiwicha-en-peru/.

Bressani, R. 1990. Chemistry, technology, and nutritive value of maize tortillas. *Food Reviews International* 6(2): 225–264. https://doi.org/10.1080/87559129009540868.

Bresson, J. L., A, Flynn, and M. Heinonen. 2009. Opinion on the safety of chia seeds (*Salvia hispanica* L.) and ground whole chia seeds as a food ingredient. *European Food Safety Authority Journal* 996: 1–26. https://doi.org/10.2903/j.efsa.2009.996.

Broughton, W. J., G. Hernández, M. Blair, S. Beebe, P. Gepts, and J. Vanderleyden. 2003. Beans (*Phaseolus* spp.) – model food legumes. *Plant and Soil* 252: 55–128. https://doi.org/10.1023/ A:1024146710611.

Bruno, M. C. 2006. A morphological approach to documenting the domestication of *Chenopodium* in the Andes. In *Documenting Domestication: New Genetic and Archaeological Paradigms*, ed. M. A. Zeder, D. G. Bradley, E. Emshwiller, and B .D. Smith, 32–45. University of California Press, Berkeley, CA.

Bruno, M. C. 2014. Beyond raised fields: Exploring farming practices and processes of agricultural change in the ancient Lake Titicaca Basin of the Andes. *American Anthropologist* 116(1): 130–145. https://doi.org/10.1111/aman.12066.

Bunsupa, S., K. Katayama, E. Ikeura, et al. 2012. Lysine decarboxylase catalyzes the first step of quinolizidine alkaloid biosynthesis and coevolved with alkaloid production in Leguminosae. *Plant Cell* 24(3): 1202–1216. https://doi:10.1105/tpc.112.095885.

Busilacchi, H., T. Qüesta, and S. Zuliani. 2015. La chía como una nueva alternativa productiva para la región pampeana. *Agromensajes* 41(2): 37–46. https://core.ac.uk/download/pdf/162568326.pdf.

Cabrera, J. C., and M. F. Cerna. 2014. Optimizaciòn de la aceptabilidad de un pan integral de chia (*Salvia hispanica* L.) mediante la metodologìa de Taguchi. *Agroindustrial Science* 4(1): 19–25. https://doi.org/10.17268/agroind.sci.

Cahill, J. P. 2001. Domestication of chia, *Salvia hispanica* L. (Lamiaceae). PhD thesis, University of California, Riverside, CA.

Cahill, J. P. 2003. Ethnobotany of chia, *Salvia hispanica* L. (Lamiaceae). *Economic Botany* 57: 604–618. https://doi.org/10.1663/0013-0001(2003)057[0604:EOCSHL]2.0.CO;2.

Cahill, J. P. 2004. Genetic diversity among varieties of chia (*Salvia hispanica* L.). *Genetic Resources and Crop Evolution* 51: 773–781. https://doi.org/10.1023/B:GRES.0000034583.20407.80.

Cahill, J. P. 2005. Human selection and domestication of chia (*Salvia hispanica* L.). *Journal of Ethnobiology* 25: 155–174. https://doi.org/10.2993/0278-0771(2005)25[155:HSADOC]2.0.CO;2.

Caicedo, C., and E. Peralta. 2000. Zonificación potencial, sistemas de producción y procesamiento artesanal de chocho (*Lupinus mutabilis* Sweet*) en Ecuador. Boletín Técnico* Nº 89. Programa Nacional de Leguminosas. Estación Experimental Santa Catalina. INIAP-FUNDACYT-P-BID-206. Quito-Ecuador. https://repositorio.iniap.gob.ec/bitstream/41000/441/4/iniapscbt89.pdf.

Caicedo, C., E. Peralta, E. Villacrés, and M. Rivera, 2001. Postcosecha y mercado del chocho (*Lupinus mutabilis* Sweet) en Ecuador. INIAP, FUNDACYT, Quito-Ecuador. https://repositorio.iniap.gob.ec/bitstream/41000/2700/1/iniapscpm105.pdf.

Caligari, P. D. S., P. Römer, M. A. Rahim, C. Huyghe, J. Neves-Martins, and E. J. Sawicka-Sienkiewicz. 2000. The potential of *Lupinus mutabilis* as a crop. In *Linking Research and Marketing Opportunities for Pulses in the 21st Century: Proceedings of the Third International Food Legumes Research Conference*, ed. R. Knight, 569–573. Springer, DordrechtNetherlands). https://doi.org/10.1007/978-94-011-4385-1_54.

Calle, E. 2004. Efecto de la cañahua (*Chenopodium pallidicaule* Aellen), trito (*Triticum aestivum* L.), soya (*Glycine max* Merr.), germinados en la alimentacion de cuyes en recria para la prevencion de escorbuto. PhD thesis, Universidad Tecnica de Oruro, Oruro, Bolivia.

Canahua, A., and P. Román. 2016. Tarwi Leguminosa andina de gran potencial. *LEISA Revista de Agroecología* 32(2): 20–22. www.canunite.org/wp-content/uploads/2016/07/Leisa_vol32n2.pdf.

Capitani, M. I., S. M. Nolasco, and M.C. Tomás. 2013. Effect of mucilage extraction on the functional properties of Chia meals. In *Food Industry*, ed I. Muzzalupo, Ch. 1, 1–19. *IntechOpen.* https://doi.org/10.5772/53171. Available from: www.intechopen.com/chapters/41676.

Cardinal-McTeague, W. M., and L. J. Gillespie. 2020. A revised sectional classification of *Plukenetia* L. (Euphorbiaceae, Acalyphoideae) with four new species from South America. *Systematic Botany* 45(3): 507–536. https://doi.org/10.1600/036364420X15935294613572.

Carrasco, D. E., and J. L. Soto. 2010. II. Importancia de los granos andinos. In *Granos Andinos. Avances, Logros y Experiencias Desarrolladas en Quinua, cañahua y Amaranto en*

Bolivia, ed. W. Rojas, J. L. Soto, M. Pinto, M. Jager, and S. Padulosi, 6–10. Bioversity International, Rome. https://cgspace.cgiar.org/bitstream/handle/10568/104701/ Granos_andinos_avances_logros_y_experiencias_desarrolladas_en_quinua_ ca%C3%B1ahua_y_amaranto_en_Bolivia_1413.pdf?sequence=3&isAllowed=y.

CGIAR-Beanfocus. 2019. Common bean: The nearly perfect food – The importance of common bean. http://ciat-library.ciat.cgiar.org/articulos_ciat/ciatinfocus/beanfo cus. pdf (accessed December 28, 2020).

Chambi, F. A. 2019. Elaboración de cup-cakes con sustitución parcial de harina de trigo con harina de quinua (*Chenopodium quinoa*), kiwicha (*Amaranthus caudatus*), cañihua (*Chenopodium pallidicaule*) y sustitución de grasa por gomas de linaza (*Linum usitatissimum*) y chia (*Salvia hispánica*). PhD thesis, Universidad Peruana Unión, Ñaña, Lima. http://repositorio.upeu.edu.pe/handle/UPEU/2941 (accessed December 3, 2020).

Chirinos, R., G. Zuloeta, R. Pedreschi, E. Mignolet, Y. Larondelle, and D. Campos. 2013. Sacha inchi (*Plukenetia volubilis*): A seed source of polyunsaturated fatty acids, tocopherols, phytosterols, phenolic compounds and antioxidant capacity. *Food Chemistry* 141: 1732–1739. https://doi.org/10.1016/j.foodchem.2013.04.078.

Chirinos-Arias, M., E. J. Jiménez, and S. L. Vilca-Machaca. 2015. Analysis of genetic variability among thirty accessions of Andean Lupin (*Lupinus mutabilis* Sweet) using ISSR molecular markers. *Scientia Agropecuaria* 6: 17–30. https://doi.org/10.17268/ sci.agropecu.2015.01.02.

CIAT (Centro Internacional de Agricultura Tropical). 2020. Conservación y uso de cultivos. https://ciat.cgiar.org/lo-que-hacemos/conservacion-y-uso-de-cultivos/?lang=es.

Ciesiolka, D., P. Gulewicz, C. Martínez-Villaluenga, R. Pilarski, M. Bednarczyk, and K. Gulewicz. 2005. Products and biopreparations from alkaloid-rich lupin in animal nutrition and ecological agriculture. *Folia Biologica (Kraków)* 53: 59–66. https://doi. org/10.3409/173491605775789443.

Cisneros, F. H., D. Paredes, A. Arana, and L. Cisneros-Zevallos. 2014. Chemical composition, oxidative stability and antioxidant capacity of oil extracted from roasted seeds of sacha-inchi (*Plukenetia volubilis* L.). *Journal of Agricultural and Food Chemistry* 62: 5191–5197. https://doi.org/10.1021/ jf500936j.

Clavijo, P. J. 1980. Resumen general de las principales características agronómicas de diferentes granos en Colombia. Instituto Interamericano de Ciencias Agrícolas. IICA, Colombia. https://repositorio.iica.int/bitstream/handle/11324/15678/CDRP210415 16e.pdf?sequence=1&isAllowed=y.

Clements, J. C., M. S. Sweetingham, L. Smith, G. Francis, G. Thomas, and S. Sipsas. 2008. Crop improvement in *Lupinus mutabilis* for Australian agriculture-progress and prospects. In *Lupins for Health and Wealth. Proceedings of the 12th International Lupin Conference, Fremantle, Western Australia, 14–18 September 2008*, ed. J. A. Palta, and J. B. Berger, 244–250. International Lupin Association, Canterbury, New Zealand. ISBN (Print) 0864761538.

Coelho, C. M. M., C. de Mattos Bellato, J. C. P. Santos, E. M. M. Ortega, and S. M. Tsai. 2007. Effect of phytate and storage conditions on the development of the 'hard-to-cook' phenomenon in common beans. *Journal of the Science of Food and Agriculture* 87: 1237–1243. https://doi.org/10.1002/jsfa.2822, www.fao.org/faostat/en/?#data/ QC (accessed December 28, 2020).

Coila, R. A. 2019. Optimización en la elaboración de galletas utilizando harina de cañihua (*Chenopodium pallidicaule*), kiwicha (*Amaranthus caudatus*) y quinua (*Chenopodium quinoa*). PhD thesis, Universidad Nacional de Trujillo, Peru. https://1library.co/document/nzwlmp1y-optimizacion-elaboracion-galletas-utilizando-chenopodium-pallidicaule-amaranthus-chenopodium.html (accessed November 24, 2020).

Contin Gomes, M. J., H. S. Duarte Martino, and E. Tako 2021. Effects of iron and zinc biofortified foods on gut microbiota in vivo (*Gallus gallus*): A systematic review. Nutrients 13(1): 189. https://doi.org/10.3390/nu13010189.

Cortez Quispe, H. A. 2016. Evaluación de cuatro niveles de polvillo de Qañäwa (*Chenopodium pallidicaule* A.) en la alimentación de Cuyes (*Cavia porcellus* L.) en crecimiento. *Revista Apthapi* 1(2): 86–95. http://ojs.agro.umsa.bo/index.php/ATP/article/view/145/145.

Cowling, W. A., B. J. Buirchell, and M. E. Tapia, 1998. Lupin. *Lupinus. Promoting the conservation and use of underutilized and neglected crops*. International Plant Genetic Resources Institute [IPGRI], Rome. www.bioversityinternational.org/fileadmin/user_upload/online_library/ publications/pdfs/Lupin_23.pdf.

Cravioto, R. O., R. K, Anderson, E. E. Lockhart, F. de P. Miranda, and R. S. Harris. 1945. Nutritive value of the Mexican tortilla. Science 102 (2639): 91–93. https://doi.org/10.1126/science.102.2639.91.

Cremer, H. D. 1983. Current aspects of legumes as a food constituent in Latin America with special emphasis on lupines: Introduction. *Plant Foods for Human Nutrition* 32: 95–100. https://doi.org/10.1007/BF01091329.

Crown, J. F. 1998. 90 years ago: The beginning of hybrid maize. *Genetics* 148(3): 923–928. www.ncbi.nlm.nih.gov/pmc/articles/PMC1460037/pdf/9539413.pdf.

De Ron, A., F. Sparvoli, J. Pueyo, and D. Bazile. 2017. Editorial: Protein crops: Food and feed for the future. *Frontiers in Plant Science* 8: 105. https://doi.org/10.3389/fpls.2017.00105.

Doxastakis, G. 2000. Lupin seed proteins. *Developments in Food Science* 41: 7–38. https://doi.org/10.1016/S0167-4501(00)80004-7.

EFSA (European Food Safety Authority). 2020. Technical Report on the notification of roasted seeds from *Plukenetia volubilis* L. as a traditional food from a third country pursuant to Article 14 of Regulation (EU) 2015/2283. *European Food Safety Authority Journal* Supporting publication 2020: EN-1817:17(3): 1–13. https://doi.org/10.2903/sp.efsa.2020.EN-1817.

Erickson, C. L. 1985. Applications of prehistoric Andean Technology: Experiments in raised field agriculture, Huatta, Lake Titicaca: 1981–1982. In *Prehistoric Intensive Agriculture in the Tropics*, ed. I. S. Farrington, 209–232. British Archaeological Reports, Oxford.

Espinoza-Moreno, R. J., C. Reyes-Moreno, J. Milán-Carrillo, J. A. López-Valenzuela, O. Paredes-López, and R. Gutiérrez-Dorado. 2016. Healthy ready-to-eat expanded snack with high nutritional and antioxidant value produced from whole amarantin transgenic maize and black common bean. *Plant Foods for Human Nutrition* 71: 218–224. https://doi.org/10.1007/s11130-016-0551-.

Estaña, W., and C. Muñoz. 2012. Variabilidad genética de la Cañihua en las provincias de Puno. Equipo Tecnico del "Proyecto Mejoramiento de Capacidades Técnico Productivas para la Competitividad de los Cultivos Andinos de Papa Nativa, Haba y Cañihua en la Region Puno", Puno, Peru. https://docplayer.es/60307065-Canihua-variabilidad-genetica-en-las-provincias-de-puno.html.

Falconí, C. 2012. Lupinus mutabilis in Ecuador with special emphasis on anthracnose resistance. PhD dissertation. [Wageningen (NL)]: Wageningen University, the Netherlands.

FAO (Food and Agriculture Organization of the United Nations). 1992. *Cultivos marginados, otra perspectiva de 1492*. Rome. www.fao.org/3/t0646s/t0646s.pdf (accessed September 12, 2020).

FAO (Food and Agriculture Organization of the United Nations). 2000. Cultivos andinos subexplotados y su aporte a la alimentación, 2nd edn. Santiago de Chile-Chile. www.fao.org/tempref/GI/Reserved/FTP_FaoRlc/old/prior/segalim/ prodalim/prodveg/cdrom/contenido/libro10/home10.htm (accessed 12 September 2020).

FAO (Food and Agriculture Organization of the United Nations). 2013. The International Year of Quinoa – A future sown thousands of years ago. www.fao.org/quinoa-2013/en/.

FAO (Food and Agriculture Organization of the United Nations). 2016. Consumo y producción de legumbres ha perdido fuerza en América Latina y el Caribe frente a cultivos más comerciales. www.fao.org/americas/noticias/ver/es/c/455947/ (accessed September 12, 2020).

FAO (Food and Agriculture Organization of the United Nations). 2018. Legumbres pequeñas semillas, grandes soluciones. Ciudad de Panamá. 292 p. ISBN 978-92-5-131129-5. www.fao.org/3/ca2597es/CA2597ES.pdf.

FAOSTAT (Food and Agriculture Organization of the United Nations). 2021. Food and agriculture data. Available at www.fao.org/faostat/en/?#data/QC (accessed December 28, 2020).

FAO/WHO (Food and Agriculture Organization of the United Nations/World Health Organization). 2016. Report joint FAO/WHO food standards programme. Codex alimentarius commission. 39th Session. Italy. www.fao.org/fao-who-codexalimentarius/sh-proxy/en/?lnk=1&url=https%253A%252F%252Fworkspace.fao.org%252Fsites%252Fcodex%252FMeetings%252FCX-718-48%252FReport%252FREP16_PRe.pdf.

Fernández, F., P. Gepts, and M. López. 1984. Etapas de desarrollo de la planta de frijol común (*Phaseolus vulgaris* L.). CIAT, Cali, Colombia. ISBN 84-89206-54-6; http://ciat-library.ciat.cgiar.org/ciat_digital/CIAT/28093.pdf (accessed December 28, 2020).

Fischer, E. F., and B. Victor. 2014. High-end coffee and smallholding growers in Guatemala. *Latin American Research Review* 49(1): 155–177. https://doi.org/10.1353/lar.2014.0001.

Flores, D. 2010. *Uso Histórico: Sacha Inchi* Plukenetia volúbilis *L.* Proyecto Perubiodiverso, Perú. https://repositorio.promperu.gob.pe/bitstream/handle/123456789/1371/Uso_historico_sacha_inchi_2010_keyword_principal.pdf?sequence=1.

Flores, D., and O. Lock. 2013. Revalorizando el uso milenario del sacha inchi (*Plukenetia volubilis* L.) para la nutrición, la salud y la cosmética. *Revista de Fitoterapia* 13(1): 23–30. www.researchgate.net/profile/Olga-Lock/publication/271523952_Revalorando_el_uso_milenario_del_sacha_inchi_Plukenetia_volubilis_L_para_la_nutricion_salud_y_cosmetica/links/54cb76460cf2598f7116ed66/Revalorando-el-uso-milenario-del-sacha-inchi-Plukenetia-volubilis-L-para-la-nutricion-salud-y-cosmetica.pdf.

Follegatti-Romero, L. A., C. A. Piantino, R. Grimaldi, and A.C. Fernando. 2009. Supercritical CO_2 extraction of omega-3 rich oil from Sacha inchi (*Plukenetia volubilis* L.) seeds. *Journal of Supercritical Fluids* 49: 323–329. https://doi.org/10.1016/j.sup flu.2009.03.010.

Fritz, G. J., M. C. Bruno, B. S. Langlie, B. D. Smith, and L. Kistler. 2017. Cultigen chenopods in the Americas: A hemispherical perspective. In *Social Perspectives on Ancient Lives from paleoethnobotanical data*, ed. M. P. Sayre, and M. C. Bruno, 55–75. Springer International Publishing, Cham, Switzerland. https://doi.org/10.1007/978-3-319-52849-6_3.

Funabashi, M. 2018. Human augmentation of ecosystems: Objectives for food production and science by 2045. *npj Science of Food* 2: 16. https://doi.org/10.1038/s41 538-018-0026-4.

Gade, D. W. 1970. Ethnobotany of cañihua (*Chenopodium-pallidicaule*), rustic seed crop of altiplano. *Economic Botany* 24(1): 55–61. https://doi.org/10.1007/bf02860637.

Gahukar, R. T. 2014. Potential of minor food crops and wild plants for nutritional security in the Developing World. *J Agr Food Inform* 15(4): 342–352. https://doi.org/10.1080/ 10496505.2014. 952429.

Galeano, C., M. Nutti, J. Ramírez-Villegas, et al. 2019. Maize for Colombia: 2030 Vision. Mexico: Centro Internacional de Mejoramiento de Maíz y Trigo (CIMMYT) and Centro Internacional de Agricultura Tropical (CIAT). https://repository.cimmyt.org/ bitstream/handle/10883/20382/ 61038.pdf?sequence=1&isAllowed=y.

Galek, R., E. Sawicka-Sienkiewicz, D. Zalewski. 2007. Evaluation of interspecific hybrids of Andean lupin and their parental forms with regard to some morphological and quantitative characters. *Fragmenta Agronomica (Poland)* 24(2): 81–87. ISSN: 0860-4088.

Galek, R., E. Sawicka-Sienkiewicz, D. Zalewski, S. Stawiński, K. and Spychała. 2017. Searching for low alkaloid forms in the Andean lupin (*Lupinus mutabilis*) collection. *Czech Journal of Genetics and Plant Breeding* 53: 55–62. https://doi.org/10.17221/71/ 2016-CJGPB.

Galluzzi, G., and I. López Noriega. 2014. Conservation and use of genetic resources of underutilized crops in the Americas—a continental analysis. *Sustainability* 6(2):980–1017. https://doi.org/10.3390/ su6020980.

Geil, P. B., and J. W. Anderson. 1994. Nutrition and health implications of dry beans: A review. *Journal of the American College of Nutrition* 13(6): 549–558. https://doi.org/ 10.1080/07315724.1994. 10718446.

Gianoli, E., I. Ramos, A. Alfaro-Tapia, Y. Valdéz, E. R. Echegaray, and E. Yábar. 2006. Benefits of a maize–bean–weeds mixed cropping system in Urubamba Valley, Peruvian Andes. *International Journal of Pest Management* 52(4): 283–289. https://doi.org/ 10.1080/09670870600796722.

Gillespie, L. J. 1993. A synopsis of Neotropical *Plukenetia* (Euphorbiaceae) including two new species. *Systematic Botany* 18(4): 575–592. https://doi.org/10.2307/2419535.

Gillespie, L. J., and W. S. Armbruster. 1997. A contribution to the Guianan Flora: *Dalechampia, Haematostemon, Omphalea, Pera, Plukenetia*, and *Tragia* (Euphorbiaceae) with notes on subfamily Acolyphoideae. *Smithson Contributions to Botany* 86: 1–48. https://doi.org/10.5479/ si.0081024X.86.

Giuliani, A., F. Hintermann, W. Rojas, et al. 2012. Biodiversity of Andean grains: Balancing market potential and sustainable livelihoods. Bioversity International, Rome. www. bioversityinternational.org/fileadmin/user_upload/online_library/publications/ pdfs/1635.pdf (accessed November 30, 2020).

González, X. 2018. Hay sembradas cerca de 500 000 hectareas de maiz en Colombia. *Agro Negocios* 12 de septiembre de 2018. www.agronegocios.co/agricultura/hay-sembra das-cerca-de-500000-hectareas-de-maiz-en-colombia-2769113.

Gorinstein. S., J. Drzewiecki, E. Delgado-Licon, et al. 2005. Relationship between dicotyledone-amaranth, quinoa, fagopyrum, soy-bean and monocots-sorghum and rice based on protein analyses and their use as substitution of each other. *European Food Research and Technology* 221(1–2): 69–77. https://doi.org/10.1007/s00217-005-1208-2.

Grobman Tversqui, A., W. S. Salhuana, R. Sevilla Panizo, and P. C. Mangelsdorf. 1961. *Races of Maize in Peru, Their Origins, Evolution and Classification*. National Academy of Sciences/National Research Council, Washington, DC.

Gross, R. 1982. *El cultivo y la utilización del Lupinus mutabilis* Sweet. FAO, Rome.

Gueguen, J., and P. Cerletti. 1994. Proteins of some legume seeds: Soybean, pea, fababean and lupin. In *New and Developing Sources of Food Proteins,* ed. B. J. F. Hudson, 145–193. Chapman & Hall, London.

Güemes-Vera, N., R. J. Peña-Bautista, C. Jiménez-Martínez, G. Dávila-Ortiz, and G. Calderón-Domínguez. 2008. Effective detoxification and decoloration of *Lupinus mutabilis* seed derivatives, and effect of these derivatives on bread quality and acceptance. *Journal of the Science of Food and Agriculture* 88(7): 1135–1143. https://doi.org/10.1002/jsfa.3152.

Guerrero, M. 1987. Algunas propiedades y aplicaciones de los alcaloides del chocho (*Lupinus mutabilis* Sweet). In *Evento de información y difusión de resultados de investigación sobre chocho y capacitación en nuevas técnicas de laboratorio,* ed. M. Guerrero, 25–28. Universidad Técnica de Ambato, Ambato, Ecuador.

Gulisano, A., S. Alves, J. Martins, and L. Trindade. 2019. Genetics and breeding of *Lupinus mutabilis*: An emerging protein crop. *Frontiers Plant Science.* https://doi.org/10.3389/fpls.2019. 01385.

Gutiérrez, L.-P., L.-M. Rosada, and A. Jiménez. 2011. Chemical composition of Sacha Inchi (*Plukenetia volubilis* L.) seeds and characteristics of their lipid fraction. *Grasas y aceites* 62(1): 76–83. https://doi.org/10.3989/gya044510.

Haileslassie, H. A., C. J. Henry, and R. T. Tyler. 2016. Impact of household food processing strategies on antinutrient (phytate, tannin and polyphenol) contents of chickpeas (*Cicer arietinum* L.) and beans (*Phaseolus vulgaris* L.): A review. *International Journal of Food Science and Technol* 51: 1947–1957. https:// doi.org/10.1111/ijfs.13166.

Hamaker, B. R., C. Valles, R. Gilman, et al. 1992. Amino acid and fatty acid profiles of the Inca peanut (*Plukenetia volubilis*). *Cereal Chemistry* 69(4): 461–463. www.cerealsgra ins.org/ publications/cc/backissues/1992/Documents/69_461.pdf.

Hawkes, J. G. 1997. Back to Vavilov: Why were plants domesticated in some areas and not in others? Paper presented at Conference "*Origins of Agriculture and Domestication of Crop Plants in the Near East*", ICARDA, Aleppo, Syria, 10–14 May 1997, 1–3. www.bioversityinternational.org/fileadmin/bioversity/   publications/Web_version/47/ch06.htm.

Hernandez Bermejo, J. E., and J. Leon. 1994. Neglected crops: 1492 from a different perspective. *FAO Plant Production and Protection Series* No.26, FAO, Rome. www.fao.org/ 3/t0646e/t0646e.pdf.

Hernández-Gómez, J. A., and S. Miranda-Colín. 2008. Morphological characterization of chia (*Salvia hispanica*). *Revista Fitotecnia Mexicana* 31(2): 105–113. www.redalyc.org/articulo.oa? id=61031203.

Hernández-López, V., M. Vargas-Vasquez, J. Maruaga-Martínez, S. Hernández-Delgado, and N. Mayek-Pérez. 2013. Origin, domestication and diversification of common beans: Advances and perspectives. *Revista fitotecnia mexicana* 32(2): 95–104. www.sci elo.org.mx/scielo.php?   script=sci_abstract&pid=S0187-73802013000200002&lng= es&nrm=iso&tlng=en.

Hernández-Pérez, T., M. E. Valverde, and O. Paredes-López, 2021. Seeds from ancient food crops with the potential for antiobesity promotion. *Critical Reviews in Food Science and Nutrition*. https://doi.org/10.1080/10408398.2021.1877107.

Horton, D. 2014. *Investigación colaborativa de granos andinos en Ecuador*. Fundación McKnight e Instituto Nacional de Investigaciones Agropecuarias. Quito, Ecuador. https://repositorio.iniap.gob.ec/ jspui/bitstream/41000/102/4/iniapsc315.pdf.

Ichuta, F., and E. Artiaga. 1986. *Relación de géneros en la producción y en la Organización Social en Comunidades de Apharuni, Totoruma, Yauricani-Ilave*. 15–17. BSc thesis in social science, Universidad Nacional del Altiplano. Puno, Peru.

INEI (Instituto Nacional de Estadisticas e Informatica). 2020. *Producción agropecuaria, según principales productos, 2013–2019*. INEI, Peru. www.inei.gob.pe/media/ MenuRecursivo/publicaciones_digitales/Est/Lib1758/cap13/ind13.htm. (accessed December 18, 2020).

Jacobsen, S.-E. 1997. Adaptation of quinoa (*Chenopodium quinoa*) to Northern European agriculture: Studies on developmental pattern. *Euphytica* 96: 41–48. https://doi.org/ 10.1023/A:1002992718009.

Jacobsen, S.-E. 2017. The scope for adaptation of quinoa in Northern Latitudes of Europe. *Journal of Agronomy and Crop Science* 203(6): 603–613. https://doi.org/10.1111/ jac.12228.

Jacobsen, S.-E., and J. L. Christiansen. 2016. Some agronomic strategies for organic quinoa (*Chenopodium quinoa* Willd.). *Journal of Agronomy and Crop Science* 202(6): 454–463. https://doi.org/10.1111/jac.12174.

Jacobsen, S.-E., and A. Mujica. 2006. El tarwi (*Lupinus mutabilis* Sweet) y sus parientes silvestres. In *Botánica Económica de los Andes Centrales*, ed. M. Moraes, B. Øllgaard, L. P. Kvist, F. Borchsenius, and H. Balslev, 458–482. Universidad Mayor de San Andrés, La Paz, Bolivia.

Jacobsen, S.-E., and A. Mujica. 2008. Geographical distribution of the Andean lupin (*Lupinus mutabilis* Sweet). *Plant Genetic Resources Newsletter* 155: 1–8. https://doi. org/http://www2. bioversityinternational.org/publications/pgrnewsletter/article. asp?lang=en&id_article=1&id_issue=155.

Jacobsen, S.-E., I. Jørgensen, and O. Stølen. 1994. Cultivation of quinoa (*Chenopodium quinoa*) under temperate climatic conditions in Denmark. *Journal of Agricultural Science* 122: 47–52. https://doi.org/10.1017/S0021859600065783.

Jacobsen, S.-E., J. L. Christiansen, and J. Rasmussen, 2010. Weed harrowing and inter-row hoeing in organic grown quinoa (*Chenopodium quinoa* Willd.). *Outlook on Agriculture* 39: 223–227. https://doi.org/10.5367/oa.2010.0001.

Jacobsen, S-E, M. Sørensen, S. M. Pedersen, and J. Weiner. 2013. Feeding the world: Genetically modified crops versus agricultural biodiversity. *Agronomy for Sustainable Development* 33(4): 651–662. https://doi.org/10.1007/s13 593-013-0138-9.

Jacobsen, S.-E., M. Sørensen, S. M. Pedersen, and J. Weiner. 2015. Using our agrobiodiversity: Plant-based solutions to feed the world. *Agronomy for Sustainable Development* 35: 1217–1235. https://doi.org/10.1007/s13593-015-0325-y.

Jamboonsri, W., T. D. Phillips, R. L. Geneve, J. P. Cahill, and D. F. Hildebrand. 2012. Extending the range of an ancient crop, *Salvia hispanica* L. – a new ω3 source. *Genetic Resources and Crop Evolution* 59(2): 171–178. https://doi.org/10.1007/s10722-011-9673-x.

Jiménez-Martínez, C., H. Hernandez, and G. Dávila-Ortíz. 2003. Lupines: An alternative for debittering and utilization in foods. In *FoodScience and Food Biotechnology*, ed. G. Gutiéerrez-López, and G. Barbosa-Cánovas, 233–252. CRC Press, Boca Raton, FL.

Katungi, E., A. Farrow, J. Chianu, L. Sperling, and S. E. Beebe. 2009. Common bean in Eastern and Southern Africa: A situation and outlook analysis. International Centre for Tropical Agriculture [IITA], Ibadan Nigeria. http://tropicallegumes.icrisat.org/wp-content/uploads/2016/02/rso-common-bean-esa.pdf.

Katwal, T. B., and D. Bazile. 2020. First adaptation of quinoa in the Bhutanese mountain agriculture systems. *PLoS ONE* 15: e0219804. https://doi.org/10.1371/journal.pone.0219804.

Khaitov, B., A. A. Karimov, K. Toderich, et al. 2020. Adaptation, grain yield and nutritional characteristics of quinoa (*Chenopodium quinoa*) genotypes in marginal environments of the Aral Sea basin. *Journal of Plant Nutrition* 44(9): 1365–1379. https://doi.org/10.1080/01904167.2020.1862200.

Kistler, L., S. Y. Maezumi, J. G. de Souza, et al. 2018. Multiproxy evidence highlights a complex evolutionary legacy of maize in South America. *Science* 362(6420): 1309–1313. https://doi.org/10.1126/science.aav0207.

Kistler, L., H. B. Thakar, A. M. VanDerwarker, et al. 2020. Archaeological Central American maize genomes suggest ancient gene flow from South America. *PNAS*: https://doi.org/10.1073/pnas.2015560117.

Kodahl, N., and M. Sørensen, 2021. Sacha inchi (*Plukenetia volubilis* L.) – an underutilized crop with a great potential. *Agronomy* 11(6): 1066. https://doi.org/10.3390/agronomy11061066.

Kulas, E. ed. 2015. *Chicha – An Andean Idenity – The History and Meaning – Chicha's Evolution and the Inca Empire*. The Ohio State University. https://u.osu.edu/chicha/chicha-and-the-inca-empire/.

Kwak, M., and P. Gepts, 2009. Structure of genetic diversity in the two major gene pools of common bean (*Phaseolus vulgaris* L., Fabaceae). *Theoretical and Applied Genetics* 118: 979–992. https:// doi.org/10.1007/s00122-008-0955-4.

Leidi, E.O., A. Monteros, G. Mercado, et al. 2018. Andean roots and tubers crops as sources of functional foods. *Journal of Functional Foods* 51: 86–93. https://doi.org/10.1016/j.jff.2018.10.007.

Li, X., and K. H. M. Siddique. 2018. *Future Smart Food Rediscovering hidden treasures of neglected and underutilized species for Zero Hunger in Asia*, Executive summary, Food and Agriculture Organization of the United Nations, Bangkok. www.fao.org/3/I9136EN/ i9136en.pdf.

Lobo Zavalia, R., M. G. Alcocer, F. J. Fuentes, W. A. Rodriguez, M. Morandini, and M. R. Devani. 2011. Desarrollo del cultivo de chia en Tucuman, Republica Argentina. *EEAOC Advance Agroindustrial* 32(4): 27–30. www.eeaoc.gob.ar/?publicacion=32-4-3.

López, A. X., A. G. Huerta, E. C. Torrez, D. M. Sangerman-Jarquín, G. O. Rosas, and M. R. Arriaga, 2017. Chia (*Salvia hispanica* L.) current situation and future trends. *Revista Mexicana de Ciencias Agrícolas* 8(7): 1619–1631. www.scielo.org.mx/scielo.php?script=sci_issuetoc&pid= 2007093420170007&lng=en&nrm=iso.

López-Marqués, R. L., A. F. Nørrevang, P. Ache, et al. 2020. Prospects for the accelerated improvement of the resilient crop quinoa. *Journal of Experimental Botany* 71(18): 5333–5347. https://doi.org/10.1093/jxb/eraa285.

Lucas, M. M., F. Stoddard, P. Annicchiarico, et al. 2015. The future of lupin as a protein crop in Europe. *Frontiers in Plant Science* 6. https://doi.org/10.3389/fpls.2015.00705.

Lundmark, C. 2007. Genetically modified maize. *BioScience* 57(11): 996. https://doi.org/10.1641/ B571115.

MAGA (Ministerio de Agricultura, Ganadería y Alimentación de Guatemala). 2017. *Informe situación del frijol a diciembre 2017*: Consumo aparente. Ciudad de Guatemala, Guatemala. www. maga.gob.gt/sitios/diplan/download/informacion_del_sector/informes_de_situacion_de_maiz_y_frijol/2017/12%20Informe%20 Situaci%C3%B3n%20Del%20Frijol%20Negro%20Diciembre%202017.pdf.

Mamani, E. 2013. *Caracterización molecular de 26 accesiones de cañihua (*Chenopodium pallidicaule *Aellen) con mayor rendimiento en grano: Altiplano – Puno*. PhD thesis. Universidad Nacional del Altiplano, Puno, Peru. http://repositorio.unap.edu.pe/han dle/UNAP/252 (accessed October 17, 2020).

Mamani, F. 2016a. *Atlas de biodiversidad genética de cañahua y quinua*. ISBN: 978-99974-65-92-4. Diseño & Impresiones FLORES. La Paz, Bolivia.

Mamani, F. 2016b. *Cultivo de cañahua (*Chenopodium pallidicaule *Aellen) para la seguridad alimentaria*. ISBN: 978-99974-65-90-0. Diseño & Impresiones FLORES. La Paz, Bolivia.

Mamani, F., S. E. Aliaga, A. Bonifacio, D. Torrico, and N. Tapia. 2018. *La cañahua grano milenario de los Andes. Arte dedicado a la producción sostenible*. ISBN: 978-99974-0-353-7. Diseño & Impresiones FLORES. La Paz, Bolivia.

Mamidi, S., M. Rossi, D. Annam, et al. 2011. Investigation of the domestication of common bean (*Phaseolus vulgaris*) using multilocus sequence data. *Functional Plant Biology* 38: 953–967. https://doi.org/10.1071/fp11124.

Mangelsdorf, P. C. 1961. Introgression in maize. *Euphytica* 10(2): 157–168. https://doi.org/ 10.1007/BF00022207.

Mangelson, H., D. E. Jarvis, P. Mollinedo, et al. 2019. The genome of *Chenopodium pallidicaule*: An emerging Andean super grain. *Applications in Plant Sciences* 7(11): e11300. https://doi.org/10.1002/ aps3.11300.

Márquez, C. 2016. La siembra de chocho es más rentable. *Revista Líderes*. www. revistalideres.ec/lideres/siembra-chocho-produccion-chimborazo.html (accessed September 12, 2020).

Martín-Cabrejas, M. A., B. Sanfiz, A. Vidal, E. Mollá, R. Esteban, F. J. López-Andréu. 2004. Effect of fermentation and autoclaving on dietary fiber fractions and antinutritional factors of beans (*Phaseolus vulgaris* L.). *Journal of Agricultural and Food Chemistry* 52: 261–266. https://doi.org/10.1021/ jf034980t.

Maurer, N. E., B. Hatta-Sakoda, G. Pascual-Chagman, and L. E. Rodriguez-Saona. 2012. Characterization and authentication of a novel vegetable source of omega-3 fatty acids, sacha inchi (*Plukenetia volubilis* L.) oil. *Food Chemistry* 134: 1173–1180. https://doi.org/10.1016/j.foodchem.

Maydana, E. 2010. *Evaluacion de la produccion de seis variedades de cañahua (*Chenopodium pallidicaule *Aellen) con participacion de agricultores en la comunidad de Pacaure del Municipio de Mocomoco*. PhD thesis, Universidad Mayor de San Andres, La Paz, Bolivia.

Mazón, N. 2018. *El chocho o tarwi como recurso genético de la región andina* [Seminario online]. Quito, Ecuador: Interaprendizaje – IPDRS. https://bit.ly/2rsbfFF (accessed September 12, 2020).

Meneses, R. 1996. *Las leguminosas en la Agricultura Boliviana*. Proyecto Rhizobiología Bolivia. CIAT-CIF-PNLG-CIFP-WALL. Cochabamba, Bolivia. pp. 209–225.

Mercado, G., J. Davalos, IPDRS, Hivos, and Cipca. 2018. *Memoria foro virtual: Los caminos del tarwi y la integración andina: Bolivia, Perú y Ecuador*. Bolivia: IPDRS. https://inter aprendizaje. ipdrs.org/images/Documentos2018/MEMORIACAMINOSDELTARWI_ 14.12.18.pdf (accessed September 12, 2020).

Michail, N. 2019. Colombia inks US$1 million deal for climate-smart corn. Food navigator-latam.com on 18 Feb. 2019. www.foodnavigator-latam.com/Article/ 2019/02/18/Maize-for-Colombia-partnership-to-develop-climate-change-resist ant-corn.

Miklas, P. N., J. D. Kelly, S. E. Beebe, and M. W. Blair. 2006. Common bean breeding for resist-ance against biotic and abiotic stresses: From classical to MAS breeding. *Euphytica* 147: 105–131. https://doi.org/10.1007/s10681-006-4600-5.

Mina, D., Q. Struelens, A. Barragán, and O. Dangles. 2018. El chocho. Un superalimento que podría convertirse en una amenaza. *Revista Nuestra Ciencia* 20: 19–21. www.puce. edu.ec/ portal/wp-content/uploads/2019/07/Nuestra-Ciencia-n.%C2%BA-20.pdf.

Miranda, F. 2014. Guía tecnica para el manejo del cultivo de la Chia (*Salvia hispánica*) en Nicaragua. *GuiagroNicaragua* 3: 130–131 [16 slides]. https://es.slideshare.net/ fpmirandasalgado/manual-de-produccion-de-chia-salvia-hispanica-40722325.

Miranda-Colin, S. 1978. Evolución de cultivares nativos de México. *Ciencia y Desarrollo* 3: 130–131.

Morais Dias, D., N. Kolba, D. Binyamin, O. Ziv, M. R. Nutti, H. S. Duarte Martino, R. P. Glahn, O. Koren, and E. Tako, 2018. Iron biofortified Carioca Bean (*Phaseolus vulgaris* L.)-based Brazilian diet delivers more absorbable iron and affects the gut microbiota in vivo (*Gallus gallus*). *Nutrients* 10(12): 1970. http://dx.doi.org/ 10.3390/nu10121970.

Moscoso-Mujica, G., A. Zavaleta, Á. Mujica, M. Santos, and R. Calixto. 2017. Fractionation and electrophoretic characterization of (*Chenopodium pallidicaule* Aellen) kanihua seed proteins. *Revista chilena de nutrición* 44(2): 144–152. http://dx.doi.org/10.4067/ S0717-75182017000200005.

Moure, A., J. Sineiro, H. Domínguez, and J. C. Parajó. 2006. Functionality of oilseed pro-tein products: A review. *Food Research International* 39: 945–963. https://doi.org/ 10.1016/j.foodres.2006.07.002.

Mujica, A., and S.-E. Jacobsen. 1998. Agrobiodiversidad de las Aynokas de Quinua (*Chenopodium quinoa* Willd.) y la seguridad Alimentaria. In *Seminario Agrobiodiversidad en la Región Andina y Amazónica. Lima, 23–24 Nov.* 28–29. CCIIA, UNALM, CLADES y RAE. Lima, Peru.

Mujica, A., and S.-E. Jacobsen. 2000. Agrobiodiversidad de las Aynokas de quinua (*Chenopodium quinoa* Willd.) y la seguridad Alimentaria. In *Proc. Seminario Taller Agrobiodiversidad en la región andina y amazónica. 23–25 noviembre 1988*, ed. C. Felipe-Morales, and A. Manrique, 151–156. NGO-CGIAR. Lima, Peru.

Muñoz, L. A., A. Cobos, O. Diaz, and J. M. Aguilera. 2013. Chia seed (*Salvia hispanica*): An ancient grain and a new functional food. *Food Reviews International* 29(4): 394–408. https://doi.org/10.1080/87559129.2013.818014.

Muziri, T., P. Chaibva, A. Chofamba, et al. 2020. Using principal component analysis to explore consumers' perception toward quinoa health and nutritional claims in Gweru, Zimbabwe. *Journal of Food Science and Nutrition* 9(2): 1025–1033. https://doi.org/10.1002/fsn3.2071.

Nájera, S. 2015. *¿Tiene la producción de chochos (L. mutabilis) el potencial de aumentar los ingresos agrícolas y contribuir a la seguridad alimentaria del Ecuador si es que sustituye a la producción de soya o maíz?* MSc thesis, School of Geociences. University of Edinburgh.

Nanduri, K. R., A. Hirich, M. Salehi, S. Saadat, and S.-E. Jacobsen. 2019. Quinoa: A new crop for harsh environments. In *Sabkha Ecosystems. Tasks for Vegetation Science, vol 49*, ed. B. Gul, B. Böer, M. Khan, M. Clüsener-Godt, and A. Hameed, 301–333. Springer, Cham. https://doi.org/10.1007/978-3-030-04417-6_19.

Norlaily, M. A., K. Y. Swee, Y. H. Wan, K. Boon, W. T. Sheau, and G. T. Soon. 2012. The promising future of chia, *Salvia hispanica* L. *Journal of Biomedicine and Biotechnology* 2012(171956): 1–9. https://doi.org/10.1155/2012/171956.

OECD (Organisation for Economic Co-Operation and Development). 2021. Chapter 1. Common Bean (*Phaseolus vulgaris*). In *Novel Food and Feed Safety, Safety Assessment of Foods and Feeds Derived from Transgenic Crops, Volume 3 Common bean, Rice, Cowpea and Apple Compositional Considerations.* www.oecd-ilibrary.org/sites/544d0 3d6-en/index.html?itemId=/content/ component/544d03d6-en.

Orona-Tamayo, D., M. E. Valverde, and Octavio Paredes-López, 2018. Bioactive peptides from selected Latin American food crops – A nutraceutical and molecular approach. *Critical Reviews in Food Science and Nutrition.* https://doi.org/10.1080/10408 398.2018.1434480.

Pascual-Villalobos, M., E. Correal, E. Molina, and J. Martínez. 1997. *Evaluación y selección de especies vegetales productoras de compuestos naturales con actividad insecticida.* Centro de Investigación y Desarrollo Agroalimentario (CIDA), Murcia, Spain. www.inia.es/sites/frontbootstrap/Pages/PageNotFoundError.aspx?requestUrl=www.inia.es/gcontrec/proyectos/resultados-97/agricola/sc94-039.pdf.

Paz Silva, L. J. 1993. Philosophy for the development of Andean Ecosystems. In *Andean Agro Ecosystem: Problems, Limitations and Perspectives.* Annals of the international workshop on Andean agro ecosystem, 30 March–2 April, Lima, Peru, pp. 11–29. https://agris.fao.org/agris-search/ search.do?recordID=QP9300027.

Pearsall, D. 1992. The origins of plant cultivation in South America. In *The Origins of Agriculture. An International Perspective*, Eds. C. Wesley Cowan & P. Jo Watson ,173–205). Washington, DC/London: Smithsonian Institution Press

Peiretti, P. G., and F. Gai, 2009. Fatty acid and nutritive quality of chia (*Salvia hispanica* L.) seeds and plant during growth. *Animal Feed Science and Technology* 148(2–4): 267–275. https://doi.org/10.1016/j.anifeedsci.2008.04.006.

Peláez, P., D. Orona-Tamayo, S. Montes-Hernández, M. Elena Valverde, O. Paredes-López, and A. Cibrián-Jaramillo, 2019. Comparative transcriptome analysis of cultivated and wild seeds of *Salvia hispanica* (chia). *Scientific Reports* 9: 9761. https://doi.org/ 10.1038/s41598-019-45895-5.

Peralta, E., and E. Villacrés, 2015. *100 recetas prácticas usando quinua, chocho y amaranto.* Publicación miscelánea N° 421. Programa Nacional de Leguminosas y Granos Andinos y Departamento de Nutrición y Calidad. Estación Experimental

Santa Catalina. Instituto Nacional de Investigaciones Agropecuarias (INIAP). Quito, Ecuador. https://repositorio.iniap.gob.ec/ bitstream/41000/2727/1/iniapscpm421.pdf.

Pérez Suárez, T. 1997. El dios del maíz en Mesoamérica. *Arqueología Mexicana* 5(25): 44–55.

Pisfil, C. A. 2017. *Optimización del nivel de sustitución de la harina de trigo por harina de quinua, cañihua y kiwicha en la elaboración de pan panini precocido.* PhD thesis, Universidad Nacional Pedro Ruiz Gallo, Peru. http://repositorio.unprg.edu.pe/handle/UNPRG/1301 (accessed 20 October, 2020).

Przybylak, J. K., D. Ciesiolka, W. Wysocka, et al. 2005. Alkaloid profiles of Mexican wild lupin and an effect of alkaloid preparation from *Lupinus exaltatus* seeds on growth and yield of paprika (*Capsicum annuum* L.). *Industrial Crops Products* 21: 1–7. https://doi.org/10.1016/j.indcrop.2003.12.001.

Quispe, E. 2004. *Morfología y variabilidad de las cañahuas cultivadas (*Chenopodium pallidicaule *Aellen).* Ing. Agr. thesis, Universidad Mayor de San Simon, Cochabamba, Bolivia.

Ramirez, S. 2016. Amido e farinha de cañihua (*Chenopodium pallidicaule*): extração, caracterização e desenvolvimento de filmes biodegradáveis. Master's dissertation, Faculdade de Zootecnia e Engenharia de Alimentos, University of São Paulo, Pirassununga, Brazil. https://doi.org/10.11606/D.74.2016.tde-12082016-104942 (accessed November 30, 2020).

Rathna Priya, T. S., and A. Manickavasagan. 2020. Common Bean. In *Pulses*, ed. A. Manickavasagan, and P. Thirunathan, 77–97. Springer International, Cham. https://doi.org/10.1007/978-3-030-41376-7_5.

Razzaghi, F., M. R. Bahadori-Ghasroldashti, S. Henriksen, A. R. Sepaskhah, and S.-E. Jacobsen. 2020. Physiological characteristics and irrigation water productivity of quinoa (*Chenopodium quinoa* Willd.) in response to deficit irrigation imposed at different growing stages – A field study from Southern Iran. *Journal of Agronomy and Crop Science* 206(3): 390–404. https://doi.org/10.1111/jac.12392.

Repo-Carrasco, R., C. Espinoza, and S.-E. Jacobsen. 2003. Nutritional value and use of the Andean crops quinoa (*Chenopodium quinoa*) and kañiwa (*Chenopodium pallidicaule*). *Food Reviews International* 19(1–2): 179–189. https://doi.org/10.1081/fri-120018884.

Repo-Carrasco-Valencia, R. 2011. Andean indigenous food crops: Nutritional value and bioactive compounds. PhD thesis, Department of Biochemistry and Food Chemistry, University of Turku, Finland. www.utupub.fi/bitstream/handle/10024/74762/Repo-Carrasco-Valencia-Diss2011.pdf? sequence=1.

Rivera-Cabrera, F., M. Medina-Valdez, C. Pelayo-Zaldívar, et al. 2017. Phytochemical composition of *Salvia hispanica* L. extracts and their satiety effect. *Revista Mexicana de Ingeniería Química* 6(1): 47–53. www.redalyc.org/pdf/620/62049878006.pdf.

Roberts, L. M., U. J. Grant, R. Ramirez E., W. H. Hatheway, and D. L. Smith. 1957. *Races of Maize in Colombia.* NAS/NRC, 510, Washington, DC www.ars.usda.gov/ARSUserFiles/50301000/Races_of_Maize/RoM_Colombia_0_Book.pdf.

Rodriguez, J. P., and F. Mamani, 2016. Participatory breeding and gender role in cañahua, an NUS Andean Grain crop in Bolivia. In *Gender, Breeding and Genomics – Case Studies.* Workshop held 18–21 October 2016, CGIAR Gender and Agriculture Research

Network, Nairobi. www.researchgate.net/publication/323053515_Participatory_breeding_and_gender_role_in_canahua_a_NUS_Andean_Grain_crop_in_Bolivia.

Rodriguez, J. P., E. Maydana, R. Flores, N. Calancha, M. Flores, and F. Mamani. 2010. Seguridad Alimentaria y revalorización de la Cañihua (*Chenopodium pallidicaule*): Caso de estudio en comunidades del Altiplano Norte de Bolivia. In *Memorias 3er Congreso Mundial de la Quinua*. FCAPV-UTO, Oruro, Bolivia.

Rodriguez, J. P., M. Aro, M. Coarite, et al. 2017. Seed shattering of Cañahua (*Chenopodium pallidicaule* Aellen). *Journal of Agronomy and Crop Science* 203(3): 254–267. https://doi.org/10.1111/jac.12192.

Rodriguez, J. P., S.-E. Jacobsen, M. Sørensen, and C. Andreasen. 2020. Cañahua (*Chenopodium pallidicaule*): a promising new crop for arid areas. In *Emerging Research in Alternative Crops, Environment & Policy 58* ed. A. Hirich, R. Ragab, R. Choukr-Allah et al., chap. 9, 221–243. Springer International, Cham. https://doi.org/10.1007/978-3-319-90472-6_9.

Rodríguez, J. P., H. Rahman, S. Thushar, and R. K. Singh, 2020. Healthy and resilient cereals and pseudo-cereals for marginal agriculture: Molecular advances for improving nutrient bioavailability. *Frontiers in Genetics*. February 27, 2020. https://doi.org/10.3389/fgene.2020.00049.

Rodríguez-Castillo, L., and X. E. Fernández-Rojas, 2003. Los frijoles (*Phaseolus vulgaris*): Su aporte a la dieta del costarricense. *Acta Médica Costarricense* 45(3): 120–125. http://actamedica.medicos.cr/ .index.php/Acta_Medica/article/view/110/93

Rodríguez del-Castillo, A.M., G. Gonzalez-Aspajo, M. F. Sánchez-Márquez, and N. Kodahl. 2019. Ethnobotanical knowledge in the Peruvian Amazon of the neglected and underutilized crop Sacha Inchi (*Plukenetia volubilis* L.). *Economic Botany* 73(2): 281–287. https://doi.org/10.1007/s12231-019-09459-y.

Rojas, W., J. L. Soto, M. Pinto, et al. 2010. *Granos andinos: avances, logros y experiencias desarrolladas en quinua, cañahua y amaranto en Bolivia*. Bioversity International, Roma.

Romeo, F. V., S. Fabroni, G. Ballistreri, S. Muccilli, A. Spina, and P. Rapisarda. 2018. Characterization and antimicrobial activity of alkaloid extracts from seeds of different genotypes of *Lupinus* spp. *Sustainability* 10(3): 788. https://doi.org/10.3390/su10030788.

Rosas, J. C. 2003. El cultivo de frijol común en América Tropical. Carrera de Ciencia y Producción Agropecuaria. Escuela Agrícola Panamericana/Zamorano. Honduras. https://bdigital.zamorano. edu/bitstream/11036/2424/1/prueba%2009.pdf.

Rossi, M., E. Bitocchi, E. Bellucci, et al. 2009. Linkage disequilibrium and population structure in wild and domesticated populations of *Phaseolus vulgaris* L. *Evolutionary Applications* 2: 504–522. https:// doi.org/10.1111/j.1752-4571.2009.00082.x.

Ruales, J., P. Pólit, and B. M. Nair. 1988. Nutritional quality of blended foods of rice, soy and lupins, processed by extrusion. *Food Chemistry* 29: 309–321. https://doi.org/10.1016/0308-8146(88)90046-5.

Salas, L. M. 2017. Produção e caracterização de filmes biodegradáveis a base do pseudocereal canihua (*Chenopodium pallidicaule*). PhD dissertation, Universidade Estadual de Campinas, Campinas, SP. www.repositorio.unicamp.br/handle/REPOSIP/330342 (accessed November 24, 2020).

Sathe, S. K., S. S. Deshpande, and D. K. Salunkhe. 1981. Functional properties of lupin seeds (*Lupinus mutabilis*) proteins and protein concentrates. *Journal of Food Science* 47: 491–502. https://doi.org/10.1111/j.1365-2621.1982.tb10110.x.

Secretaría de Economía. 2012. Análisis de la cadena de valor del frijol. Dirección General de Indrustrias Básicas, Estados Unidos Mexicanos. www.economia.gob.mx/files/comunidad_negocios/industria_comercio/analisis_cadena_valor_frijol.pdf (accessed December 28, 2020).

Sellami, M. H., C. Pulvento, and A. Lavini. 2021. Agronomic practices and performances of quinoa under field conditions: A systematic review. *Plants* 10(1): 72. https://doi.org/10.3390/plants10010072.

Serrano, R. 2012. Distribución de la diversidad genética y etnobotánica de cañahua (*Chenopodium pallidicaule* Aellen) en las comunidades del Altiplano Norte. PhD thesis. Facultad de Agronomía, Universidad Mayor de San Andres, La Paz, Bolivia. https://repositorio.umsa.bo/bitstream/handle/ 123456789/8002/T-1652.pdf?sequence=1&isAllowed=y.

Singh, S. P., R. Nodari, and P. Gepts. 1991. Genetic diversity in cultivated common bean: I. Allozymes. *Crop Science* 31(1): 19–23. https://doi.org/10.2135/cropsci1991.0011183X003100010004x.

Skarbø, K., and K. van der Molen. 2016. Maize migration: Key crop expands to higher altitudes under climate change in the Andes. *Climate and Development* 8(3): 245–255. https://doi.org/10.1080/17565529.2015.1034234.

Sosa-Baldivia, A., G. Ruiz-Ibarra, R. R. Robles de la Torre, R. Robles Lopez, and A. Montufar Lopez, 2018. The chia (*Salvia hispanica*): Past, present and future of an ancient Mexican crop. *Australian Journal of Crop Science* 12(10): 1626–1632. https://doi.org/10.21475/ajcs.18.12.10.p1202.

Spehar, C. R., and R. L. D. Santos. 2002. Quinoa BRS Piabiru: Alternative for diversification of cropping systems. *Pesquisa Agropecuária Brasileira* 37(6): 889–893. https://doi.org/10.1590/s0100-204x2002000600020.

Srichamnong, W., P. Ting, P. Pitchakarn, O. Nuchuchua, and P. Temviriyanukul. 2018. Safety assessment of *Plukenetia volubilis* (Inca peanut) seeds, leaves, and their products. *Food Science and Nutrition* 6(4): 962–969. https://doi.org/10.1002/fsn3.633.

Staller, J. E. 2010. Maize cobs and cultures: History of *Zea mays* L. Springer, Berlin. https://doi.org/10.1007/978-3-642-04506-6.

Staller, J. E. 2016. High altitude maize (*Zea mays* L.) cultivation and endemism in the Lake Titicaca Basin. *Journal of Botany Research* 1(1): 8–21. https://doi.org/10.36959/771/556.

Stokes, P., and P. Rowley-Conwy. 2002. Iron Age Cultigen? Experimental return rates for fat hen (*Chenopodium album* L.). *Environmental Archaeology* 7(1): 95–99. https://doi.org/10.1179/env.2002.7.1.95.

Strozycki, P. M., A. Szczurek, B. Lotocka, M. Figlerowicz, and A. B. Legocki. 2007. Ferritins and nodulation in *Lupinus luteus*: Iron management in indeterminate type nodules. *Journal of Experimental Botany* 58: 3145–3153. https://doi.org/10.1093/jxb/erm152.

Švec, I., M. Hruskova, R. Kapacinskaite, and T. Hofmanova. 2019. Effect of quinoa (*Chenopodium quinoa*) and cañahua wholemeals (*Chenopodium pallidicaule*) on pasting behaviour of wheat flour. *Advances in Food Science and Engineering* 3(1): 1–8. https://dx.doi.org/10.22606/afse.2019.3100.

Tapia, M. 1982. Proceso agroindustrial del tarwi (*Lupinus mutabilis*). In *Actas de la Conferencia Internacional del Lupino,* 58–62. Asociación Internacional del Lupinu. Torremolinos, Spain. Servicio de Publicaciones Agrarias, Ministerio de Agricultura, Pesca y Almentación, Madrid, Spain.

Tapia, M. E. 2015. *El tarwi, lupino Andino. Tarwi, tauri o chocho.* Corporación gráfica Universal, Lima, Peru. http://fadvamerica.org/wp-content/uploads/2017/04/TARWI-espanol.pdf.

Tapia, M. E., and A. M. Fries, 2007. Guía de campo de los cultivos andinos. FAO, ANPE-Peru. https://runamaqui.fr/wp-content/uploads/2020/07/FAO-Los-cultivos-andinos-documento-completo.pdf.

Tapia, M. E., S.A. Mujica, and A. Canahua. 1980. Origen, distribución geográfica y sistemas de producción de la quinua. In *I Reunión sobre genética y fitomejoramiento de la quinua,* A1–A8. PISCA-UNTA-IBTA-IICA-CIID, Publicacion – Universidad Nacional Tecnica del Altiplano (Peru). Puno, Peru.

Tiwari, B. K., and N. Singh. 2012. *Pulse Chemistry and Technology.* Royal Society of Chemistry, Cambridge, UK. ISBN 978-1-84973-331-1.

Triana-Maldonado, D. M., S. A. Torijano-Gutiérrez, and C. Giraldo-Estrada. 2017. Supercritical CO_2 extraction of oil and omega-3 concentrate from Sacha inchi (*Plukenetia volubilis* L.) from Antioquia, Colombia. *Grasas y Aceites* 68(1): 1–11. https://doi.org/10.3989/gya.0786161.

Valdivia-López, M. Á., and A. Tecante, 2015. Chia (*Salvia hispanica*): A review of native Mexican seed and its nutritional and functional properties. *Advances in Food and Nutrition Research* 75: 53–75. https://doi.org/10.1016/bs.afnr.2015.06.002.

Vašek, J., P. H. Čepková, I. Viehmannová, M. Ocelák, D. Cachique Huansi, and P. Vejl. 2017. Dealing with AFLP genotyping errors to reveal genetic structure in *Plukenetia volubilis* (Euphorbiaceae) in the Peruvian Amazon. *PLoS ONE* 12(9): e0184259. https://doi.org/10.1371/journal.pone.0184259.

Vavilov, N.I. 1935. Theoretical basis for plant breeding. Vol. 1. Moscow. Origin and Geography of Cultivated Plants. In *The Phytogeographical Basis for Plant Breeding,* ed. Y. A. Ovchinikov (trans. D. Love). Vol. 1, 316–366. Cambridge University Press, Cambridge, UK.

Villacrés, E., C. Caicedo, and E. Peralta. 2000. Diagnóstico del procesamiento artesanal, comercialización y consumo del chocho. In *Zonificación Potencial, Sistemas de Producción y Procesamiento Artesanal del Chocho (Lupinus mutabilis Sweet) en Ecuador,* 24–41. Quito, Ecuador.

Villacrés, E., E. Peralta, and M. Alvarez. 2003. Chochos en su punto. In *Chochos. Recetarios. Disfrute Cocinando con Chochos,* ed. E. Peralta, Publicación Miscelánea 118: 1–53. INIAPFUNDACYT, Quito, Ecuador.

Villarino, E. 2018. Here's how to do bean breeding the climate-smart way. https://bigdata.cgiar.org/heres-how-to-do-bean-breeding-the-climate-smart-way/ (accessed December 28, 2020).

Williams, J. T. ed. 1993. *Underutilized Crops: Pulses and Vegetables.* Chapman and Hall, London.

Wink, M. 1991. Plant breeding: Low or high alkaloid content. In *Proceedings of the 6th International Lupin Conference, 25–30 Nov 1990,* ed. D. von Bayer, 326–334. Temuco, Pucón, Chile.

Xiu-shi, Y., Q. Pei-you, G. Hui-min, and R. Gui-xing. 2019. Quinoa industry development in China. *Ciencia e Investigación Agraria* 46: 208–219. https://doi.org/10.7764/rcia.v46i2.2157.

Yanapa Velasquez, L. L. 2018. *Elaboración de biopelículas para envasado de alimentos a partir de quitosano y cañihua (*Chenopodium pallidicaule*).* Thesis. Universidad Nacional del Altiplano. Puno, Peru. http://repositorio.unap.edu.pe/handle/UNAP/11044 (accessed December 5, 2020).

Zegarra, S. I. 2018. *Elaboración de un pan apto para celiacos a base de harina de* Chenopodium pallidicaule *y evaluación de su aceptabilidad sensorial.* PhD dissertation, Universidad San Ignacio de Loyola, Peru. http://dx.doi.org/10.20511/USIL.thesis/3023 (accessed December 3, 2020).

Zielinska-Dawidziak, M. 2015. Plant ferritin–a source of iron to prevent its deficiency. *Nutrients* 7(2): 1184–1201. https://doi.org/10.3390/nu7021184.

Chapter 2

Genotype and Environment as Key Factors Controlling Seed Quality in Latin-American Crops

Luisa Bascuñán-Godoy[1], María Reguera[2], Ángel Mujica[3],
Néstor Fernández del Saz[1], Carolina Sanhueza[1],
Catalina Castro[1], José Ortiz[1], Gabriel Barros[1], José
Delatorre-Herrera[4], Karina B. Ruiz[5], Teodoro Coba de la Peña[6],
Enrique Ostria[7], Enrique Martínez[7], and Susana Fischer[8]
[1]Universidad de Concepción, Concepción, Chile.
[2]Universidad Autónoma de Madrid, Madrid, Spain.
[3]Universidad Nacional Altiplano, Puno, Peru.
[4]Universidad Arturo Prat, Iquique, Chile.
[5]Universidad Arturo Prat, Iquique, Chile.
[6]Centro de Estudios Avanzados en Zonas
Áridas (CEAZA), La Serena, Chile.
[7]Centro de Estudios Avanzados en Zonas
Áridas (CEAZA), La Serena, Chile.
[8]Universidad de Concepción, Concepción, Chile.

CONTENTS

DOI: 10.1201/9781003088424-2

53

2.1 LATIN AMERICAN CROPS

Plants respond to adverse biotic and abiotic conditions within their local environment, whereby they use their remarkable ability to adjust their physiology to cope with and acclimate to changing growth conditions.

Climate change threatens agriculture, increasing the loss of soil and reducing areas suitable for crop production systems; by 2050 it will be necessary to increase plant productivity by 70% to satisfy the needs of an ever-increasing world population (FAO 2019). Crop yield is largely determined by climate conditions, therefore minor deviations from optimal conditions can strongly penalize yield potential in terms of quantity, but also quality. Considering the agricultural challenges we face for the next coming decades, a deeper understanding of the effect of environmental factors on crop growth and development could significantly reduce yield losses and improve quality, ensuring food security worldwide.

The development of new varieties with high and stable genetic potential for high density crop production is the main challenge for breeding and modern agronomy. However, there are contrasting views to accomplish this goal. It has been highlighted that modern cultivars usually achieve higher yields with plenty of soil nutrients and optimal climate conditions. However, there are several reports indicating a better performance of landraces (genotypes adapted to local regions) compared to modern genetic materials, which generally may present low to medium yields. It has been accepted that the main contributions of landraces are traits related to adaptations to stressful environments, which can be used in plant breeding programs (Dwivedi et al. 2016).

Since the 1960s, there has been plenty of evidence of a slowdown in yield improvement rates of major food crops (including rice, wheat and maize) (FAO 2016), and current yield strategies are not enough to meet future requirements (Lenné and Wood 2011). Considering that climate change will affect the potential of traditional crops, searching for ancestral crops and landraces with high genetic diversity and food potential is mandatory. Thus, landraces and wild relatives will be important genetic sources for developing new varieties with higher yields (in terms of quantity and quality) and enhanced stress tolerance.

Latin-American crops comprise high diversity of plant species as they are adapted to a wide range of environments; this has resulted in large genetic pools and, therefore one of the richest repositories worldwide (Bazile et al. 2014). Several cultivars from this region have been widely explored, and many hybrids and improved varieties have been produced. This is the case of *Zea mays* (maize) and *Phaseolus vulgaris* (pea). However, other crops yet remain to be further explored, such as *Amaranthus* sp (amaranto or kiwicha), *Chenopodium pallidicaule* (cañahua), *Chenopodium quinoa* (quinoa), *Salvia hispanica* (chia), *Plukenetia volubilis* (inchi) and *Lupinus mutabilis* (lupino andino or tarwi), ones that show great food potential and an unexploited capacity to resist different abiotic stresses that make them suitable to grow in marginal environments. Besides, wild accessions of these species will remain a viable source of germplasm, not only to maintain high productivity, but also to increase seed quality (FAO 2016).

2.2   MAJOR ABIOTIC STRESSES AFFECTING SEED QUALITY

Many works have suggested that low protein content, the scarcity of micronutrients or other nutritional relevant compounds such as antioxidants, which are consumed from food products coming from staple cultivars, have resulted in unhealthy diets that are linked to different human pathologies (Kucek et al. 2015; Fan et al. 2008). On the other hand, ancient plant food resources could be a good source of healthy and beneficial nutrients (Arzani and Ashraf 2017; Peñas et al. 2014). Nonetheless, how abiotic stresses (such as extreme temperatures, low water availability, high salts and mineral deficiencies) have shaped the production quality of ancient crops is something that requires further investigation.

The presence of stress at any growth stage can impact crop yield and its nutritional quality. These effects include processes such as impaired gametogenesis, fertilization, embryogenesis, altered nutrient assimilation and mobilization, and a reduction in the accumulation of reserves in the endosperm, affecting seed development and composition (Waqas, et al. 2021). A brief description of the effect of the three main abiotic stresses enhanced by global warming, which are able to change plant productivity (in terms of seed quality and quantity) is presented in this section. Also, at this point it is necessary to clarify that the terms seed and grain will be distinguished throughout the text, although the fact they may be related terms, they are different from a botanical standpoint. While the term seed (consisting of seed coat, endosperm and embryo) will be referred to as an ovule containing an embryo, grain (consisting of bran, endosperm and germ) will be referred to as fusion of the seed coat and fruit tissues. Additionally, the term seed quality will be referred to aspects related to human nutritional consumption rather than their ability to germinate.

2.2.1 Drought

Climate change generates considerable uncertainty about future water availability. Drought affects 45% of the world's agricultural land, and forecasts predict increasing frequency of insufficient precipitation and consequent aridity in many parts of the world (Bates et al. 2008; IPCC 2018). There is strong scientific consensus that one of the primary physiological targets of water stress in plants is photosynthesis (Lawlor and Cornic 2002). Primarily, drought decreases photosynthetic rates through the reduction of stomatal conductance, photooxidation and enzyme damage, thereby decreasing the number of assimilates available for sink tissues (Sehgal et al. 2018). Besides, drought stress during seed filling induces embryo abortion, decreases the mobilization of reserves and reduces the number of amyloplasts. Thus, carbohydrate synthesis (assimilation and metabolism) in plants is severely altered, affecting quality and yield (Çakir 2004).

Interestingly, drought can induce increased content of molecules with antioxidant capacity in several plant species, as will be discussed throughout this chapter (Bascunan-Godoy et al. 2016; Fischer et al. 2017; Laxa et al. 2019). The physiological basis of this response may vary among species, but in general, it involves an increase in nutrient remobilization and the induction of antioxidant metabolism. Within this context, the control of water availability could be used as a strategy to improve the nutritional quality of several seeds, grains and fruits.

2.2.2 High Temperatures

The world is facing a gradual temperature rise, along with a higher frequency of heat waves leading to warmer days and nights. Indeed, it is projected that temperatures will have increased by 2°C by the end of the 21st century (IPCC 2021). Major effects of high temperature on plants (usually $\geqslant 30$°C) include the reduction of life cycles, pollen abortion, kernel shrinkage, reduction of seed reserves, anther indehiscence and reduced development of the pollen tube. Physiologically, high temperatures alter processes including photosynthesis and respiration impairment interfering with enzyme activities, resulting in chlorophyll degradation, or causing general damage to photosystem II, electron transfer and energy balance, which is crucial to maintaining cell functioning (Shah and Paulsen 2003; Prasad et al. 2006; Li et al. 2020). All these events result in reduced crop yield (Rezaei et al. 2015).

Reproductive stages are sensitive to elevated temperatures in plants. Simulated high temperature waves during the vegetative and reproductive stages in maize showed that the latter stages of development are more sensitive to elevated temperatures. Thus, high temperatures during pre- and post-anthesis reduce CO_2 exchange rates (~17%), growth (17–29%), grain number (7–45%),and grain yield (10–45%) (Waqas et al. 2021). Furthermore, during flowering, elevated day and/or night-time temperatures negatively affect floret number, silk number, and grain development and quality in rice, canola and maize (Matsui et al. 2001; Cicchino et al. 2010; Pokharel et al. 2020). Air temperature above 35°C suppresses rice ovary fertilization and inhibits the grain filling process, related to a detrimental effect on grain yield (Zhang et al. 2007).

It should be noted that normally in the field, plants are exposed to a combination of stresses (i.e. heat stress is often associated with drought). This circumstance worsens the crop's stress response resulting in larger yield penalties compared to a single stress situation (Dreesen et al. 2012). Interestingly, heat and drought stress can disrupt the accumulation of various seed constituents, primarily starch and proteins, by inhibiting the enzymatic processes of synthesis (Behboudian et al. 2001; Farooq et al. 2018).

2.2.3 Salinity

Salinity is one of the most deleterious environmental factors limiting crop productivity (Shrivastava and Kumar 2015). The main causes of salinity in crops are irrigation with highly salted groundwater, disintegration, and release of seawater in coastal areas, and accumulation of salts in arid/semiarid regions due to insufficient leaching of ions (Chinnusamy et al. 2005). Nowadays, nearly 20% of irrigated agricultural lands are considered saline. Adverse effects of salinity on plant growth are related to the effects of osmotic stress (due to lower water availability, similar to what happens under drought and high temperatures) and ionic stress inducing homeostasis imbalance (Epstein et al. 1980). These two major effects are accompanied by those caused by oxidative stress and nutrient imbalance that eventually trigger cytotoxicity and plant growth impairment (Abobatta 2020).

Salt resistance involves adaptation in order to maintain physiological and biochemical homeostasis, including structural and molecular adaptation and salt exclusion mechanisms towards minimizing salt concentrations in cells. Interestingly, among salt-tolerant species, there are halophytes that naturally grow and complete their life cycle under salinity conditions, whereas most crops fail in growth and development and consequently cannot produce yield (Ruiz et al. 2016).

Several crops, including many from the Andean region such as quinoa, kiwicha or cañahua, can increase the velocity and total percentage of germination under moderate (100 to 150 mM NaCl, equivalent to 10 to 15 dS m^{-1}) to high salinity conditions (Delatorre and Pinto 2009; Razzaghi et al. 2011). However, growing at high salinity conditions reduces seed value by both, changing the protein profile and/or protein content (Aloisi et al. 2016; Ruiz et al. 2016; Fischer et al. 2017). This could be explained, at least partially, because salinity induces nutritional plant deficiencies or imbalances due to competition with sodium (Na$^+$) or chloride (Cl$^-$), resulting in decrease in the uptake and transportation of nutrients including nitrogen, phosphorus, potassium and calcium (Hu and Schmidhalter 2005).

2.3   GENOTYPE AND ENVIRONMENT INTERACTION ON CROP PRODUCTIVITY AND SEED QUALITY

While much literature has centered on studying the impact of abiotic stress on crop yield and the underlying mechanisms, there is scarce information on the relationship between abiotic stress and grain/seed quality (even for staple crops). This section attempts to present an update of the available information regarding how environmental conditions affect seed quality in the most relevant and promissory Latin-American crops.

2.3.1   *Chenopodium pallidicaule* and *Chenopodium quinoa*

Chenopodium quinoa Willd. (quinoa, quinua) (Figure 2.1a, b) and *Chenopodium pallidicaule* Aellen (also known as cañahua or cañihua) (Figure 2.1c, d) are two related annual crops that belong to the family Amaranthaceae. Both species are native to the Andean altiplano, a high plateau situated at 3500–4200 m above sea level between the western and eastern areas of the Andean Cordillera. The cultivation of cañahua and quinoa dates back more than 7000 years when it was established in the area as a staple crop by ancient Incan and pre-Incan societies. After the Spanish conquest, the cultivation of ancient crops was likely discouraged due to their association with indigenous cultures and ignorance of their benefits (Aellen and Just 1943), and became marginal crops for farmers' subsistence in the Andean region.

The edible seeds of cañahua and quinoa (dicot plants) are not grains, rather they are achenes (fruits), composed of a single seed enclosed by an outer pericarp (Abdelbar 2018). Seeds of both plants have unique nutritional profiles suitable for human consumption. Seeds from *C. pallidicaule* and *C. quinoa* are gluten free, good sources of dietary fiber (Repo-Carrasco et al. 2009, 2019) and contain

Figure 2.1   Latin-American crops.

Notes: (a) *Chenopodium quinoa* Willd (quinoa) at seed filling stage (picture taken in the northern part of Chile [Vicuña, La Serena, Chile]); (b) dry seeds of a red cultivar of quinoa; (c) *Chenopodium pallidicaule* Aellen (cañihua) at flowering stage (note the very small flowers); (d) dry seeds of cañihua ecotype Kello; (e) *Amaranthus* ssp. (kiwicha) at flowering stage; (f) dry seeds of *Amaranthus caudatus* (variety Oscar Blanco); (g) *Phaseolus vulgaris* (beans) at reproductive stage (fresh pods in the plant); (h) dry seeds of beans; (i) *Zea mays* (maize) during flowering in an established plantation in Peru; (j) diversity of corn cobs colors including yellow, red and white; (k) *Salvia hispanica* (chia) at flowering; (l) variegated seeds of chia, cultivar Oaxaca from Mexico; (m) fresh capsules of *Plukenetia volubilis (*sacha inchi); (n) dry capsules of sacha inchi with testa; (o) plantation of *Lupinus mutabilis* (tarwi) during flowering in the experimental station of Camacani (Puno [Peru]); (p) seeds of tarwi, variety SLP-4. **Source: Pictures courtesy of Dr. Enrique Martinez (Center of Advanced Studies in Arid zones [CEAZA, Chile]), Dr. Alberto Pedreros (Universidad de Concepcion [Chile), Dr. Grandez de Iquitos (University of Altiplano [Peru]), Dra. Nancy Chasquibol Silva (Centro de Estudios e Innovación de Alimentos Funcionales (CEIAF), Universidad de Lima [Peru]) and Dr. Angel Mujica (University of Altiplano (Peru) and Amaranth Promotion Network [Mexico]).**

Notes: Figure 2.1 shows 16 panels with pictures of different plants and seeds.

A) Panel A shows a close look at many star-shaped seeds that form a panicle in quinoa with a reddish color.

Figure 2.1   (continued)

B) Panel B has two hands showing quinoa seeds belonging to Pandela landrace with different colors including white and reddish.

C) Panel C is a close look at cañihua plants in green and dark color in the field.

D) Panel D shows seeds of cañihua in an orange color.

E) Panel E shows a dark red panicle of a kiwicha plant (in green) growing in the field.

F) Panel F shows kiwicha whitish seeds.

G) Panel G shows pods of a bean plant.

H) Panel H shows oval shaped white beans with dark spots.

I) Panel I shows maize plants in the field.

J) Panel J shows different types of maize ears dispersed in five rows and in orange to dark red color.

K) Panel K shows chia flowers in a plant with small blue petals.

L) Panel L shows oval shaped chia seeds in a brownish color and darker spots.

M) Panel M shows a Sacha inchi plant with four or five lobes of fruit capsules.

N) Panel N shows mature, star-shaped, five lobe Sacha inchi fruit.

O) Panel O shows tarwi plants with panicles with many pink to purple flowers in the field.

P) Panel P shows bean-like tarwi seeds of white and black colors and a label in the center of the picture indicating that these are tarwi seeds.

about 20% protein, with a complete set of essential amino acids (Penarrieta et al. 2008). In addition to high-quality protein, both seeds contain a wide variety of other health-promoting compounds, including antioxidants such as phenols and flavonoids (Repo-Carrasco-Valencia et al. 2010), and lipids, including the fatty acids Oleic (C18:1) and Linoleic (C18:2) (Khaitov et al. 2020). Both species are considered Andean superfoods (Mangenson et al. 2019; Repo-Carrasco-Valencia et al. 2019) and quinoa has been chosen as a crop that might contribute to global food security during the next century (FAO 2019).

Cañahua is a poorly studied species and is considered a partially domesticated crop whose cultivation is mainly restricted to the Andean region. Non-uniform seed ripening and small seed size are the principal agronomic issues that have prevented a more extensive cultivation of cañahua (Mujica 1994). In contrast, *C. quinoa*, with different available genotypes and a higher size of seed, is a more popular crop, cultivated in many areas of the world, with varieties already developed in North America and Europe (Alandia et al. 2020).

There is little information available regarding the genetic diversity of cañahua (Mangelson et al. 2019). In fact, the scientific community is worried about its *in situ* preservation, which is crucial for future genetic improvement (plant breeding). Nevertheless, efforts to preserve the genetic diversity of cañahua have resulted in the creation of a collection of cañahua germplasm in two banks located in Peru and Bolivia (Flores 2006; IPGRI et al. 2005).

In contrast, there is much more information about the genetic diversity of quinoa. Worldwide, there are more than 6000 landraces of quinoa cultivated by farmers. According to their adaptation capacity to specific agro-ecological conditions, cultivars can be classified into five ecotypes: Highlands (or Altiplano type); Inter-Andean valleys; Yungas (grown under tropical conditions); Salares (grown at high altitude salt lake areas and with limited volume of annual rainfall [150–300 mm]); and coastal/lowlands (rainfall ranges from 500 to 1500 mm annually) (Martinez et al. 2009; Bazile et al. 2016). In spite of the difficulties in sequencing a tetraploid organism such as quinoa, its genome sequence was recently published using a Chilean coastal ecotype (Jarvis et al. 2017) and the variety "Real" from the altiplano (Zhou et al. 2017).

Importantly, among quinoa landraces, coastal/lowlands are of particular importance due to their widely longitudinal geographic range, which allows greater photoperiod adaptation that makes them highly suitable for quinoa cultivation into different climatic zones (Jacobsen 1997; Bendevis et al. 2014).

The genotype and environment interaction in quinoa has been studied in different published works. Bertero et al. 2004 tested 24 cultivars of quinoa at 14 sites across three continents during two growing seasons. They found great differences in seed yield and composition among genotypes through different environmental growing conditions. The great variability of quinoa seed quality was supported by the work of Reguera et al. 2018, in which three varieties of quinoa were adapted to specific agroecological conditions –Salcedo-INIA (developed by INIA-Peru), Titicaca (developed by the University of Copenhagen, Denmark) and Regalona (developed by Baer, Chile) – were grown in different environmental conditions in Spain, Peru and Chile. The results revealed that protein content, amino acid profiles, mineral composition and phytate amount in seeds varied depending on cultivation conditions and genotype used, while other parameters, such as saponin or fiber were stable across locations. Interestingly all genotypes studied presented higher antioxidant contents when they were cultivated in the north of Chile.

When analyzing quinoa abiotic stress responses, it should be considered that one of the major limitations found in quinoa cultivation in the field is the combined effect of drought and high temperatures. Quinoa is especially sensitive to elevated temperatures at the reproductive stage, i.e., from the beginning of flowering (pre-anthesis) to the end of flowering (post-anthesis), with the seed filling period also being a sensitive stage) (Geerts et al. 2008). In general, several reports have stated that seed yield potential and the content and quality of proteins in quinoa seeds are reduced under water deficits (Fisher et al. 2013; Bascunan-Godoy et al. 2016). Working with two lowland quinoas cultivars, Bascunan-Godoy et al. (2016) found that water stress increased the accumulation of metabolites that belong to the ornithine pathway (N metabolism) in both genotypes, Faro and BO78; however, after a few days of re-irrigation, only Faro was able to recover metabolite levels to those found in control conditions. Remarkably, this genotype showed higher drought tolerance and the ability to store betacyanins, which are nitrogen-containing pigments with high antioxidant power (Bascunan-Godoy et al. 2018a). Interestingly, Fisher et al. 2013 reported an increase in antioxidant capacity measured using DPPH (2,2-diphenyl-2-picryl-hydrazyl) while hydric restriction in Chilean lowland genotypes increased. The authors agree that it is possible to produce seeds with higher nutritional value when subjecting plants to controlled water limitation, avoiding penalties on seed yield.

When referring to temperature stress, Lesjak and Calderini 2017, using a lowland Chilean quinoa genotype, Regalona, in an experiment performed under controlled conditions, found that the increase of 4°C during the night reduced grain yield, biomass and grain number. In another experiment, the effect of temperature on quinoa seed quality was approached under field conditions to try different sowing dates, therefore modifying the photo-thermal conditions during seed-filling stage (Curti et al. 2020). Differences in sowing dates induced changes in the lipid content, accompanied by significant variation in fatty acid concentration. However, the general decrease in lipid content did not similarly affect the major unsaturated fatty acids among cultivars. A decrease in oleic and α-linolenic concentration was observed for almost all cultivars, whereas linoleic concentration remained unchanged for some cultivars or even increased in others. This result again suggests that cultivar-specific responses to photo-thermal conditions during the seed-filling period are involved in quinoa (Curti et al. 2020). More recently, work performed by Matias et al. (2021) showed that when quinoa, grown under Mediterranean field conditions, suffered heat stress, it eventually affected yields and seed quality in different ways depending on the length of the cultivar life cycle. Thus, those early maturing varieties are, potentially, better adapted genotypes for areas that are more susceptible to suffering heat waves when aiming at preserving yields and seed nutritional qualitative-related parameters.

Even though quinoa is considered a facultative halophyte (Boindi et al. 2015), there is a broad gradient of saline tolerance according to the cultivar´s origin (Delatorre and Pinto 2009; Razzaghi et al. 2011). Quinoa coming from the altiplano or those coming from lowlands, two representative quinoa ecotypes, present contrasting salinity resilience (Jacobsen and Stølen 1993; Delatorre-Herrera and Pinto 2009; Ruiz-Carrasco et al. 2011). Many quinoa trials tested worldwide using salinity irrigation have been performed (Hussain et al. 2018, 2020; Roman et al. 2020; Adolf et al. 2012; Cocozza 2013; Shabala et al. 2013). These reports found significant interaction between the genotype used and the irrigation conditions tested for seed yield, biomass and different agronomic traits. This highlights the great genotypic plasticity of quinoa and also points to the need to assess the genotypic performance under particular growing conditions. Furthermore, the impact of salinity on the nutritional quality of quinoa seeds has started being analyzed in recent years. For instance, Fischer et al. 2017 showed that salinity is correlated with reduced content of proteins and antioxidants in seeds. Ruiz et al. (2016), using two contrasting landraces, salares and coastal, revealed interesting responses to salinity in that belonging to the latter ecotype in terms of growth, yield and seed quality expressed in enhanced total polyphenol content and antioxidant activities in seeds, accompanied by superior ability to germinate under saline conditions; however, the total protein content decreased slightly under salinity. In contrast, in another study conducted by Wu et al. 2016, in which four quinoa cultivars were grown under six salinity treatments and two levels of fertilization, little variation was found in the protein contents, but changes were detected in seed density and hardness. Nevertheless, further studies should be conducted to target the genotypic dependence on salinity stress response in quinoa.

2.3.2   *Amaranthus* ssp.

The genus *Amaranthus* comprises about 400 species of annual or short-lived perennial plants collectively known as amaranth. They belong to the family Amaranthaceae, as does quinoa or cañahua. The historical evidence depicted that amaranthus domestication and cultivation started about 8000 years ago by pre-Columbian civilizations in South and Central America. Amaranths (also known as Huahtli or Kiwicha) (Figure1e, f) were a staple food for the Aztecs, Mayans and Incas civilizations (Sauer 1950a, 1950b; Pal and Khoshoo 1972; Alvarez-Jubete et al. 2010a). However, after the arrival of and further colonization by Europeans, amaranth crops were nearly eradicated and confined to very few indigenous communities.

Some species produce edible leaves (e.g., *Amaranthus tricolor*) and gluten free grains (*Amaranthus hypochondriacus, Amaranthus caudatus, Amaranthus cruentus*), being a rich and inexpensive source of daily required proteins, antioxidants and bioactive compounds (Bressani 1994; Gimplinger et al. 2007; Alvarez-Jubete et al. 2010a, 2010b). Additionally, Amaranthus seeds show an exceptional nutritional profile which includes crude protein 12–19%, fats 5–8%, starch 62–69%, total sugars 2–3%, dietary fiber 4–20% and ash 3–4% (Alvarez-Jubete et al. 2009). Notably, seeds are highly enriched in essential amino acids, such as lysine (363 – 421 mg/g N). Regarding fatty acids, amaranth seeds have high concentrations of the unsaturated fatty acids linoleic and oleic acids (62 and 20% respectively). Other important fatty components are tocopherol (vitamin E) and tocotrienols. Besides, amaranth seeds have the highest content of minerals, especially Ca, Mg, Zn and Fe compared with other pseudocereals and cereals (Alvarez-Jubete et al. 2010a).

Amaranth grows well in temperate and tropical regions. It shows high resilience to various abiotic stresses, either imposed by climate or soil conditions, especially those encountered in semiarid biomes. Amaranth possesses a C4 photosynthetic pathway, which is an evolved mechanism for CO_2 fixation, reducing water loss (Sage et al. 2007). Hence, amaranth plants can perform well during the vegetative and reproductive stages without significant growth and yield penalties under heat and drought stress. Flowering starts about 4 to 8 weeks after sowing, but phenological stages vary significantly among species/cultivars and soil nitrogen content.

Due to their high resilience plus their nutritional and pharmaceutical value, there is a large germplasm collection of amaranth crop species. For example, the United States Department of Agriculture (USDA) and the National Botanical Research Institute (NBRI) collections include approximately 3000 and 2500 accessions, respectively. However, non-domesticated species are underrepresented in seed banks worldwide (Achigan-Dako et al. 2014).

Different works have highlighted the outstanding ability of acclimatization and adaptation to different abiotic stresses of *Amaranthus* (Liu and Stützel 2004; Moreno et al. 2017). Under drought stress, several genotypes of amaranth showed osmotic adjustment capacity and the ability to control water loss by reducing stomatal conductance and leaf expansion, thus, conferring resistance to severe drought (Liu and Stützel 2002). However, resistance depends on the genotype. For example, Liu and Stützel (2004) evaluated the effects of drought stress on water use efficiency and economic spectrum traits in four genotypes of amaranth. The authors found differences in biomass partitioning between roots and shoots depending on the genotype, but without detecting a significant effect on water use efficiency. However, Tsutsumi et al. (2017) reported that the water use efficiency of amaranth crops can vary significantly with genotype, mainly due to variations in structural, biochemical and physiological components. Probably, both the physiological C4 pathway and morphological traits together finely modulate drought resistance in *Amaranthus* species. Seed resilience against water deficit is remarkable, as shown by Pulvento et al. (2021). The seed yield components of *Amaranthus hypochondriacus* were not significantly affected by reduced irrigation, nor was seed quality. Although more research is needed, this is a promising feature projecting intensive use of amaranth plants for food security in a climate changing world.

Amaranth plants are not only suitable crops for drought prone regions, but also grow rapidly under high temperatures. Indeed, full genotypic potential is reached at day/night temperatures above 25/15°C, respectively, with optimal growth at about 30°C (Achigan-Dako et al. 2014). Nevertheless, much more research work is needed to fully understand the physiological mechanism of *Amaranthus* to cope with heat stress.

Salinity is the third major environmental constraint for plant growth, which is exacerbated by climate change. Amaranth species are recognized as glycophytes (Wang et al. 1999). However, amaranth can tolerate salt concentrations up to 150 mM NaCl. Saucedo et al. (2017) reported that under salt stress (0.4 to 0.8 M NaCl), *Amaranthus cruentus* and wild relatives accumulated protective proteins (LEA) in leaves, stems and roots. Agronomical practices such as seed priming have demonstrated significant improvement in *Amaranthus caudatus* seed germination under severe salt stress (up to 200 mM NaCl) (Moreno et al. 2017).

Responses to combined abiotic stress are far from being well understood in *Amaranthus* species. Pulvento et al. (2021) studied drought and salinity effects in grains of *Amaranthus hypochondriacus* under field conditions during 3 years in Italy (Pulvento et al. 2021). Individually, differential water supply (irrigation time, irrigation volume) did not induce changes in yield and quality, but salinity reduced seed yield by 55%, compared to control conditions. However, it was remarkable that the quality of amaranth seeds was preserved under combined drought-salinity treatments and no significant differences in starch, protein and lipids were found (Pulvento et al. 2021). Interestingly, salinity induced a significant and remarkable increase in β-cyanin, β-xanthin, betalain, total carotenoids, β-carotene, ascorbic acid, total polyphenolic content, total flavonoid content and total antioxidant capacity in leaves of *Amaranthus tricolor*. Hence, it is highly probable to find similar results in Latin-American *Amaranthus* crop species.

There is no doubt about the nutritional quality and the potential of discovering new bioactive compounds in *Amaranthus*. Nowadays, *Amaranthus* presents a great research opportunity to properly understand the genetic control of physiological responses and resilience to environmental stresses. Also, plant biologists have the challenge of studying the gaps in biological processes against combined stresses, enhanced yield and harvesting index, while keeping the nutritional quality of *Amaranthus*.

2.3.3   *Phaseolus vulgaris*

Phaseolus vulgaris L. (beans) (Figure 1 g, h) belongs to the Fabaceae family, which comprises species displaying a wide variety of forms: trees, shrubs and herbs, including many with a climbing growth habit. Common beans are the most frequently produced and consumed worldwide legume, with high commercial value (Broughton et al. 2003). Like other legumes, common beans can fix atmospheric nitrogen (N_2) through the symbiotic fixation process with rhizobia (Lugtenberg and Kamilova 2009), thus allowing a reduced use of chemical fertilizers and promoting more sustainable agriculture.

Although the dried seeds of *P. vulgaris* are low in methionine and cyst-eine, these are an important source of vegetal protein for millions of people, supplementing those amino acids which are lacking in diets based on maize, rice or other cereals (Broughton et al. 2003; Wortmann 2006). Beans are especially rich in lysine and tryptophan, iron, copper and zinc; and beneficial antioxidants and flavonoids (Wortmann 2006). Due to extensive plant-breeding efforts, *P. vulgaris* is cultivated in several agroecological environments and comprises numerous cultivars that differ in their growth habitat as well as seed size and color (Purseglove 1968; Singh et al. 1991a).

The wild bean was domesticated several times in the pre-Colombian era in both Mesoamerica (Mexico) and the Andean zone, but phylogenetic and popula-tion structure analyses have confirmed the origin of *P. vulgaris* in Mesoamerica (Bitocchi et al. 2012). Mesoamerican and Andean populations are distinguished by yield potential, morphology (Singh et al. 1991a,c), isozymes (Singh et al. 1991b), DNA molecular markers (Nodari et al. 1993), as well as by physiological traits related to photosynthesis (Castonguay et al. 1991). Additionally, there are other important diversity centers in Brazil, North America, Africa, the Middle East and Europe. The world collection of cultivated and wild *P. vulgaris* is held in the germplasm bank of CIAT (International Center of Tropical Agriculture) in Cali, Colombia. In 1997, 364 genotypes were identified among the accessions held in the CIAT collection, and among commonly grown national or regional cultivars (Beebe et al. 1997).

Many studies considering genotype and environmental factors have been conducted (Sozen et al. 2018; Suarez et al. 2018; Rainey et al. 2005), con-cluding that climate changes between years affect yield parameters in dry bean genotypes.

Regarding abiotic stress, water scarcity is one of the main limiting factors for common bean crops, negatively affecting seed yield and quality, which might be related to the fact that the symbiotic nitrogen fixation is rapidly inhibited under these conditions (Sinclair and Serraj 1995). Local farming practices, domesti-cation and worldwide spread has entailed the development of a wide variety of common bean genotypes with a wide range of resistance to water scarcity. Lopez et al. 2020 studied the molecular basis of differential drought tolerance in two accessions of common beans (PMB-0220 and PHA-0683). They found that differential regulation of ABA synthesis and signaling related genes among the two genotypes and control of drought-induced senescence makes a relevant contribution to the higher drought resistance level of PHA-0683 accession. The effect of water stress on the physical and chemical qualities of the common bean seeds (variety BRS Realce) was studied in Planaltina, Brazil (Silva et al. 2020). In this work, water deficit increased the protein and reduced the carbohydrate and ash contents. Additionally, they found that water stress reduced macro- and micro-mineral content in the grains and changed the physical quality of the seeds. In spite of the results reported here, other studies have reported a reduction in protein content under drought (Khalil et al. 2010), but different proteins respond in various ways.

Under field conditions it is difficult to separate the effect of drought and high temperatures. Controlled condition experiments conclude that the effect of high temperatures on bean seeds were more detrimental to seed quality than was limited rainfall (Muasya et al. 2008). Exposure to high temperatures during two reproductive growth stages, namely flower bud formation and pod-filling, resulted in severe damage (Shonnard et al. 1994) inducing abscission in numerous flowers and pods (Konsens 1991), reducing the number of pollen grains, number of viable seeds and yield components (Da Silva et al. 2020).

Suarez et al. 2020 intended to identify genotypic differences in 91 bean genotypes (derived from interspecific and intraspecific crosses) to high temperature under the environmental conditions of a tropical dry forest ecosystem. They found that a few Andean lines presented good adaptation capacity to heat stress. The superior performance under higher temperatures of the identified genotypes was related to greater ability to partition dry matter, high pollen viability and early growth maturity, resulting in greater grain filling.

Something interesting was the effect of altitude on *P. vulgaris*, promoting the content of antioxidants, total phenols, essential amino acids such as valine, phenylalanine and isoleucine content. The effect of altitude can be linked to the higher temperature and different light quality, creating a greater stressful condition for the crop that responded by increasing the synthesis of antioxidant compounds (Nicoletto et al. 2019). The information reported so far must also be related to the genotype.

P. vulgaris is considered an extremely salt-sensitive species. Several authors have approached the effect of sodium chloride (NaCl) on productivity and yield in contrasting salt-sensitive genotypes. Assimakopoulou et al. (2015) found that 'Corallo', is a salinity tolerant genotype due to its capacity to sequester Na in the roots and maintain appropriate K/Na and Ca/Na ratios, restricting the levels of toxic ions in growing shoots. The higher sensitivity to salt of 'Romano' cultivar was related to its higher level of Na in the leaves and the reduction of leaf number and leaf water content. Taibi et al. 2016, on the other hand, also working in contrasting salinity-sensitive genotypes, found that the only qualitative difference among genotypes was the level of phenolic compounds in leaves, therefore, the high-yielding genotype may have increased the activity of antioxidant enzymes that give better protection against oxidative damage throughout higher flavonoid and ascorbic acid contents stored under high salinity. Regarding seed quality, Farooq et al. (2017) found that seed protein content declines under salt stress, which can be linked with changes in the symbiotic relationship observed under salinity in common beans (Abdi et al. 2015). These works give some clues for selectioning genotypes under saline stress conditions.

2.3.4 *Zea mays*

Maize (*Zea mays* L.) (Figure1i, j) is one of the leading cereal grains worldwide, along with wheat and rice (Ali et al. 2014). Maize production areas yield about 850 million tons and the average grain yield is about 5200 Kg ha^{-1} (FAO 2016). Mexico is recognized as the center of origin and domestication of maize. In fact, maize domestication and growth were already underway in the Tehuacán valley (Mexico) about 5000 years ago (Vallebueno-Estrada et al. 2016) and has generated great genetic diversity through farming and environmental selection (Perales and Golicher 2014). The huge genetic diversity of maize is reflected in more than 5500 accessions that exist worldwide (from more than 40 countries) (Prasanna 2012). However, it should be noted that commercial lines present a narrow genetic base (Kasoma et al. 2021). The modest genetic diversity of commercial genotypes has limited variability for value-added traits. Commercial maize genotypes are poor in terms of protein content, essential amino acids – lysine and tryptophan – and also have low levels of biological healthy molecules (Gavicho-Uarrota et al. 2011). Teosinte (*Z. mays* ssp. parviglumis Iltis & Doebley) and tripsacum are the two wild relatives of maize that have been extensively characterized (Matsuoka et al. 2002). Unlike many hybrid varieties, maize landraces present high levels of phytochemical compounds such as phenolic compounds (flavonoids and anthocyanins), carotenoids, xanthophylls, vitamins and dietary fiber, that contribute to human health (Gavicho-Uarrota et al. 2011). Nowadays, new lines of maize hybrids have been improved with relevant traits for quality and yielding through gene introgression from landraces or wild relatives (White et al. 2007).

Numerous articles report tolerance to abiotic stress, including drought, high temperatures and salinity of different populations, landraces and inbred lines. However, the fact that about 80% of the total area of maize sown in Mexico is produced through subsistence farming, indicates that it is a species capable of withstanding extreme conditions (Bellon et al. 2011). Drought affects various morpho-physiological processes, including plant biomass, root length, shoot length and photosynthesis in maize (Ali et al. 2014, Yamori et al. 2014). The occurrence of drought stress during the growth period may lead to significant yield losses due to nutrition and water imbalances (Ashraf et al. 2016). Several studies involving non-drought-tolerant maize hybrids reveal a reduction in total phenols, hydroxycinnamic acids and antioxidant capacity, while works on drought-tolerant maize populations reported the maintenance or increased level of healthy molecules such as oil, fatty acid composition, protein, and tryptophan content (Ignjatovic-Micic et al. 2015).

Drought is the major abiotic stress that hinders crop productivity across the world (Aslam et al. 2015). The impact of drought largely depends on the intensity and duration of water shortage. Mi et al. (2018) studied different levels of progressive drought on yield formation of maize, reporting that grain yield was significantly reduced by either the vegetative (18.6–26.2%) or reproductive stage (41.6–46.6%), which was largely caused by the decrease in kernels per ear. Also, the effect of drought on yield and seed quality is highly dependent on the sensitivity or genotype tolerance (Anjum et al. 2017). In more drought tolerant genotypes, osmoprotectans' accumulation might play a key role in their physiological performance and grain biomass trait (Salgado-Aguilar et al. 2020). Development of tassel and ear, pollination, fertilization, embryo development, endosperm development and grain filling are seriously affected by drought stress in maize (Aslam et al. 2015). Moderate drought produced a decrease in the grain yield of 33.7% and 62.3% in a drought-tolerant and a drought-sensitive maize variety; respectively (Qi et al. 2010). Additionally, a decrease in soil water content negatively affects the physiological quality of the maize grains produced, affecting germination and post-germination performance (Aslam et al. 2015; Machado et al. 2020).

Previous reports indicated that high temperatures affect the persistence and productivity of leaves, and reduced both whole plant dry mass accumulation and grain yield by a shorter duration of grain filling (Wilson et al. 1973). Thus, high-temperature stress during the reproductive stage has been the main obstacle for increasing maize productivity in many places worldwide (Tao et al. 2016). Despite the negative effect of diurnal warming, higher night temperatures caused an increased gain in kernel weight, resulting from the remobilization of stored dry mass (Badu-Apraku et al. 1983).

Warming may affect maize growth and production due to the direct effect on its structure, biochemical properties, and gas exchange of maize leaves, resulting in increased net photosynthesis, C:N ratio and soluble sugars (Zheng et al. 2013). However, at rising temperatures pollen viability and corn silk receptivity resulted in poor seed set and decreased yield (Hussain et al. 2006). Heat-stressed maize plants are unable to convert photosynthates into starch in pollen because key enzymes involved in starch biosynthesis are inhibited at both transcriptional and post-transcriptional levels (Boehlein et al. 2019). The reduction in pollen number contributes to distorting the fertilization process (Sánchez et al. 2014). Additionally, the development of grain filling is accelerated, reducing the period of amyloplast biogenesis and endosperm cell division reducing grain size (Yang et al. 2017; Waqas et al. 2019). To overcome the negative effect of higher temperatures on grain yield, hybrid breeding is recommended because hybrid plants have shown a higher capacity to tolerate heat stress in the field (Hussain et al. 2006).

The effect of salinity on maize yield is also genotype-dependent, thus determination of the salt tolerance level in large genetic resources and breeding populations will be very important to solving the salinity problem (Konuskan et al. 2017). Exclusion of excessive Na amounts or its storage in vacuoles, is an important adaptive strategy for maize under salt stress (Farooq et al. 2015). The response of maize to salinity also varies depending on the developmental stage, with germination and stand establishment more sensitive than at later stages (Fortmeier and Schubert 1995; Radic et al. 2007; Farooq et al. 2015; Konuskan et al. 2017). Despite this, Wang et al. (2016) and Kang et al. (2010) worked with moderate salinity irrigation (2–3 g NaCl L^{-1}) in maize, resulting in normal growing and greater productivity. In fact, Li et al. (2019) reported that moderate salinity (5 g NaCl·L^{-1}) maintains oil, crude fiber and ash contents in grains. However, at increased levels of salinity, the excessive uptake of Cl and Na ions by maize roots leads to severe interference with other essential mineral elements, inducing nutritional imbalance (Hasegawa et al. 2000; Karimi et al. 2005; Turan et al. 2010). The salinity effect on germination and vegetative growth is very important in determining yield (Konusan et al. 2017). In some maize genotypes, yield is significantly reduced under salt stress affecting both grain weight and number (Farooq et al. 2015). In addition, sink limitations and reduced acid invertase activity in developing grains produce poor grain setting under saline (Farooq et al. 2015). In fact, Li et al. 2019 working in the Neidan 314, which is considered a vigorous maize genotype (Ye et al. 2020), found that grain moisture and starch content decreased under salinity conditions; however, oil and ash were maintained, and protein and fiber increased. Still, further studies are needed for in-depth understanding of the effect of salinity stress on maize quality.

2.3.5 *Salvia hispanica*

Chia (*Salvia hispanica* L.) (Figure1 k, l) is an annual crop that belongs to the Lamiaceae family. The area from western Mexico to East-Central Mexico (with altitudes between 1400 and 2200 m a.s.l.) has been identified as the center of its genetic origin (Cahill 2004). Recent contributions using single nucleotide polymorphisms (SNPs) support this information and revealed that most of the genetic diversity variation of chia remains in wild populations (Pelaez et al. 2019).

In pre-Columbian Mesoamerica, chia was cultivated for its edible seeds with nutritional and therapeutic properties (Cahill 2003); however, after the Spanish conquest it was dramatically eradicated (Ayerza and Coates 2005a). Chia is a good source of polyunsaturated fatty acids and has been pointed out as the oil-seed plant with the highest omega 3 fatty acid content (Cahill 2003, 2004; Ayerza and Coates 2005a, 2005b, 2009). Additionally, chia contains high amounts of proteins (up to 26%), dietary fiber (33.9–39.9%) and high antioxidant activity because of high levels of tocopherols, phytosterols, carotenoids and phenols such as chlorogenic and caffeic acids, myricetin, quercetin and kaempferol (Amato et al. 2015; Reyes-Caudillo et al. 2008). Additionally, their leaves present active compounds of nutraceutical, antioxidant and antimicrobial value (Amato et al. 2015; Elshafie et al. 2018).

Chia is now cultivated and commercialized due to its highly nutritional seeds; however, genetic studies suggest a slight loss of diversity accompanying domestication and a near lack of diversity in modern commercial varieties (Cahill 2004).

Chia is a summer crop characterized as a short-day plant and intolerant to frost (Jamboonsri et al. 2012; Baginsky et al. 2014). The characteristics of short-day flowering species have been considered a problem to expanding their cultivation to a wider range of latitudes. This is because flowers are too late in the season, and seeds are not able to end maturation before frosting. Nevertheless, breeding efforts have produced longer-day flowering genotypes (early flowering) to extend cultivation of this crop to other climate areas (Jamboonsri et al. 2012). Most of these new lines are mutants capable of inducing flowering in light cycles of 13 and 16 h of day length and a few are day-length insensitive. Certainly, the in-depth study of this germplasm including yield, quality and the identification of seed metabolites has been a main goal.

Genotype and environment (GX E) interaction on chia quality was studied in a pioneer experiment performed by Ayerza and Coates (2009). They investigated how seed yield, protein content, oil content and fatty acid composition varies among three chia selections when planted in three inter-Andean valleys of Ecuador, which differ in elevation (affecting chia's length of growing period). They found that location mainly affected seed yield, and, to a lesser degree, the quality (protein and oil contents as well as fatty acid composition). Additionally, no differences in protein, oil, fiber, amino acids and antioxidant content composition were found between Ecuadorian chia genotypes that differed in seed color (Ayerza 2013).

However, while no great variation in seed quality has been reported in Mesoamerica, important differences have been observed in studies conducted in Europe (Italy) comparing commercial and long day/ vs. early flowering genotypes. Studies carried out by de Falco et al. 2017, 2018 compared the composition of commercial chia genotypes (two black, and one white) and long day/ early flowering genotypes: G3, G8, G17 (mutant genotypes obtained through treatment with gamma radiation (Jamboonsri et al. 2012)) and W13.1 (derived from a cross between G8 and a white-flowered, white-seeded commercial chia (wild type)) through nuclear magnetic resonance (NMR) spectroscopy and metabolomics analysis. The metabolic fingerprinting showed great differences among these populations. Most of the detected metabolites showed larger variations in the seeds of early flowering genotypes, compared to commercial seeds. In particular, early flowering and white genotypes were reduced in carbohydrates compared to black genotypes. Interestingly, the highest content of omega-3 fatty acids was found in white commercial chia. The early flowering genotypes showed the highest content of antioxidant metabolites including caffeoyl derivatives caffeic, chlorogenic and rosmarinic acid; flavonoids genistein and quercetin, as well as carbohydrates such as sucrose and raffinose (but not glucose). These results suggest great potential for different purposes (related to the food industry) among commercial and early flowering genotypes. Interestingly, a recent study was performed aiming at identifying single nucleotide polymorphisms (SNPs) and simple sequence repeat (SSRs) markers which can contribute highly to breeding chia towards generating more nutritious seeds (Pelaez et al. 2019).

Nonetheless, there is little information about the interaction GX E in this crop and how it affects chia seed quality, which should be further explored in future experiments. In its zone of origin, chia grows under rainfed or irrigated conditions (Coates and Ayerza 1996). Silva et al. 2016 worked in the semiarid conditions in the north of Chile, and using two genotypes black and white (commercial accessions from Bolivia) found that chia plants induce strategies related to controlling water loss under water shortage. This is consistent with the results observed by Lovelli et al. 2019 using both commercial and early flowering genotypes under controlled conditions. There is little information on the effect of drought on seed composition, but Silva et al. 2016 reported a decrease in total oil, omega-3 and linoleic acid in seeds. Besides, changes in fatty acid compositions were reported by Ayerza et al. 1995, who performed experiments in five of Argentina's northwestern locations, differentially influenced by temperature. This work reported that seed composition, including total oil content and oleic, linoleic, and linolenic fatty acid concentrations, varied significantly with location. The effect of drought and temperatures was similarly observed, when Heuer et al. 2002 showed that salinity reduced the oil content and changed the fatty acid composition (palmitic, stearic, oleic, linoleic and linolenic) of chia seeds. Additionally, Paiva et al. 2018 showed that conductivity levels higher than 4.5 dS m^{-1} combined with temperatures ranging between 20 and 30°C, negatively affected germination, growth and biochemical components, such as chlorophylls and proteins of chia seedlings. More studies that involve metabolic changes during the reproductive stage are necessary to establish a more direct relationship regarding the changes in the plant and seed composition.

2.3.6 *Plukenetia volubilis*

Plukenetia volubilis L. (family Euphorbiaceae) or their name in Quechuan 'sacha inchi' (Figure1 m, n) is a native plant from the Amazon cultivated as early as 3000–5000 years ago (Bernal and Correa 1992; Brack 1999).

In America, *P. volubilis* grows in countries such as Peru, Bolivia, Venezuela, Colombia, Ecuador and Brazil in tropical or subtropical climates, with temperatures ranging from 10 to 26°C, relative humidity of 78% and altitudes of at least 1490 m.a.s.l. (Arfini and Antonioli 2013).

Recently, *P. volubilis* has attracted increasing attention, because of its oleaginous seeds that contain an unusual and outstanding nutritional composition (Kodahl 2020). Depending on extraction methods, most of these studies report a lipid content of 45–50%, with a very high proportion of polyunsaturated fatty acid, greater than 80% (Castaño et al. 2012), including essential fatty acids such as α-linolenic and linoleic acids (Hamaker et al. 1992) and an omega-3: omega-6 ratio close to 1 (Chirinos et al. 2013). Furthermore, the seeds have high protein content (between 22 and 30%) and are a good source of antioxidants (Hamaker et al. 1992; Gutierrez et al. 2011; Ruiz et al. 2013). Kodahl (2020) compared the content of omega-3, omega-6, and antioxidant capacities reported in different plant species and found that they were significantly higher in *P. volubilis* compared to others typically found in the healthy market.

Studies analyzing the genetic diversity of this crop reveal that, unlike other Latin-American species, *P. volubilis* maintains a high degree of diversity (Corazon-Guivin et al. 2009). Rodríguez et al. (2010) reported high variability in morphology, plant production and oil content. Differences among cultivars in the chemical and phytochemical composition of seeds were also established by Chirinos et al. (2013) who worked with 16 genotypes from Peru. Results showed that content of phytochemicals and fatty acid profiles varied according to genotype.

To date, there is scarce information regarding the effects of environmental stresses on the productivity and quality of sacha inchi. However, several works have shown sensitivity to drought and chilling stress in this crop (Luo et al. 2014; Lei et al. 2014; Tian et al. 2013), and it seems that in association with mycorrhizal fungi it increased drought tolerance (Tian et al. 2013). Further studies should focus on elucidating the impact of environmental stresses on the seed quality of this crop to shed light on variations that might suffer the same nutritional characteristics.

2.3.7 *Lupinus mutabilis*

The genus *Lupinus* includes almost 200 species, but only four of them play an important role in agriculture: *L. albus, L. angustifolius, L. luteus* and *L. mutabilis*

(Gresta et al. 2017). *Lupinus mutabilis* Sweet (Andean lupin, tauri, tarwi, tarhui, chocho or kirku) (Figure 1o, p), is an important crop originally from the Central Andes, with high seed crude protein (32–53%), and oil (13–25%) content (Gulisano et al. 2017).

It seems that *L. mutabilis* was separated from its wild progenitor *L. piurensis* about 2600 years ago (Gulisano et al. 2019). Andean lupin was first domesticated in the Cajamarca region (Peru) and cultivated about 1800 years ago in the plains of South America (Gulisano et al. 2019), being an important plant used in crop rotation and contributing to soil fertility through nitrogen fixation and phosphorus mobilization (Kurlovich et al. 2002).

Germplasm collections are held mainly in Peru, Bolivia and Ecuador, and smaller collections are also present in Chile and Argentina, among other countries. Overall, these gene banks hold more than 3000 genotypes of *L. mutabilis* (Gulisano et al. 2019 and references therein). The genetic diversity of the Andean lupin has been addressed by different authors. Guilengue et al. 2020 performed a phenotypic analysis on the yield component of 23 Andean lupin accessions. They found that the productivity of primary branches was an important component of total yield accession, as well as the ability to produce large seeds. The authors highlighted one accession (LM268) that achieved these characteristics reaching the highest seed production. The genetic diversity revealed in this study, however, prompts further breeding opportunities outside the altiplane. On the other hand, Berru et al. 2021 compared seed characteristics and composition in 33 Andean lupine ecotypes (Peruvian) with other lupine species, including *L. albus*, *L. angustifolius* and *L. luteus*. The authors found that the Andean lupines had higher protein, lipid and tocopherol content than *L. albus* and *L. angustifolius* and presented similar values to those presented in *L. luteus*.

Seeds of *L. mutabilis* are not directly consumed due to the presence of quinolizidinic alkaloids that give it a bitter flavor. The alkaloid in higher concentrations are lupanine, followed by tetrahydrorombifoline, 4-hydroxilupanine, sparteine and 13-hydroxilupanine (Gross et al. 1988; Ortega-David et al. 2010; Castañeda et al. 2008). These substances are important for the plant since they protect it from phytopathogens and herbivores (Keeler 1976). Cortes-Avendano et al. 2020, working in ten ecotypes, found that alkaloids were influenced by geographical location, likely due to the different climatic conditions. They evaluated the effect of the aqueous debittering process of seeds in the profile and levels of quinolizidine alkaloids by gas chromatography and mass spectrometry (Jacobsen and Mujica 2006). From eight alkaloids identified before debittering, only small amounts of lupanine and sparteine remained in the seeds, and no other alkaloids were identified. The debittering of Andean lupines reduced the level of alkaloids to levels far below the maximal level allowed by international regulations. These results suggest that *L. mutabilis* harbors nutritional characteristics suitable for modern food trends.

Regarding the necessary climate conditions for its growth, it is known that the plant is susceptible to excess humidity and scarce water availability (Jacobsen and Mujica 2006). Optimal precipitations range from 350 to 850 mm during their growing cycle; and drought during flowering might stimulate early maturation and cause considerable biomass and seed yield losses (Hardy et al. 1997). Additionally, it does not tolerate frosts at early stages (Jacobsen and Mujica 2006; Simioniuc et al. 2021).

The effect of water stress on *L. mutabilis* was analyzed by Carvalho et al. 2004, 2005. The water deficit was imposed 15 days after anthesis, for 20 days. They found that water stress had little effect on total biomass, but strongly reduced plant water status, gas exchange and leaf area. Water stress changed the composition of all organs in the plant, increasing sugars and oils in stems and sugars in leaves. Regarding seed quality, water stress increased more than twice the level of total carbohydrates (from 50 to 130 g kg-1 of seed dry weight), modifying the sugar profile, increasing sucrose and reducing raffinose. The water stress also halved oil content, while protein content was maintained. It would be interesting to know how the quality of the oils and proteins were affected; however, these results suggest an impairment in nutrient translocation to seeds, affecting their final quality.

Another interesting abiotic factor that impacts Andean lupin growth is freezing. This legume plant species shows low tolerance to spring frost, which has limited its adaptation to temperate environments (Simioniuc et al. 2021). Recently, Simioniuc et al. 2021 compared three genotypes of *L. mutabilis* with *L. albus* in a range of freezing temperatures (from -2°C to -10°C) at different seedling stages. They found greater frost damage effect in the first stage studied (when cotyledons were breaking through the soil surface). Additionally, they reported that the LIB222 genotype displayed the highest frost resistance, related to a higher level of anthocyanin in its tissues. The anthocyanins are involved in cold and frozen resistance in many plant species due to their high antioxidant activity (Chalker-Scott 1999; Ahmed et al. 2015). However, it is not known so far if these results found at early growing stages could be extrapolated to other phases of growth (such as flowering or seed filling stage), or how the quality of the seed may vary under low temperatures (Simioniuc et al. 2021).

Considering that *L. mutabilis* is a subsistence species with potential for growth on marginal lands, more research efforts should be focused on identification of genotypes with early maturation and high frost and drought tolerance, while conserving the high protein and oil content in seeds.

2.4 PERSPECTIVES OF CROP PRODUCTION UNDER STRESS CONDITIONS

Current domesticated crop plants have resulted from a combination of years of evolution and human selection. It is estimated that more than 30,000 plant species have been cultivated throughout human history, and of these, approximately 7000 are crop species (Jacobsen et al. 2015; Kew 2016; Khoshbakht and Hammer 2008). Nevertheless, nowadays, fewer than 20 species provide most of the world's food, with only three crops – rice, wheat and maize – providing 60 percent of the world's food energy intake (Lenné and Wood 2011; FAO 2016). The majority of the edible plant species in the world are non-staple foods, often native species which are still underutilized. In other words, they are used locally or regionally and very little is known about their biology, or their nutritional and nutraceutical characteristics (Sogbohossou et al. 2018).

While research efforts are mainly focused on the major staple crops, exploring underutilized species would reveal untapped potential and may contribute not only to improving human and animal nutrition, but also to reducing the loss of biodiversity, contributing to sustainable agriculture in terms of utilization of resources, mainly soil and water (Kahane et al. 2013; Mayes et al. 2012; Sogbohossou et al. 2018; Ulian et al. 2020). This would increase our possibilities when aiming at coping with climate change through a more environmentally friendly agriculture that includes exploring resilient species against an ample range of stresses. In line with this, Latin-American crops, including their wild relatives and landraces, developed and adapted to marginal areas and extreme environments, could make a very valuable contribution. Only if the current modern breeding technologies are combined with a deeper awareness/knowledge about our agrobiodiversity will we have better chances of meeting the current food security challenge of feeding 10 billion people by 2050, within the context of climate change.

ACKNOWLEDGMENTS

This work was supported by grant laValSe-Food-CYTED (Ref. 119RT0567), Ministerio de Ciencia e Innovación (MICINN, Spain) (PID2019-105748RA-I00) and ANID Fondecyt Regular N° 1211473.

REFERENCES

Abdelbar, O.H. 2018. Flower vascularization and fruit developmental anatomy of quinoa (*Chenpodium quinoa* Willd) Amaranthaceae. *Annals of Agricultural Sciences* 63: 67–75.

Abdi, N., I. Hmissi, M. Bouraoui, B. L'taief, and B. Sifi. 2015. Effect of salinity on Common bean (*Phaseolus vulgaris* L.)-Sinorhizobium strain symbiosis. *Journal of New Sciences, Agriculture and Biotechnology* 16: 559–566.

Abobatta, W.F. 2020. Plant responses and tolerance to combined salt and drought stress. In *Salt and Drought Stress Tolerance in Plants. Signaling and Communication in Plants*, edited by M. Hasanuzzaman and M. Tanveer. Cham: Springer. https://doi.org/10.1007/978-3-030-40277-8_2.

Achigan-Dako, E.G., O.E.D. Sogbohossou, and P. Maundu. 2014. Current knowledge on *Amaranthus* spp.: Research avenues for improved nutritional value and yield in leafy amaranths in sub-Saharan Africa. *Euphytica* 197: 303–317.

Adolf, V.I., S. Shabala, M.N. Andersen, F. Razzaghi, and S.E. Jacobsen. 2012. Varietal differences of quinoa's tolerance to saline conditions. *Plant and Soil* 357: 117–129.

Aellen, P. and T. Just. 1943. Key of the American species of the genus Chenopodium L. *American Midland Naturalist* 30: 46–76.

Ahmed, N.U., J.I Park, H.J. Jung, Y. Hur, and I.S. Nou. 2015. Anthocyanin biosynthesis for cold and freezing stress tolerance and desirable color in *Brassica rapa. Functional & Integrative Genomics* 15: 383–394.

Alandia, G., J.P Rodriguez, S.E. Jacobsen, D. Bazile, and B. Condori. 2020. Global expansion of quinoa and challenges for the Andean region. *Global Food Security* 26: 100429.

Ali, Q., A. Ali, M. Waseem, A. Muzaffar, S. Ahmed, S. Ali, K.S. Bajwa, M.F. Awan, T.R. Samiullah, A.I. Nasir, and T. Husnain. 2014. Correlation analysis for morpho-physiological traits of maize (*Zea mays* L.). *Life Science Journal* 11(12s): 9–13.

Aloisi, I., L. Parrotta, K. B. Ruiz, C. Landi, L. Bini, G. Cai, S. Bioondi, and S. Del Duca. 2016. New insight into quinoa seed quality under salinity: Changes in proteomic and amino acid profiles, phenolic content, and antioxidant activity of protein extracts. *Frontiers in Plant Science* 7: 656.

Alvarez-Jubete, L., E.K. Arendt, and E. Gallagher. 2009. Nutritive value and chemical composition of pseudocereals as gluten free ingredients. *International Journal of Food Science and Nutrition* 60: 240–257.

Alvarez-Jubete, L., E.K. Arendt, and E. Gallagher. 2010a. Nutritive value of pseudocereals and their increasing use as functional gluten-free ingredients. *Trends in Food Science and Technology* 21: 106–113.

Alvarez-Jubete. L., H.H. Wijngaard, E.K. Arendt, and E. Gallagher. 2010b. Polyphenol composition and in-vitro antioxidant activity of amaranth, quinoa and buckwheat as affected by sprouting and bread baking. *Food Chemistry* 119: 770–778.

Amato, M., Caruso, M.C., F. Guzzo, F. Galgano, M. Commisso, R. Bochicchio, R. Labella, and F. Favati. 2015. Nutritional quality of seeds and leaf metabolites of Chia (*Salvia hispanica* L.) from Southern Italy. *European Food Research Technology* 241: 615–625.

Anjum, S.A., U. Ashraf, M. Tanveer, I. Khan, S. Hussain, B. Shahzad, A. Zohaib, F. Abbas, M.F. Saleem, I. Ali, and L.C. Wang. 2017. Drought induced changes in growth, osmolyte accumulation and antioxidant metabolism of three maize hybrids. *Frontiers Plant Science* 8: 69.

Arfini, F., and F. Antonioli. 2013. *Sacha inchi. Research about the Conditions for Recognition of Geographical Indications in Peru.* Lima: CRED.

Arzani, A. and M. Ashraf. 2017. Cultivated ancient wheats (*Triticum spp.*): A potential source of health-beneficial food products. *Comprehensive Reviews in Food Science and Food Safety* 16: 477–488.

Ashraf, U., M.N. Salim, A. Sher, S.R. Sabir, A. Khan, S. Pan, and X. Tang. 2016. Maize growth, yield formation and water-nitrogen usage in response to varied irrigation and nitrogen supply under semi-arid climate. *Turkish Journal of Field Crops* 21: 88–96.

Aslam, M., M.A. Maqbool, and R. Cengiz. 2015. *Drought Stress in Maize (Zea mays L.) Effects, Resistance Mechanisms, Global Achievements and Biological Strategies for Improvement.* Springer Briefs in Agriculture ISBN 978-3-319-25442-5.

Assimakopoulou, A., I. Salmas, K. Nifakos, and P. Kalogeropoulos. 2015. Effect of salt stress on three green bean (*Phaseolus vulgaris* L.) cultivars. *Notulae Botanicae Horti Agrobotanici Cluj-napoca.* 43: 113–118.

Ayerza, R. 1995. Oil content and fatty-acid composition of chia (*Salvia hispanica* L.) from 5 northwestern locations in Argentina. *Journal of the American Oil Chemists Society* 72: 1079–1081.

Ayerza, R. 2013. Seed composition of two chia (*Salvia hispanica* L.) genotypes which differ in seed color. *Emirates Journal of Food and Agriculture* 25: 495–500.

Ayerza, R. and W. Coates. 2005a. *Chia: Rediscovering a Forgotten Crop of the Aztecs.* Tucson, AZ: University of Arizona Press.

Ayerza, R. and W. Coates. 2005b. Effect of ground chia seed and chia oil on plasma total cholesterol, LDL, HDL, triglyceride content, and fatty acid composition when fed to rats. *Nutrition Research* 11: 995–1003.

Ayerza, R. and W. Coates. 2009. Influence of environment and genotype on crop cycle and yield; seed protein, oil, and α-linolenic ω-3-fatty acid content of chia (*Salvia hispanica* L.). *Industrial Crops and Products* 30: 321–324.

Badu-Apraku, B., R.B. Hunter, and M. Tollenaar. 1983. Effect of temperature during grain filling on whole plant and grain yield in maize (*Zea mays* L.). *Canadian Journal Plant Science* 63 :357–363.

Baginsky, C., J. Arenas, H. Escobar, M. Garrido, D. Valero, D. Tello, L. Pizarro, L. Morales, and H. Silva. 2014. *Determinación de fecha de siembra óptima de chia en zonas de clima desértico y templado mediterráneo semiárido bajo condiciones de riego en Chile.* Universidad de Chile, Facultad de ciencias Agronómicas, Escuela de pregrado. http://repositorio.uchile.cl/bitstream/handle/2250/152812/Efecto-de-la-fecha-de-siembra-en-el-rendimiento-en-grano-de-chia-%28Salvia-hispanica-L%29-y-su-relac ion-con-el-crecimiento-y-desarrollo.pdf?sequence=1&isAllowed=y

Bascunan-Godoy, L., M. Reguera, Y.M. Abdel-Tawab, and E. Blumwald. 2016. Water deficit stress-induced changes in carbon and nitrogen partitioning in *Chenopodium quinoa* Willd. *Planta* 243: 591–603.

Bascunan-Godoy, L., C. Sanhueza, C.E. Hernández, L. Cifuentes, K. Pinto, R. Álvarez, M. González-Teuber, and L.A. Bravo. 2018. Nitrogen supply affects photosynthesis and photoprotective attributes during drought-induced senescence in quinoa. *Frontiers in Plant Science* 9: 994.

Bates, B.C., Z.W. Kundzewicz, S. Wu, and J.P. Palutikof. 2008. Climate change and water. Technical Paper of the Intergovernmental Panel on Climate Change, IPCC Secretariat, Geneva. *The American Midland Naturalist*, 168.

Bazile, D., E.A. Martínez, and F. Fuentes. 2014. Diversity of quinoa in a biogeographical island: A review of constraints and potential from arid to temperate regions of Chile. *Notulae Botanicae Horti Agrobotanici Cluj-Napoca* 42: 289–298.

Bazile, D., S.E. Jacobsen, and A. Verniau. 2016. The global expansion of quinoa: Trends and limits. *Frontiers in Plant Science* 7: 622.

Beebe, S.E., O. Toro, A.V. González, M.I. Chacon, and D.G. Debouck. 1997. Wild–weed–crop complexes of common bean (*Phaseolus vulgaris* L, Fabaceae) in the Andes of Peru and Colombia, and their implications for conservation and breeding. *Genetic Resources and Crop Evolution* 44: 73–91.

Behboudian, M.H., Q. Ma, N.C. Turner, and J.A. Palta. 2001. Reactions of chickpea to water stress: Yield and seed composition. *Journal of the Science Food and Agriculture* 81: 1288–1291.

Bellon, M.R., D. Hodsonb, and J. Hellin. 2011. Assessing the vulnerability of traditional maize seed systems in Mexico to climate change. *Proceedings of the National Academy of Sciences* 108(33): 13432–13437.

Bendevis, M.A., Y. Sun, E. Rosenqvist, S. Shabala, F. Liu, and S.E. Jacobsen. 2014. Photoperiodic effects on short-pulse 14C assimilation and overall carbon and nitrogen allocation patterns in contrasting quinoa cultivars. *Environmental and Experimental Botany* 104: 9–15.

Bernal, Y and J. Correa. 1992. Especies vegetales promisorias del convenio Andrés Bello. *SECAB.* 7: 577–596.

Berru, L.B., P. Glorio-Paulet, C. Basso, A. Scarafoni, F. Camarena, A. Hidalgo, and A. Brandolini. 2021. Chemical composition, tocopherol and carotenoid content of seeds from different Andean Lupin (*Lupinus mutabilis*) ecotypes. *Plant Foods Human Nutrition* 76: 98–104.

Bertero, H.D., A.J. De la Vega, G. Correa, S.E. Jacobsen, and A. Mujica. 2004. Genotype and genotype-by-environment interaction effects for grain yield and grain size of quinoa (*Chenopodium quinoa* Willd.) as revealed by pattern analysis of international multi-environment trials. *Field Crops Research* 89: 299–318.

Biondi, S., K.B. Ruiz, E.A. Martinez, A. Zurita-Silva, F. Orsini, F. Antognoni, G. Dinelli, I. Marotti, G. Gianquinto, S. Maldonado, H. Burrieza, D. Bazile, V.I. Adolf, and S.E Jacobsen. 2015. Tolerance to saline conditions. In *State of the Art Report of Quinoa in the World in 2013*, Chap. 2.3. Paris: FAO and CIRAD.

Bitocchi, E., L. Nanni, E. Bellucci, M. Rossi, A. Giardini, P.S. Zeuli, G. Logozzo, J. Stougaard, P. McClean, G. Attene, and R. Papa. 2012. Mesoamerican origin of the common bean (*Phaseolus vulgaris* L.) is revealed by sequence data. *Proceedings of the National Academy of Sciences* 109 : E788–E796.

Boehlein, S.K., P. Liu, A. Webster, C. Ribeiro, M. Suzuki, S. Wu, J.C. Guan, J.D. Stewart, W.F. Tracy, A.M. Settles, D.R. McCarty, K.E. Koch, L.C. Hannah, T.A Hennen-Bierwagen, and A. Myers. 2019. Effects of long-term exposure to elevated temperature on *Zea mays* endosperm development during grain fill. *The Plant Journal* 99: 23–40.

Brack, A. 1999. *Diccionario enciclopédico de plantas útiles del Perú.* Cuzco: Centro de Estudios Regionales Andinos – Bartolomé de las Casas, p. 400.

Bressani, R. 1994. Composition and nutritional properties of amaranth. In *Amaranth-Biology, Chemistry and Technology*, edited by Octavio Paredes-Lopez, 185–205. London: CRC Press.

Broughton, W.J., G. Hernandez, M.W. Blair, S. Beebe, P. Gepts, and J. Vanderleyden. 2003. Beans (*Phaseolus* spp.) – model food legumes. *Plant and Soil* 252: 55–128.

Cahill, J.P. 2003. Ethnobotany of chia, *Salvia hispanica* L. *Economy Botany* 57: 604–618.

Cahill, J.P. 2004. Genetic diversity among varieties of chia (*Salvia hispanica* L.). *Genetic Resources of Crop Evolution* 51: 773–781.

Çakir, R. 2004. Effect of water stress at different development stages on vegetative and reproductive growth of corn. *Field Crops Research* 89: 1–16.

Carvalho, I.S., C.P. Ricardo, and M. Chaves. 2004. Quality and distribution of assimilates within the whole plant of lupines (*L. albus* and *L. mutabilis*) influenced by water stress. *Journal of Agronomy and Crop Science* 190: 205–210.

Carvalho, I.S., M. Chaves, and C.P. Ricardo. 2005. Influence of water stress on the chemical composition of seeds of two lupins (*Lupinus albus* and *Lupinus mutabilis*). *Journal of Agronomy and Crop Science* 191: 95–98 doi: 10.1111/j.1439-037X.2004.00128.x

Castañeda, B., R. Manrique, F. Gamarra, A. Muñoz, F. Ramos, F. Lizaraso, and J. Martínez. 2008. Probiótico elaborado en base a las semillas de *Lupinus mutabilis* Sweet (chocho or tarwi) seeds. *Acta Médica Peruana* 25: 210–215.

Castaño, D.L., M. Valencia, E. Murillo, J.J. Mendez, and J.E. Joli. 2012. Fatty acid composition of Inca peanut (*Plukenetia volubilis* Linneo) and its relationship with vegetal bioactivity. *Revista Chilena Nutricion* 39: 45–52.

Castonguay, Y. and A.H Markhart III.1991. Saturated rates of photosynthesis in water-stressed leaves of common bean and tepary bean. *Crop Science* 31: 1605–1611.

Chalker-Scott, L. 1999. Environmental significance of anthocyanins in plant stress responses. *Photochemistry and Photobiology* 70: 1–9.

Chinnusamy, V., A. Jagendorf, and J.K. Zhu. 2005. Understanding and improving salt tolerance in plants. *Crop Science* 45:437–448.

Chirinos, R., G. Zuloeta, R. Pedreschi, E. Mignolet, Y. Larondelle, and D. Campos. 2013. Sacha inchi (*Plukenetia volubilis*): A seed source of polyunsaturated fatty acids, tocopherols, phytosterols, phenolic compounds and antioxidant capacity. *Food Chemistry* 141: 1732–1739.

Cicchino, M., J.I. Rattalino-Edreira, M. Uribelarrea, and M.E. Otegui. 2010. Heat stress in field-grown maize: Response of physiological determinants of grain yield. *Crop Science* 50:1438–1448.

Coates, W., and R. Ayerza. 1996. Production potential of chia in Northwestern Argentina. *Industrial Crops and Products* 5: 229–233.

Cocozza, C., C. Pulvento, A. Lavini, M. Riccardi, R. d'Andria, and R. Tognetti. 2013. Effects of increasing salinity stress and decreasing water availability on ecophysiological traits of Quinoa (*Chenopodium quinoa* Willd.) grown in a Mediterranean-type agroeco-system. *Journal of Agronomical Crop Science* 199: 229–240.

Corazon-Guivin, M., D. Castro-Ruiz, W. Chota-Macuyama, Á. Rodríguez, D. Cachique, E. Manco, D. Del-Castillo, J.F. Renno, and C. García-Dávila. 2009. Caracterización genética de accesiones SanMartinenses del banco nacional de germoplasma de sacha inchi *Plukenetia volubilis* L. (E.E. El Porvenir—INIA). *Folia Amazónica* 18: 23–31.

Cortés-Avendaño, P., M. Tarvainen, S. Jukka-Pekka, P. Glorio-Paulet, B. Yang, and R. Repo-Carrasco-Valencia. 2020. Profile and content of residual alkaloids in ten ecotypes of Lupinus mutabilis Sweet after aqueous debittering process. *Plant Foods for Human Nutrition* 75: 184–191.

Curti, R.N., M.D. Sanahuja, S.M. Vidueiros, C.A. Curti, A.N. Pallaro, and H.D. Bertero. 2020. Oil quality in sea-level quinoa as determined by cultivar-specific responses to temperature and radiation conditions. *Journal of the Science of Food and Agriculture* 100: 1358–1361.

Da Silva, D.A., C.A.F. Pinto-Maglio, E.C.de Oliveira, R.L.D. dos Reis, S.A.M. Carbonell, and A.F. Chiorato. 2020. Influence of high temperature on the reproductive biology of dry edible bean (*Phaseolus vulgaris* L.) *Scientia Agricola* 77. doi: 10.1590/ 1678-992X-2018-0233

de Falco, B., G. Incerti, R. Bochicchio, T.D. Phillips, M. Amato, and V. Lanzotti. 2017. Metabolomic analysis of *Salvia hispanica* seeds using NMR spectroscopy and multivariate data analysis. *Industrial Crops and Products* 99: 86–96.

de Falco, B., A. Fiore, R. Rossi, M. Amato, and V. Lanzotti. 2018. Metabolomics driven analysis by UAEGC-MS and antioxidant activity of chia (*Salvia hispanica* L.) commercial and mutant seeds. *Food Chemistry* 254: 137–143.

Delatorre-Herrera, J. and M. Pinto. 2009. Importance of ionic and osmotic components of salt stress on the germination of four quinua (*Chenopodium quinoa* Willd.) selections. *Chilean Journal of Agricultural Research* 69: 477–485.

Dreesen, F.E., H.J. De Boeck, I.A. Janssens, and I. Nijs. 2012. Summer heat and drought extremes trigger unexpected changes in productivity of a temperate annual/biannual plant community. *Environmental and Experimental Botany* 79: 21–30.

Dwivedi, S.L., S. Ceccarelli, M.W. Blair, H.D. Upadhyaya, A.K. Are, and R. Ortiz. 2016. Landrace germplasm for improving yield and abiotic stress adaptation. *Trends in Plant Science* 21: 31–42.

Elshafie, H.S., L. Aliberti, M. Amato, V. De Feo, and I. Camele. 2018. Chemical composition and antimicrobial activity of chia (*Salvia hispanica* L.) essential oil. *European Food Research and Technology* 244: 1675–1682. doi: 10.1007/s00217-018-3080-x

Epstein, E., J.D. Norlyn, D.W. Rush, R.W. Kingsbury, D.B. Kelly, G.A. Cunningham, and A.F. Wrona. 1980. Saline culture of crops: A genetic approach. *Science* 210: 399–404.

Fan, M.S., F.J. Zhao, S.J. Fairweather-Taitc, P.R. Poultona, S.J. Dunhama, and S.P. McGrath. 2008. Evidence of decreasing mineral density in wheat grain over the last 160 years. *Journal of Trace Elements in Medicine and Biology* 22: 315–324.

FAO (Food and Agriculture Organization). 2016. *Save and Grow in Practice: Maize, Rice, Wheat. A Guide To Sustainable Cereal Production.* www.fao.org/ag/save-and-grow/ MRW/index_en.html.

FAO (Food and Agriculture Organization). 2019. *FAO: Challenges and Opportunities in a Global World.* Rome. Licence: CC BY-NC-SA 3.0 IGO.

Farooq, M., M. Hussai, A. Wakeel, and K.H.M. Siddique. 2015. Salt stress in maize: effects, resistance mechanisms, and management. A review. *Agronomy for Sustainable Development* 35: 461–481.

Farooq, M., N. Gogoi, M. Hussain, S. Barthakur, S. Paul, N. Bharadwaj, H.M. Migdadi, S.S. Alghamdi, and K.H.M. Siddique. 2017. Effects, tolerance mechanisms and management of salt stress in grain legumes. *Plant Physiology and Biochemistry* 118: 199–217.

Farooq, M., M. Hussain, M. Usman, S. Farooq, S.S. Alghamdi, and K.H.M. Siddique. 2018. Impact of abiotic stresses on grain composition and quality in food legumes. *Journal of Agricultural and Food Chemistry* 66: 8887–8897.

Fischer, S., R. Wilckens, J. Jara, and M. Aranda. 2013. Controlled water stress to improve functional and nutritional quality in quinoa seed. *Boletin Latinoamericano y del Caribe de Plantas Medicinales y Aromáticas* 12: 457–468.

Fischer, S., R. Wilckens, J. Jara, M. Aranda, W. Valdivia, L. Bustamante, F. Graf, and I. Obal. 2017. Protein and antioxidant composition of quinoa (*Chenopodium quinoa* Willd.) sprout from seeds submitted to water stress, salinity and light conditions. *Industrial Crops and Products* 107: 558–564.

Flores, R. 2006. *Evaluación preliminar agronómica y morfológica del germoplasma de cañahua (Chenopodium pallidicaule Aellen) en la Estación Experimental Belén.* Tesis de Grado. La Paz, Bolivia: UMSA.

Fortmeier, R., and S. Schubert. 1995. Salt tolerance of maize (*Zea mays* L.): The role of sodium exclusion. *Plant Cell & Environment* 18: 1041–1047.

Gavicho-Uarrota, V., R. Brasil-Severino, and M. Maraschin. 2011. Maize landraces (*Zea mays* L.): A new prospective source for secondary metabolite production. *International Journal of Agricultural Research* 6 :218–226.

Geerts, S., D. Raes, M. Garcia, J. Vacher, R. Mamani, J. Mendoza, R. Huanca, B. Morales, R. Miranda, J. Cusicanqui, and C. Taboada. 2008. Introducing deficit irrigation to stabilize yields of quinoa (*Chenopodium quinoa* Willd.). *European Journal of Agronomy* 28: 427–436.

Gimplinger, D.M., G. Dobos, R. Schönlechner, and H.P. Kaul. 2007. Yield and quality of grain amaranth (*Amaranthus* sp.) in eastern Austria. *Plant Soil and Environment* 53: 105–112.

Gresta, F., M. Wink, U. Prins, M. Abberton, J. Capraro, A. Scarafoni, and G. Hill. 2017. Lupins in European cropping systems. *Legumes in Cropping System* 88–108.

Gross, R., E. Von Baer, R. Koch, L. Marquard, L. Trugo, and M. Wink. 1988. Chemical composition of a new variety of the Andean lupin (*Lupinus mutabilis* cv. Inti) with low alkaloid content. *Journal of Food Composition and Analysis* 1: 353–361.

Guilengue, N., S. Alves, P. Talhinhas, and J. Neves-Martins. 2020. Genetic and genomic diversity in a Tarwi (*Lupinus mutabilis* Sweet) germplasm collection and adaptability to Mediterranean climate conditions. *Agronomy* 10: 21.

Gulisano, A., S. Alves, J.N. Martins and L.M. Trindade. 2019. Genetics and breeding of *Lupinus mutabilis:* An emerging protein crop. *Frontier in Plant Science* 10: 1385.

Gutiérrez, L.F., L.M. Rosada, and A. Jiménez. 2011. Chemical composition of Sacha Inchi (*Plukenetia volubilis* L.) seeds and characteristics of their lipid fraction. *Grasas y Aceites* 62: 76–83.

Hamaker, B.R., C. Valles, R. Gilman, R.M. Hardmeier, D. Clark, H.H. Garcia, A.E. Gonzales, I. Kohlstad, M. Castro, R. Valdivia, T. Rodriguez, and M. Lescano. 1992. Amino acid and fatty acid profiles of the inca peanut (*Plukenetia volubilis*). *Cereal Chemistry* 69: 461–463.

Hardy, A., C. Huyghe, and J. Papineau. 1997. Dry matter accumulation and partitioning, and seed yield in indeterminate Andean lupin (*Lupinus mutabilis* Sweet). *Australian Journal of Agricultural Research* 48: 91–102.

Hasegawa, P.M., R.A. Bressan, J.K. Zhu, and H.J. Bohnert. 2000. Plant cellular and molecular response to high salinity. *Annual Review of Plant Physiology and Plant Molecular Biology* 51: 463–499.

Heuer, B., Z. Yaniv, and I. Ravina. 2002. Effect of late salinization of chia (*Salvia hispanica*), stock (*Matthiola tricuspidata*) and evening primrose (*Oenothera biennis*) on their oil content and quality. *Industrial Crops and Products* 15: 163–167.

Hu, Y. and U. Schmidhalter. 2005. Drought and salinity: A comparison of their effects on mineral nutrition of plants. *Journal of Plant Nutrition and Soil* 168: 541–549.

Hussain, M.I., A.J. Al-Dakheel, and M.J. Reigosa. 2018. Genotypic differences in agro-physiological, biochemical and isotopic responses to salinity stress in quinoa (*Chenopodium quinoa* Willd.) plants: Prospects for salinity tolerance and yield stability. *Plant Physiology and Biochemistry* 129: 411–420.

Hussain, M.I., A. Muscolo, M. Ahmed, M.A. Asghar, and A.J. Al-Dakheel. 2020. Agro-morphological, yield and quality traits and interrelationship with yield stability in quinoa (*Chenopodium quinoa* Willd.) Genotypes under saline marginal environment. *Plants* 9: 1763.

Hussain, T., I.A. Khan, M.A. Malik, Z. Ali. 2006. Breeding potential for high temperature tolerance in corn (*Zea mays* L.). *Pakistan Journal of Botany* 38: 1185–1195.

Ignjatovic-Micic, D., J. Vancetovic, D. Trbovic, Z. Dumanovic, M. Kostadinovic, and S. Bozinovic. 2015. Grain nutrient composition of maize (*Zea mays* L.) Drought-tolerant populations. *Journal of Agricultural Food Chemistry* 63: 1251–1260.

IPCC (Intergovernmental Panel on Climate Change). 2018. *Progress Report of the Special Report on Global Warming of 1.5°C*. www.ipcc.ch/site/assets/uploads/2018/04/13022 0180459-INF.6-ReportSR-15.pdf (accessed February 25, 2019).

IPCC (Intergovernmental Panel on Climate Change). 2021. *Climate Change 2021: The Physical Science Basis*. www.ipcc.ch/report/ar6/wg1/ (accessed August 19, 2021.

IPGRI (International Plant Genetic Resources Institute), PROINPA (Promocion e Investigacion de Productos Andinos), and IFAD (International Fund for Agricultural Development). 2005. *Descriptores para cañahua (Chenopodium pallidicaule Aellen)*. Rome/La Paz: IFAD; IPGRI; Fundación PROINPA.

Jacobsen, S.E. 1997. Adaptation of quinoa (*Chenopodium quinoa*) to Northern European agriculture: Studies on developmental pattern. *Euphytica* 96: 41–48.

Jacobsen, S.E., and A. Mujica. 2006. *El tarwi (*Lupinus mutabilis *Sweet) y sus parientes silvestres*. Universidad Mayor de San Andrés, La Paz, 458–482.

Jacobsen, S.E., and O. Stølen. 1993. Quinoa-morphology, phenology and prospects for its production as a new crop in Europe. *European Journal of Agronomy* 2: 19–29.

Jacobsen, S.E, M. Sørensen, S.M. Pedersen, and J. Weiner. 2015. Using our agrobiodiversity: Plant-based solutions to feed the world. *Agronomy for. Sustainable Development* 35: 1217–1235.

Jamboonsri, W., T.D. Phillips, R.L. Geneve, J.P. Cahill, and D.F. Hildebrand. 2012. Extending the range of an ancient crop, Salvia hispanica L. – a new ω3 source. *Genetic Resources and Crop Evolution* 59: 171–178.

Jarvis, D., Y. Ho, D. Lightfoot, S. Schmöckel, B. Li, J. Theo, A. Borm, H. Ohyanagi, K. Mineta, C.T. Michell, N. Saber, N.M. Kharbatia, R.R. Rupper, A.R. Sharp, N. Dally, B.A. Boughton, Y.H. Woo, G. Gao, G.W.M. Schijlen, A.A. Momin, S. Negrao, S. Al-Babili, C. Gehring, U. Roessner, C. Jung, K. Murphy, S.T. Arold, T. Gojobori, C.G. van der Linden, E.N. van Loo, E.N. Jellen, P.J. Maughan, and M. Tester. 2017. The genome of *Chenopodium quinoa*. *Nature* 542: 307–312.

Kahane, R., T.Hodgkin, H. Jaenicke, C. Hoogendoorn, M. Hermann, J.D.H. Keatinge, J.A Hughes, S. Padulosi, and N. Looney. 2013. Agrobiodiversity for food security, health and income. *Agronomy for Sustainable Development* 33: 671–693.

Kang, Y.H., M. Chen, and S.Q. Wan. 2010. Effects of drip irrigation with saline water on waxy maize (*Zea mays* L. var. ceratina Kulesh) in North China Plain. *Agricultural Water Management* 97: 1303–1309.

Karimi, G., M. Ghorbanli, H. Heidari, R.A. Khavarinejadand, and M.H. Assareh. 2005. The effects of NaCl on growth, water relations, osmolytes and ion content in Kochia prostrate. *Biologia Plantarum* 49: 301–304.

Kasoma, C., Shimelis, H., Laing, M.D., Shayanowako, A.I.T. and Mathew, I. 2021. Revealing the genetic diversity of maize (*Zea mays* L.) populations by phenotypic traits and DArTseq markers for variable resistance to fall armyworm. *Genetic Resources and Crop Evolution* 68: 243–259.

Keeler, R.F., Cronin, E.H., and Shupe, J.L. 1976. Lupin alkaloids from teratogenic and nonteratogenic lupines. IV –Concentration of total alkaloids, and the teratogen anagyrine as a afunction of plant part and stage of growthand their relationship to crooked calf disease. *Journal of Toxicology Environmental and Health* 1: 899–908.

Kew, R.B.G. 2016. *State of the World's Plants Report—2016*. London: Royal Botanic Gardens, Kew.

Khaitov, B., A.A. Karimov, K. Toderich, Z. Sultanova, A. Mamadrahimov, K. Allanov, and S. Islamov. 2020. Adaptation, grain yield and nutritional characteristics of quinoa (*Chenopodium quinoa*) genotypes in marginal environments of the Aral Sea basin. *Journal of Plant Nutrition* 44: 1365–1379.

Khalil, S.E. and E.G. Ismael. 2010. Growth, yield and seed quality of *Lupinus termisas* affected by different soil moisture levels and different ways of yeast application. *Journal of American Science* 6: 141–153.

Khoshbakht, K., and K. Hammer. 2008. How many plant species are cultivated?. *Genetic Resources and Crop Evolution* 55: 925–928.

Kodahl, N. 2020. Sacha inchi (*Plukenetia volubilis* L.)—from lost crop of the Incas to part of the solution to global challenges? *Planta* 251: 80.

Konsens, I., M. Ofir, and J. Kigel. 1991. The effect of temperature on the production and abscission of flowers and pods in snap-bean (*Phaseolus vulgaris* L.). *Annals of Botany* 67: 391–399.

Konuşkan, Ö., H. Gözübenli, I. Atiş, and M. Atak. 2017. Effects of salinity stress on emergence and seedling growth parameters of some maize genotypes (*Zea mays* L.). *Turkish Journal of Agriculture* 5: 1668–1672.

Kucek, L.K., L.D. Veenstra, P. Amnuaycheewa, and M.E. Sorrells. 2015. A grounded guide to gluten: How modern genotypes and processing impact wheat sensitivity. *Comprehensive Reviews in Food Science and Food Safety* 14: 285–302.

Kurlovich, B.S., A.K. Stankevich, and S.I. Stepanova. 2002. The history of lupin domestication. In *Lupins: Geography, Classification, Genetic Resources and Breeding*, edited by B.S. Kurlovich, 147–164. St. Petersburg: OY International Express.

Lawlor, D.W. and G. Cornic. 2002. Photosynthetic carbon assimilation and associated metabolism in relation to water deficits in higher plants. *Plant, Cell and Environment* 25: 275–294.

Laxa, M., M. Liebthal, W. Telman, K. Chibani, and K. J. Dietz. 2019. The role of the plant antioxidant system in drought tolerance. *Antioxidants* 8: 94.

Lei, Y., Y. Zheng, and K. Dai. 2014. Different responses of photosystem I and photosystem II in three tropical oilseed crops exposed to chilling stress and subsequent recovery. *Trees* 28: 923–933.

Lenné, J.M. and D. Wood. 2011. *Agrobiodiversity Management for Food Security*. 12–26. Wallingford, Oxon: CABI.

Lesjak, J. and D.F. Calderini. 2017. Increased night temperature negatively affects grain yield, biomass and grain number in Chilean quinoa. *Frontiers in Plant Science* 8: 352.

Li, J., J. Chen, J. Jin, S. Wang, and B. Du. 2019. Effects of irrigation water salinity on maize (*Zea may* L.): Emergence, growth, yield, quality, and soil salt. *Water* 11: 2095.

Li, Y.T., W.W. Xu, B.Z. Ren, B. Zhao, J. Zhang, P. Liu, and Z.S. Zhang. 2020. High temperature reduces photosynthesis in maize leaves by damaging chloroplast ultrastructure and photosystem II. *Journal of Agronomy and Crop Science* 206: 548–564.

Liu, F. and H. Stützel. 2002. Leaf expansion, stomatal conductance, and transpiration of vegetable amaranth (Amaranthus sp.) in response to soil drying. *Journal of the American Society of Horticultural Science* 127: 878–883.

Liu, F. and H. Stützel. 2004. Biomass partitioning, specific leaf area, and water use efficiency of vegetable amaranth (*Amaranthus* spp.) in response to drought stress. *Scientia Horticulturae* 102: 15–27.

Lopez, C.M., M. Pineda, and J.M. Alamillo. 2020. Differential regulation of drought responses in two *Phaseolus vulgaris* genotypes. *Plants* 9: 1815.

Lovelli, S., M. Valerio, T.D. Phillips, and M. Amato. 2019. Water use efficiency, photosynthesis and plant growth of Chia (*Salvia hispanica* L.): A glasshouse experiment. *Acta Physiologiae Plantarum* 41: 3.

Lugtenberg, B. and F. Kamilova. 2009. Plant-growth-promoting rhizobacteria. *Annual Reviews of Microbiology* 63: 541–556.

Luo, Y.L., Z.L. Su, T.J. Bi, X.L. Cui, and Q.Y. Lan. 2014. Salicylic acid improves chilling tolerance by affecting antioxidant enzymes and osmoregulators in sacha inchi (*Plukenetia volubilis*). *Brazilian Journal of Botany* 37: 357–363.

Machado, F.B., S. De, A.M. David, S.R. Dos Santos, J.C. Figueiredo, C.D. Da Silva, and D.A. Nobre. 2020. Physiological quality of maize seeds produced under soil water deficit conditions. *Revista Brasileira de Engenharia Agricola e Ambiental* 24: 451–456.

Mangelson, H., D.E. Jarvis, P. Mollinedo, O.M. Rollano-Penaloza, V.D. Palma-Encinas, L.R. Gomez-Pando, E.N. Jellen, and P.J. Maughan. 2019. The genome of *Chenopodium pallidicaule*: An emerging Andean super grain. *Applications in Plant Sciences* 7(11): e11300.

Martínez, E.A., E. Veas, C. Jorquera, R. San Martín, and P. Jara. 2009. Re-introduction of quinoa into arid Chile: Cultivation of two lowland races under extremely low irrigation. *Journal of Agronomy and Crop Science* 195: 1–10.

Matías, J., M.J Rodríguez, V. Cruz, P. Calvo, and M. Reguera. 2021. Heat stress lowers yields, alters nutrient uptake and changes seed quality in quinoa grown under Mediterranean field conditions. *Journal of Agronomy and Crop Science*. https://doi.org/10.1111/jac.12495.

Matsui, T., K. Omasa, and T. Horie. 2001. The difference in sterility due to high temperatures during the flowering period among japonica-rice varieties. *Plant Production Science* 4: 90–93.

Matsuoka, Y., Y. Vigouroux, M.M. Goodman, J. Sanchez, E. Buckler, and J. Doebley. 2002. A single domestication for maize shown by multilocus microsatellite genotyping. *Proceedings of the National Academy of Sciences* 99: 6080–6084. d

Mayes, S., F.J. Massawe, P.G. Alderson, J.A. Roberts, S.N. Azam-Ali, and M. Hermann. 2012. The potential for underutilized crops to improve security of food production. *Journal of Experimental Botany* 63: 1075–1079.

Mi, N., F. Cai, Y.S Zhang, R.P. Ji, S.J. Zhang, and Y. Wang. 2018. Differential responses of maize yield to drought at vegetative and reproductive stages. *Plant Soil and Environment* 64: 260–267.

Moreno, C., C.E. Seal, and J. Papenbrock. 2017. Seed priming improves germination in saline conditions for Chenopodium quinoa and *Amaranthus caudatus*. *Journal of Agronomy and Crop Science* 204: 40–48.

Muasya, R.M., W.J.M. Lommen, C.W. Muui, and P.C. Struik. 2008. How weather during development of common bean (Phaseolus vulgaris L.) affects the crop's maximum attainable seed quality. *NJAS – Wageningen Journal of Life Sciences* 56 : 85–100.

Mujica, A. 1994. Andean grains and legumes. In *Neglected Crops: 1492 from a Different Perspective*, edited by J. E. Hernández, and J. León, 141–148. vol. FAO Plant Production and Protection, Series, no. 26. Rome: Food and Agriculture Organization of the United Nations.

Nicoletto, C., G. Zanin, P. Sambo, and D. L. Costa. 2019. Quality assessment of typical common bean genotypes cultivated in temperate climate conditions and different growth locations. *Scientia Horticulturae* 256: 108599.

Nodari, R.O., S.M. Tsai, R.L. Gilbertson, and P. Gepts.1993. Towards an integrated linkage map of common bean: 2. Development of an RFLP-based linkage map. *Theoretical and Applied Genetics* 85: 513–520.

Ortega-David, E., A. Rodriguez, A. David, and A. Zamora-Burbano. 2010. Caracterización de semillas de *Lupinus mutabilis* sembrado en los andes de Colombia. *Acta agronómica* 59: 111–118.

Paiva, E.P., S.B. Torres, T.R.C. Alves, F.V. da S. Sá, M. de S. Leite, and J.L.D. Dombroski. 2018. Germination and biochemical components of *Salvia hispanica*; L. seeds at different salinity levels and temperatures. *Acta Scientiarum. Agronomy* 40: e39396.

Pal, M. and T.N. Khoshoo. 1972. Evolution and improvement of cultivated Amaranths (v. inviability, weakness and sterility in hybrids). *J Heredity* 63: 78–82.

Peláez, P., D. Orona-Tamayo, S. Montes-Hernández, M. Valverde, O. Paredes-López, and A. Cibrian- Jaramillo. 2019. Comparative transcriptome analysis of cultivated and wild seeds of Salvia hispanica (chia). *Scientific Reports* 9: 9761.

Peñarrieta, J.M., J.A. Alvarado, B. Åkesson, and B. Bergenståhl. 2008. Total antioxidant capacity and content of flavonoids and other phenolic compounds in canihua (*Chenopodium pallidicaule*): An Andean pseudocereal. *Molecular Nutrition and Food Research* 52: 708–717.

Peñas, E., F. Uberti, C. di Lorenzo, C. Ballabio, A. Brandolini, and P. Restani. 2014. Biochemical and immunochemical evidences supporting the inclusion of quinoa (*Chenopodium quinoa* Willd.) as a gluten-free ingredient. *Plant Foods for Human Nutrition* 69: 297–303.

Perales, H. and D. Golicher. 2014. Mapping the diversity of maize races in Mexico. *Plos One* 9: e114657.

Pokharel, M., A. Chiluwal, M. Stamm, D. Min, D. Rhodes, and S.V.K. Jagadish. 2020. High night-time temperature during flowering and pod filling affects flower opening, yield and seed fatty acid composition in canola. *Journal of Agronomy and Crop Science* 206: 579–596.

Prasad, P., K. Boote, L. Allen Jr., J. Sheehy, and J. Thomas. 2006. Species, ecotype and cultivar differences in spikelet fertility and harvest index of rice in response to high temperature stress. *Field Crops Research* 95: 398–411.

Prasanna, B.M. 2012. Diversity in global maize germplasm: Characterization and utilization. *Journal of Biosciences* 37: 843–855.

Pulvento, C., M.H. Sellami, and A. Lavini. 2021. Yield and quality of *Amaranthus hypochondriacus* grain amaranth under drought and salinity at various phenological stages in southern Italy. *Journal of the Science of Food Agriculture.* doi: 10.1002/jsfa.11088.

Purseglove, J.W. 1968. *Tropical Crops: Dicotyledons.* London: Longmans, pp. 346–381.

Qi, W., J.W. Zhang, K.J. Wang, P. Liu, and S.T. Dong. 2010. Effects of drought stress on the grain yield and root physiological traits of maize varieties with different drought tolerance. *The Journal of Applied Ecology* 21: 48–52.

Radić, V., D. Beatović, and J. Mrđa. 2007. Salt tolerance of corn genotypes (*Zea mays* l.) during germination and later growth. *The Journal of Agricultural Science* 52: 115–120.

Rainey, K.M., and P.D. Griffiths. 2005. Differential response of common bean genotypes to high temperature. *Journal of the American Society for Horticultural Science* 130: 18–23.

Razzaghi, F., S.H. Ahmadi, V.I. Adolf, C.R. Jensen, S.E. Jacobsen, and M.N. Andersen. 2011. Water relations and transpiration of quinoa (*Chenopodium quinoa* Willd.) under salinity and soil drying. *Journal of Agronomy and Crop Science* 197: 348–360.

Reguera, M., C.M. Conesa, A. Gil-Gómez, C.M. Haros, M.Á. Pérez-Casas, V. Briones-Labarca, L. Bolaños, I. Bonilla, R. Álvarez, K. Pinto, Á. Mujica, and L. Bascuñán-Godoy. 2018. The impact of different agroecological conditions on the nutritional composition of quinoa seeds. *Peer J* 6: e4442.

Repo-Carrasco-Valencia, R. and J.M. Vidaurre-Ruiz. 2019. Quinoa and Other Andean Ancient Grains: Super Grains for the Future. *Cereal Foods World* 6: 1–10.

Repo-Carrasco-Valencia, R., A.A. de La Cruz, J.C. Alvarez, and H. Kallio. 2009. Chemical and functional characterization of Kaiwa (*Chenopodium pallidicaule*) grain, extrudate and bran. *Plant Foods for Human Nutrition* 64: 94–101.

Repo-Carrasco-Valencia, R., J.K. Hellström, J.M. Pihlava, and P.H. Mattila. 2010. Flavonoids and other phenolic compounds in Andean indigenous grains: Quinoa (*Chenopodium quinoa*), kañiwa (*Chenopodium pallidicaule*) and kiwicha (*Amaranthus caudatus*). *Food Chemistry* 120: 128–133.

Reyes-Caudillo, E., A. Tecante, M.A. Valdivia-López. 2008. Dietary fibre content and antioxidant activity of phenolic compounds present in Mexican chia (*Salvia hispanica* L.) seeds. *Food Chemistry* 107: 656–663.

Rezaei, E.E., S. Siebert, and F. Ewert. 2015. Intensity of heat stress in winter wheat—Phenology compensates for the adverse effect of global warming. *Environmental Research Letters* 10: 024012.

Rodríguez, Á., M. Corazon-Guivin, D. Cachique, K. Mejia, D. Castillo, J.F. Renno, and C. Garcia-Dávila. 2010. Diferenciación morfológica y por ISSR (Inter simple sequence repeats) de especies del género *Plukenetia* (Euphorbiaceae) de la Amazonía peruana: propuesta de una nueva especie. *Revista Peruana de Biologia* 17: 325–330.

Roman, V.J., L.A. den Toom, C.C. Gamiz, N. van der Pijl, R.G. Visser, E.N. van Loo, and C.G. van der Linden. 2020. Differential responses to salt stress in ion dynamics, growth and seed yield of European quinoa varieties. *Environmental and Experimental Botany* 177: 104146.

Ruiz, C., C. Díaz, J. Anaya, and R. Rojas. 2013. Análisis proximal, antinutrientes, perfil de ácidos grasos y de aminoácidos de semillas y tortas de 2 especies de Sacha inchi (*Plukenetia volubilis* y *Plukenetia huayllabambana*). *Revista de la Sociedad Química del Perú* 79: 29–36.

Ruiz, K.B., I. Aloisi, V. Canelo, S. Del Duca, P. Torrigiani, H. Silva, and S. Biondi. 2016. Salt flat versus coastal ecotypes of quinoa: salinity responses in Chilean landraces from contrasting habitats. *Plant Physiology and Biochemistry* 101: 1–13.

Ruiz-Carrasco, K., F. Antognoni, A.K. Coulibaly, S. Lizardi, A. Covarrubias, E.A. Marínez, M.A. Molina-Montenegro, S. Biondi, and A. Zurita-Silva. 2011. Variation in salinity tolerance of four lowland genotypes of quinoa (*Chenopodium quinoa* Willd.) as assessed by growth, physiological traits, and sodium transporter. *Plant Physiology and Biochemistry* 49: 1333–1341.

Sage, R.F., T.L. Sage, R.W. Pearcy, and T. Borsch. 2007. The taxonomic distribution of C4 photosynthesis in Amaranthaceae sensu stricto. *American Journal of Botany* 94: 1992–2003.

Salgado-Aguilar, M., T. Molnar, J.L. Pons-Hernández, J. Covarrubias-Prieto, J.G. Ramírez-Pimentel, J.C. Raya-Pérez, S. Hearne, and G. Iturriaga. 2020. Physiological and biochemical analyses of novel drought-tolerant maize lines reveal osmoprotectant accumulation at silking stage. *Chilean Journal of Agricultural Research* 80: 241–252.

Sánchez, B., A. Rasmussen, and J.R. Porter. 2014. Temperatures and the growth and development of maize and rice: A review. *Global Change Biology* 20: 408–417. doi: 10.1111/gcb.12389.

Saucedo, A.L., E. Hernández-Domínguez, L. Luna-Valdez, A. Guevara-García, A. Escobedo-Moratilla, E. Bojorquéz-Velázquez, F. Río-Portilla, D. Fernández-Velasco, and A.P. Barba de la Rosa. 2017. Insights on the structure and function of a Late Embryogenesis Abundant protein from *Amaranthus cruentus*: An intrinsically disordered protein involved in protection against desiccation, oxidant conditions, and osmotic stress. *Frontiers in Plant Sciences* 8: 497.

Sauer, J.D. 1950a. Amaranths as dye plants among the pueblo peoples. *South-west. Journal of Anthropological Sciences* 6: 412–415.

Sauer, J.D. 1950b. The grain amaranths: A survey of their history and classification. *Annals of the Missouri Botanical Garden* 37: 561–632.

Sehgal, A., K. Sita, K.H.M. Siddique, R. Kumar, S. Bhogireddy, R.K. Varshney, B. HanumanthaRao, R.M. Nair, P.V.V. Prasad, and H. Nayyar. 2018. Drought or/and heat-stress effects on seed filling in food crops: Impacts on functional biochemistry, seed yields, and nutritional quality. *Frontiers in Plant Science* 9: 1705.

Shabala, S., Y. Hariadi, and S.E. Jacobsen. 2013. Genotypic difference in salinity tolerance in quinoa is determined by differential control of xylem Na^+ loading and stomatal density. *Journal of Plant Physiology* 170(10): 906–914.

Shah, N. and G. Paulsen. 2003. Interaction of drought and high temperature on photosynthesis and grain-filling of wheat. *Plant and Soil* 257 : 219–226.

Shonnard, G.C. and P. Gepts. 1994. Genetics of heat tolerance during reproductive development in common bean. *Crop Science* 34: 1168–1175.

Shrivastava, P. and R. Kumar. 2015. Soil salinity: A serious environmental issue and plant growth promoting bacteria as one of the tools for its alleviation. *Saudi Journal of Biological Sciences* 22: 123–131.

Silva, A.N., G.M.L. Ramos, W.Q. Ribeiro, E. Rodrigues, P. Carvalho, C. Andrea, C. Cleo, and M.A. Vanderlei. 2020. Water stress alters physical and chemical quality in grains of common bean, triticale and wheat. *Agricultural Water Management* 231: 106023.

Simioniuc, D.P., V. Simioniuc, D. Topa, M. van den Berg, U. Prins, P.J. Bebeli, and I. Gabur. 2021. Assessment of Andean lupin (*Lupinus mutabilis*) genotypes for improved frost tolerance. *Agriculture* 11: 155.

Sinclair, T.R., and R. Serraj. 1995. Legume nitrogen fixation and drought. *Nature Cell Biology* 378: 344.

Singh, S.P., P. Gepts, and D.G Debouck. 1991a. Races of common bean (*Phaseolus vulgaris*, Fabaceae). *Economic Botany* 45: 379–396.

Singh, S.P., R. Nodari, and P. Gepts. 1991b. Genetic diversity in cultivated common bean. I. Allozymes. *Crop Science* 31: 19–23.

Singh, S.P., J.A. Gutiérrez, A. Molina, C. Urrea, and P. Gepts. 1991c. Genetic diversity in cultivated common bean. II. Marker–based analysis of morphological and agronomic traits. *Crop Science* 31: 23–29.

Sogbohossou, E.O.D., E.G. Achigan-Dako, P. Maundu, S. Solberg, E.M.S. Deguenon, R.H. Mumm, I. Hale, A.V. Deynze, and M.E. Schranz. 2018. A roadmap for breeding orphan leafy vegetable species: A case study of *Gynandropsis gynandra* (Cleomaceae). *Horticulture Research* 5: 2.

Sozen, O., U. Karadavut, H. Ozcelik, H. Bozoglu, and M. Akcura. 2018. Genotype x environment interaction of some dry bean (*Phaseolus vulgaris* L.) genotypes. *Legume Research* 41: 189–195.

Suarez, J.C., J.A. Polanía, A.T. Contreras, L. Rodriguez, L. Machado, C. Ordoñez, S. Beebe, and I M. Rao. 2020. Adaptation of common bean lines to high temperature conditions: Genotypic differences in phenological and agronomic performance. *Euphytica* 216: 1–20.

Suarez-Salazar, J.C., J.A., Polania, A.T. Contreras-Bastidas, L. Rodríguez Suárez, S. Beebe, and I.M. Rao. 2018. Agronomical, phenological and physiological performance of common bean lines in the Amazon region of Colombia. *Theoretical and Experimental Plant Physiology* 30: 303–320.

Taibi, K., F. Taibi, L.A. Abderrahim, A. Ennajah, M. Belkhodja, and J.M. Mulet. 2016. Effect of salt stress on growth, chlorophyll content, lipid peroxidation and antioxidant defence systems in *Phaseolus vulgaris* L. *South African Journal of Botany* 105: 306–312.

Tao, Z.Q., Y.Q. Chen, C. Li, J.X. Zou, P. Yan, S.F. Yuan, X. Wu, and P. Sui. 2016. The causes and impacts for heat stress in spring maize during grain filling in the North China Plain - A review. *Journal of Integrative Agriculture* 15 : 2677–2687.

Tian, Y.H., Y.B. Lei, Y. Zheng, and Z.Q. Cai. 2013. Synergistic effect of colonization with arbuscular mycorrhizal fungi improves growth and drought tolerance of *Plukenetia volubilis* seedlings. *Acta Physiologiae Plantarum* 35: 687–696.

Turan, M.A., A.H.A. Elkarim, N. Taban, and S. Taban. 2010. Effect of salt stress on growth and ion distribution and accumulation in shoot and root of maize plant. *African Journal of Agricultural Research* 5: 584–588.

Tsutsumi, N., M. Tohya, T. Nakashima, and O. Ueno. 2017. Variations in structural, biochemical, and physiological traits of photosynthesis and resource use efficiency in *Amaranthus* species (NAD-ME-type C4). *Plant Production Science* 20: 300–312.

Ulian T., M. Diazgranados, S. Pironon, S. Padulosi, L. Udayangani, L. Davies, M.J.R. Howes, J.S Borrell, I. Ondo, O.A. Perez-Escobar, S. Sharrock, P. Ryan, D. Hunter, M.A. Lee, C. Barstow, L. Luczaj, E. Mattana, et al. 2020. Unlocking plant resources to support food security and promote sustainable agriculture. *Plants, People, Planet* 2: 421–445.

Vallebueno-Estrada, M., I. Rodríguez-Arévalo, A. Rougon-Cardoso, J. Martínez González, A. García Cook, F. Montiel, and J.P. Vielle-Calzada. 2016. The earliest maize from San Marcos Tehuacán is a partial domesticate with genomic evidence of inbreeding. *Proceedings of the National Academy of Sciences of the United States of America* 113: 14151–14156.

Wang, Q.M., Z.L. Huo, L.D. Zhang, J.H. Wang, and Y. Zhao. 2016. Impact of saline water irrigation on water use efficiency and soil salt accumulation for spring maize in arid regions of China. *Agricultural Water Management* 163: 125–138.

Wang, Y., Y. Meng, H. Ishikawa, T. Hibino, Y. Tanaka, N. Nii, and T. Takabe. 1999. Photosynthetic adaptation to salt stress in three-color leaves of a C4 *Amaranthus tricolor*. *Plant and Cell Physiology* 40: 668–674.

Waqas, M.A., X. Wang, S.A. Zafar, M.A. Noor, H.A. Hussain, A.M. Nawaz, and M. Farooq. 2021. Thermal stresses in maize: Effects and management strategies. *Plants* 10: 293.

White, P.J., L.M. Pollak, and S.A. Duvick. 2007. Improving the fatty acid composition of corn oil by using germplasm introgression. *Lipid Technology* 19: 35–38.

Wilson, J.H., M.S.J. Clowes, and J.C.S. Allison. 1973. Growth and yield of maize at different altitudes in Rhodesia. *Annals of Applied Biology* 73: 11–84.

Wortmann, C.S. 2006. *Phaseolus vulgaris* L. (common bean). In *PROTA 1: Cereals and Pulses/Céréales et légumes secs,* edited by M. Brink and G. Belay. Wageningen, Netherlands: PROTA. https://uses.plantnet-project.org/f/index.php?title=Phaseolu s_vulgaris_-_haricot_sec_(PROTA)&mobileaction=toggle_view_desktop) Accessed December 07, 2022.

Wu, G., A.J. Peterson, C.F. Morris, and K.M. Murphy. 2016. Quinoa seed quality response to sodium chloride and sodium sulfate salinity. *Frontiers in Plant Science* 7: 790.

Yamori, W., K. Hikosaka, and D.A. Way. 2014. Temperature response of photosynthesis in C3, C4, and CAM plants: Temperature acclimation and temperature adaptation. *Photosynthesis Research* 119: 101–117.

Yang, H., T. Huang, M. Ding, D. Lu, and W. Lu. 2017. High temperature during grain filling impacts on leaf senescence in waxy maize. *Agronomy Journal* 109: 906–916.

Ye, X., M. Zhang, M. Zhang, and Y Ma. 2020. Assessing the performance of maize (*Zea mays* L.) as trap crops for the management of sunflower broomrape (*Orobanche cumana* Wallr.). *Agronomy* 10: 100.

Zhang, G.L., L.Y. Chen, S.T. Zhang, G.H. Liu, W.B. Tang, Z.Z. He, and M. Wang. 2007. Effects of high temperature on physiological and biochemical characteristics in flag leaf of rice during heading and flowering period. *Scientia Agricultura Sinica* 40: 1345–1352.

Zheng, Y., M. Xu, R. Shen, and S. Qiu. 2013. Effects of artificial warming on the structural, physiological, and biochemical changes of maize (*Zea mays* L.) leaves in northern China. *Acta Physiologia Plantarum* 35: 2891–2904.

Zou, C., A. Chen, L.H. Xiao, H.M. Muller, P. Ache, G. Haberer, M.L. Zhang, W. Jia, P. Deng, R. Huang, D. Lang, F. Li, D.L. Zhan, X.Y. Wu, H. Zhang, J. Bohm, R.Y. Liu, S. Shabala, R. Hedrich, J.K. Zhu, and H. Zhang. 2017. A high-quality genome assembly of quinoa provides insights into the molecular basis of salt bladder-based salinity tolerance and the exceptional nutritional value. *Cell Research* 27: 1327–1340.

Chapter 3

The Outlook for Latin-American Crops

Challenges and Opportunities

Nieves Fernández-García[1], Inmaculada Román-García[1], and Enrique Olmos[1]
[1]Centro de Edafología y Biología Aplicada del Segura - Consejo Superior de Investigaciones Científicas (CEBAS-CSIC). Espinardo, Murcia, Spain.

CONTENTS

DOI: 10.1201/9781003088424-3

3.1 INTRODUCTION

There is no single, easy solution to world hunger and poverty. Proposals for the future of the European Commission's common agricultural and food policy stress the reduction in meat consumption and production as a starting point (Schiermeier, 2019), given that a growing share of grain production is used for animal feed.

Of the thousands of known edible plant species, only three cereals—maize, wheat and rice, as the main cultivated plants in the world—contribute to nearly 60% of the calories and proteins consumed by humans from vegetables (FAO, 2000). These three crops are considered the backbone of agriculture, playing a critical role in global food provision (FAO, 2000, 2016, 2017).

This dependence on just a few crops makes current cropping systems highly vulnerable to projected impacts of climate change and accompanying extreme weather events. Climate change poses a considerable threat to global food security, so expanding our food sources by integrating so-called emerging or alternative crops can be a sustainable solution. Thousands of neglected and underutilized species with balanced amino acid and micro-nutrient profiles are known of, which can help address climate change and reduce hunger. These crops include quinoa, amaranth, chia, sacha inchi, tarwi and kaniwa. Implementation of these underutilized crops could provide us with a more diversified agricultural system and improve nutritional quality and climate resilience of future crop systems. Currently, these emerging crops are considered "the food of the future for humanity" by the Food and Agriculture Organization of the United Nations (FAO) and the World Health Organization (WHO), as they are able to replace animal proteins and provide bioactive compounds; they also contain gluten-free proteins (FAO, 2017).

Latin-American crops such as quinoa, amaranth and chia are alternative, profitable and strategic crops. Healthy lifestyles and healthy eating are on the rise, and we are increasingly concerned about what we eat, filling our shopping baskets with more real food and less ultra-processed food. This trend, which is currently associated with healthy and organic diets, has led to a notable increase in the sale of these "superfoods" of Latin-American origin in recent years. These foods are highly appreciated for their nutritional properties,

Population (billion)

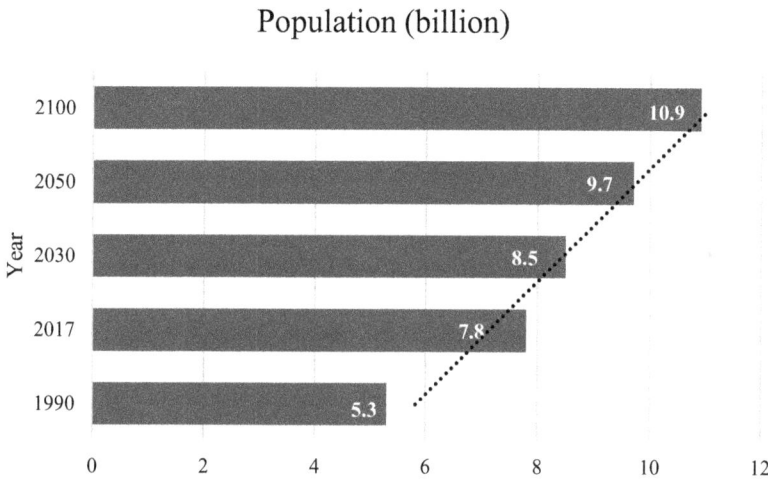

Figure 3.1 World Population 2019: Wall Chart (ST/ESA/SER.A/434). Revision Produced by United Nations Department of Public Information. **Source: United Nations, Department of Economic and Social Affairs, Population Division (2019). World Population Prospects - Population Division - United Nations. https://population. un.org/wpp/.**

and some are consumed as alternatives to animal products like milk (OECD/ FAO, 2019).

On the other hand, the FAO estimates that by 2050 there will be 2 billion more inhabitants on the planet, and that this increase will be accompanied by a progressive but rapid loss of arable land (Figure 3.1). What agrifood measures are on the horizon to meet global food demand? Could Latin-American food systems feed the world (Pastor-Pazmiño, 2017)?

In addition, the world's potential land for food production is threatened by land degradation, affected by soil erosion, depletion and contamination due to both climate change and human misuse (EEA, 2019). This situation represents a major loss that will translate into a loss of global capacity to produce food and supply our growing world population. In order to be able to go into the future with the certainty that we will have food for the entire population, the vast majority of specialists considers precision agriculture to be necessary. When and where to sow, and what to sow, are two of the essential questions yet to be answered (Castillo, 2015).

These degraded lands are likely to end up in a process of desertification. For this reason, there is an obligation to find alternative crops that are able to both cover nutritional demand and grow in adverse soils. Crops like quinoa, amaranth and chia are able to grow in adverse areas with conditions such as saline soils, high temperatures, etc. That is why we can ask ourselves the question: Could Latin-American crops feed the world?

The outstanding properties of Andean crops are the main reason why their consumption has been maintained for millennia in certain civilizations, from ancient times to their expansion to global recognition today. Many of these crops were underestimated by Spanish conquerors, who considered them as irrelevant foods, and were thus replaced by crops with which they were already familiar, such as wheat (Atchison, 2016).

Within the Latin-American crops considered "superfoods", in addition to quinoa, maize, chia and amaranth, we also find crops that are less well known, such as sacha inchi, black bean or tarwi. These lesser known crops are also valued for their high percentage of proteins, high levels of essential amino acids and high concentrations of antioxidants, and some of them are also gluten free.

3.2   QUINOA (*CHENOPODIUM QUINOA* WILLD)

Quinoa (*Chenopidium quinoa* Willdenow), of the Chenopodiaceae family and known as quinua (Quechua), jopa (Aymara), suba (Chibcha), quinhua (Mapuche) and quinoa (Spanish), is native to the Andes. It originated in Peru and Bolivia, around Lake Titicaca, and was probably domesticated in several areas simultaneously (Vietmeyer, 1989). With approximately 7,000 years of cultivation, quinoa is one of the oldest crops in the Andean region, where ancient civilizations such as the Tiahuanacota and Inca contributed to its domestication and conservation throughout time (Jacobsen, 2003). Quinoa and potatoes seem to have been the staple foods of many ancient highland civilizations. On the arrival of the Spaniards, quinoa was replaced by cereals despite being a local staple food at the time.

On archeological digs, quinoa has been discovered in tombs in Tarapacá, Calama and Arica in Chile, and in different regions of Peru (Tapia, 2014). At the time of the Spanish arrival, quinoa was well developed technologically and was widely distributed within and beyond Inca territory. The first Spaniard to note the cultivation of quinoa was Pedro de Valdivia who, on noticing the planted crops around Concepción, recorded that the native Indians also sowed quinoa among other plants for food (Tapia, 2014).

Quinoa is an annual broad-leaved, dicotyledonous herb usually standing about 1-3 m high (Mujica, 1992). The woody central stem is either branched or unbranched depending on the genotype, sowing density and environmental conditions, and it may be green, red or purple. The inflorescences emerge from the higher part of the plant or from the axils of the stem, and they have a central axis from which a secondary (amarantiform) and/or tertiary axis with flowers (glomeruliform) emerges. Depending on the cultivar, the seeds are on average about 2 mm in diameter and they come in a range of different colors; they can be white, red or even black (Rojas, 2003).

3.2.1 Quinoa Production and Market

The consumption of quinoa has increased significantly in recent years, mainly in China and EU countries such as France, Germany and Spain. According to a Euromonitor study (Mascaraque, 2018), consumption of quinoa and other Andean grains is mainly trend driven and health motivated.

Quinoa production showed a remarkable increase after 2013, which was declared the year of quinoa by the FAO (Bazile et al., 2015). Worldwide quinoa production increased steadily until 2015, then declined and was maintained in the following years, possibly due to the fall in the price of quinoa seeds in 2015 (Figure 3.2).

The main producers worldwide are Peru, Bolivia and Ecuador (Figure 3.3). Strikingly, quinoa production in Spain has grown exponentially in recent years: the country went from no production 10 years ago to being the main producer in Europe by 2019 (17,217 ton). Currently, according to the Spanish Ministry of Agriculture (MAPA, 2020), 6,638 ha are dedicated to quinoa cultivation in Spain, and Andalusia is the main production region (93% of the total).

3.2.2 Present and Future Uses of Quinoa

Archaeological and ethnobotanical data show that quinoa has been cultivated in the Andean regions of South America for thousands of years. Quinoa remains have been found in various contexts, including burial grounds, hearths, storage structures, human digestive tracts and coprolites (Lopez et al., 2011). This indicates that quinoa was not used only for food, but also in religious rites.

Quinoa is traditionally consumed in three different forms: (1) as a whole seed, (2) in soups or (3) as *pitu*; a kind of toasted refined flour. It is mainly consumed in the form of seed and soup in the Andean regions and as *pitu* in the Plateau. Both the seed and quinoa flour have also traditionally been used for bread making and fried or cooked products (Angeli et al., 2020). Another ancient way of consuming quinoa is in puffed form. To make puffed quinoa, the seed is treated through an extrusion process using high temperatures and pressure. Flour can also be used to make pasta, chips and tortillas (flatbreads). Moreover, in recent years, the consumption of sprouts has increased exponentially. Sprouting seeds are now common on supermarket shelves, and they are appreciated as a rich source of

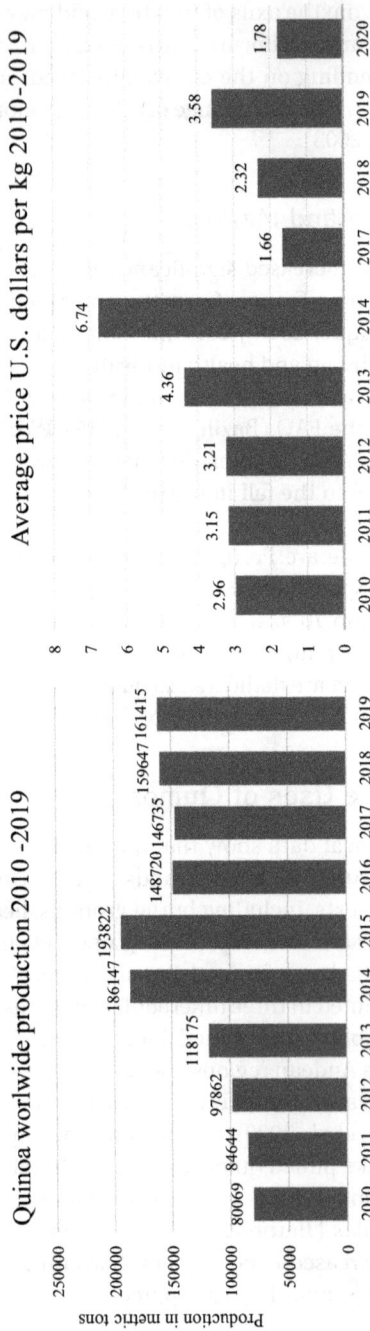

Figure 3.2   (left panel) Worldwide evolution of production). Source: FAO, 2019.
Figure 3.2   (right panel) Price of quinoa between 2010 and 2019. Source: Prospectiva, 2020; FAO Worldwide, 2010–2019; Tridge blog
©Statista, 2021.

Quinoa production worldwide 2010-2019, by country

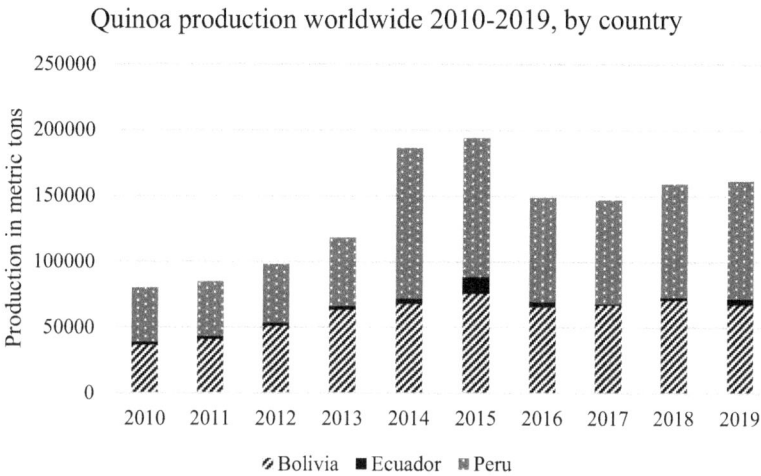

Figure 3.3   Main quinoa producers worldwide: Peru, Bolivia and Ecuador. **Source: Worldwide; FAO, 2010 to 2019. Data were obtained from Statista. www.fao.org/faos tat/en/#data/QCL.**

healthy phytochemicals and for their sensory characteristics. They are rich in nutrients like minerals, amino acids and vitamins, in addition to nutraceuticals, and they contain low levels of anti-nutritional substances like tannin, lectin and galactoside compared to non-germinated seeds. The most common form of sprout consumption is in salads as a food supplement.

Nowadays, a large number of sprouting seeds are consumed, such as sun-flower, broccoli and chia. Nutritional analysis of sprouts shows that sprouting quinoa seeds are being consumed in countries like Peru. On the other hand, quinoa seeds can be fermented to obtain beer, which is gluten-free and therefore suitable for consumption by people with coeliac condition. The food industry is producing more and more quinoa-derived products, including pasta and breakfast products rich in quinoa, as well as snacks, energy bars and ready-to-eat meals.

New uses for quinoa include the extraction of starch granules and their application in emulsifying systems. The starch granule is a biopolymer that is abundant in the seeds and inexpensive to extract. It is both biodegradable and biocompatible, which makes it a suitable candidate to replace synthetic and inor-ganic particles used as Pickering stabilizers. Among the starch-based Pickering emulsions, octenylsuccinate quinoa starch (OSQS) shows interesting emulsi-fying properties, mainly due to the small size of the starch granule compared to that obtained from other species (Li et al., 2020).

3.3   AMARANTHUS: KIWICHA (*AMARANTHUS CAUDATUS* L.), AND AMARANTH (*A. CRUENTUS* L. AND *A. HYPOCHONDRIACUS* L.)

The genus Amaranthus is considered an ancient crop that has been cultivated from 5,000–7,000 years ago (Sauer, 1967). It comprises approximately 60 species that grow in many regions of the world (Saunders and Becker, 1984). Most amaranth species are native to America, mainly to Mesoamerica and South America, with only 15 species originating in Asia, Africa, Australia and Europe (Martinez-Lopez et al., 2020). The main species cultivated for the characteristic of the seeds are *A. caudatus*, *A. cruentus* and *A. hypochondriacus*, which have been domesticated in the Andean region of South America and Mesoamerica, respectively (Sauer, 1967). These amaranth species are herbaceous annual plants that were domesticated in prehistoric times (Sauer, 1976). In Tehuacan, Puebla, Mexico, archaeological remains were found indicating that *A. cruentus* was already being cultivated around 4000 BC, and that *A. hypochondriacus* was grown around AD 500 (Sauer, 1976; Jacoben and Mujica, 2003). The oldest archaeological evidence of *A. caudatus* dates back 2,000 years and was found in northern Argentina (Hurnziker and Planchuelo, 1971). The inflorescence consists of large branches of solid red, yellow, green, or variable colors. Likewise, the seeds are extremely variable in color. Amaranth has small seeds of between 1 and 1.5 mm in diameter.

Interest in amaranth has increased progressively since the 1980s, when the US National Academy of Sciences developed a research program entitled "Underexploited Tropical Plants with Promising Economic Value", which included 36 promising species with the greatest potential for cultivation (Berghofer and Schoenlechner, 2002; Caselato-Sousa and Amaya-Farfan, 2012; Orona-Tamayo and Paredes-López, 2017; Soriano-Garcia et al., 2018). Since then, a major effort has been made to study the nutritional and antioxidant characteristics of amaranth and its cultivation possibilities in different parts of the world (Park et al., 2020).

3.3.1   Present and Future Uses of Amaranthus

Amaranth is a versatile seed, and is considered one of the most nutritious pseudo-cereals that is high in gluten-free protein. Amaranth is also used in sweets or drinks, in fresh leaf salads, popped seeds, and as a source of natural red dye, animal feed and for obtaining oils and cosmetic products (López-Mejía et al., 2014; Repo-Carrrasco-Valencia, 2011). The Amaranth seed can be roasted or milled into flour, and it can be consumed on its own or included in other crop products such as bread, cakes, muffins, pancakes, biscuits, dumplings, noodles, pasta and crackers (Cardenas-Hernandez, 2016).

Moreover, the incorporation of amaranth flour in pasta processing has been found to produce a low level of stickiness that is more desirable for pasta products (Chillo et al., 2008; Fiorda et al., 2013). Guardianelli et al. (2021) have studied mixing amaranth flour from germinated and non-germinated seeds with wheat flour for bread. Wheat flour withstood the inclusion of 25% germinated amaranth seeds without substantial changes in bread quality. Additionally, roasted amaranth flour has the potential to replace wheat flour in the preparation of acceptable gluten-free noodles (Beniwal et al., 2019). Singh and Liu (2021) compared noodles made from amaranth flours with wheat flour noodles and demonstrated that roasting improved cooking results and the texture of cooked noodles.

Amaranth is also consumed as a vegetable throughout Asia, Africa, the Caribbean, Greece and Mexico. Amaranth leaves are usually picked fresh for use as greens in salads or steamed, boiled or fried in oil. Cooked greens can be used as a side dish, in soups or as an ingredient (FAO, 2020).

3.3.2 New Regions of Cultivation

Currently, amaranth is widely cultivated and consumed throughout India, Nepal, China, Indonesia, Malaysia and the Philippines in Asia; throughout the whole of Central America and Mexico; and in South and East Africa. Different species of amaranth are cultivated in different countries. *A. hypochondriacus* is cultivated in some countries of South America, Mexico and Asia, while *A. caudatus* is cultivated in Argentina, Peru and Bolivia, and *A. cruentus* is cultivated in Guatemala.

The African continent represents one of the poorest and most malnourished regions on the planet. African nations in this continent need to develop new strategies in order to eradicate poverty and child malnutrition in the future. One of the possibilities is to pay greater attention to underutilized crop species that have great potential for influencing and improving food security for African nations and promoting sustainable rural growth and development (Aderibigbe et al., 2020). Amaranth is an underutilized crop that has great potential in Africa, given its stress-tolerant characteristics and ability to grow under extreme situations, which can help to mitigate the effects of climate change (Aderibigbe et al., 2020; Alemayehu et al., 2015; Dizyee et al., 2020). This crop was introduced in sub-Saharan Africa in the 20th century, mainly *A. hypochondriacus*, *A. cruentus*, *A. caudatus*, and *A. dubius* (Dinssa et al., 2016). Farmers in this part of the world use it for both its leaf and seed. Available data from Tanzania and Kenya show that 25,548 ha were cultivated in 2017, with a production of approximately 70,000 tons (Ochieng et al., 2019).

The area under kiwicha (amaranth) cultivation in Peru ranged from 2,300 ha in 2015 to 1,400 ha in 2017, with an average yield of 1,885 kg ha^{-1} (Valvas et al., 2020). Recently, the Ministry of Agriculture and Irrigation of Peru (MINAGRI) has developed a new variety of kiwicha, INIA 442 LA FRONDOSA, resistant to different pests and able to grow at altitudes of 2,500 to 3,000 m above sea level. This new variety will increase production to 3.5 ton per ha, increasing the profitability of the crop (MINAGRI). In Mexico, amaranth production in 2019 was about 6,000 tons, much less than those 7,115 produced in 2018 (Panorama Alimentario, 2020). The main producing area in the country is Puebla, accounting for about half of the amaranth produced. The area dedicated to amaranth cultivation in Mexico in 2016 was 4,545 ha, and Puebla was the main state with the largest area dedicated to this crop.

3.4   CHIA (*SALVIA HISPÁNICA* L.)

Chia (*Salvia hispánica* L.), of the Lamiaceae family, is native to the central valleys of Mexico and northern Guatemala, where species of the Labiatae family are concentrated. Chia seeds began to be used in human food around 3500 BC and acquired importance as a staple crop in central Mexico between 1500 and 900 BC (Cahill, 2003; Pelaez et al., 2019). It was one of the main crops of pre-Columbian societies, and Aztecs and Mayas used the seeds for the preparation of various medicines, foods and paintings. The Aztecs received chia seeds as annual tributes from the people under their rule, and it was used as an offering to the gods in religious ceremonies (Beltran-Orozco et al., 2003; Muñoz et al., 2013).

Carl Linnaeus classified chia in 1753. *Salvia* is a Spanish word meaning "cure," and in Latin, *hispanica* means Spanish; *Salvia hispánica* thus means "Spanish plant used to cure or save" (Urbina, 1887; Sharma and Mogra, 2019). The word "chia" (meaning "oily") is a Spanish adaptation of *chian*, or *chien* in its plural form, which has its origin in the Nahuatl language of the Aztecs. Chia has been cultivated since ancient times on the banks of the Grijalva River, the ancient territory of the Nahuatl Chiapan, which is now the state of Chiapas in modern Mexico; the plant took its name from the Grijalva River, which means "river of chia" (Munoz et al., 2013).

Chia (*Salvia hispanica* L.) is an annual herb that blooms during the summer months (Alfaro and Silva, 2013; Munoz et al., 2013) and grows up to 1–2 m in height. The seeds are oval, smooth and shiny, and mottled with brown, gray, dark red and white; they are generally found in groups of four. The plant has ribbed and hairy quadstems (Ayerza and Coates, 2005; Di Sapio et al., 2012; Pelaez et al., 2019;).

3.4.1   State of the Art in Chia Crops

The chia seed has great nutritional qualities. It is rich in antioxidants and an excellent source of fiber; it has a high protein content and high levels of omega-3,

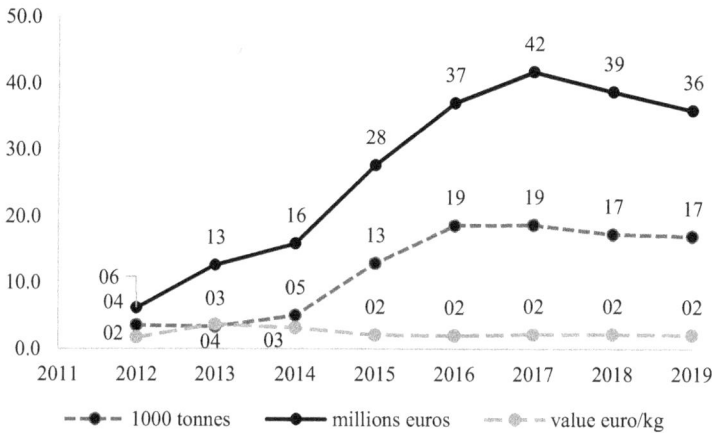

Figure 3.4   Indicative chia imports to Europe from the main producing and supplying countries: Paraguay, Bolivia, Argentina, Peru, Mexico, Uganda, Chile and Australia. **Source: Eurostat, Market Access Database; The European market potential for chia seeds CBI.**

and is rich in minerals, among other features (Kulczynski et al., 2019; Orona-Tamayo et al., 2019). Although the health-promoting and biologically active characteristics of chia have been known for many years, it is only in the last few years that interest in chia consumption has grown from a scientific and commercial point of view (Teoh et al., 2018). The chia market in Europe has been increasing significantly since 2012. As can be seen in Figure 3.4, chia imports have grown fourfold in the last eight years.

In recent years, Paraguay has been the main producer of chia in the world with more than 20,000 tons, which represents almost 50% of the world's production. Other important chia producers are Mexico, Bolivia, Argentina, Peru, Chile, Uganda and Australia. The world's largest importer is the United States (Figure 3.5). Europe's largest importers of chia are the Netherlands and Germany, accounting for importations of over 50% (Figure 3.5).

Chia has been considered a novel food by the EU, following the regulation stipulating that any food that was not consumed to a significant degree within the European Union before May 15, 1997 is to be considered novel. Since January 1, 2018, this region has drawn up a list of authorized novel foods and their requirements, including chia seeds and chia-derived oil (Kulczynski et al., 2019). The following uses for chia are permitted in the EU: (1) as an ingredient in food in pastry, bread products, yogurts, fruit jellies and dressings; (2) as a nutritional supplement; and (3) as a base for beverages.

Chia seeds import value in 2020

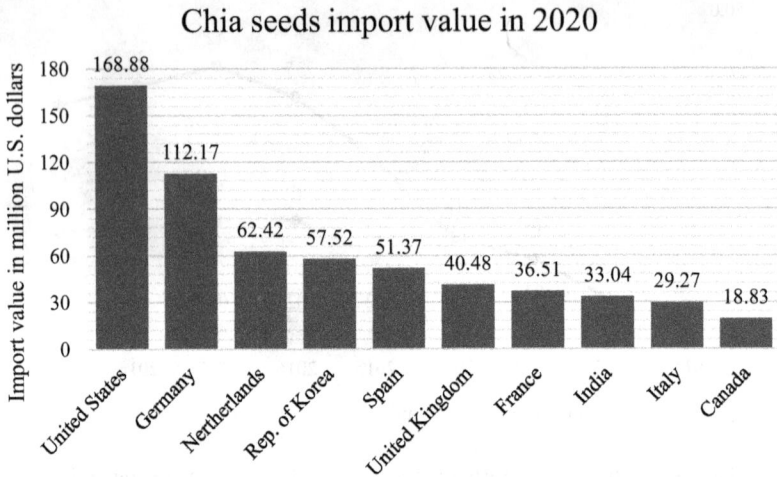

Figure 3.5   Chia seeds import value in 2020. **Source: Worldwide; UN Comtrade, 2020. https://comtrade.un.org/data/.**

3.4.2   New Regions of Cultivation

Chia is a short-day flowering plant, sensitive to frost at all stages of development (Bochicchio et al., 2015). The induction of flowering and the transition between vegetative and generative phases of plant growth are highly dependent on the length of the day (<12 hours). These characteristics mean that chia cultivation is limited to latitudes below 25°, close to the equator. Chia grows naturally in tropical and subtropical areas, with optimum growth between 400 and 2,500 m above sea level; below 200 m, it does not develop adequately (Orozco et al., 2014). Chia requires temperatures between 11 and 36°C, and its optimum range is 16–26°C (Ayerza and Coates, 2009). Chia grows best in sandy, well-drained soils with moderate salinity and with a pH ranging from 6 to 8.5 (Yeboah et al., 2014). Currently, the main cultivation areas for chia are Paraguay, Mexico, Argentina, Bolivia, Australia, Colombia, Guatemala and Peru. Chia cultivation is extending to other regions, however, and in Europe, trials are being carried out in Greece, France, Germany and Italy (Bochicchio et al., 2015; Gravé et al., 2019; Grimes et al., 2020; Tabaxi et al., 2016;). Although the results are interesting, it is clear that there is a need to develop new varieties that can grow in days with a longer light cycle. Breeders have recently succeeded in developing varieties with low day-length dependence, allowing for growth in 12–15h light conditions (Gravé et al., 2019). This opens up possibilities for cultivation in regions such as Europe (Grimes et al., 2020). According to our knowledge, a new variety, named ORURO, has been introduced in southwestern France at the Regional Centre for Organic Agriculture, in Auch. A stable and homogeneous variety selected to bloom in long summer days, ORURO chia seeds flower early, which allows for production in temperate regions (Gravé et al., 2019). In other Latin-American countries, such as Chile, the possibility of introducing chia cultivation is also being evaluated (Baginsky et al., 2016; Silva et al., 2018).

3.4.3   Present and Future Uses of Chia

Nowadays, chia seeds are used whole, ground, and in the form of gel and oil (Muñoz-Gonzalez et al., 2019; Zettel and Hitzmann, 2018;). In recent years, chia gel has been used in the production of various food products, such as low-fat sausages (Camara et al., 2020; Pintado et al., 2020). Chia gel has also been used as an egg substitute in chocolate cakes or as partial fat replacement in pork burgers (Gallo et al., 2020; Lucas-Gonzalez et al., 2020). The use of chia gels in these food products not only reduces the fat and calorie content, but also increases the amounts of beneficial fatty acids contained in chia oils, such as omega-3. However, this substitution should not exceed 25% of the oil or egg in cakes, as it may affect different product qualities like texture, color or flavor (Borneo et al., 2010).

Chia flour has also been used to partially replace wheat flour. Borneo et al. (2010) observed that replacing 7.5% of wheat flour with chia flour was the most appropriate proportion in terms of nutritional capabilities and organoleptic qualities. Constatini et al. (2014) observed that replacing 10% of wheat flour with chia flour had no effect on the final flavonoid content, but it did increase fat, moisture, polyphenols and dietary fiber, while reducing carbohydrate content. Other authors have observed that the substitution of 7.5% wheat flour with ground chia seeds negatively affects the sensory qualities of bread and reduces consumer acceptance (Kowalski et al., 2020). However, 5% substitution with ground chia seeds was not found to affect the quality of the bread produced, which had a higher polyunsaturated fatty acid to saturated fatty acid ratio than regular wheat bread. The addition of 5% chia flour to white bread can increase the dietary fiber content by up to 50% (Švec and Hrušková, 2015). Chia flour can also be used to increase the nutritional quality of gluten-free foods such as bread, as these tend to be poorer in minerals and fats (Pellegrini and Agostoni, 2015). Chia flour has also been used in the production of pasta. Oliveira et al. (2015) observed that the addition of chia flour in pasta production significantly improved the protein, fiber and mineral content. Similarly, Cota-Gastelum et al. (2019) observed that the preparation of pasta with a mixture of wheat semolina (25%), chia flour (10%) and chickpea flour (35%) improved the nutritional qualities of the pasta as well as its antioxidant properties, increasing consumer acceptability.

The inclusion of chia in dairy products has also been increasing in recent years. For example, ice cream has been obtained by partially replacing the milk fat with the olein fraction of chia oil (Ullah et al., 2017). The addition of chia enhanced the concentration of omega-3 fatty acids and improved the antioxidant perspectives of ice cream. Chia seed mucilage has also been used as a stabilizer and emulsifier in ice cream (Campos et al., 2016). Chia mucilage has been added to yogurts, as well, replacing fats and increasing nutritional value (Ribes et al., 2021). The addition of 2% chia oil to natural yogurts resulted in higher amounts of unsaturated fatty acids, especially linoleic and alpha-linolenic acids, and a high phytosterol content (Derewiaka et al., 2019).

This seed is also a promising source of bioactive peptides which exert biological functions to promote health, as treatment for or prevention of some diseases (improvement of the digestive system, stronger bones and muscles, reduction of risk of heart disease) (Grancieri et al., 2019; Orona-Tamayo et al., 2019; Quintal-Bojórquez et al., 2021). Benefits for the human body of bioactive peptides are mainly due to the composition of amino acids present in each peptide and the different reactivity. Actually, the most studied properties of bioactive peptides in human health are about anticancer, antimicrobial, antihypertensive, antioxidant, anticholesteolemic, anti-inflammatory and immunomodulatory diseases (Orona-Tamayo et al., 2019; Santiago-Lopez et al., 2016).

3.5 TARWI (*LUPINUS MUTABILIS*)

Tarwi, a leguminous species (*Lupinus mutabilis* Sweet) of the Fabaceae family, also known as tauri, chocho or Andean Lupin, is native to the Andes of Peru, Bolivia and Ecuador. It has been cultivated in the Andean area since pre-Incan times. The main characteristics are the high protein, fat contents and its bitter taste due to anti-nutritional alkaloids that make processing necessary before consumption in seeds; and the plant capacity of fixing atmospheric nitrogen in the soil due to the association with rhizobia and its ability to adapt to adverse climatic conditions with minimal demands on soil quality (Gutierrez, 2016).

The wide genetic diversity of tarwi means that there is great variability in its adaptation to different environmental conditions (soil, temperature, precipitation, etc.). This genetic diversity also means that the seeds' concentration of proteins, alkaloids, oils, among other components, also vary widely (Jacobsen and Mujica, 2006).

Tarwi is considered to be the "Andean soybean" thanks to its high levels of proteins and oils, macronutrients that can be used in the formulation of various functional foods to prevent metabolic diseases. Proteins and oils represent more than half of the seed's weight. Some of the fatty acids present in tarwi seeds are essential fatty acids for our organism, i.e., we cannot synthesize them ourselves, so must acquire them in our diet (Jacobsen and Mujica 2006).

Tarwi is an excellent potential source for obtaining protein isolates and bioactive compounds for the functional food and nutraceutical industry (Jacobsen and Mujica, 2006; Chirinos-Arias, 2015; Tapia, 2015). Protein is the most abundant and important macronutrient in this legume (46–48%), with levels similar to those found in soybeans; moreover, their concentration can increase after defatting, reaching values of 47–64% and a digestibility of over 90% (Schoeneberger et al., 1982; Intiquilla-Quispe, 2018). This extract can be used to obtain protein concentrates and isolates for use in gastronomy, the food industry, livestock feed, and others (Tapia, 2015). Tarwi seeds are raw materials with a high industrial potential in the production of food products such as pastes, flours and oils. Tarwi paste is used to make drinks, yogurt and jams, while the flour is mainly used in the production of products like breads, cakes, biscuits, cookies, noodles and instant soups.

Seed from this crop also has a high fat content, which makes it viable for industrial oil production; its use in human consumption makes it necessary to obtain treated grains free of the toxic substances they contain (Ruiz and Sotelo, 2001; Resta et al., 2008).

Flours after protein and oil extraction can be a basic component for the manufacture of products for the food industry and an efficient use of the crop (Sosa, 2000; Rosell et al., 2009; Morales et al., 2012).

Unfortunately, this species has not been given the importance it deserves, partly due to the scarce dissemination of its nutraceutical properties and the presence of alkaloids that give the seeds a bitter taste, limiting its consumption. For such reasons, a future objective should be to disseminate the nutraceutical and medicinal properties of "Andean Lupin" in order to arouse the interest of other countries in its study and diffusion.

3.6   SACHA INCHI (*PLUKENETIA VOLUBILIS* L.)

Sacha inchi (*Plukenetia volubilis* L.), of the Euphorbiaceae family, is also known as sacha peanut, mountain peanut, Inca nut or Inca peanut (Wang et al., 2018). Sacha inchi is native to the Peruvian Amazon, and it is also found in the Colombian, Bolivian and Brazilian Amazon. The fruits of the plant are starshaped where seeds are contained (Kumar et al., 2017, Vasquez-Osorio et al., 2017). These seeds show high levels of fat, 41–43%, and more calories than sugar; they are also a good source of vitamins, proteins and minerals; better than most meat products, so they could be a good substitute for meat consumption (Chirinos et al., 2013; Alvarado, 2014).

Andean populations have long consumed this seed in traditional preparations such as soups, biscuits and children's food. This is especially true in Peru, where they obtain oil and flour by various methods, which are used to prepare food, drinks and snacks that provide omega-3 and omega-6 (Vásquez-Osorio et al., 2017).

Oil has long been used for food by various ethnic groups in Latin America; it is now obtaining international recognition for its healthy properties, especially as an extraordinary source of polyunsaturated fatty acids, one of the most prospective nutraceutical sources to promote healthy nutrition with extraordinary benefits for human health. Currently, the protein isolate obtained from the residue after oil extraction has high protein levels (of ~59%); this residue may be used in other potential products and it can be further used in a high number of products. Using waste or by-products obtained in the preparation of sacha inchi oil to produce protein isolates can contribute to human health in the fight against malnutrition, and it can also contribute to global knowledge on underutilized crops.

3.7   MAIZE (*ZEA MAYS* L., VARIETY PURPLE)

Maize, or corn (*Zea mays* L.) is native to the Americas. It began its journey *c.*10,000 years ago someplace in Mesoamerica, and from there it spread to the whole continent. During the 16th and 17th centuries, maize cultivation was widespread throughout the world, as Spaniards and other Europeans exported maize from the Americas to Europe. At present, this crop is the third most important crop (after wheat and rice), and it is produced in most countries of the world. Maize has great variability in the color of the grain, including white, yellow, black, purple, blue, red and orange in different varieties (Arteaga et al., 2016; Kistler et al., 2018).

Zea mays L., a purple variety (purple corn or maize), originated in the Andean region of what is now Peru, and it has been widely cultivated and consumed throughout the Andean region of South America, mainly in Peru, Ecuador, Bolivia and Argentina (Lao et al., 2017). Purple corn has purple episperm seeds (grains) and cobs, which gives the pigments special characteristics: between 1.5% and 6.0% of the pigments are anthocyanins, which belong to the group of flavonoids including cianidine-3-Glucoside (Guillén-Sánchez et al., 2014). Although it has been consumed in Andean regions for millennia for its many properties, purple corn is not well known in Europe. It is considered a superfood due to its high anthocyanin content, and it is grown sustainably and exported all over the world. Purple corn not only possesses high concentrations of antioxidant pigments, but also of fiber. Overall, this crop is nutritionally and compositionally rich, being similar in the mineral and vitamin content to the yellow corn but showing higher levels of easily digestible carbohydrates (Ai and Jane, 2016; Lao et al., 2017). It should be noted that the presence of anthocyanins and other phenolic compounds that differentiate purple maize from other conventional maize varieties make it stand out as a good food source with healthy benefits. However, the information currently available on its benefits is still limited and more research would be needed to understand its contribution to human health (Lao et al., 2017).

Consumers can find it canned or vacuum-packed, although it can also be consumed in the form of flour or in food supplements.

3.8   KANIWA (*CHENOPODIUM PALLIDICAULE* AELLEN)

Kaniwa (*Chenopodium pallidicaule* Aellen), of the Chenopodiaceae family, is also known as kañiwa, kañawa and cuchi-quinoa, among other names. The origin of this pseudo-cereal is uncertain, but it is almost certainly an Andean native. Kaniwa is closely related to quinoa, and it was considered a variety of quinoa until 1929 (Repo-Carrrasco-Valencia, 2011). Kaniwa is cultivated in the Andes of Peru and Bolivia and is characterized by its tolerance to low temperatures; it grows in extreme environments where other species such as wheat and maize cannot grow. This crop grows in the extreme altiplano environment, in the highest areas of Peru and Bolivia on the banks of the Titicaca River, at an altitude of between 3,800 and 4,100 m above sea level where the average temperature is less than 10ºC. Here, other staple crops considered essential for global food security, such as wheat, barley, even quinoa, do not grow as reliably as this plant (Popenoe, 1989). It is a very important crop for highland farmers; when other crops fail due to frost, kaniwa still provides food (Repo-Carrasco-Valencia, 2011).

Unlike quinoa, kaniwa has very low saponin levels, and it can be used directly without any prior treatment (Repo-Carrasco-Valencia et al., 2009). Its seed does not contain gluten, so it is of interest to people with coeliac disease; it is also highly nutritious and rich in nutrients like calcium and iron, and high in protein. Fortunately, it has a healthy amino acid profile as well.

In kaniwa, proteins (15–19%) are the second most abundant nutrient after carbohydrates, and they are of better quality than the proteins found in other cereals due to a balanced portion of amino acids and the presence of those is recommended for children and adults (Repo-Carrasco-Valencia, 2011). Moreover, these proteins are particularly rich in lysine (5–6%), isoleucine and tryptophan (3.4 and 0.9 g/ 100 g of protein, respectively) (Repo-Carrasco-Valencia et al., 2003; Gallego-Villa et al., 2014). They are the perfect complement to daily protein intake.

The high content of proteins, and outstanding levels of some key amino acids of these seeds, provide a basis for considering kaniwa as an important crop for current and future food security

3.9 COMMON BEAN (*PHASEOLUS VULGARIS*)

Studies about the origin of the common bean indicate a Mesoamerican origin for these species from which different migratory events extended the distribution of *Phaseolus vulgaris* towards South America. Their hypothesis on the origin of the common bean indicates that, from a central area on the western slopes of the Andes in northern Peru and Ecuador, wild beans dispersed northwards (Colombia, Central America and Mexico) and southwards (southern Peru, Bolivia and Argentina), giving rise to Mesoamerican and Andean genetic variability, respectively (Bitocchi et al., 2003; Chacón-Sanchez et al., 2005; Kaplan, 1965).

The common bean, of family Fabaceae, which has been an important source of livelihood in the Americas since it was first cultivated more than 7,000 years ago, is now considered one of the most important food legumes, mainly in Africa and Latin America (Broughton et al., 2003). This legume is often grown by small-scale farmers under limiting conditions. Among these limiting conditions is abiotic stress caused by drought, which currently affects more than 50% of the world's bean-growing areas and can cause production losses of between 10% and 100%. However, according to Polania et al. (2016), a genetic improvement for the environment for drought resistance (the main abiotic stress factor limiting bean cultivation) would be in the selection of morphophysiological characteristics in outstanding genotypes.

On the other hand, this legume – being a source of bioactive compounds with the potential to contribute to healthier diets – is considered a nutraceutical or functional food (Bressani, 1993). Its antioxidant capacity is determined by the content of compounds such as polyphenols (including flavonoids), anthocyanins, and condensed tannins (which contribute to the color of the bean) (Bressani et al., 1961; Chávez-Mendoza et al., 2017; Singh et al., 2021).

Although beans are traditionally included in the diet in many Latin-American countries and are beneficial for health and disease prevention, a considerable amount of research is needed to characterize the very diverse genotypes in terms of their content of bioactive compounds and their effect on consumer health (Acevedo et al., 1994; Calderon et al., 1992; Suarez-Martinez et al., 2016). Chavez-Mendoza et al. (2017) stressed the need to study wild-type genotypes for their nutraceutical qualities to be used as a valuable resource in breeding programs and strategies to promote their consumption.

Common beans are considered one of the main sources of protein in the diets of many countries. Cultivation of this legume is mainly carried out by smallholders under limiting environmental conditions (Buruchara et al., 2011; Beebe et al., 2013; Keller et al., 2020: Mukankusi et al., 2018). In addition to seed yield losses related to the challenging environmental conditions where beans are cultivated, storage of common beans under adverse conditions (including high temperatures and high humidity) renders them susceptible to a hardening phenomenon, also known as the hard-to-cook (HTC) defect (Bressani, 1993; Reyes-Moreno et al., 1993). The HTC defect diminishes the nutritional properties of this seed due to extension of cooking time affecting the complementary nutritional role of this legume as well as in diets when it is combined with cereals. This effect is especially relevant in countries where diet is based on legumes (as are common beans) and cereals (maize, wheat, or rice). Processing methods such as soaking, germination, fermentation, and cooking have been reported as detoxifying the legume seed. Also, soaking prior to cooking softens the seeds, significantly reducing cooking time (Bressani, 1993; Reyes-Moreno et al., 1993).

The importance of considering the cooking time and its relationship with bean quality has resulted in the development of breeding programs aimed at improving this qualitative aspect (Carvalho et al., 2017). Incorporating cooking quality into cultivar selection will allow improvement in the length of cooking time, thus preserving the nutritional quality of the seeds (Ribeiro et al., 2014).

3.10 CONCLUSIONS

Climate change is a threat to our world pantry, making food security at risk. It is therefore necessary to transform our food systems by unlocking underutilized plant resources to support and improve food security, promoting sustainable agriculture and affordable healthy diets for all.

In the next few years, preservation of biodiversity will be essential. In this regard, sustainable agricultural productivity is going to be required to meet growing demand for food that should possess high nutritional quality.

Most of the agricultural species described in this book are not considered to be among the main staple crops in the world and could be categorized as neglected and underutilized species. Establishment of these underutilized crops may contribute to the FAO's goal of a "world free from hunger and malnutrition, where food and agriculture contribute to improving the standard of living for all" in an economical, social, and environmentally sustainable manner, improving food security and providing essential nutrients thus complementing the current staple crops (wheat, corn, and rice).

In this chapter, we have tried to highlight the agronomic, technological, and health aspects of some Latin-American crops such as chia, amaranth, sacha inchi, tarwi, common bean, maize, kaniwa, and quinoa which could offer natural food resources with high nutritional value. Indeed, their contribution to food security has been increasingly recognized worldwide, for present and future generations. These species are underutilized food crops that indigenous people have protected and preserved using their traditional knowledge and practices.

ACKNOWLEDGMENTS

This work was supported by "Proyectos estrategicos Ris3Mur (Ref: 2I18SAE00057); Consejeria de Empleo, Universidades, Empresa y Medio Ambiente de la Región de Murcia".

REFERENCES

Acevedo, E., Velazquez-Coronado, L. and Bressani, R. 1994. Changes in dietary fiber content and its composition as affected by processing of black beans (*Phaseolus vulgaris*, Tamazulapa variety). *Plant Foods for Human Nutrition* 46: 139–145.

Aderibigbe, O.R., Ezekiel, O.O., Owolade, S.O., Korese, J.K., Sturm, B. and Henseld, O. 2020. Exploring the potentials of underutilized grain amaranth (*Amaranthus* spp.) along the value chain for food and nutrition security: A review. *Critical Reviews in Food Science and Nutrition* DOI: 10.1080/10408398.2020.1825323.

Ai, Y. and Jane, J. 2016. Macronutrients in corn and human nutrition. *Comprehensive Reviews in Food Science and Food Safety* 15 : 581–598.

Alemayehu, F.R., Bendevis, M.A. and Jacobsen, S.E. 2015. The potential for utilizing the seed crop amaranth (*Amaranthus* spp.) in East Africa as an alternative crop to support food security and climate change mitigation. *Journal of Agronomy and Crop Science* 201: 321–329.

Alfaro, F. and Silva, H. 2013. *Determinación de umbrales de respuestas fisiológicas e identificación de mecanismos de tolerancia al déficit hídrico en cuatro accesiones de Chía (*Salvia hispánica *L.).* Santiago:

Alvarado-Quiroz, K.D. 2014. Obtención, caracterización fisicoquímica, caracterización electroforética y digestibilidad del aislado proteico del residuo agroindustrial de *Plukenetia volubilis* (Sacha Inchi). Tesis de grado Arequipa-Perú, Universidad Católica de Santa María.

Angeli, V., Silva, P.M., Massuela, D.C., Khan, M.W., Hamar, A., Khajehei, F., Graeff-Hönninger, S. and Piatti, C. 2020. Quinoa (*Chenopodium quinoa* Willd.): An overview of the potentials of the "Golden Grain" and socio-economic and environmental aspects of its cultivation and marketization. *Foods* 9: 216.

Arteaga, M.C., Moreno-Letelier, A., Mastretta-Yanes, A., Vázquez-Lobo, A., Breña-Ochoa, A., Moreno-Estrada, A., Eguiarte, L.E. and Piñero, D. 2016. Genomic variation in recently collected maize landraces from Mexico. *Genomics Data* 7: 38–45.

Atchison, G.W., Nevado, B., Eastwood, R.J., Contreras-Ortiz, N., Reynel, C., Madriñán, S., Filatov, D.A. and Hughes, C.E. 2016. Lost crops of the Incas: Origins of domestication of the Andean pulse crop tarwi, Lupinus mutabilis. *American Journal of Botany* 103(9): 1592–1606.

Ayerza, R. and Coates, W. 2005. Chia: Rediscovering a forgotten crop of the Aztecs. Tucson, AZ: University of Arizona. www.jstor.org/stable/j.ctv29sfps7.

Ayerza, R. and Coates, W. 2009. Influence of environment on growing period and yield, protein, oil and α-linolenic content of three chia (*Salvia hispanica* L.) selections. *Industrial Crops and Products* 30: 321–324.

Baginsky, C., Arenas, J., Escobar, H., Garrido, M., Valero, N., Tello, D., Pizarro, L., Valenzuela, A., Morales, L. and Silva, H. 2016. Growth and yield of chia (*Salvia hispanica* L.) in the Mediterranean and desert climates of Chile. *Chilean Journal of Agricultural Research* 76: 255–264.

Bazile, D. and Baudon, F. 2015. The dynamics of the global expansion of quinoa growing in view of its high biodiversity. In *State of the Art Report of Quinoa in the World in 2013*, edited by D. Bazile, D. Bertero and C. Nieto, 42–55. Rome: FAO & CIRAD.

Beebe, S., Rao, I., Mukankusi, C., and Buruchara, R. 2013. Improving resource use efficiency and reducing risk of common bean production in Africa, Latin America and the Caribbean. In *Eco-Eefficiency: From Vision to Reality*, edited by C. Hershey and P. Neate, 117–134. Colombia: CIAT.

Beltrán-Orozco, M.C. and Romero, M.R. 2003. *La Chía, Alimento Milenario*. Mexico: Departamento de Graduados e Investigación en Alimentos, www.jstor.org/stable/j.ctv29sfps7. Escuela Nacional de Ciencias Biológicas. (ENCB), Instituto Politécnico Nacional (IPN).

Beniwal, S.K., Devi, A. and Sindhu, R. 2019. Effect of grain processing on nutritional and physico-chemical, functional and pasting properties of amaranth and quinoa flours. *Indian Journal of Traditional Knowledge* 18: 500–507.

Berghofer, E. and Schoenlechner, R. 2002. Grain Amaranth. In *Pseudocereals and Less Common Cereals*, 219–260. Berlin: Springer.

Bitocchi, E., Laura Nannia, L., Belluccia, E., Rossia, M., Giardinia, A., Zeulib, P.S., Logozzob, G., Stougaardc, J., McCleand, P., Attenee, G. and Papaa, R. 2003. Mesoamerican origin of the common bean (*Phaseolus vulgaris* L.) is revealed by sequence data. *Proceedings of the National Academy of Sciences* 109(14): E788–796.

Bochicchio, R., Rossi, R., Labella, R., Bitella, B., Permiola, M. and Amato, M. 2015. Effect of sowing density and nitrogen top-dress fertilization on growth and yield of Chia (*Salvia hispanica* L.) in a Mediterranean environment. *Italian Journal of Agronomy* 10: 163–166.

Borneo, R., Aguirre, A. and León, A.E. 2010. Chia (*Salvia hispanica* L) gel can be used as egg or oil replacer in cake formulations. *Journal of the Academy of Nutrition and Dietetics* 110: 946–949.

Bressani, R. 1993. Grain quality of common beans. *Food Reviews International* 9(2): 237–297.

Bressani, R., Elias, L.G. and Navarrete, D.A. 1961. Nutritive value of Central American beans. IV. The essential amino acid content of samples of black beans, red beans, rice beans, and cowpeas of Guatemala. *Journal of Food Science* 26(5): 457–555.

Broughton, W.J., Hernández, G., Blair, M., Beebe, S., Gepts, P.and Vanderleyden, J. 2003. Beans (*Phaseolus* spp.) – model food legumes. *Plant and Soil* 252: 55–128.

Buruchara, R., Chirwa, R., Sperling, L., Mukankusi, C., Rubyogo, J.C., Muthoni, R. and Abang, M.M. 2011. Development and delivery of bean varieties in Africa: The Pan-Africa Bean Research Alliance (PABRA) model. *African Crop Science Journal* 19(4): 227–245.

Cahill, J. 2003. Ethnobotany of chia, *Salvia hispanica* L. (Lamiaceae). *Economic Botany* 57: 604–618.

Calderon, E., Velasquez, L. and Bressani, R. 1992. Comparative-study of the chemical-composition and nutritive-value of scarlet beans (*Phaseolus-Coccineus*) and common beans (*Phaseolus-Vulgaris*). *Archivos Latinoamericanos de Nutricion* 42(1): 64–71.

Camara, A.K.F.I., Vidal, V.A.S., Santos, M., Bernardinelli, O.D., Sabadini, E. and Pollonio, M.A.R. 2020. Reducing phosphate in emulsified meat products by adding chia (*Salvia hispanica* L.) mucilage in powder or gel format: A clean label technological strategy. *Meat Science* 163: 108085.

Campos, B.E., Ruivo, T.D., da Silva Scapim, M.R., Madrona, G.S. and Bergamasco, R.C. 2016. Optimization of the mucilage extraction process from chia seeds and application in ice cream as a stabilizer and emulsifier. *Lebensmittel-Wissenschaft und Technologie [Food Science and Technology]* 65: 874–883.

Cardenas-Hernandez, A., Beta, T., Loarca-Pina, G., Castano-Tostado, E., Nieto-Barrera, J.O. and Mendoza S. 2016. Improved functional properties of pasta: enrichment with amaranth seed flour and dried amaranth leaves. *Journal of Cereal Science* 72: 84–90.

Carvalho, B.L., Patto-Ramalho, M.A., Cunha-Vieira, I. and Barbosa-Abreu, A. F. 2017. New strategy for evaluating grain cooking quality of progenies in dry bean breeding programs. *Crop Breeding and Applied Biotechnology* 17: 115–123.

Caselato-Sousa, V.M. and Amaya-Farfan, J. 2012. State of knowledge on amaranth grain: A comprehensive review. *Journal of Food Science* 77(4): 93–104.

Castillo, J. 2015. *Quién nos va a alimentar en el 2050?* Foro Económico Mundial [World Economic Forum] (weforum.org).

Chacón-Sánchez, M.I., Pickersgill, B. and Debouck, D.G. 2005. Domestication patterns in common bean (*Phaseolus vulgaris* L.) and the origin of the Mesoamerican and Andean cultivated races. *Theoretical Applied Genetics* 110(3): 432–444.

Chávez-Mendoza, C. and Sánchez, E. 2017. Bioactive compounds from Mexican varieties of the common bean (*Phaseolus vulgaris*): Implications for health. *Molecules* 22(8): 1360.

Chillo, J., Laverse, P., Falcone, M. and Del Nobile A. 2008. Quality of spaghetti in base amaranthus wholemeal flour added with quinoa, broad bean and chick pea. *Journal of Food Engineering* 84(1): 101–107.

Chirinos, R., Zuloeta, G., Pedreschi, R., Mignolet, E., Larondelle, Y. and Campos, D. 2013. Sacha inchi (*Plukenetia volubilis*): A seed source of polyunsaturated fatty acids, tocopherols, phytosterols, phenolic compounds and antioxidant capacity. *Food Chemistry* 141: 1732–1739.

Chirinos-Arias, M.C. 2015. Andean Lupin (*Lupinus mutabilis* Sweet) a plant with nutraceutical and medicinal potential Tarwi (*Lupinus mutabilis* Sweet) una planta con potencial nutritivo y medicinal. *Revista Bio Ciencias* 3 (3): 163–172.

Constantini, L., Lukšič, L., Molinari, R., Kreft, I., Bonafaccia, G., Manzi, L. and Merendino, N. 2014. Development of gluten-free bread using tartary buckwheat and chia flour rich in flavonoids and omega-3 fatty acids as ingredients. *Food Chemistry* 165: 232–240.

Cota-Gastelum, A.G., Salazar-Garcia, M.G., Espinoza-Lopez, A., Perez-Perez, L.M., Cinco-Moroyoqui, F.J., Martinez-Cruz, O., Wong-Corral, F.J. and Del-Toro-Sanchez, C.L. 2019. Characterization of pasta with the addition of Cicer arietinum and Salvia hispanica flours on quality and antioxidant parameters. *Italian Journal of Food Science* 31: 626–643.

Derewiaka, D., Stepnowska, N., Brys, J., Ziarno, M., Ciecierska, M. and Kowalska, J. 2019. Chia seed oil as an additive to yogurt. *Grasa y Aceites* 70: e302.

Di Sapio, O., Bueno, M., Busilacchi, H., Quiroga, M. and Severin, C. 2012. Caracterización morfoanatómica de hoja, tallo, fruto y semilla de *Salvia hispánica* L. (Lamiaceae). *Latin American and Caribbean Bulletin of Medicinal and Aromatic Plants* 11: 249–268.

Dinssa, F.F., Hanson, P., Dubois, T., Tenkouano, A., Stoilova, T., Hughes, J. and Keating, J.D.H. 2016. AVRDC—the World Vegetable Center's women-oriented improvement and development strategy for traditional African vegetables in sub-Saharan Africa. *European Journal of Horticultural Science* 81: 91–105.

Dizyee, K., Baker, D., Herrero, M., Burrow, H., McMillan, L., Sila, D.N. and Rich, K.M. 2020. The promotion of amaranth value chains for livelihood enhancement in East Africa: A systems modelling approach. *African Journal of Agricultural and Resource Economics* – AFJARE 15: 81–94.

EEA (European Environment Agency). 2019. Soil, land and climate change. www.eea.eur opa.eu/signals/signals-2019-content-list/articles/soil-land-and-climate-change.

FAO (Food and Agriculture Organization). 2000.Website for Agrobiodiversity: www.fao. org/biodiversity/index.asp?lang=en.

FAO (Food and Agriculture Organization). 2010–2019. *Worldwide*. Rome: FAO. www.stati sta.com/statistics/486442/global-quinoa-production/.

FAO (Food and Agriculture Organization). 2016. *Save and Grow in Practice: Maize, Rice and Wheat*. Rome: FAO.

FAO (Food and Agriculture Organization). 2017. *The Future of Food and Agriculture – Trends and Challenges*. Rome: FAO.

FAO (Food and Agriculture Organization). 2020. *The Best Thing about Fruits and Vegetables? Their Diversity! Five Lesser Known but Surprisingly Nutritious Fruits and Vegetables.* www.fao.org/fao-stories/article/en/c/1364251.

Fiorda, F.A., Soares Jr., M.S., da Silva, F.A., Grosmann, M.V. and Souto, L.R. 2013. Microstructure, texture and colour of gluten-free pasta made with amaranth flour, cassava starch and cassava bagasse. *Food Science and Technology* 54(1): 132–138.

Gallego-Villa, D.Y., Russo, L., Kerbab, K., Landi, M. and Rastrelli, L. 2014. Chemical and nutritional characterization of *Chenopodium pallidicaule* (cañihua) and *Chenopodium quinoa* (quinoa) seeds. *Emirates Journal of Food and Agriculture* 26(7): 609.

Gallo, L.R.D., Botelho, R.B.A., Ginani, V.C., de Oliveira, L.D., Riquette, R.F.R. and Leandro, E.D. 2020. Chia (*Salvia hispanica* l.) gel as egg replacer in chocolate cakes: Applicability and microbial and sensory qualities after storage. *Journal of Culinary Science and Technology* 18: 29–39.

Grancieri, M., Martino, H.S.D., and Gonzalez de Mejia, E. 2019. Chiaseed (*Salvia hispanica* L.) as a source of proteins and bioactive pep-tides with health benefits: A review. *Institute of Food Technologists* 18(2): 480–.499.

Gravé, G., Mouloungui1, Z., Poujaud, F., Cerny, M., Pauthe, C., Koumba, I.S: Diakaridja, N. and Merah, O. 2019. Accumulation during fruit development of components of interest in seed of Chia (*Salvia hispanica* L.) cultivar Oruro© released in France. *Oilseeds and Fats Crops and Lipids* 26(50): 1–7.

Grimes, S.J., Capezzone, F., Nkebiwe, P.M. and Grae -Hönninger, S. 2020. Characterization and evaluation of *Salvia hispanica* L. and Salvia columbariae Benth. Varieties for their cultivation in Southwestern Germany. *Agronomy* 10: 2012.

Guardanielli, L.M., Salinas, M.V. and Puppo, M.C. 2021. Quality of wheat breads enriched with flour from germinated amaranth seeds. *Food Science and Technology International*. https://doi.org/10.1177/10820132211016577.

Guillén-Sánchez, J., Mori-Arismendi. S. and Paucar-Menacho, L.M. 2014. Características y propiedades funcionales del maíz morado (*Zea mays* L.) var. subnigroviolaceo Characteristics and functional properties of purple corn (*Zea mays* L.) var. *Subnigroviolaceo*. *Scientia Agropecuaria* 5: 211–217.

Gutierrez, A., Infantes, M., Pascual, G. and Zamora, J. 2016. Evaluación de los factores en el desamargado de tarwi (*Lupinus mutabilis* Sweet). *Agroindustrial Science* 6: 145–149.

Hunziker, A. T. and Planchuelo A. M. 1971. Sobre un nuevo hallazgo de *Amaranthus caudatu* en tumbas de Argentina. *Kurtziana* 6: 63–67.

Intiquilla-Quispe, A., Flores-Fernández, C., Jiménez Aliaga, K. and Iris-Zavaleta, A. 2018. Capítulo 4: Potencial biotecnológico de las semillas de tarwi. In Lupinus mutabilis (*Tarwi) Leguminosa andina con gran potencial industrial*, edited by Iris Zavaleta, 89–121. Lima: Fondo Editorial de la Universidad Nacional Mayor de San Marcos.

Jacobsen, S.E. 2003. The worldwide potential for quinoa (*Chenopodium quinoa* Willd.). *Food Reviews International* 19: 167–177.

Jacobsen, S.E. and Mujica, A. 2003. The genetic resources of Andean grain amaranths (*Amaranthus caudatus* L, *A. cruentus* L. and *A. hipochondriacus* L) in America. *Plant Genetic Resources Newsletter* 133: 41–44.

Jacobsen, S.E and Mujica, A. 2006. El tarwi (*Lupinus mutabilis* Sweet.) y sus parientes silvestres. 25. In *Botánica Económica de los Andes Centrales*, edited by R. Moraes, B. Øllgaard, L. P. Kvist, F. Borchsenius and H. Balslev, 458–482. La Paz: Universidad Mayor de San Andrés.

Kaplan, L. 1965. Archeology and domestication in American Phaseolus (beans). *Economic Botany* 19: 358–368.

Keller, B., Ariza-Suarez, D., de la Hoz, J., Aparicio, J.S., Portilla-Benavides, A.E., Buendia, H.F., Mayor, V.M., Studer, B. and Raatz, B. 2020. Genomic prediction of agronomic traits in common Bean (*Phaseolus vulgaris* L.) under environmental stress. *Frontiers in Plant Science* 11: 1001.

Kistler, L., Maezumi, S.Y., de Souza, J.G., Przelomska, N.A.S., Costa, F.M., Oliver Smith, Hope Loiselle, Ramos-Madrigal, J., Wales, N., Ribeiro, E.R., Morrison, R.R., Grimaldo, C., Prous, A.P., Arriaza, B., Gilbert, M.T.P., Freitas, F.O. and Allaby, R.G. 2018. Multiproxy evidence highlights a complex evolutionary legacy of maize. *South America Science* 362(6420): 1306–1313.

Kowalski, S., Mikulec, A. and Pustkowiak, H. 2020. Sensory assessment and physico-chemical properties of wheat bread supplemented with chia seeds. *Polish Journal of Food and Nutrition Sciences* 70: 387–397.

Kulczynski, B., Kobus-Cisowska, J., Taczanowski, M. and Kmiecik, D. and Gramza-Michałowska, A. 2019. The chemical composition and nutritional value of chia seeds—Current state of knowledge. *Nutrients* 11: 1242.

Kumar, B., Smita, K., Cumbal, L. and Debut, A. 2017. Sacha inchi (Plukenetia volubilis L.) shell biomass for synthesis of silver nanocatalyst. *Journal of Saudi Chemical Society* 21(1): 293–298.

Lao, F., Sigurdson, G.T. and Giust, M. 2017. Health Benefits of Purple Corn (*Zea mays* L.) phenolic compounds. *Comprehensive Reviews in Food Science and Food Safety* 16(2): 234–236. DOI.org/10.1111/1541-4337.12249.

Li, S., Zhang, B., Li, C., Fu, X. and Huang, Q. 2020. Pickering emulsion gel stabilized by octenylsuccinate quinoa starch granuleas lutein carrier: Role of the gel network. *Food Chemistry* 305: 125476.

Lopez, M.L., Capparelli, A. and Nielsen A. 2011. Traditional post-harvest processing to make quinoa grains (*Chenopodium quinoa* Willd) apt for consumption in Northern Lipez (Potosı´, Bolivia): Ethnoarchaeological and archaeobotanical analyses. *Archaeological and Anthropological Sciences* 3: 49–70.

López-Mejía, O.A., López-Malo, A. and Palou, E. 2014. Antioxidant capacity of extracts from amaranth (*Amaranthus hypochondriacus* L.) seeds or leaves. *Industrial Crops and Products* 53: 55–59.

Lucas-Gonzalez, R., Roldan-Verdu, A., Sayas-Barbera, E., Fernandez-Lopez, J., Perez-Alvarez, J.A. and Viuda-Martos, M. 2020. Assessment of emulsion gels formulated with chestnut (*Castanea sativa* M.) flour and chia (*Salvia hispanica* L) oil as partial fat replacers in pork burger formulation. *Journal of the Science of Food and Agriculture* 100: 1265–1273.

MAPA (Ministerio de Agricultura, Pesca y Alimentación). 2020. *Encuesta sobre superficies y rendimientos de cultivos*. www.mapa.gob.es/es/estadistica/temas/estadisticas-agrar ias/totalespanayccaa2020_tcm30-553610.pdf.

Martinez-Lopez, A., Millan-Linares, M.C., Rodriguez-Martin N.M., Millan, F. and Montserrat-de la Paz, S. 2020. Nutracetical value of kiwicha (*Amaranthus caudatus* L.). *Journal of Functional Foods* 65: 103735.

Mascaraque, M. 2018. *Ancient Grains: From Traditional Staple Food to Superfood*. London: Euromonitor International. https://blog.euromonitor.com/ancient-grains-from-traditional-staple-food-to-superfood/.

Morales, P., Rodrigo, V., González M., González, E., Tapia O., Sanhueza, C. and Valenzuela, B. 2012. Nuevas fuentes dietarias de ácido alfa-linolénico: una visión crítica. *Revista Chilena de Nutrición* 39(3): 79–87.

Mujica, A. 1992. *Granos y leguminosas andinas: cultivos marginados*. Roma: Organización de la Naciones Unidas para la Agricultura y la Alimentación FAO.

Mukankusi, C., Raatz, B., Nkalubo, S., Berhanu, F., Binagwa, P., Kilango, M., Williams, M., Enid, K., Chirwa, R. and Beebe, S. 2018. Genomics, genetics and breeding of common bean in Africa: A review of tropical legume project. *Plant Breeding* 138(4): 401–414.

Muñoz, L.A., Cobos, A., Diaz, O. and Aguilera, J.M. 2013. Chia seed (*Salvia hispanica*): An ancient grain and a new functional food. *Food Reviews International* 29: 394–408.

Muñoz-Gonzalez, I., Merino-Alvarez, E., Salvador, M., Pintado, T., Ruiz-Capillas, C., Jimenez-Colmenero, F. and Herrero, AM. 2019. Chia (*Salvia hispanica* l.) a promising alternative for conventional and gelled emulsions: Technological and lipid structural characteristics. *Gels* 5: 19.

Ochieng, J., Schreinemachers, P., Ogada, M., Dinssa, F.F., Barnos, W. and Mndiga, H. 2019. Adoption of improved amaranth varieties and good agricultural practices in East Africa. *Land Use Policy* 83: 187–194.

OECD/FAO (Organisation for Economic Co-operation and Development/Food and Agriculture Organization). 2019. *Latin American Agriculture: Prospects and Challenges*, OECD-FAO Agricultural Outlook 2019-2028. Paris: OECD Publishing.

Oliveira, M.R., Novack, M.E., Santos, C.P., Kubota, E. and Rosa, C.S. 2015. Evaluation of replacing wheat flour with chia flour (*Salvia hispánica* L.) in pasta. *Semina: Ciencias Agrárias* 36: 2545–2553.

Orona-Tamayo, D. and Paredes-López, O. 2017. Amaranth Part 1—*Sustainable crop for the 21st century: Food properties and nutraceuticals for improving human health. Sustainable Protein Sources,* 239–256. Cambridge, MA: Academic Press.

Orona-Tamayo, D., Valverde, M.E. and Paredes-López, O. 2019. Bioactive peptides from selected Latin America crops – A nutraceutical and molecular approach. *Critical Reviews in Food Science and Nutrition* 59(12): 1949.

Orozco, G., Durán, N., González, D., Zarazúa, P., Ramírez, G. and Mena, S. 2014. Proyecciones de cambio climático y potencial productivo para Salvia hispanica L. en las zonas agrícolas de México. *Revista Mexicana de Ciencias Agrícolas* 10: 1831–1842.

Panorama Alimentario 2020, SIAP. www.inforural.com.mx/wp-content/uploads/2020/11/Atlas-Agroalimentario-2020.pdf.

Park, S.J. Sharma, A. and Lee, H.J. 2020. A review of recent studies on the antioxidant activities of a third-millennium food: Amaranthus spp. *Antioxidants* 9: 1236.

Pastor-Pazmiño, C., Chonheiro, L. and Wahren, J. 2017. *Agriculturas alternativas en latinoamérica. Tipología, alcances y viabilidad para la transformación social-ecológica.* Mexico: Fundación Friedrich Ebert.

Pelaez, P., Orona-Tamayo, D., Montes-Hernandez, S., Valverde, M.E., Paredes-Lopez, O.and Cibrian-Jaramillo, A. 2019. Comparative transcriptome analysis of cultivated and wild seeds of *Salvia hispanica* (chia). *Scientific Reports* 9: 9761.

Pellegrini, N. and Agostoni, C. Nutritional aspects of gluten-free products. *Journal of the Science of Food and Agriculture* 95: 2380–2385.

Pintado, T., Ruiz-Capillas, C., Jimenez-Colmenero, F.and Herrero, A.M. 2020. Impact of culinary procedures on nutritional and technological properties of reduced-fat longanizas formulated with chia (*Salvia hispanica* l.) or oat (*Avena sativa* l.) emulsion gel. *Foods* 9: 1847.

Polania, J.A.; Poschenrieder, C.; Beebe, S.and Rao, I.M. 2016. Effective use of water and increased dry matter partitioned to grain contribute to yield of common bean improved for drought resistance. *Frontiers in Plant Science* 7: article 660.

Popenoe, H., King, S.R., Leon, J. and Kalinowski, L.S. 1989. *Lost Crops of the Incas: Little-Known Plants of the Andes with Promise for Worldwide Cultivation.* Washington, DC: National Research Council.

Quintal-Bojórquez, N.D.C., Carrillo-Cocom, L.M., Hernández-Álvarez, A.J. and Segura-Campos, MR. 2021. Anticancer activity of protein fractions from chia (*Salvia hispanica* L.). *Journal Food Science* 86: 2861–2871.

Repo-Carrasco-Valencia, R. 2011. *Andean Indigenous Food Crops: Nutritional Value and Bioactive Compounds.* University of Turku, Turku Finlad ISBN 978-951-29-4605-1.

Repo-Carrasco-Valencia, R., Espinoza, C. and Jacobsen, S-E. 2003. Nutritional value and use of the Andean crops quinoa (*Chenopodium quinoa*) and Kañiwa (*Chenopodium pallidicaule*). *Food Reviews International* 19(1–2): 179–189.

Repo-Carrasco-Valencia, R., Acevedo de La Cruz, A., Icochea Alvarez, J.C. and Kallio, H. 2009. Chemical and functional characterization of kañiwa (*Chenopodium pallidicaule*) grain, extrudate and bran. *Plant Foods for Human Nutrition* 64(2): 94–101.

Resta, D., Boschin, G., D'Agostina, A. and Arnoldi, A. 2008. Evaluation of total quinolizidine alkaloids content in lupin flours, lupin-based ingredients, and foods. *Molecular Nutrition and Food Research* 52(4): 490–495.

Reyes-Moreno, C. and Paredes-Lopez, O. 1993. Hard-to-cook phenomenon in common beans — A review. *Critical Reviews in Food Science and Nutrition* 33(3): 227–286.

Ribeiro, N.D., Rodrigues, J. de A., Prigol, M., Nogueira, C.W., Storck, L., Gruhn, E.M. 2014. Evaluation of special grains bean lines for grain yield, cooking time and mineral concentration. *Crop Breeding and Applied Biotechnology* 14(1): 15–22.

Ribes, S., Pena, N., Fuentes, A., Talens, P. and Barat, J.M. 2021. Chia (*Salvia hispanica* L.) seed mucilage as a fat replacer in yogurts: Effect on their nutritional, technological, and sensory properties. *Journal of Dairy Science* 104: 2822–2833.

Rojas, W. 2003. Multivariate analysis of genetic diversity of Bolivian quinoa germplasm. *Food Reviews International* 19: 9–23.

Rosell, C.M., Cortez, G. and Repo-Carrasco R. 2009. Breadmaking use of Andean crops quinoa, kañiwa, kiwicha, and tarwi. *Cereal Chemistry* 86(4): 386–392.

Ruiz, M.A. and Sotelo, A. 2001. Chemical composition, nutritive value, and toxicology evaluation of Mexican wild lupins. *Journal of Agricultural and Food Chemistry* 49: 5336–5339.

Santiago-Lopez, L., Hernandez-Mendoza, A., Vallejo-Cordoba, B., Mata-Haro V. and Gonzalez-Cordova, A. F. 2016. Food-derived immunomodulatory peptides. *Journal of the Science of Food and Agriculture* 96 (11): 3631–3641.

Sauer, J.D. 1967. The grain amaranths and their relatives: A revised taxonomic and geographic survey. *Annals of the Missouri Botanical Garden* 54(2): 103–137.

Sauer, J.D. 1976. Grain amaranths Amaranthus spp.(Amaranthaceae). In *Evolution of Crop Plants*, edited by N.W. Simmonds, chap. 2: 4–6. London and New York: Longman.

Saunders, R.M. and Becker, R. 1984. Amaranthus: A potential food and feed resource. In *Advances in Cereal Science and Technology*, edited by Y. Pomeranz, 6, 357–396. St. Paul, MN: American Association of Cereal Chemists

Schiermeier, Q. 2019. Eat less meat: UN climate-change report calls for change to human diet. *Nature* 572: 291–292.

Schoeneberger, H., Gross, R., Cremer, H. and Elmadfa, I. 1982. Composition and protein quality of Lupinus mutabilis. *The Journal of Nutrition* 112: 70–76.

Sharma, V. and Mogra, R. 2019. A comprehensive review on chia seeds (*Salvia hispanica* L.). *International Journal of Chemical Studies* 7: 57–61.

Silva, H., Arriagada, C., Campos-Saez, S., Baginsky, C., Castellaro-Galdames, G.and Morales-Salinas, L. 2018. Effect of sowing date and wáter availability on growth of plants of chia (*Salvia hispanica* L) established in Chile. *PLoS ONE* 13(9): e0203116.

Singh, J.P., Singh, B. and Kaur A. 2021. Bioactive compounds of legume seeds. In *Bioactive Compounds in Underutilized Vegetables and Legumes*, Reference Series in Phytochemistry, edited by H.N. Murphy and K.Y. Pack, 645–665. Cham: Springer.

Singh, M. and Liu, S.X. 2021. Evaluation of amaranth flour processing for noodle making. *Journal of Food Processing and Preservation* 45 : 1–9.

Soriano-García, M., Arias-Olguín, I.I., Montes, J.P.C., Ramírez, D.G.R., Figueroa, J.S.M., Flores-Valverde, E. and Valladares-Rodríguez, M.R. 2018. Nutritional functional value and therapeutic utilization of Amaranth. *Journal of Analytical & Pharmaceutical Research* 7: 596–600.

Sosa, C.A. 2000. Influencia de dos métodos de extracción de un aislado proteico de lupino (*Lupinus mutabilis*) en sus propiedades funcionales, Tesis Magis-tral, Lima: Biblioteca Agrícola Nacional, Universidad Nacional Agraria La Molina.

Suarez-Martinez, S.E., Ferriz-Martínez, R.A., Campos-Vega, R., Elton-Puente, J.E., Carbota, K.T. and García-Gasca, T. 2016. Bean seeds: Leading nutraceutical source for human health [Semillas del frijol: fuente líder de nutracéuticos para la salud humana]. CyTA – *Journal of Food* 14(1): 131–137.

Švec, I. and Hrušková, M. 2015. Hydrated chia seed effect on wheat flour and bread technological quality. *Agricultural Engineering International: CIGR Journal* Special Issue: 259–263.

Tabaxi, I., Zervas, G., Tsiplakou, E., Travlos, I.S., Kakabouki, I. and Tsioros, S. 2016. Chia (*Salvia hispanica*) fodder yield and quality as affected by sowing rates and organic fertilization AU – bilalis, dimitrios. *Communications in Soil Science and Plant Analysis* 47: 1764–1770.

Tapia, M.E. 2014. El largo camino de la quinoa: ¿Quiens escribieron su historia? In Estado del arte de la quinua en el mundo en 2013, 3–10. Santiago de Chile/ Montpellier, France: FAO and CIRAD.

Tapia, M.E. 2015. El tarwi, lupino andino – La Fondazione L'Albero della Vita. Fondo Ítalo Peruano http://fadvamerica.org/wp-content/uploads/2017/04/TARWI-espanol.pdf.

Teoh, S.L., Lai, N.M., Vanichkulpitak, P., Vuksan, V., Ho, H. and Chaiyakunapruk, N. 2018. Clinical evidence on dietary supplementation with chia seed (*Salvia hispanica* L.): A systematic review and meta-analysis. *Nutrition Reviews* 76: 219–242.

Ullah, R., Nadeem, M. and Imran, M. 2017. Omega-3 fatty acids and oxidative stability of ice cream supplemented with olein fraction of chia (*Salvia hispanica* l.) oil. *Lipids in Health and Disease* 16: 34.

United Nations, 2019). *World Population Prospects*. New York: United Nations Population Division. https://population.un.org/wpp/.

Urbina M. 1887. La chía y sus aplicaciones. La Naturaleza. Sociedad Mexicana de Historia *Natural Tomo* 1: 27–36.

Valvas, R.L.M., Gomez-Pando, L. and Pinedo-Taco, R. 2020. Sustainability of the production units of the amaranto crop (*Amaranthus caudatus* L.). *Ecosistemas y Recursos Agropecuarios* 7: e2483.

Vásquez-Osorio, D., Hincapié-Llanos, G.A., Cardona, M., Jaramillo, D.I. and Vélez Acosta, L. 2017. Formulación de una colada empleando harina de Sacha Inchi (*Plukenetia Volubilis* L.) proveniente del proceso de obtención de aceite. *Perspectivas En Nutrición Humana* 19(2): 167–179.

Vietmeyer, N.D. 1989. *Lost Crops of the Incas*. National Research Council. Washington, DC: National Academy Press.

Wang, S., Zhub, F., Kakudac, Y. 2018. Sacha inchi (*Plukenetia volubilis* L.): Nutritional composition, biological activity, and uses. *Food Chemistry* 265: 316–328.

Yeboah, T., Owusu, L., Arhin, A. and Kumi, E. 2014. Fighting poverty from the street: Perspectives of some female informal sector workers on gendered poverty and livelihood portfolios in Southern Ghana. *Journal of Economic and Social Studies* 5(1): 239–267.

Zettel, V. and Hitzmann, B. 2018. Applications of chia (*Salvia hispánica* L.) in food products. *Trends in Food Science and Technology* 80: 43–50.

Chapter 4

Structure and Composition of Latin-American Crops

Barbara Borczak[1], José María Coll Marqués[2],
Octavio Paredes-López[3], and Claudia M. Haros[1]
[1]University of Agriculture, Krakow, Poland
[2]Instituto de Agroquímica y Tecnología de
Alimentos (IATA-CSIC), Valencia, Spain
[3]Centro de Investigación y de Estudios Avanzados (CINVESTAV),
Instituto Politécnico Nal., Unidad Irapuato, Guanajuato, Mexico

CONTENTS

4.1 INTRODUCTION

Latin-American indigenous seeds constitute cereals, pseudo-cereals, oilseeds, and legumes. The selected crops are maize from the Latin-American region

DOI: 10.1201/9781003088424-4

(*Zea mays*), amaranth (*Amaranthus* spp), quinoa (*Chenopodium* spp), kañiwa (*Chenopodium pallidicaule*), chia (*Salvia hipsanica* L.), sacha inchi (*Plukentia volubilis*), as well as legumes such as black turtle bean (*Phaseolus vulgaris*) and tarwi (*Lupinus mutabilis*). It is pertinent to mention that some important additional crops also originated in Latin America and are widely distributed around the world, such as cocoa, sunflower, peanuts, and other type of maize. Below, a short description is shown of the first eight specific crops described above.

Cereals are mainly represented by maize (Table 4.1). The words maize and corn are used indiscriminately to refer to the same crop and seed. Based on the origin of maize (*Teocintle,* Mexico) and the process of its domestication, two different cultivars have been developed: Mesoamerican and Andean (Salvador-Reyes et al., 2021). There are about 260 very well-known races, of which 98 are found in Mesoamerica, while 146 have been found in the Andes (Colombia, Ecuador, Peru, Brazil, Chile, Argentina, and Bolivia) (Boege, 2009; Serratos, 2012). The remaining races are found in some regions of South and Central America.

Zea mays is one of the most important staple crops found in the highlands of the Andean regions of Ecuador, Peru, and Bolivia, between 1,700 and 3,500 m above sea level and is one of the main sources of energy in the diet of the inhabitants (Zambrano et al., 2021). Zambrano et al. (2021) indicated that in Ecuador, there are 29 races of maize, 17 found in the highlands, 6 of which are not well defined (Yánez et al., 2003). In the north of this country, traditional races called *chauchos, huandangos,* and *mishcas* are generally cultivated (soft yellow kernels). In the central zone, preferably floury white races of maize are sown, such as *blanco blandito, chazo,* and *guagal.* In the south, a type of maize called *zhima* (white semi-dent kernel) is mostly grown (Zambrano et al., 2021). In Peru, there are 52 races of maize, 32 are grown in the highlands, and the vast majority corresponds to floury maize. In Bolivia, 40 races were found of which 31 are cultivated in the highlands (Salhuana, 2004a). There are a large number of traditional varieties due to the diverse uses for human consumption. The most important native cultivars in this region are *karapampa, kellu, uchupilla, hualtaco, huillcaparu, pasakalla, chuspillo checchi,* and *kulli chunkula* (Ortiz, 2012). The floury maize is consumed by humans in different forms: as fresh maize (*choclo*), dry toasted kernels (*tostado, cancha*), soups (*chuchuca*), drinks (*chicha, colada, maicena*), breads, and other traditional preparations (Villacrés et al., 2016). The grain is a great source of energy and amino acids due to its starch, protein, and oil content. It also contains eight minerals (K, Fe, Zn, Ca, P, Si, Cu and Mg), nine vitamins (A, B1, B2, B3, B4, B6, B9, C, and E), some phenolic compounds and anthocyanin, in the case of black, red, and blue maize (Prasanna et al., 2020; Zambrano et al., 2021) (Figure 4.1).

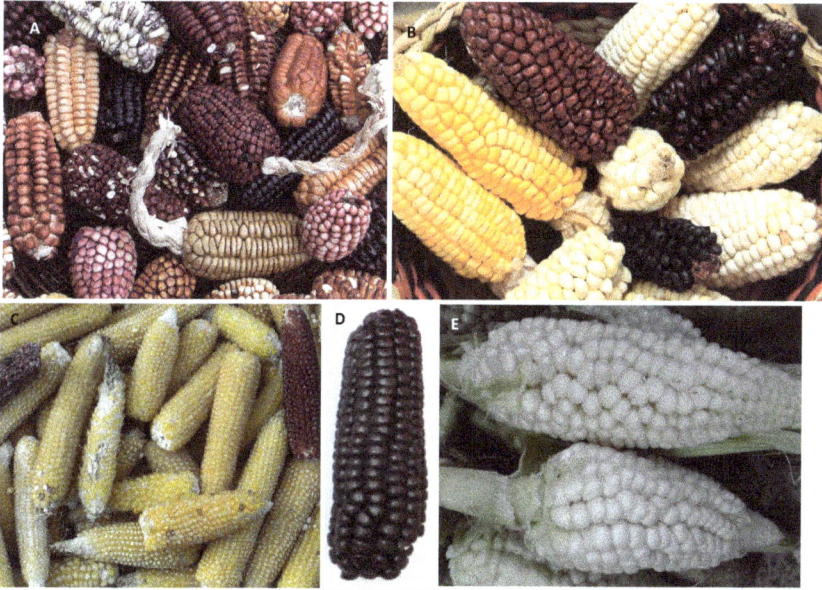

Figure 4.1 Races of maize from the Latin-American region.
Notes: There are five images. The top left image shows many different Andean maize ear color types, in which the grains also vary in color within an ear. The top right photo shows many different Andean maize ear color types (white, yellow, purple, and red). The bottom left image shows pop-corn maize called "maíz pisingallo" of an orange-yellow color. The bottom middle image shows one purple corn. The bottom right image shows two white maize ears of "Choclo" maize. **Source: https://es.wikipedia.org/wiki/Archivo:Peruvian_corn.jpg (Source: www.flickr.com/photos/jennifrog/57252925/, Author: Jenny Mealing);**

Image B: https://es.wikipedia.org/wiki/Archivo:D%C3%ADa_Internacional_de_los_Pueblos_Ind%C3%ADgenas_(7852553230).jpg (www.flickr.com/photos/10021639@N05/7852553230, Author: Cancillería Ecuador);

Image C: https://es.wikipedia.org/wiki/Palomitas_de_ma%C3%ADz#/media/Archivo:PopcornCobs2007.jpg;

Image D: https://es.wikipedia.org/wiki/Zea_mays#/media/Archivo:Maizmorado.png;

Image E. https://es.wikipedia.org/wiki/Variedades_peruanas_de_ma%C3%ADz#/media/Archivo:(Choclo)_from_Quito.JPG.

Pseudo-cereals of the Latin-American regions are mainly composed of amaranth, quinoa, and kañiwa. They were commonly consumed by Aztecs, Maya, Incas, and other pre-Columbian civilizations during pre-Hispanic times. However, after the Spanish conquest, their cultivation likely disappeared due to their association with the religions of indigenous cultures (De la Cruz Torres et al., 2008; Reguera and Haros, 2017; Mangelson et al., 2019; Orona-Tamayo et al., 2019). The genus *Amaranthus* L. contains about 50–70 species and 400 varieties (Joshi et al., 2018; Aderibigbe et al., 2022). The main species of amaranth cultivated for seeds and most commonly used for human consumption are *A. cruentus* in Guatemala, *A. caudatus* in Peru, and *A. hypochondriacus* in Mexico (Bressani, 2003; Reguera and Haros, 2017; Aderibigbe et al., 2022). Quinoa is a part of the *Amaranthaceae* family and closely related with amaranths. Quinoa (*Chenopodium quinoa*) was a fundamental daily food of the ancient South American Andean cultures, and is now grown mainly in the Andean countries of Peru and Bolivia (Reguera and Haros, 2017). Similarly, kañiwa/cañahua is a close relative of quinoa. However, unlike quinoa, kañiwa is less bitter because it contains a smaller amount of saponin (Huamani et al., 2020). As of now, there are about 30 cultivated and wild kañiwa varieties known (Mangelson et al., 2019). Kañiwa is grown in Andean highland plateaus at over 4,000 m.a.s.l. and is very resistant to strict climatic conditions. At present, kañiwa is grown mainly in the Peruvian and Bolivian altiplano by families for their own consumption and used as *kañiwaco*, toasted kañiwa flour. This nutty-tasting flour is mixed with water or milk and eaten as breakfast cereal as a great source of calories and nutritive proteins (Repo-Carrasco-Valencia et al., 2009).

Oleaginous seeds include sacha inchi (*Plukenetia volubilis* L.) and chia (*Salvia hispanica* L.). The first is a plant of the *Euphorbiaceae* family with large seeds (Wang et al., 2018) (Table 4.1) and used to be described as "wild peanut", "mountain peanut", "Inca peanut", or "Inca inchi". It grows in the Amazonian forest (Gutiérrez et al., 2011) in Surinam, Venezuela, Colombia, Ecuador, Peru, Bolivia, and Brazil. The genus *Plukenetia* comprises more than 25 species (Kodahl and Sørensen, 2021), including *P. brachybotrya, P. polyadenia, P. loretensis, and P. huayllabambana*. However, their morphological and physicochemical properties differ from *P. volubilis*. Cultivation of sacha inchi dates from pre-Hispanic times and most probably constituted part of Incans' diets (3,000–5,000 years ago) (Kodahl and Sørensen, 2021). It grows at altitudes ranging from 200 to 1,500 m (Wang et al., 2018).

Seeds of *P. volubilis* have been used in traditional cooking in Peru (*inchacapi, lechona api, pururuca, cutacho, inchi cucho, tamales, turron*, and *chichi*). They can also be consumed in roasted and salted form, in chocolate, pressed as oil, ground to a buttery substance, or as cooked leaves (Kodahl and Sørensen, 2021). At the same time, chia (*Salvia hispanica* L.) is a herbaceous plant belonging to the *Lamiaceae* family (Palma-Rojas et al., 2017; Hernández-Pérez et al., 2021) (Table 4.1). The genus *Salvia* consists of approximately 900 species (Knez Hrnčič et al., 2020). The plants can grow to 1–1.5 m high and the fruits are indehiscent achene. It originates from west Mexico and northern Guatemala (Pelaez et al., 2019), although today it is also cultivated in Australia, Bolivia, Columbia, Peru, Argentina, the USA, and Europe (Palma-Rojas et al., 2017) with Mexico being the world's largest chia producer (Hrnčič et al., 2020). Chia as an oilseed has great economic potential since it is a source of oil with remarkable nutritional and nutraceutical characteristics; thus, chia has a brilliant future for industrial and human purposes (Ayerza, 2009; Orona-Tamayo et al., 2019; Hernández-Pérez et al., 2021).

The legume family (*Leguminosae*) comprises 643 genera (18,000 species) which are grouped into 40 tribes and cultivated in either tropical or temperate climates (Broughton et al., 2003). Legumes native to Latin America that we are going to deal with in this chapter are *Lupinus mutabilis* Sweet and *Phaseolus vulgaris* also known as black turtle bean, among many others. *L. mutabilis* is a plant variously known as *ullush, talwish, tauri, tarwi, chocho, lupino* or *ccquella* (Chirinos et al., 2015). It belongs to genus *Lupinus* (Table 4.1) with 85–100 distinct species and more than 500 specie names (Atchison et al., 2016). It is a relative of *Lupinus albus, Lupinus luteus*, and *Lupinus angustifolius*, three crucial pulses collected worldwide (Córdova-Ramos et al., 2020). This legume grows in the Andes at 2,000 to 3,800 m.a.s.l and is distributed from Colombia to the north of Argentina (Tapia, 2015, Chirinos et al., 2020). The seeds were found in the Nazca cultural remains in Peru as possible evidence of their consumption by the ancient Peruvian cultures Tiahuanaco and Chavin (Castañeda, 1988; FAO, 2010).

Tarwi has a considerable genetic variability within the Andean region, with different morphological types reported (Gross, 1982) as follows: (1) small plants of about 60 cm high, no branches, presented in the region of Potosí (Bolivia) and in the regions above 3,500 m.a.s.l.); (2) branched and tall plants (in the Andean valleys of Bolivia and Southern Peru), as well as (3) highly branched plants, taller than 1.8 m and growing in Colombia, Ecuador, and northern Peru (Neves Martins et al., 2016). Use of tarwi might be limited because of high alkaloid content (Jacobsen and Mujica, 2006), which results in a bitter taste which is simultaneously toxic. However, with the soaking and washing of seeds in water, the bitter compounds are removed.

The tribe Phaseoleae containing, among others, common beans (*P. vulgaris*) is among the most important economic group of legumes in the world. Basal species (*P. microcarpus, P. vulgaris*) are part of a complex species that includes *P. acutifolius, P. coccineus*, and *P. polyanthus* (Broughton et al., 2003). The black bean, also known as black Spanish bean, Tampico bean, and Venezuelan bean, has its origin in South America with two centers of domestication (Gepts and

TABLE 4.1   BOTANICAL CLASSIFICATION OF LATIN-AMERICAN CROPS AND WHEAT FOR COMPARISON PURPOSES

	CEREALS		PSEUDO-CEREALS	
Name	**WHEAT**	**CORN**	**AMARANTH**	**QUINOA**
Class	*Monocotyledoneae*		*Dicotyledoneae*	
Owpder	*Poales*		*Caryophyllales*	
Family	*Poaceae*		*Amaranthaceae*	*Chenopodiaceae*
Subfamily	*Pooide*	*Bambusoideae*	*Amaranthoideae*	*Chenopodioideae*
Tribe	*Triticeae*	*Andropogoneae*	*Amarantheae*	*Chenopodieae/ Chenopodiea*
Genus	*Triticum*	*Zea*	*Amaranthus*	*Chemopodium*
Species	*T. aestivum, T. aestivum spp. spelta, T. durum*	*Z. mays*	*A. caudatus, A. cruentus, A. hypochondriacus*	*Ch. quinoa Willd, Ch. pallidicaule Aellen (kanigua/kanihua/canihua), Ch. nuttalia Safford*

Source: Adapted from Carvajal-Larena (2019), Chirinos-Arias (2015), Moreno et al. (2014), Berghofer and Schöenlechner (2007), Saleem et al. (2016), Kodahl (2020), Kodahl and Sørensen (2021), Coelho and Salas-Mellado (2014), Jamshidi et al. (2019), Knez Hrnčič et al. (2020).

KAÑIWA	LEGUMES			OLEAGINOUS SEEDS	
	BLACK TURTLE BEAN	**TARWI**		**CHIA**	**SACHA INCHI**
	Dicotyledoneae			*Magnoliopsida*	
		Fabales		*Lamiales*	
	Fabaceae or Leguminosae			*Lamiaceae/ Labiate*	*Euphorbiaceae*
		Papilionoideae		*Nepetoideae*	*Acalyphoideae*
	Phaseolae	*Genisteae*		*Mentheae*	*Plukenetieae (Benth) Hutch., subtribe Plukenetiinae Benth.*
Chenopodium pallidicaule Aellen	*Phaseolus*	*Lupinus*		*Salvia*	*Plukentia L.*
	P. acutifolius, P. coccineus, P. Polyanthus, P.vulgaris	*L.mutabilis and four species as possible close relatives (L. ellsworthianus C.P.Sm., L. piurensis C.P.Sm., L. praestabilis C.P.Sm., and L. semperflorens); Mediterranean species (Lupinus albus L. and Lupinus luteus L. and Lupinus angustifolius L.)*		*S. hispanica*	*ca. 25 species P. volubilis, P. brachybotrya, P. polyadenia, P. loretensis, and P. huayllabambana.*

Dpbouk, 1991). The small seed varieties originated from a small-seeded wild type (*P.vulgaris var. aborigineus* (Burk.) Baudet) in Central America, while the large-seeded varieties are from another type (*P. vulgaris var. aborigineus*) in the Andean region of South America. Black turtle bean is traditionally used in Mexican, Caribbean, and South American dishes (Priya and Manickavasagan, 2020) in the form of burrito, feijoada (stew of beans, beef, and pork), gallo pinto or pabellón criollo (rice, shredded beef, and stewed black beans) (Lim, 2012). The Andean mountain range in South America is a significant location for crop domestication intended for human nutrition, even though growing conditions are demanding with high and varied altitudes (1,500–4,200 m.a.s.l.) as well as accompanied by changing climates. It is pertinent to mention that indigenous people of the Andes domesticated several dozens of separate crop species (Repo-Carrasco-Valencia, 2020; Hernández-Pérez et al., 2020; Hernández-Pérez et al., 2021).

The botanical classification and physicochemical characteristic of Latin-American crops are summarized in Tables 4.1, 4.2, and 4.3.

4.2 GROSS STRUCTURAL FEATURES

Latin-American seeds are classified in this report into different plant crops (cereals, pseudo-cereals, legumes, and oilseeds) and they differ between each other in biological quality proteins, amino acids, polyunsaturated fatty acids, dietary fiber, and starch content, among other components and traits.

The main parts of maize in general, include: pericarp, aleurone, endosperm, and germ (Salvador-Reyes et al., 2021). The outer layer of kernels may be yellow (*Chullpi*), black with white spots (*Piscorunto*), pale yellow (*Giant Cuzco*), red–orange with white stripes (*Sacsa*), and an intense and homogeneous purple (*Purple*). Differences in color are determined by climatic conditions and genetic factors. High altitudes together with lower temperatures induce natural pigment development, such as anthocyanins (determining blue, purple, and black colorations). On the other hand, warm climates promote the concentration of carotenoids (leading to red, yellow, and orange colorations) (Figures 4.1–4.2). *Piscorunto, Giant Cuzco, Sacsa*, and *Purple* varieties are classified as floury maizes, because their endosperm is composed exclusively of soft starch. At the same time, *Chullpi* is considered a sweet type of maize since it presents part of its endosperm in a vitreous state (Paliwal et al., 2000). There is also a higher ratio of starch:protein presented in their endosperm (11:1) compared to vitreous (6:1) (Wang and Wang, 2019).

In amaranth, quinoa, and kañiwa the bran fraction is higher than traditional cereals (Bressani, 2003; Reguera and Haros, 2017). They also present greater content of protein and fat compared with cereals, whereas the starch content is lower (Figure 4.3, Table 4.3). Oleaginous plants occur in the form of fruits (sacha inchi) or grain-like seeds (chia). The kernels are only covered with a thin layer in the form of shell and contain a higher amount of protein than cereals and pseudo-cereals together with fat and dietary fiber being the greatest among all Latin-American crops (Figure 4.4, Table 4.3). The main parts of legumes (tarwi and black turtle bean) are seed coat, radicle, and cotyledon with the latest having the greatest share in the whole seed (Figure 4.5). The protein content of tarwi is remarkable, being one of the greatest among all Andean crops.

4.3   PHYSICAL PROPERTIES

Knowledge about engineering features, mainly the physical properties of crops, including cereals, pseudo-cereals, legumes and oilseeds is important for engineers and technicians involved in several harvest and postharvest practices. Physical characteristics are crucial in investigating the maintenance of crops in handling operations (Mohsenin, 1986; Suleiman, 2019). Data on physical properties of Latin-American crops, including their morphology and size distribution are essential to proposing equipment and amenities for the following processes: harvesting, handling, cleaning, conveying, grading, separation, drying, storing, and processing of crops (Sacilik et al., 2003; Vilche et al., 2003; Reguera and Haros, 2017). The most important physical and thermal properties of the grain, oilseed, and other agricultural commodities are seed dimension (length, width, and thickness), arithmetic and geometric mean diameter, surface area, volume, sphericity ,and aspect ratio. They also may include the weight of 1,000 grains, bulk, and kernel density, fractional porosity, the static and dynamic coefficient of friction against different materials, as well as the angle of repose, heat capacity, thermal conductivity and diffusivity, and latent heat of vaporization (Mohsenin, 1986; Suleiman, 2019).

For example, bulk density influences the capacity of storage and transport systems, while information about true density is needed for separation and classification equipment. Seed porosity determines the resistance to airflow during the aeration and drying of kernels (Reguera and Haros, 2017). Table 4.2 presents a data sheet of known physical properties of Latin-American crops compared to wheat. These properties vary widely, depending on moisture content, variety, or cultivar in general, year and region of cultivation, as well as temperature of grains, legumes, and oilseeds (Mohsenin, 1986; Vilche et al., 2003; Reguera and Haros, 2017; Suleiman, 2019). Latin-American crops may differ in size and shape (Figures 4.1–4.2), and as expected they show different physical and chemical properties (Tables 4.2–4.3).

TABLE 4.2   PHYSICAL PROPERTIES AND GEOMETRIC DIMENSIONS OF LATIN-AMERICAN CROPS. (WHEAT WAS PRESENTED FOR COMPARISON PURPOSES)

Parameter	Units	Wheat[a] varieties S/D/N***	Maize[b]	*Amaranthus cruentus*[c-d]	*Chenopodium quinoa* Willd[c-e] L/M/S*
Moisture	% dm	15.8/15.8/ 16.4	8.40/14.44/ 12.75/NR/ 12.02 and 11.1	7.7-43.9[c] 9.5-43.6[d]	4.6-25.8[c] 15.0[d]
1000-seed mass	G	46.2/45.9/ 40.0	324.28- 1340.63	0.79[c]/1.2[c'']	2.5-3.1[c]
Seed weight	G	NR	50.06-61.56	NR	NR
Hectoliter weight	kg/ 100L	NR	NR	NR	NR
Specific Volume	m³/kg	NR	NR	0.78-1.10 x 10⁻³[c]	NR
Bulk density	kg/m³	791/789/ 732	NR	840-720[c] 820-867[c']	747-667[c]
True density[a,c,d]/ compacted density	kg/m³	1104/ 1151/1076	NR	1390-1320[c]	928-1188[c]
Porosity	%	0.283/ 0.315/ 0.320	NR	0.40-0.45[c]	0.194-0.438[c]
L*(brightness or white-ness),		NR	66.73/ 57.53/ 76.31/ 52.50/14.72	NR	NR
a*(red-ness and green-ness)		NR	10.98/1.62/ 1.64/17.71/ 2.82	NR	NR
b*(yellow-ness and blue-ness)		NR	39.72/ 11.58/ 23.83/ 21.41/0.59	NR	NR

Chenopodium pallidicaule [f-g]	*Salva hispanica L.* [f-h]	*Plukentia volubilis* [i]	*Lupinus mutabilis* [j-k]	*Phaseolus vulgaris* [l-m]
10.37/11.79[f] 10.37/9.61/9.79/ 10.39 [f'] 10.7/10/.7/10.7/ 7.7/8.4 [f"]	7.0±0.4[h]/ 6.2 ± 0.517	3.3-8.32[i]	9.99 ± 0.14 (8.19–13.80)[k]	11.02[l]; 9.25[l"]
2.5-4.0[g]	2.0-3.5[f']/1.267-1.387[h]	NR	153-320[k]	256.2/ 175.9/ 246.1/ 196.3/ 224.4[l]
NR	NR	NR	0.25±0.06 (0.184-0.302)[k]	NR
NR	NR	0.70-1.22[f']	NR	NR
NR	NR	NR	NR	NR
958-904[g]	729-458[f']/722[g]	NR	NR	2.4-10 [m']
NR	937-1075[g]	NR	NR	NR
NR	22.9-32.8[g]	NR	NR	NR
29.8/30.5/29.8[g]	42/41.9/41.30[g]	NR	61.21±0.10[k"]/ (67.2-87.5)[k]/ 84.59 ± 0.90[j]	69.9 ± 2.5[m]
11.3/11.1/11.9[g]	3.2/3.8/3.8[g]	NR	2.22±0.10[k"]/ (0.3-(-3.4))[k]/ −2.10 ± 0.44[j]	0.6 ± 0[j]
12.9/14.1/14.8[g]	7.4/8.6/8.6[g]	NR	11.47±0.04[k"]/ (22.0-33.3)[k]/ 20.32 ± 1.55[j]	5.9±0[m]

(*continued*)

TABLE 4.2 (CONTINUED)

Parameter	Units	Wheat[a] varieties S/D/N***	Maize[b]	*Amaranthus cruentus*[c-d]	*Chenopodium quinoa* Willd[c-e] L/M/S*
Length	mm	5.46/5.37/ 5.38 6.51–8.66[a']	18.23-20.36	1.35-1.50[d]	2.045/1.889/ 1.691[d]*
Width	mm	2.56/2.47/ 2.62 3.026– 3.952[a']	7.93-17.34	1.22-1.37[d]	2.015/1.885/ 1.689[d]*
Thickness	mm	2.12/2.18/ 2.43 2.25–2.79[a']	5.17-6.13	0.81-0.93[d]	0.930/0.980/ 0.973[d]*
Equivalent diameter	mm	3.09/3.06/ 3.24	NR	1.10-1.24[d]	1.394-1.607[c]
Sphericity		0.57/ 0.57/0.60 57.40– 69.43%[a']	NR	0.81-0.83[d]	0.77-0.80[c]
Volume	mm³	NR	NR	0.55-0.76[d]	NR
Equivalence sphere area	mm²	30.07/ 29.64/33.2	NR	3.26-4.04[d]	NR
Oblate spheroid area	mm²	0	NR	3.59-4.43[d]	NR
Solid of revolution area	mm²	NR	NR	3.60-4.47[d]	NR
Arithmetic mean dia	mm	NR	NR	NR	NR
Geometric mean dia,	mm	3.36–4.24[a']	NR	NR	NR
Surface area S,	mm²	47.92– 51.65[a']	NR	NR	NR
Aspect ratio R	%	NR	NR	NR	NR

Chenopodium pallidicaule [f-g]	*Salva hispanica L.* [f-h]	*Plukentia volubilis* [i]	*Lupinus mutabilis* [j-k]	*Phaseolus vulgaris* [l-m]
1-1.2[f]and 1.14[g] (min.0.13 max.1.45)	1.99 (min.0.26 max.2.49) [f]/ 1.76–2.42[g]	7.49-8.92[i'];15-18[i'']	9.64±0.11 (8.86-10.99)[k]/ 9.98±0.64	10.93/9.64/ 10.60/9.39/ 10.37[l]; 9.1±0.9[r]
0.92[g] (min.0.12max.1.20)	1.21 (min.0.14-1.68) [f]/1.22-1.36[g]	16-20[i']	7.91±0.11 (6.76-9.39)[k]/ 8.39 ± 0.41	7.05/6.21/ 6.94/6.63/ 6.97[l]; 6.0±0.5[r]
0.78[g] (min0.12max.1.08)	1.01 (min.0.12 max.1.25) [f]/ 0.77-0.84[g]	7-8[i']	4.95±0.11 (3.26-5.72)[k]/ 6.02 ± 0.35	4.73/4.22/ 5.02/4.64/ 4.41[l]; 4.5±0.6[r]
NR	1.32-1.39[g]	13.91-20.13[i']	NR NR	NR NR
0.24 (min.0.01 max.0.40)[g]	0.41[f] (min.0.01max.0.70)	NR	NR	NR
0.18 (min.0.29 max.0.80)[g]	0.88 (min.0.13max.0.83) [f]/ 1.19-1.42[g]	NR	NR	NR
NR	NR	NR	NR	NR
NR	NR	NR	NR	NR
NR	NR	NR	NR	NR
0.95[g]	1.40[f]	NR	NR	NR
0.27[g]	0.81[f]/ 1.1-1.69[g]	NR	NR	NR
2.05[g]	5.29[f]/ 3.78-7.01[g]	NR	NR	NR
80.70[g] (min.82.75 max.92.3)	60.56 (min.55.3max.67.2)[f]/ 48.0-81.8[g]	NR	NR	NR

(*continued*)

TABLE 4.2 (CONTINUED)

Parameter	Units	Wheat[a] varieties S/D/N[***]	Maize[b]	*Amaranthus cruentus*[c-d]	*Chenopodium quinoa* Willd[c-e] L/M/S[*]
Coefficient of friction		0.41 ± 0.0[a']	NR	NR	0.211-0.265[c] NR
Ply-wood		0.36/0.33/0.36	NR	NR	0.145-0.240[c]
Galvanized iron[e]/steel/ metal sheet[a]		NR	NR	NR	NR
Concrete		NR	NR	NR	NR
Angle of repose	°	NR	NR	22.7-30.6[c]	18-25[c]
Thermal velocity	m/s	NR	NR	NR	0.6-1.02[c]
Thermal diffusivity	x10⁻³ mm²/s	NR	NR	NR	NR
Electrical conductivity	µS/cm	NR	NR	NR	NR

Notes: (a) Babić et al. (2011); (a') Shafaei and Kamgar (2017).

(b) Data for five varieties of Peruvian maize grown in the Andean region: *Chullpi* (Ch), *Piscorunto*(P), *Giant Cuzco*(GC), *Sacsa* (S) and *Purple* (P) varieties and yellow corn (Y) from Salhuana (2004) and Salvador-Reyes and Clerici (2020).

(c) Abalone et al. (2004); (c') Ogrodowska et al. (2014); (c") Brust et al. (2014).

(d) Abugoch et al. (2009).

(e) Vilche et al. (2003).

(f) Data for two varieties of kaniwa (*Cupi, Ramis*) from Repo-Carrasco-Valencia et al. (2009); (f ') Data for four ecotypes of kaŻiwa from the Agronomical Experimental Station-INIA Salcedo, Puno, Peru (Kello//Wila/ Guinda/Ayara) from Repo-Carrasco-Valencia et al. (2010); (f ") Data for Peruvian ecotypes *Roja/Blanca/Amarilla/Illpa-Inia* from La Rosa et al. (2016).

(g) Data from Suleiman et al. (2019) for kaŻiwa and chia as moisture content increased from 10 to 20% d.b.

(h) Ixtaina et al. (2008); Coelho and Salas-Mellado (2014).

(i) Gutiérrez et al. (2011) and Wanga et al. (2018); (i')Data for 25 of sacha inchi, *Plukenetia volublilis* from Rodrigues et al. (2018); (i") Data from Kodahl and Sørensen (2021).

Chenopodium pallidicaule [f-g]	*Salva hispanica L.* [f-h]	*Plukentia volubilis* [i]	*Lupinus mutabilis* [j-k]	*Phaseolus vulgaris* [l-m]
NR	NR	NR	NR	NR
NR	0.28±0.01 [g]	NR	NR	NR
NR	NR	NR	NR	NR
NR	NR	NR	NR	NR
NR	16-18 [g]	NR	NR	NR
NR	NR	NR	NR	NR
92.0/91.0/91.0 [g]	93.0/105/93.0 [g]	NR	NR	NR
NR	NR	NR	NR	75.9/91.6/ 80.5/82.2/ 80.9 [l]

(j) Data for three bitter Lupinus mutabilis genotypes originating from different regions of Peru (Altagracia from Ancash, Andenes from Cusco, and Yunguyo from Puno) from Córdova-Ramos et al. (2020).

(k) Data for 33 lupin ecotypes from different Peruvian regions from Berru et al. (2021) and from Miano et al. (2015); (k')Neves Martins et al. (2016); (k")Córdova-Ramos et al. (2020); Mohamed and Rayas (1995).

(l) Data for five black bean cultivars (Campeiro/ Esplendor/Grafite/ Supremo/Valente) from Vanier et al. (2019); (l') Data from USDA (2010); Lim (2012); (l") Berrios et al. (1999).

(m) Data for Black beans (*Phaseolus vulgaris*) var. "San Luis" (BB) from de la Rosa-Millán et al. (2019); (m')Data from Audu and Aremu (2015).

*L/M/S: Large>2.0 mm (27.4%)/Medium 1.7-2.0 mm (72.3%)/Small<1.7 mm (0.3%);***S/D/N: Simonida/Dragana/NS40S varieties of *Triticum aestivum*, NR: not reported.

4.4   KERNEL STRUCTURES

As indicated before, structurally, the seeds of Latin-American crops are composed of different parts as they belong to different categories (cereals, pseudo-cereals, legumes, and oleaginous); these crops are shown in Figure 4.2. Wheat is included for comparison purposes.

Maize kernel, as indicated before, is divided into layers: pericarp, aleurone, endosperm, and germ, which reflect in general the structure of common cereals and pseudo-cereals (Figure 4.2). Different Latin-American races of maize differ in shape and size. *Chullpi* and *Piscorunto* kernels present an elongated shape and low thickness (width 7.93–9.54 mm, length 18.28–18.47 mm, and thickness

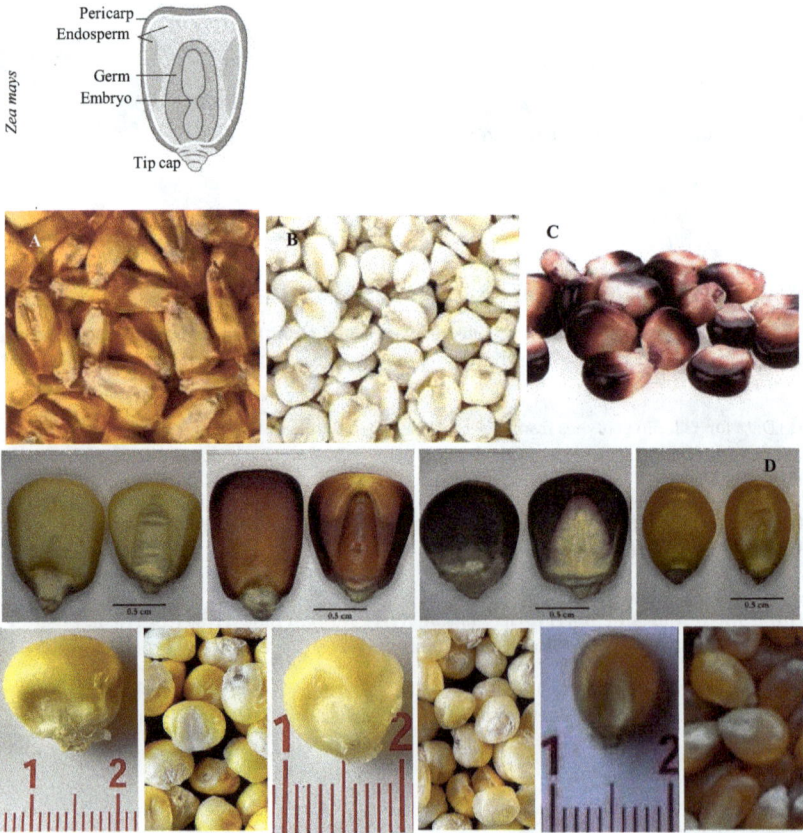

Figure 4.2   A. *Piscorunto,*[a] **B.** *Chullpi,*[a] **C.** *Giant Cuzco,*[a] **D.** *Sacsa,*[a] **E.** *Purple maize,*[a] F. Latin-American maize. **Source: (A). Adapted from www.shutterstock.com/image-photo/peruvian-chulpe-maiz-popcorn-unpopped-676249492.**

Figure 4.2   (continued)

(B). Adapted from www.shutterstock.com/image-photo/giant-white-corn-backgro und-uses-166618307.

(C). Adapted from www.shutterstock.com/image-photo/purple-corn-maize-isola ted-on-white-1446030731.

(D). Authors' own work.

Notes: The figure is composed of four parts, in the following order: structure of the grain, up close images of three types of maize grains (A, B, and C), details of four different Latin-American maize grains (each one from a frontward and backward perspective (D)) and details of three types of Latin-American grains with their sizes (E).

Common maize grain illustration of its structure, seen from a cross-section of the seed. The image shows a tooth-like shape surrounded by a structure called the pericarp. Moreover, the grain shows a tip cap at the bottom. In the upper part of the tip cap there is a vertical oval shape section called germ and inside of it the embryo is found. Finally, in the area between the outside of the germ and the pericarp is the endosperm.

A) Up close chullpi maize seeds with an elongated tooth-like-shape in a yellow color.

B) Up close image of giant white maize seeds with a rounded tooth-like-shape.

C) Up close purple maize seeds with a rounded tooth-like-shape; the top half being purple and the bottom part white.

D) Four different color types of Latin-American maize (beige, caramel, purple, and an orange-yellow color) showing both their frontward and backward views. They have an approximate range of 0.6 cm–1 cm.

E) There are three types of maize. The first one has a ball-like shape, range of 1.2 cm, in a yellow color. The second one has a ball-like shape, range of 1 cm, in a deeper yellow color. The third one has a rounded tooth-like shape, range of 0.7 cm, and in an orange color.

5.17–5.74 mm). Meanwhile, *Giant Cuzco, Sacsa*, and *Purple* have circular shapes and different sizes (width 11.37–17.34 mm, length 12.62–20.36 mm, and thickness 5.18–6.13 mm) (Salvador-Reyes et al., 2021). It has been suggested that the differences may be linked to evolutionary factors (Paliwal et al., 2000) and that the evolved races (*Giant Cuzco, Sacsa*, and *Purple*) present larger grains and more homogeneous shapes than the primitive *Chullpi* and *Piscorunto*, whose kernels are smaller and irregularly shaped. The hectoliter weight (50.06–61.56 kg/100 L) of Latin-American maize was lower in comparison to the values for commercial corn (69 to 75 kg/100 L) (UNALM/MINAGRI, 2014). This lower density might be dependent on the strict environmental conditions, plus genetic factors, of the Andean highlands (i.e., temperature variation, water availability), which might decrease the period of filling the grain, reducing its final weight and the amount of cells accumulating starch and proteins (Salvador-Reyes et al., 2021), giving as a result a lower protein content.

The endosperm, embryo, and seed coat (pericarp) are the fundamental structures of pseudo-cereals (amaranth, quinoa, and kañiwa). Starch and proteins are mainly stored in the endosperm, while oil, some proteins, and minerals occur in embryo. In the pericarp, deposition of mainly cellulose and hemicellulose, together with protein and lignin, takes place (Baltensperger, 2003; Reguera and Haros, 2017).

The color of amaranth's pericarp differs depending on the species and can be cream, yellow, white, pale brown, dark brown, black, red, or pink (Aderibigbe et al., 2022). Its surface is glossy, smooth, and a bit straighter with a lens-like shape. The length of kernels might be between 1.3 and 1.7 mm, width in the range of 0.9 to 1.3 mm, while its weight is usually in the frame of 0.6 to 1.0 mg (Reguera and Haros, 2017). The kernels of amaranth are smaller than cereal grains (wheat or maize), some pseudo-cereals (quinoa), Andean lupins, or oleaginous seeds (sacha inchi) (Figure 4.3–4.5, Table 4.2). Quinoa seeds are yellow,

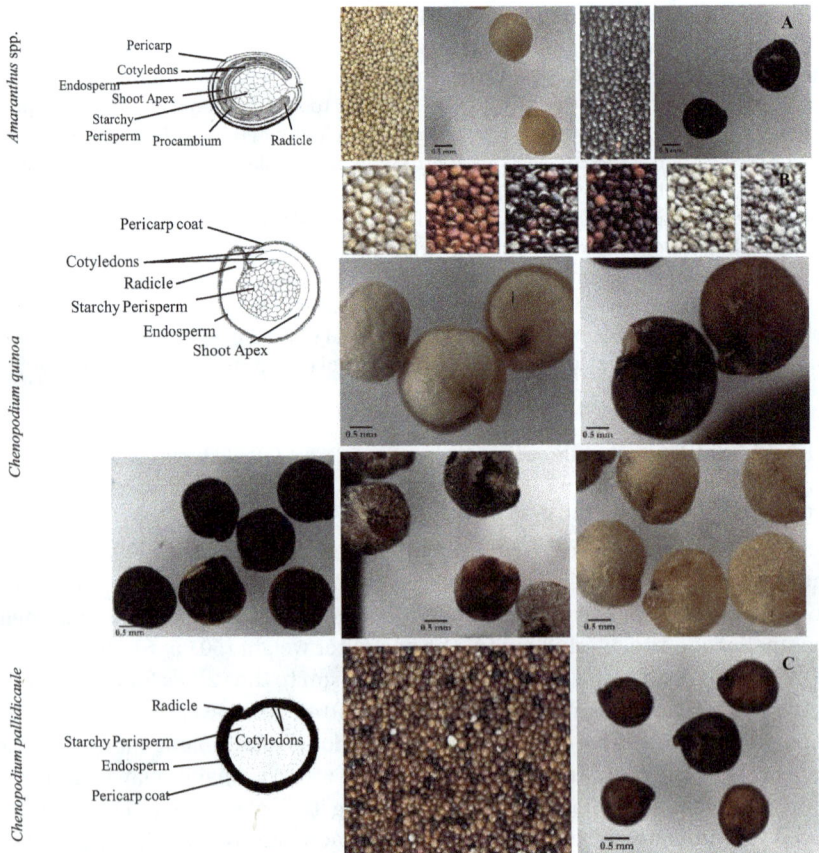

Figure 4.3   Pseudocereals from Latin-America: **A.** Amaranth;[a] **B.** Quinoa;[a] **C.** Cañahua, Cañihua o Kañiwa.[b] **Source: Seed structure adapted from Valcárcel-Yamani and Lannes (2012) and [b]Bruno et al. (2018).**
Images A, B, C: Authors' own work.

Figure 4.3   (continued)

Notes: The figure is composed of three parts, the structure (seen from a cross-section of the seed) and images of amaranth (A), quinoa (B) and kañiwa(C).

A)

Illustration of the structure of the pseudocereal amaranth. It has an oval shape, surrounded by a pericarp, followed by another layer called endosperm, followed by the germ with two cotyledons and an inner starchy perisperm. Next to the structure, there is an upclose image and a detailed image (the rounded shape can be observed and the germ surrounding it) of both white amaranth (approximately 1.1mm) and black amaranth (approximately 0.94-1.3 mm).

B)

Illustration of the structure of the pseudocereal quinoa. It has a round shape, surrounded by a pericarp coat, followed by another layer called endosperm, followed by the germ with two cotyledons and an inner starchy perisperm. Next to the structure, there are five up close images of quinoa and their detailed images: white (approximately 2.5 mm), red (approximately 2.4 mm), black (approximately 1.3 mm), grey (approximately 1.75 mm) and beige (approximately 2 mm). Their rounded shape can be observed and the germ surrounding it.

C)

Illustration of the structure of the pseudocereal kañiwa. It has a round shape, surrounded by a pericarp coat, followed by another layer called endosperm, followed by the germ with two cotyledons and an inner starchy perisperm. Next to the structure, there is an up close image of kañiwa and a detailed image. Their rounded shape can be observed and the germ surrounding it, they have an approximate range of 1-1.25mm.

white, orange, red, brown, or black depending on the crop, especially in the case of wild species; and the pericarp takes on a black tint (Figure 4.3; Saturni et al., 2010; Reguera and Haros, 2017). Quinoa is a small, dry, one-seeded fruit with round-shaped seeds and diameters that vary between 1.0 and 2.6 mm, together with 250–500 seeds per gram (Vilche et al., 2003; Valencia-Chamorro, 2003; Reguera and Haros, 2017). The longitudinal section of the seed is round, because the length and width are roughly equal (Vilche et al., 2003; Reguera and Haros, 2017; Table 4.2).

Kañiwa is not a true cereal equivalent to quinoa. This annual, herbaceous plant is 0.2–0.7 m tall and its seeds are 1.0–1.2 mm long, 0.12–1.2 mm wide, and 0.12–1.08 mm thick (Repo-Carrasco-Valencia et al., 2009; Figure 4.3, Table 4.2). The color of pericarp varies depending on the ecotype and can be yellow, brown, gray, or black (Repo-Carrasco-Valencia et al., 2010; La Rosa et al., 2016).

Sacha inchi is characterized by a star-shaped fruit capsule (3–5 cm). With fruit maturation, the color changes from green to blackish brown (Wang et al., 2018). The fruit capsules contain edible dark brown oval seeds, approximately 1.8×0.8×1.6 cm in size, and the testa is hard and brown, with dark brown markings (Figure 4.4) (Kodahl and Sørensen, 2021). Chia seeds are generally very small. They are about 2 mm long by 1.5 mm wide and less than 1 mm thick, oval in shape and shiny (Figure 4.4). The color changes from black, gray, or black spotted to white (Hernández-Gómez et al., 2008; Knez Hrnčič et al., 2020).

The main parts of *L. mutabilis* are pericarp (seed coat), cotyledon, radicle, hilar rin, and hilar fissure (Figure 4.5; Miano et al., 2015). The broad variation exists for pericarp color, ranging from white to cream to brown and to black (Berru

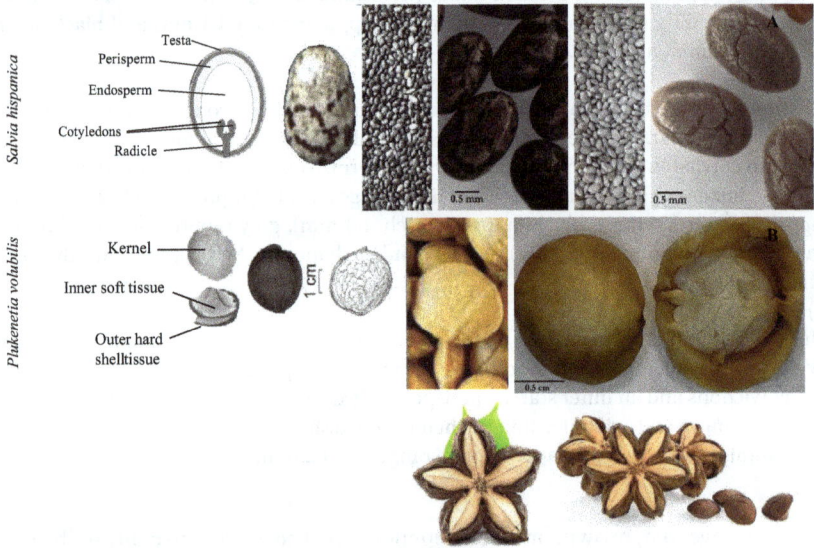

Figure 4.4   Latin-American oilseeds: **A.** Chia;[a] **B.** Sacha inchi.[b] **Source: Seed structure adapted from www.ladobe.com.mx/2011/09/las-semillas-exploradores-solitarios/ and** [b]**https://es.wikipedia.org/wiki/Plukenetia_volubilis (*). www.shutterstock.com/ image-photo/sacha-inchi-peanut-seed-isolate-on-1460624984 and www.shutterst ock.com/image-photo/sacha-inchi-isolated-on-white-background-625267805.**

Images A and B: Authors' own work.

Notes: The figure is composed of two parts, the structure (seen from a cross-section of the seed) and images of chia (A) and sacha inchi (B).

A) Illustration of the structure of chia. It has an oval shape, surrounded by the testa, followed by another layer called perisperm in the inner section of which is found the endosperm. At the bottom the two cotyledons are found. Next to the structure, there are two up-close images of white (approximate height of 1.85 mm and width of 1.2 mm) and black chia (approximate height of 2.2 mm and width of 1.4 mm) along with a detailed image for each.

B) Illustration of the structure of sacha inchi. It has a circular shape, surrounded by outer hard shell tissue and an inner soft tissue. There are dark brown seeds with oval ribs of 1.5–2 cm on the outer layer. Next to the structure, there is an up-close image along with a cross-section of the image (1.3–1.5 cm).

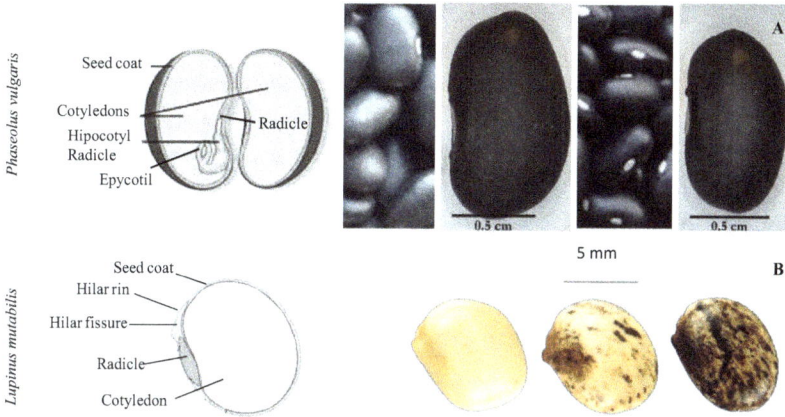

Figure 4.5 Latin-American legumes: **A.** Black Turtle Beans;[a] **B.** Lupin, *Tarwi*, *Altramuz* or Chocho.[b] **Source: [a]Seed structure adapted from https://botanicalgarden.berkeley. edu/glad-you-asked/seeds-p1 and [b]Miano et al. (2015).**

Image A: Authors' own work.

Image B: adapted from https://en.wikipedia.org/wiki/Lupinus_mutabilis.

Notes: The figure is composed of two parts, the structure and images of (A) black turtle beans (A) and (B) lupin.

A) Illustration of a cross-section of the structure of black turtle bean. It has a common bean shape, surrounded by a seed coat, followed by two cotyledons. Next to the structure, there are two up-close images of black turtle beans and a detailed image. The first bean has an approximate height of 1.2 cm and width of 0.7 cm; the second darker bean has an approximate height of 1.1 cm and width of 0.7 cm.

B) Illustration of the structure of lupin. It has a more rounded bean shape, surrounded by a seed coat, and one cotyledon (there are two cotyledons, but as it is not a cross-section, only one of them can be seen). Next to the structure, there are three different colored detailed images of tarwi (white, white with dark spots, and mostly dark spots). Tarwi has an approximate height of 0.9 cm and width of 0.7 cm.

et al., 2021). The color of the pericarp influences the color of the corresponding dish which may have some impact on the processing industry/consumer preference. The seed coat has three cell layers (Miano et al., 2015). The external layer resistant to water is formed by macrosclereid cells (palisade tissue) containing lignin, polysaccharides, pectin, calose, quinones, suberin, cutin, and phenols. The second layer is formed by osteosclereid cells, while the third layer is a sclerified parenchyma.

The seeds of *P. vulgaris* are small, glossy, with a dense meaty flavour similar to that of mushrooms and are nutritionally rich especially in antioxidants (Lim,

2012). They are mottled black, plump, oblong, or kidney-shaped: 9.1–10.93 (length) × 6–7.05 (width) × 4.22–5.0 mm (thick) (Table 4.2; Berrios et al., 1999; Vanier et al., 2019). Some structural characteristics, external and internal, of *P. vulgaris* are shown in Figure 4.5 A

4.5 CHEMICAL COMPOSITION OF KERNELS

The proximate composition of Latin-American seeds is shown in Table 4.3. The chemical composition of proteins, lipids, carbohydrates, fiber, and bioactive compounds are well covered in other chapters.

4.5.1 Proteins

The nutritional value of cereals, pseudo-cereals, oleaginous seeds, and legumes is strongly correlated with their protein content and its composition (Schöenlechner et al., 2008; Reguera and Haros, 2017). Pseudo-cereals, oilseeds, and legumes are extraordinarily valuable sources of protein, having a well-balanced amino acid set, with a mostly high content of lysine and a sufficient amount of sulfur-containing amino acids (Reguera and Haros, 2017). It is worth mentioning that the amino acid profile of cereals, like maize, plays a remarkably nutritional complementary role with that of legumes, like common beans (Paredes-López et al., 1984). The proteins in pseudo-cereals (quinoa, amaranth, kañiwa), chia, and legumes predominantly consist of globulins and albumins, and comprise lower content of, or even lack of storage of prolamin proteins, which constitute the main proteins in cereals, and the detrimental ones in the case of celiac disease (Alvarez-Jubete et al., 2010; Sandoval-Oliveros and Paredes-Lopez, 2013; Reguera and Haros, 2017).

Table 4.3 shows variations in the levels of protein content in the different cultivars of maize crops; interestingly, maize samples from the Andean region were found to have high levels of protein as reported by Salvador- Reyes et al. (2021). Changes in protein content, and in composition in general, is a function of the specific cultivar, genetically modified crops, nutrients available in the soil, and environmental influences (Rascon-Cruz et al., 2004).

Some studies have reported that the *Chullpi* variety showed the highest protein content among all kinds included in this work, which might be explained by the greater amount of protein that vitreous endosperm has compared to the floury samples (Salvador-Reyes et al., 2020). All essential amino acids in Latin-American maize varieties are presented, with leucine being the most abundant. The limiting amino acid, as expected, was lysine in all varieties together with valine and isoleucine in *Cuzco* and *Sacsa*. Significant amounts of tryptophan were detected in *Chullpi* and *Purple* maizes, giving them an advantage over traditional maize, in which this amino acid is at a low level (Rascon-Cruz et al. 2004;

Badui, 2006). Chapter 8 reports the further and more detailed protein composition of maize from the Latin-American region.

Commonly, amaranth is characterized by higher protein, also lysine, content than cereals (i.e., wheat, maize) or other pseudo-cereals (quinoa or kañiwa) but when compared with some Latin-American crops (oleaginous seeds, legumes), it may exhibit considerably lower amounts (Table 4.3). The exogenous amino acid content of amaranth is superior and its mutual proportions are more favorably balanced than in most cereals (Ballabio et al., 2011; Reguera and Haros, 2017; Joshi et al., 2018; Segura-Nieto et al., 1994). Globulin and glutelins are the main protein fractions, followed by albumin and prolamins. These proteins have important bioactive peptides which may act as modulators of metabolism and possess other outstanding biological activities, such as antihypertensive and antioxidant properties (Orona-Tamayo and Paredes-López, 2017). Proteins are influenced by growing factors and genotype (Schöenlechner et al., 2008; Reguera and Haros, 2017). Mostly, proteins are deposited in the germ and seed coat (~65%), while the rest are located in the endosperm (35%) (Saunders and Becker, 1984; Reguera and Haros, 2017). The protein amount of quinoa is greater than cereals (Table 4.3) and characterized by high biological value (83%), similar to that of milk protein (Ranhotra et al., 1993; Reguera and Haros, 2017) delivering considerable amounts of all essential amino acids (Ruales and Nair, 1992; Abugoch James, 2009; Gonzalez et al., 2012).

Compared to cereal grains, quinoa proteins are particularly rich in lysine, a limiting amino acid in most cereals. Their balance of essential amino acids is perfect due to a wider range of amino acids than that found in grains and legumes, which includes not only higher lysine amount but also higher methionine (Ruales and Nair, 1993; Abugoch James, 2009). Similarly, the protein content of kañiwa (14.0–15.7%) is close to quinoa. The protein quality of kañiwa is exceptional with a balanced essential amino acid composition, including 5–6% lysine, which is typically limiting in cereal grain crops (Mangelson et al., 2019; Peñarrieta et al., 2008). The protein composition and amino acid profile of amaranth, quinoa, and kañiwa are also extensively presented in Chapter 8.

Oleaginous seeds are a great source of protein, which seems even better than that of cereals and pseudo-cereals. The protein content of sacha inchi ranged between 24.2 and 27.0% (Hamaker et al., 1992; Gutiérrez et al., 2011). According to Hamaker et al. (1992), leucine (64 mg/g of protein) is the most abundant essential amino acid; next are tyrosine, isoleucine, lysine, threonine and valine (55, 50, 43, 43, and 40 mg/g of protein, respectively). These proteins have sulfur amino acids (methionine+cysteine) by 37 mg/g and phenylalanine by 9 mg/g (Hamaker et al., 1992). Albumin (43.7%) is the predominant aqueous soluble protein, followed by glutelin (31.9%), globulin (27.3%), and prolamin (3.0%) (Sathe et al., 2002).

Chia seeds are rich in protein (16.0–28.6% d.b.) and the essential amino acids of seed flour exhibited in general a relatively good balance of them, especially methionine and cysteine (Sandoval-Oliveros and Paredes-López, 2013). The main protein fraction corresponded to globulins (52%) which are also a good source of aromatic amino acids (Sandoval-Oliveros and Paredes-López, 2013). The exogenous aminoacids are arginine, leucine, phenylalanine, valine, and lysine; and the endogenous amino acids are glutamic and aspartic acids, alanine, serine, and glycine. The content of amino acid serine is 1.05 g/100 g, glutamic acid 3.50 g/100 g, glycine 0.95 g/100 g, alanine 1.05 g/100 g, lysine 0.97 g/100 g, and histidine 0.53 g/100 g of seeds (Ullah et al., 2016). Composition has been found to be affected by factors such as genotype, environment, and agronomical input (Jamshidi et al., 2019). Chia seeds do not contain the proteins that make up gluten, which makes them very valuable for people suffering from celiac disease. The protein composition and amino acid profile of oleaginous seeds are also extensively presented in Chapter 8.

The protein content of tarwi is the highest among all Latin-American crops, and reaches up to 52.6% d.b. (Table 4.3). It is rich in lysine and cysteine and comprises two protein classes – globulins (80% of total protein) and albumins (Brücher, 1989; Chirinos et al., 2021). Its seeds have low amounts of sulphured amino acids (methionine and cysteine) when considering human nutrition; however, its proteins show higher biological values than those of other species of Lupinus (*L. albus, L. angustifolius* and *L. cosentinii*) and are comparable to soybean proteins. The proteins digestibility is high (>90%) (Neves-Martins et al., 2016).

The amount of protein (up to 25% d.b.) (Priya and Manickavasagan, 2020) in black turtle bean is higher than in cereals (*Zea mays*), pseudo-cereals (kañiwa, quinoa, amaranth), and comparable to plants of oleaginous seeds (chia, sacha inchi) but lower than in tarwi seeds (Table 4.3). The amino acid lysine (1.483% d.b.) is greater than in other crops, such as maize. However, the proteins are poor in methionine (0.325% d.b.) and tryptophan (0.256% d.b.) (Lim, 2012). It is pertinent to mention that several of the Latin-American cultivars (e.g., amaranth, chia, common beans) contain peptides of different lengths and sequence which once hydrolyzed from proteins by different techniques show a wide variety of remarkable nutraceutical and medicinal properties, inlcuding roles as vegetable vaccines, with great commercial potential in the near future (Orona-Tamayo and Paredes-López, 2017). The composition and amino acid profile of Latin-American legumes are also extensively described in Chapter 8.

4.5.2 Carbohydrates

The most typical simple sugars of grains are glucose, fructose, arabinose, and xylose, together with sucrose and maltose (Reguera and Haros, 2017), which amount is rather small in cereals and pseudo-cereals (Berghofer and Schöenlechner, 2007; Reguera and Haros, 2017). Starch, non-starch polysaccharides and resistant starch are the main polysaccharides in cereals, whereas non-starch polysaccharides found in kernels are predominantly composed of β-glucans, cellulose, and hemicelluloses, which are included as part of dietary fiber together with resistant starch (Reguera and Haros, 2017). The amount of dietary fiber in pseudo-cereals and maize is similar and found to be like that in other cereals (Table 4.3), together with starch as the most crucial plant carbohydrate occurring in different shapes and sizes (Valcárcel-Yamani et al., 2012; Reguera and Haros, 2017). Oleaginous seeds of chia and sacha inchi contain mainly complex carbohydrates in the form of insoluble fractions (72.4% and 93%, respectively) (Wang et al., 2018; Jamshidi et al., 2019; Kodahl, 2020). Latin-American legumes are predominantly composed of complex carbohydrates but differ in their composition. Black turtle bean contains starch and dietary fiber in almost equal amounts, while tarwi is composed mainly of oligosaccharides. The common feature of both legumes is that they are rich in dietary fiber and contain small amounts of soluble sugars.

Maize is characterized by high amounts of starch, which is the main carbohydrate in its endosperm (Table 4.3). The quantity is higher than in other cereals and in other Latin-American crops (pseudo-cereals, oleaginous seeds, legumes). Starch granules differ in their shape, size, and distribution depending on the cultivar – *Chullpi*, *Sacsa*, and *Purple* starches are spherical, whereas those of *Piscorunto* and *Giant Cuzco* are irregular and polyhedral. *Sacsa* had a bimodal distribution presenting large granules type-A (>10 μm) and small granules type-B (<10 μm), whereas in other varieties only type-A granules were observed. Starch granules of *Giant Cuzco*, *Purple*, *Sacsa*, and *Chullpi* were larger than those of *Piscorunto* (Salvador-Reyes et al., 2021).

The starch percentage of pseudo-cereals varies between 38.03 and 70.40 (Table 4.3). In amaranth, starch is the main carbohydrate component, but its content is lower than in cereals (Valcárcel-Yamani et al., 2012; Reguera and Haros, 2017; Table 4.3). It is situated in the perisperm. The granules of amaranth starch (1-3 μm) are smaller than in other cereals (Berghofer and Schöenlechner, 2007; Valcárcel-Yamani et al., 2012; Reguera and Haros, 2017), and with polygonal, circular, and elliptical shapes according to their species (López et al., 1994). Amaranth starch is composed of smaller amylose amounts (0.1–11.1%) than those found in other cereals and characterized by normal and waxy types of starch contained in the same species of amaranth (Stone and Lorenz, 1984; Schöenlechner et al., 2008; Reguera and Haros, 2017). It has also been reported that cultivation and environmental conditions of amaranth affect the amylose/amylopectin ratio (Stone and Lorenz, 1984). Similarly, in quinoa, starch is also the most abundant carbohydrate (Table 4.3) with a polygonal shape and a diameter of 1.5–3.0 μm, being smaller than starch of typical grains (Koziol, 1992; Vega-Gálvez et al., 2010; Reguera and Haros, 2017). The amount of amylose in quinoa grains is smaller (11.0–12.4%) than that reported in wheat, rice, or maize (Koziol, 1992; Reguera and Haros, 2017).

Starch is the most abundant component, not only in quinoa and amaranth, but also in kañiwa (Repo-Carrasco-Valencia et al., 2010). Granules of starch kañiwa are in polygonal form with a diameter between 0.7 and 1.3 μm (Luna-Mercado and Repo-Carrasco-Valencia, 2021), being smaller when compared to cereal grains and pseudo-cereals (amaranth, quinoa). Kañiwa starch is found mainly in the perisperm surrounded by the embryo and is strongly associated with protein and other components of the grain. It was found that the amylose content of kañiwa starch variety Illpa Inia was 6.48% (Luna-Mercado and Repo-Carrasco-Valencia, 2021). Higher values of amylose in kañiwa starch were also reported elsewhere (10.7–17.44%) (Steffolani et al., 2013). More descriptions of the starch characteristics of maize and pseudo-cereal starch are explained in Chapter 6.

The carbohydrate composition of oleaginous seeds included in this report needs more study. Information about the carbohydrate content and composition of sacha inchi (*P. volubilis*) seeds is very limited (Kodahl, 2020). According to Wang et al. (2018), the carbohydrate amount ranged from 12.1–30.9% and results were obtained by simply subtracting the moisture, lipid, protein, and ash amounts from the total weight. Takeyama and Fukushima (2013) reported that the carbohydrate fraction of sacha inchi consists of 18.7% of glucides (Carbohydrate – (IDF + SDF), 72.3% water insoluble dietary fiber (IDF), and 9.0% soluble dietary fiber (SDF)). However, the exact composition of the fiber and the existence of starch in the seeds needs further research (Wang et al., 2018; Kodahl, 2020). Even less is known on carbohydrates composition in chia seeds (16.5–42.12% d.b.) with no starch detected (Vrancheva et al., 2019). Most of the literature data deals about dietary fiber (18–37.5% d.b.) (Table 4.3; Ayerza, 1995; Ayerza and Coates, 2004; Ixtaina et al., 2011; USDA, 2011; Da Silva Marineli et al., 2014; Coelho and Salas-Mellado, 2014; Amato et al., 2015; García-Salcedo et al., 2018; Carillo et al., 2018; Jamshidi et al., 2019; Knez Hrnčič et al., 2020; Hernández-Pérez et al., 2020). The dietary fiber of chia is composed mainly of insoluble fractions (>93%) and the rest (7%) is soluble form (Jamshidi et al., 2019). Interesting is the presence of a polysaccharide with a molecular weight of 0.8–2×10^6 Da (tetrasaccharide with 4-O-methyl-α-D-glucoronopyranosyl residues), which is situated in the fruit exocarp and has the ability to form a highly hydrophilic mucilage (Bochicchio et al., 2015). White and black-spotted chia samples from Salta, Argentina showed a mucilage composition with mannose (1.48 and 1.20%), galactose (3.26 and 3.16%), galacturonic acid (4.60 and 3.58%), glucose (7.41 and 5.88%), arabinose (9.48 and 9.80%), glucuronic acid (12.74 and 13.72%), xylose (61.04–63.07%), and with water capacity absorption 54.24 and 54.03, respectively (Muñoz et al., 2021).

Latin-American legumes differ in the amount and composition of carbohydrates. Black turtle bean is mainly composed of complex carbohydrates (up to 65% d.b.) (Priya and Manickavasagan, 2020). Nearly as much starch (18.17–28.06% d.b.) as dietary fiber (17.63–22.83% d.b.) may be found depending on varieties (Table 4.3; USDA, 2010; Lim, 2012; Vanier et al., 2019). Among total sugars (2.12% d.b.) (Lim, 2012), monosaccharides, disaccharides, and oligosaccharides are found. The common oligosaccharides are raffinose, stachyose, and verbascose. The black turtle starch is granular, round and oval in shape with

TABLE 4.3   CHEMICAL COMPOSITION OF LATIN-AMERICAN CROPS AND WHEAT (FOR COMPARISON PURPOSES)

CROPS	Protein, % d.b.	Lipids, % d.b.	Starch, % d.b.	Dietary fiber, % d.b.	Ash, % d.b.
CEREALS					
Wheat	11.6 (f5.70)[c], 11.7[e], 14.3(f6.25)[h], 13.20[a]	1.7[c], 2.0[e], 2.3[h]; 2.7[a]	61.0[e], 78.4[h]	2.8[h], 6.5[g]; 12.20[a]	1.4[e], 2.2[h]
maize	7.80/7.27/5.10/7.38 and 5.7(f6.25)[i]	2.70/3.93/3.52/4.50 and 4.8[i]	74.60/69.37/ 74.80/70.53 and 72.00[i]	4.30/1.86/1.12/ 1.87 and 3.8[i]	1.30/1.46/1.38/ 1.50 and 1.40[i]
PSEUDO-CEREALS					
Amaranthus cruentus	14.0(f5.85)[b], 14.0-14.8(f5.85)[g], 14.6(f5.8)[f], 14.9(f5.70)[c], 15.2[e], 13.2-18.2(f5.85)[e], 16.5(f5.85)[a], 13.80-21.50; 15.00-16.60; 13.10-21.00[a]	5.6[c], 5.7[a], 5.9-6.0[g], 6.0[h], 8.0[e], 8.8[f]; 5.60-8.10, 6.10- 7.30; 5.80-10.90[a], 6.3-8.1[s]	55.1[f], 61.4[a], 67.3[e]	11.1[f], 20.6[a]; 3.10-4.20, 4.90-5.00, 2.70- 4.90[a], 3.6-4.4[s]	2.8[a], 2.4[b], 2.9[c], 3.3[f], 2.4-2.6[g], 2.8-3.9[s]
Chenopodium quinoa	11.0(f5.70)[d], 12.8-13.5(f5.77)[g], 13.3[e], 13.8(f5.8)[f], 16.5[b], 9.1-15.7[b]	4.1-5.8[g], 5.0[f], 5.2[a], 6.3[h], 7.5[d], 7.5[e]; 4.0-7.6[h]	64.2[d], 66.9-70.4[g], 67.4[f], 69.0[e]; 48.5-69.8[h]	3.8[h], 6.72[d], 12.9[f], 14.2[a], 14.6-19.7[g]; 8.8-14.1[h]	2.7[a], 2.7[d], 3.8[h], 3.3[f], 2.3-2.5[g]; 2.0-7.7[h]
Chenopodium pallidicaule (kañiwa)	14.41/14.88[i]; 15.38/13.29/ 14.72/14.38[f], 15.3/15.5/ 14.7[f*]; 15.4/15.4/15.7/13.8 [k] (f6.25)	5.88/6.96[i]; 7.36/ 6.87/4.46/6.66[f]; 8.5/8.0/7.6[f*]; 7.5/7.8/7.5/3.9 [k]	52.4/57.47[j]; 54.07/52.78/ 54.04/38.03[f]; 60.4/58.5/62.0[f*]	11.24/8.1[i]; 5.33/7.52/7.46/ 14.37[f] 5.6/7.0/6.0[f*];	5.03/4.33[m], 3.56/3.67/3.38/ 3.13[m]; 4.6/4.0/3.7[m*] 3.7/3.7/3.5/4.2[k]

(continued)

TABLE 4.2 (CONTINUED)

CROPS	Protein, % d.b.	Lipids, % d.b.	Starch, % d.b.	Dietary fiber, % d.b.	Ash, % d.b.
OLEAGINOUS SEEDS					
Salva hispanica L. (chia)	28.56 ± 0.235[t] (f:6.25)/ 19.78±0.015[t']/ 16-26[t']/ 18.3 ± 1.613[t'], 22.7±0.7[t]	7.58 ± 0.028[t]/ 16.06±0.038[t']/ 20.30 -38.60[t']/ 32.4±0.214[t'], 32.5±2.7[t]	(CHO)17.887[t]/ 31.46±0.062[t']/ 42.12[t']/ 16.5±1.628[t']/ 3.1[t] (other carbohydrates)	39.87 ± 0.757[t] (soluble fiber)/ 27.88±0.021[t']/ 32.4-37.50[t']/ 22.2 ± 0.323[t']/ 33.5±2.7[t]	0.103 ± 0.015[t]/ 4.82±0.025[t']/ 4.3±0.035[t']/ 3.7±0.3[t]
Plukentia volubilis (sacha inchi)	24.2-27 [m],[n] (f:6.25)	33.4-54.3[m],[n']	12.1-30.9 [m],[n'],[m''] (CHO)	72.4 [m]	2.7-6.46 [m]
LEGUMES					
Phaseolus vulgaris (black turtle bean)	23.08/25.36/23.81/26.00/ 23.89 (f:6.25)[q]; 21.60[r]; 25.93 (f:6.25)[r'']	1.23/1.37/1.22/1.68/ 1.22[q]; 1.42[r]; 1.95[r'']	25.50/18.17/ 28.06/24.97/ 21.17[r] / 62.36 (CHO) [r]/ 48.59/51.67/ 50.56/46.76/ 50.74 (CHO)[r];	22.83/17.63/ 20.32/21.15/ 19.93[r]; 15.2[r']	4.25/3.94/4.06/ 4.39/4.20[q]; 3.60[r]; 4.65[r'']
Lupinus mubilis (tarwi)	40.87 ± 0.40 (32.03–46.90)[o] (f:6.25) 41.4–47.7[o] 34.6–50.2[o'] 32.0–52.6[p] 47.36±2.80 [p] (f:6.25)	16.12 ± 0.14(13.60–18.55)[o] 15.0–20.1[o] 14.3–23.6[o'] 13.0–24.6[p] 16.19 ± 1.03[p]	29.45 ± 0.43 (24.85–33.90)[t] 31.65 ± 2.47 [p]	8.2[q]	3.58 ± 0.06 (2.70–4.40)[t] 2.4–5.2[p] 4.80 ± 0.21[p]

Notes: d.m. dry basis; f: nitrogen to protein conversion factor used. [Continued]

(a) Alvarez-Jubete et al.(2010); [a'] Data for *A.cruentus, A.hypochondriacus, A.caudatus,* respectively from Joshi et al. (2018).

(b) Sanz-Penella et al. (2013) of *A. cruentus.*

(c) García-Mantrana et al. (2014) of *A. cruentus* and *T. aestivum* L.

(d) Iglesias-Puig et al. (2015).

(e) Souci et al. (2000) of *A. cruentus,* abrased quinoa, *F. escutentum* and *T. aestivum* L.

(f) Valcárcel-Yamani et al. (2012).

(g) Own measurements from *Amaranthus* spp., Real quinoa and wheat.

(h) Koziol (1992). [h'] Data for different varieties, cultivars and ecotypes of quinoa from Nowak et al. (2016);

(i) Salhuana (2004): Salvador-Reyes and Clerici (2020) of *Chullpi (Ch), Piscorunto(P), Giant Cuzco(GC), and Purple* (P) varieties and *yellow corn* (Y) as control.

(j) Data for two Peruvian varieties of kaniwa *(Cupi/ Ramis)* from Repo-Carrasco-Valencia et al. (2009); [f'] Data for four ecotypes of kaŽiwa from the Agronomical Experimental Station-INIA Salcedo, Puno, Peru (Kello/ Wila/ Guinda/Ayara) from Repo-Carrasco-Valencia et al. (2010); [j']three Peruvian ecotypes *(Chilliwa/Purple plant/ Red kañiwa Condorsaya)* from Huamaní et al. (2020).

(k) Data for Peruvian ecotypes *Roja/Blanca/Amarilla/Illpa-Inia* from de La Rosa et al. (2016).

(l) García-Salcedo et al. (2018); [l'] Data for Ecuadorian chia seeds from Carillo et al. (2018). [l''] Data from Ayerza (1995); Ayerza and Coates (2004); Ixtaina et al. (2011); USDA, 2011; Da Silva Marineli et al. (2014); Amato et al. (2015); Jamshidi et al. (2019); Coelho and Salas-Mellado (2014).

(m) Gutiérrez, et al. (2011), Wang et al. (2018); (m') Ruiz et al. (2013); (m'') Takeyama, Fukushima (2013).

(n) Hamaker et al. (1992); (n')Kodahl and Sørensen (2021).

(o) Data for 33 Andean lupin ecotypes from different Peruvian regions from Berru et al. (2021); [o'] Data for in Ecuadorian *L. mutabilis* from Schoeneberger et al. (1982); [o'] Data in Peruvian ecotypes from Caligari et al. (2000).

(p) Carvajal-Larenas et al. (2016); (p') Data for three bitter Lupinus mutabilis genotypes originating from different regions of Peru (Altagracia, from Ancash, Andenes, from Cusco, and Yunguyo, from Puno) from Córdova-Ramos et al. (2020).

(q) Carvajal-Larenas et al. (2016).

(r) Data for five black bean cultivars (Campeiro/Esplendor/Grafite/Supremo/Valente) (Embrapa Rice and Beans, in the city of Santo António de Goiás, State of Goiás, Brazil) from Vanier et al. (2019); (r')Data from USDA (2010); Lim (2012); (r'') Data from Berrios et al. (1999).

(s) Seguera-Nieta et al. (2018).

(t) Hernández-Pérez, et al. (2020).

indentation. The surface is smooth without fissure (Du et al., 2014). The particle size varies between 10.0 and 60.3 μm, with a mean granule diameter of 25.3 μm. The reported amylose content was high and equals 45.4±0.8% (Du et al., 2014). In the seeds of tarwi, starch is usually absent, and the carbohydrates (29.45–31.65%) (Table 4.3.) are mainly oligosaccharides (e.g., stachyose and raffinose) and cell wall storage polysaccharides (Trugo et al., 2003; Berru et al., 2021). The reported mean content of soluble sugars in three Andean genotypes (*Altagracia*, *Andenes*, *Yunguyo*) from different regions in Peru were as follows: glucose (0.30%), fructose (0.52%), maltose (0.14%), reducing sugars (0.84%), and saccharose (4.34% d.b.) (Córdova-Ramos et al., 2020).

4.5.3 Lipids

Maize is characterized by higher lipid content than wheat and lower than the rest of Latin-American crops. Lipids are mainly deposited in the germ, but they are minor compounds of endosperm as for common cereals and take a great part in the assembly of the starch–protein matrix of grains which strongly affect the texture of kernels (hardness). Depending on the variety/ecotype of common maize, the endosperm might be of vitreous type, richer in free fatty acids than triacylglycerol or floury structure with no differences between those two components (Gayral et al., 2015).

The amount of lipids in the Latin-American maize samples used in this study by Salvador-Reyes et al. (2021), is higher than wheat and lower than yellow maize, and other Latin-American crops (Table 4.3). The composition of fat in Latin-American maize is interesting and contains in average 15% of saturated fatty acids (mainly palmitic acid), 30% of monounsaturated (oleic) acid, and 55% of polyunsaturated, including the essential fatty acids linoleic (53%) and α-linolenic (1.2%) (Salvador-Reyes et al., 2021). A 100 g portion of whole maize fulfills the daily recommendation for linoleic essential fatty acid for adults (FAO, 2010). The *Chullpi* variety presented traces of behenic acid (<1%), not detected in the other four varieties of Andean maize. The presence of this acid in the human diet is associated with increased cholesterol concentration in the blood (Cater and Denke, 2001).

Pseudo-cereal kernels have an analogous arrangement of lipids in the form of fat droplets. In amaranth, lipid bodies are situated in the embryo and the endosperm, while in the case of quinoa in the embryo and perisperm (Reguera and Haros, 2017). Lipid amounts in quinoa, amaranth, and kañiwa are approximately 2–3 times greater than wheat and maize (Table 4.3; Alvarez-Jubete et al., 2010; Reguera and Haros, 2017). Generally, pseudo-cereal lipids are characterized by a high content of unsaturated fatty acids (between 75 and 86%) (Valcárcel-Yamani et al., 2012; Reguera and Haros, 2017; Huamani et al., 2020). Linoleic acid is the most important fatty acid of pseudo-cereals (47.5–47.8, 48.2–56.0; 46.9–48.5%), followed by oleic acid (23.7–32.9, 24.5–26.7, and 24.8–25.9%), and palmitic acid (12.3–20.9, 9.7–11.0; 12.9–13.5%), for amaranth, quinoa, and kañiwa, respectively (Becker, 1994; Valcárcel-Yamani et al., 2012; Salas et al. 2015; Huamani et al., 2020; Repo-Carrasco-Valencia, 2020). Squalene, an open-chain triterpene with many unsaturated bonds is the biochemical precursor of the entire steroid family presented in amaranth (1.9–11.2% in its oil). The presence of 3.4–5.8% of squalene was also found in quinoa seeds (Valcárcel-Yamani et al., 2012; Reguera and Haros, 2017). Despite the high fat content and level of unsaturation, pseudo-cereal lipids are generally resistant to oxidation resulting from the presence of antioxidants such as tocopherols (Álvarez-Jubete et al., 2010; Reguera and Haros, 2017). Pseudo-cereals could be part of a balanced diet regimen as a result of their low saturated fatty acid content (Becker, 1994).

The oleaginous seeds such as sacha inchi and chia are of great interest, because of their exceptionally high lipid content in the form of oil: 35–60% and 25–38%, respectively with unique unsaturated fatty acids composition (Ixtaína et al., 2008; Gutiérrez et al., 2011). Common legumes, such as black turtle bean (<2% d.b.) are generally low in fat. There are however some exceptions which include, among others tarwi characterized by high lipid content (19–24% d.b.) (Table 4.3). Legumes are comprised mainly of mono- and polyunsaturated fatty acids, with no cholesterol and saturated fatty acids (Maphosa and Jideani, 2017).

The lipid content of sacha inchi is very high (33.4–54.3%) (Gutiérrez et al., 2011; Wang et al., 2018; Kodahl and Sørensen, 2021) and its composition is unique (neutral lipids – 97.2%, free fatty acids – 1.2% and phospholipids -0.8%) (Gutiérrez et al., 2011). Approximately 77.5–84.4% polyunsaturated fatty acids (PUFAs), 8.4–13.2% monounsaturated fatty acids (MUFAs), and 6.8–9.1% saturated fatty acids (SFAs) were found in seeds (Follegatti-Romero et al., 2009; Maurer et al., 2012; Chirinos et al., 2013). The predominant fatty acid is α-linolenic acid (ALA, ω-3) (46.8–50.8%), then linoleic acid (ω-6, 33.4–36.2%) and oleic acid (ω-9, 8.7–9.6%) (Guillén et al., 2003; Follegatti-Romero et al., 2009; Chirinos et al., 2013). Very low amounts of myristic acid (C14:0), eicosanoic acid (C20:0), and gadoleic acid (C20:1, ω-11, 0.16%) were also reported in sacha inchi (Follegatti-Romer et al., 2009; Chandrasekaran and Liu, 2015). Lipids in chia seeds constitute a significant amount (up to 38.60% d.b.) (Table 4.3) (Ayerza, 1995; Ixtaína et al., 2011; Ayerza and Coates, 2004; Da Silva Marineli et al., 2014; Amato et al., 2015). The main fatty acids presented by chia are linolenic acid (18:3), linoleic acid (18:2), stearic acid (18:0), palmitic acid (16:0), and oleic acid (18:1) (Ayerza, 1995; Ayerza and Coates, 2004; Hernández-Pérez et al., 2020). ALA constitutes the greatest share (>60%), making chia one of the best plant-based sources of omega-3 (Ayerza and Coates, 2001).

Together with proteins, lipids comprise (up to 24.6% d.b.) (Table 4.3) more than half of the seed's weight of *L. mutabilis* and are composed of some essential fatty-acids like oleic, linoleic, and linolenic that represent the 40.4%, 37.1%, and 2.9% from the total, respectively (Chirinos, 2015). The oil does not have any erucic (toxic) acid as in *L. albus* (Neves-Martins et al., 2016). Tarwi is also the only species capable of containing 18% oil content (the minimum for industrial extraction) (Neves-Martins et al., 2016). The content of lipids in black turtle bean is very low (1.2–1.95% d.b.) (Berrios et al., 1999; USDA, 2010; Lim, 2012; Vanier et al., 2019), being the lowest among all Latin-American crops and comparable to wheat grain (Table 4.3). The fat composition differs depending upon the bean variety and growth conditions (Priya and Manickavasagan, 2020). Total saturated fatty acids constitute 0.366%, including myristic (0.001%), palmitic (0.343%), and stearic (0.022%). At the same time, total monounsaturated fatty acids, mainly in the form of oleic (0.123%) and total polyunsaturated fatty acids – 0.610%, above all linoleic (0.332%) and linolenic (0.278%) are also present (Lim, 2012).

The composition of lipids in Latin-American crops is extensively described in Chapter 8.

4.6   CONCLUSIONS

Latin-American seeds belong to different types of crops: cereals, pseudo-cereals, oleaginous seeds, and legumes. They differ between each other not only in terms of physicochemical properties but also because of their exceptional nutritional characteristics. They are a great source of high quality proteins (pseudo-cereals, oleaginous seeds, and legumes), an excellent source of unsaturated fatty acids (pseudo-cereals, oleaginous seeds, legumes), and abundant in dietary fiber (oleaginous seeds, black turtle bean). Because of great functional and pro-healthy properties, Latin-American crops constitute a promising future alternative to already well-known and cultivated crops intended for human nutrition worldwide. Despite the above, some data in the literature need completion and further research. In other words, more thorough research on physicochemical and functional properties, better knowledge about the content and composition of carbohydrates and fiber of different types, and minor components fundamental to the daily diet present in all Latin-American genetic materials are still required.

In addition to agronomic types of studies, including traditional and molecular agricultural techniques, it is necessary to pursue extensive *in vitro* and *in vivo* investigation – following strict scientific strategies – to assess the remarkable nutritional and nutraceutical potential of the Latin-American crops involved in this report. It is pertinent to underline that several of these cultivars have been classified by official international organizations as key food sources of the 21st century.

ACKNOWLEDGMENTS

This work was financially supported by the Integrated Program of the University of Agriculture in Krakow and co-financed by European Union Funds (No. POWR.03.05.00-00-z222/17), Food4ImNut Food4ImNut PID2019-107650RB-C21 funded by MCIN/AEI/10.13039/501100011033, Spain.

REFERENCES

Abalone, R., A. Cassinera, A. Gastón, and M. A. Lara. 2004. Some physical properties of amaranth seeds. *Biosystems Engineering* 89: 109–117.

Abugoch, J. L. E. 2009. Quinoa (*Chenopodium quinoa* Willd.): Composition, chemistry, nutritional, and functional properties. *Advances in Food and Nutrition Research* 58: 1–31.

Aderibigbe, O. R., O. O. Ezekiel, S. O. Owolade, J. K. Korese, B. Sturm, and O. Hensel. 2022. Exploring the potentials of underutilized grain amaranth (*Amaranthus* spp.) along the value chain for food and nutrition security: A review. *Critical Reviews in Food Science and Nutrition* 62(3): 656–669.

Alvarez-Jubete, L., E. K. Arendt, and E. Gallagher. 2010. Nutritive value of pseudocereals and their increasing use as functional gluten-free ingredients. *Trends in Food Science & Technology* 21: 106–113.

Amato, M., M. C. Caruso, F. Guzzo, F. Galgano, M. Commisso, R. Bochicchio, R. Labella, and F. Favati. 2015. Nutritional quality of seeds and leaf metabolites of Chia (*Salvia hispanica* L.) from Southern Italy. *European Food Research Technology* 241: 615–625.

Atchison, G. W., B. Nevado, R. J. Eastwood, N. Contreras-Ortiz, C. Reyne, S. Madriñán, D. A. Filatov, and C. E. Hughes. 2016. Lost crops of the Incas: Origins of domestication of the Andean pulse crop tarwi, *Lupinus mutabilis*. *American Journal of Botany* 103(9): 1592–1606.

Audu, S. S., and M. O. Aremu. 2015. Effect of domestic processing on the levels of some functional parameters in black turtle bean (*Phaseolus vulgaris* L). *Food Science and Quality Management* www.iiste.org, ISSN 2224-6088 (Paper) ISSN 2225-0557), 38.

Ayerza, R. 1995. Oil content and fatty acid composition of chia (*Salvia hispanica* L.) from five northwestern locations in Argentina. *Journal of the American Oil Chemists' Society* 72: 1079–1081.

Ayerza, R. 2009. The seed's protein and oil content, fatty acid composition and growing cycle length of a single genotype of Chia as affected by environmental factors. *Journal of Oleo Science* 58(7): 347–354.

Ayerza, R., and W. Coates. 2001. Omega-3 enriched eggs: the influence of dietary α-linolenic fatty acid source on egg production and composition. *Canadian Journal of Animal Science* 81: 355–362.

Ayerza, R., and W. Coates. 2004. Protein and oil content, peroxide index and fatty acid composition of chia (*Salvia hispanica* L.) grown in six tropical and subtropical ecosystems of South America. *Tropical Science* 44: 131–135.

Babić, Ljiljana., Mirko Babić, Jan Turan, Snežana Matić-Kekić, Milivoj Radojčin, Sanja Mehandžić-Stanišić, Ivan Pavkov, Miodrag Zoranović 2011. Physical and stress–strain properties of wheat (*Triticum aestivum*) kernel. *Journal of the Science of Food and Agriculture* 91: 1236–1243. https://doi.org/10.1002/jsfa.4305.

Badui, S. 2006. Química de los alimentos (4th ed.). Mexico: Editorial Alhambra Mexicana, pp.140–145.

Ballabio, C., F. Uberti, C. Di Lorenzo, A. Brandolini, E. Penas, and P. Restani. 2011. Biochemical and immunochemical characterization of different varieties of amaranth (*Amaranthus* L. ssp.) as a safe ingredient for gluten-free products. *Journal of Agricultural and Food Chemistry* 59: 12969–12974.

Baltensperger, D. D. 2003. Cereal grains and pseudo-cereals. In *Encyclopedia of Food and Culture*, edited by S.H. Katz and W.W. Weaver. New York: Scribner, URL: www.encyclopedia.com/doc/1G2-3403400119.html. Accessed 03.06.15.

Becker, R. 1994. Amaranth oil: Composition, processing, and nutritional qualities. In *Amaranth Biology, Chemistry, and Technology*, edited by O. Paredes-López, 133–141. Boca Raton, FL: CRC Press.

Berghofer, E., and R. Schoenlechner. 2007. Pseudocereals – An overview. IFS Workshop: Traditional grains for low environmental impact and good health, Pretoria.

Berriosa, J. D. J., B. G. Swansonb, and W. A. Cheong. 1999. Physico-chemical characterization of stored black beans (*Phaseolus vulgaris* L.). *Food Research International* 32: 669–676.

Berru, L. B., P. Glorio-Paulet, C. Basso, A. Scarafoni, F. Camarena, A. Hidalgo, and A. Brandolini. 2021. Chemical composition, tocopherol and carotenoid content of seeds from different Andean Lupin (*Lupinus mutabilis*) Ecotypes. *Plant Foods for Human Nutrition* 76: 98–104.

Bochicchio, R., T. D. Philips, S. Lovelli, R. Labella, F. Galgano, A. Di Marisco, M. Perniola, and M. Amato. 2015. Innovative crop productions for healthy food: The case of chia (*Salvia hispanica* L.). In *The Sustainability of Agro-Food and Natural Resource Systems in the Mediterranean Basin*, edited by A. Vastola. New York: Spriger. DOI 10.1007/978-3-319-16357-4_3.

Boege, E. 2009. Centros de origen, pueblos indígenas y diversificación del maíz. *Ciencias* 92–93: 18–28.

Bressani, R. 2003. Amaranth. In *Encyclopedia of Food Sciences and Nutrition*, edited by B. Caballero, 166–173. Oxford: Academic Press.

Broughton, W. J., G. Hernández, M. Blair, S. Beebe, P. Gepts, and J. Vanderleyden. 2003. Beans (*Phaseolus* spp.) – model food legumes. *Plant and Soil* 252: 55–128.

Brücher, H. 1989. *Lupinus mutabilis* Sweet. In *Useful Plants of Neotropical Origin and Their Wild Relatives*, edited by R. Gros, 80–84. New York: Springer.

Bruno, M. C., M. Pinto, and W. Rojas. 2018. Identifying domesticated and wild kañawa (*Chenopodium pallidicaule*) in the archeobotanical record of the Lake Titicaca Basin of the Andes. *Economic Botany* 72(2):137–149.

Brust, J., W. Claupein, and R. Gerhards. 2014. Growth and weed suppression ability of common and new cover crops in Germany. *Crop Protection* 63: 1–8.

Caligari, P. D. S., P. Römer, M. A. Rahim, C. Huyghe, J. Neves-Martins, and E. J. Sawicka-Sienkiewicz. 2000. The potential of *Lupinus mutabilis* as a crop. In *Linking Research and Marketing Opportunities for Pulses in the 21st Century*, 569–573. Dordrecht: Springer.

Carrillo, W., M. Cardenas, C. Carpio, D. Morales, M. Álvarez, and M. Silva. 2018. Content of nutrients component and fatty acids in chia seeds (*Salvia hispanica* L.) cultivated in Ecuador. *Asian Journal of Pharmaceutical and Clinical Research* 11(2): 1–4.

Carvajal-Larenas, F. E. 2019. Nutritional, rheological and sensory evaluation of *Lupinus mutabilis* food products – a review. *Czech Journal of Food Sciences* 37(5): 301–311.

Carvajal-Larenas, F. E., A. R. Linnemann, M. J. R. Nout, M. Koziol, and M. A. A. J. van Boekel. 2016. *Lupinus mutabilis*: Composition, uses, toxicology, and debittering. *Critical Reviews in Food Science and Nutrition* 56: 1454–1487.

Castañeda, M. 1988. *Estudio comparativo de diez variedades de tarwi (Lupinus mutabilis Sweet.) conducidas en dos ambientes de la Sierra norte y centro del Perú*. Tesis Ingeniero Agrónomo). Universidad Nacional Agraria La Molina, Lima.

Cater, N. B., and M. A. Denke. 2001. Behenic acid is a cholesterol-raising saturated fatty acid in humans. *The American Journal of Clinical Nutrition* 73(1): 41–44.

Coelho, M. A., and M. de las M. Salas-Mellado. 2014. Chemical characterization of chia (*Salvia hispanica* L.) for use in food products. *Journal of Food and Nutrition Research* 2(5): 263–269.

Córdova-Ramos, J. S., P. Glorio-Paulet, F. Camarena, A. Brandolini, and A. Hidalgo. 2020. Andean lupin (*Lupinus mutabilis* Sweet): Processing effects on chemical composition, heat damage, and *in vitro* protein digestibility. *Cereal Chemistry* 97: 827–835.

Chandrasekaran, U., and A. Liu. 2015. Stage-specific metabolization of triacylglycerols during seed germination of Sacha Inchi (*Plukenetia volubilis* L.). *Journal of the Science of Food and Agriculture* 95: 1764–1766.

Chirinos-Arias, M. C. 2015. Andean Lupin (*Lupinus mutabilis* Sweet) a plant with nutraceutical and medicinal potential. *Revista Bio Ciencias* 3(3): 163–172.

Chirinos, R., G. Zuloeta, R. Pedreschi, E. Mignolet, Y. Larondelle, and D. Campos. 2013. Sacha inchi (*Plukenetia volubilis*): A seed source of polyunsaturated fatty acids, tocopherols, phytosterols, phenolic compounds and antioxidant capacity. *Food Chemistry* 141: 1732–1739.

Chirinos, R., E. Cerna, R. Pedreschi, M. Calsin, A. Aguilar-Galvez, and D. Campos. 2021. Multifunctional *in vitro* bioactive properties: Antioxidant, antidiabetic and antihypertensive of protein hydrolyzates from tarwi (*Lupinus mutabilis* Sweet) obtained by enzymatic biotransformation. *Cereal Chemistry* 98: 423–433.

Da Silva Marineli, R. É. A. Moraes, S. A. Lenquiste, A. T. Godoy, M. N. Eberlin, and M. R. Maróstica Jr. 2014. Chemical characterization and antioxidant potential of Chilean chia seeds and oil (*Salvia hispanica* L.). *LWT Food Science and Technology* 59: 1304–10.

De la Cruz Torres, E., C. Mapes Sánchez, A. Laguna Cerda, J. M. García Andrade, A. López Monroy, J. González Jiménez, and T. Falcón Bárcenas. 2008. Ancient underutilised pseudocereals – potential alternatives for nutrition and income generation. In *New Crops and Uses: Their Role in a Rapidly Changing World*, edited by J. Smartt and N. Haq, 186–203. Chichester, UK: RPM Print & Design.

de la Rosa-Millán, J., E. Heredia-Olea, E. Perez-Carrillo, D. Guajardo-Flores, S. Román, O. Serna-Saldívar. 2019. Effect of decortication, germination and extrusion on physicochemical and *in vitro* protein and starch digestion characteristics of black beans (*Phaseolus vulgaris* L.). *LWT - Food Science and Technology* 102: 330–337.

Du, Shuang-Kui, Hongxin Jiang, Yongfeng Ai, Jay-Lin Jane. 2014. Physicochemical properties and digestibility of common bean (*Phaseolus vulgaris* L.) starches. *Carbohydrate Polymers* 108(8): 200–205.

FAO (Food and Agriculture Organization of the United Nations). 2010. *Fats and Fatty Acids in Human Nutrition, Report of an Expert Consultation*. FAO Food and Nutrition Paper 91 (Final report).Rome: FAO. www.fao.org/3/a-i1953e.pdf.

Follegatti-Romero, L. A., C. A. Piantino, R. Grimaldi, and A. C. Fernando. 2009. Supercritical CO_2 extraction of omega-3 rich oil from Sacha inchi (*Plukenetia volubilis* L.) seeds. *Journal of Supercritical Fluids* 49: 323–329.

García-Mantrana, I., V. Monedero, and M. Haros. 2014. Application of phytases from bifidobacteria in the development of cereal-based products with amaranth. *European Food Research and Technology* 238: 853–-862.

García-Salcedo, A. J., O. L. Torres-Vargas, amd H. Ariza-Calderón. 2018. Agroindustria y Ciencia de los Alimentos/Agroindustry and Food Science. Physical-chemical characterization of quinoa (*Chenopodium quinoa* Willd.), amaranth (*Amaranthus caudatus* L.), and chia (*Salvia hispanica* L.) flours and sedes. *Acta Agronomica* 67(2): 215–222.

Gayral, M., B. Bakan, M. Dalgalarrondo, K. Elmorjani, C. Delluc, S. Brunet, L. Linossier, M. H. Morel, and D. Marion. 2015. Lipid partitioning in maize (*Zea mays* L.) endosperm highlights relationships among starch lipids, amylose, and vitreousness. *Journal of Agriculture and Food Chemistry* 63: 3551–3558.

Gepts, P., and D. Dpbouk. 1991. Origin, domestication, and evolution of the common bean (*Phaseolus vulgaris* L.). In *Common Beans: Research for Crop Improvement*, edited by A. Van Schoonhoven and O. Voyset, 7–53. Wallingford: CAB International.

González, J. A., Y. Konishi, M. Bruno, M. Valoy, and F. E. Prado. 2012. Interrelationships among seed yield, total protein and amino acid composition of ten quinoa (*Chenopodium quinoa*) cultivars from two different agroecological regions. *Journal of the Science of Food and Agriculture* 92: 1222–1229.

Gross, R. 1982. *El cultivo y utilización del tarwi*, L. mutabilis. Rome: FAO.

Guillén, M. D., A. Ruiz, N. Cabo, R. Chirinos, and G. Pascual. 2003. Characterization of Sacha Inchi (*Plukenetia volubilis* L.) oil by FTIR spectroscopy and H-1 NMR. Comparison with linseed oil. *Journal of the American Oil Chemists Society* 80: 755–762.

Gutiérrez, L.-P., L.-M. Rosada, and A. Jiménez. 2011. Chemical composition of Sacha Inchi (*Plukenetia volubilis* L.) seeds and characteristics of their lipid fraction. *Grasas y Aceites* 62(1): 76–83.

Hamaker, B. R., C. Valles, R. Gilman, R. M. Hardmeier, D. Clark, H. H. García, A. E. Gonzáles, I. Kohlstad, M. Castro, R. Valdivia, T. Rodriguez, and M. Lescano. 1992. Amino acid and fatty acid profiles of the Inca peanut (*Plukenetia volubilis* L.). *Cereal Chemistry* 69: 461–463.

Hernández-Gómez, J. A., S. M. Colín, and A. Penae Lomelí. 2008. Natural outcrossing of chia (*Salvia hispanica* L.). *Revista. Chapingo Serie Horticultura* 14: 331–337.

Hernández-Pérez, T., M. E. Valverde, D. Orona-Tamayo, O. Paredes-López. 2020. Chia (*Salvia hispanica*): Nutraceutical properties and therapeutic applications. *Proceedings* 53. doi:10.3390/proceedings2020053017.

Hernández-Pérez, T., M. E. Valverde, and O. Paredes-López. 2021. Seeds from ancient food crops with the potential for antiobesity promotion. *Critical Reviews in Food Science and Nutrition*. https://doi.org/10.1080/10408398.2021.1877107.

Huamaní, F., M. Tapia, R. Portales, V. Doroteo, C. Ruiz, and R. Rojas. 2020. Proximate analysis, phenolics, betalains, and antioxidant activities of three ecotypes of kañiwa (*Chenopodium pallidicaule* aellen) from Peru. http://pharmacologyonline.silae.it. ISSN: 1827-8620, pp. 229–236.

Iglesias-Puig, E., V. Monedero, and M. Haros. 2015. Bread with whole quinoa flour and bifidobacterial phytases improve contribution to dietary mineral intake and their bioavailability without substantial loss of bread quality. *LWT-Food Science and Technology* 60: 71–77.

Ixtaína, V. Y., S. M. Nolasco, and M. C. Tomás. 2008. Physical properties of chia (*Salvia hispanica* L.) seeds. *Industrial Crops and Products* 28: 286–293.

Ixtaína, V. Y., M. L. Martínez, V. Spotorno, C. M. Mateo, D. M. Maestri, B. W. K. Diehl, S. M. Nolasco, and M. C. Tomás. 2011. Characterization of Chia seed oils obtained by pressing and solvent extraction. *Journal of Food Composition and Analysis* 24: 166–74.

Jacobsen, E., and A. Mujica. 2006. El tarwi (*Lupinus mutabilis* Sweet) y sus parientes silvestres. In *Botánica Económica de los Andes Centrales*, 458–482. La Paz: Universidad Mayor de San Andrés.

Jamshidi, A. M., A. Ahmadi, R. Bochicchio, and R. Rossi. 2019. Chia (*Salvia hispanica* L.) as a novel forage and feed source: A review. *Italian Journal of Agronomy* 14: 1–18.

Joshi, D. C., S. Sood, R. Hosahatti, L. Kant, A. Pattanayak, A. Kumar, D. Yadav, and M. G. Stetter. 2018. From zero to hero: The past, present and future of grain amaranth breeding. *Theoretical and Applied Genetics* 131: 1807–1823.

Knez Hrnčič, M. K., M. Ivanovski, D. Cör, and Ž. Knez. 2020. Chia seeds (*Salvia hispanica* L.): An overview—phytochemical profile, isolation methods, and application. *Molecules* 25(1): 11.

Kodahl, N. 2020. Sacha inchi (*Plukenetia volubilis* L.)—from lost crop of the Incas to part of the solution to global challenges? *Planta* 251: 80.

Kodahl, N., and M. Sørensen. 2021. Sacha inchi (*Plukenetia volubilis* L.) is an underutilized crop with a great potential. *Agronomy* 11: 1066.

Koziol, M. J. 1992. Chemical composition and nutritional evaluation of quinoa (*Chenopodium quinoa* Willd.). *Journal of Food Composition and Analysis* 5: 35–68.

La Rosa, R., E. Anaya, Z. Flores, M. Bejarano, L. Brito, and E. Pérez. 2016. Germinación de *Chenopodium pallidicaule* aelle "kañiwa" bajo diferentes condiciones de salinidad y temperatura. *The Biologist* (Lima) 14(1): 5–10.

Lim, T. K. 2012. Edible medicinal and non-medicinal plants. *Fruits* 2: 815–848.

Lopez, M. G., L. A. Bello-Perez, and O. Paredes-López. 1994. Amaranth carbohydrates. In *Amaranth Biology, Chemistry, and Technology*, edited by O. Paredes-López, 107–131. Boca Raton, FL: CRC Press.

Luna-Mercado, G. I., and R. Repo-Carrasco-Valencia. 2021. Gluten-free bread applications: Thermo-mechanical and techno-functional characterization of kañiwa flour. *Cereal Chemistry* 98: 474–481.

Mangelson, H., D. E. Jarvis, P. Mollinedo, O. M. Rollano-Penaloza, V. D. Palma-Encinas, R. L. Gomez-Pando, E. N. Jellen, and P. J. Maughan. 2019. The genome of *Chenopodium pallidicaule*: An emerging Andean super grain. *Applications in Plant Sciences* 7(11): e11300.

Maphosa, Y., and V. A. Jideani. 2017. The role of legumes in human nutrition. In *Functional Food – Improve Health Through Adequate Food*, edited by Hueda M. Chavarri, 103–122. Zagreb: IntechOpen Online ISBN 978-953-51-3439-8.

Maurer, N. E., B. Hatta-Sakoda, G. Pascual-Chagman, and L. E. Rodríguez-Saona. 2012. Characterization and authentication of a novel vegetable source of omega-3 fatty acids, sacha inchi (*Plukenetia volubilis* L.) oil. *Food Chemistry* 134: 1173–1180.

Miano, A. C., J. A. García, and P. E. Duarte Augusto. 2015. Correlation between morphology, hydration kinetics and mathematical models on Andean lupin (*Lupinus mutabilis* Sweet) grains. *LWT - Food Science and Technology* 61: 290–298.

Mohamed, A. A., and P. Rayas-Duarte. 1995. Composition of *Lupinus albus*. *Cereal Chemistry* 72: 643–647.

Mohsenin, N. N. 1986. *Physical Properties of Plant and Animal Materials* (2nd ed.). New York, NY: Gordon and Breach Science.

Moreno, M. L, I. Comino I, and C. Sousa 2014. Alternative grains as potential raw material for gluten-free food development in the diet of celiac and gluten-sensitive patients. *Austin Journal of Nutrition and Food Sciences* 2(3): 9.

Muñoz, L. A., N. Vera, M. C. Zúñiga, M. Moncada, and C. M. Haros. 2021. Physicochemical and functional properties of soluble fiber extracted from two phenotypes of chia (*Salvia hispanica* L.) seeds. *Journal of Food Composition and Analysis* 104: 104138.

Neves Martins, J. M., P. Talhinhas, and P. R. de Sousa, 2016. Yield and seed chemical composition of *Lupinus mutabilis* in Portugal. *Revista de Ciências Agrárias* 39(4): 518–525.

Nowak, V., J. Du, and U. Ruth. 2016. Charrondière. Assessment of the nutritional composition of quinoa (*Chenopodium quinoa* Willd.). *Food Chemistry* 193: 47–54.

Ogrodowska, D., R. Zadernowski, M. Tanska, and S. Czaplicki. 2011. Physical properties of *Amaranthus cruentus* seeds from different cultivation regions in Poland. *Zywnosc-Nauka Technologia Jakosc* 18: 91–104.

Orona-Tamayo D., O. Paredes-López. 2017. Amaranth Part 1 - Sustainable crop for the 21st century: Food properties and nutraceuticals for improving human health. In *Sustainable Protein Sources*, edited by S. R Nadathur, J. P. D Wanasundara, and L. Scanlin, 239–256. Cambridge, MA: Academic Press.

Orona-Tamayo, D., M. E. Valverde, and O. Paredes-López. 2019. Bioactive peptides from selected Latin American food crops – A nutraceutical and molecular approach. *Critical Reviews in Food Science and Nutrition*, 59(12): 1949–1975.

Ortiz, A. 2012. *Los Maíces en la Seguridad Alimentaria de Bolivia*. La Paz: Centro de Investigación y Promoción del Campesinado (CIPCA).

Paliwal, R. L., G. Granados, H. R. Lafitte, and A. D. Violic. 2000. El maíz en los trópicos: Mejoramiento y producción. www.fao.org/d ocrep/003/X7650S/x7650s02.htm.

Palma-Rojas, C., C. González, B. Carrasco, H. Silva, and H. Silva-Robledo. 2017. Genetic, cytological and molecular characterization of chia (*Salvia hispanica* L.) provenances. *Biochemical Systematics and Ecology* 73 : 16–21.

Paredes López, O., C. Ordorica-Falomir, and F. Guevara-Lara. 1984. Las proteínas vegetales: presente y futuro en la alimentación. In *Prospectiva de la Biotenología en México*, edited by R. Quintero Ramírez, 331–350. Mexico: Conacyt.

Peláez, P., D. Orona Tamayo, S. Montes Hernández, M. Valverde, O. Paredes López, and A. Cibrián Jaramillo. 2019. Comparative transcriptome analysis of cultivated and wild seeds of *Salvia hispanica* (chia). *Scientific Reports* 9: 9761–9767.

Peñarrieta, J. M., J. A. Alvarado, B. Åkesson, and B. Bergenståhl. 2008. Total antioxidant capacity and content of flavonoids and other phenolic compounds in canihua (*Chenopodium pallidicaule*): An Andean pseudocereal (en línea). *Molecular Nutrition and Food Research* 52: 708–717.

Prasanna, B. M., N. Palacios-Rojas, F. Hossain, V. Muthusamy, A. Menkir, T. Dhliwayo, T. Ndhlela, F. S. Vicente, S. K. Nair, B. S. Vivek, X. Zhang, M. Olsen, and X. Fan. 2020. Molecular breeding for nutritionally enriched maize: Status and prospects. *Frontiers in Genetics* 10: 1392.

Priya, T. R. S., and A. Manickavasagan. 2020. Common bean. In *Pulses Processing and Product Development*, edited by A. Manickavasagan and T. Praveena, 77–98. New York: SpringerLink. ISBN 978-3-030-41375-0 ISBN 978-3-030-41376-7 (eBook) https://doi.org/10.1007/978-3-030-41376-7.

Ranhotra, G. S., J. A. Gelroth, B. K. Glaser, K. J. Lorenz, and D. L. Johnson. 1993. Composition and protein nutritional quality of quinoa. *Cereal Chemistry* 70: 303–305.

Rascon-Cruz, Q., Y. Bohorova, J. Osuna-Castro, and O. Paredes-López. 2004. Accumulation, assembly and digestibility of amarantin expressed in transgenic tropical maize. *Theoretical and Applied Genetics* 108(2): 335–342.

Reguera, M., and C. M. Haros. 2017. Structure and composition of kernels. In *Pseudocereals. Chemistry and Technology* (1st ed.), edited by C. M. Haros and R. Schoenlechner, 28–48. Oxford: Wiley-Blackwell.

Repo-Carrasco-Valencia, R. 2020. Nutritional value and bioactive compounds in Andean ancient grains. *Proceedings* 53(1): 1.

Repo-Carrasco-Valencia, R., A. Acevedo de la Cruz, J. C. Icochea Álvarez, and H. Kallio. 2009. Chemical and functional characterization of kañiwa (*Chenopodium pallidicaule*) grain, extrudate and bran. *Plant Food for Human Nutrition* 64: 94–101.

Repo-Carrasco-Valencia, R., J. K. Hellström, J. M. Pihlava, and P. H. Mattila. 2010. Flavonoids and other phenolic compounds in Andean indigenous grains: Quinoa (*Chenopodium quinoa*), kañiwa (*Chenopodium pallidicaule*) and kiwicha (*Amaranthus caudatus*). *Food Chemistry* 120: 128–133.

Rodrígues, H. S., A. Borém, M. S. F. Valente, M. T. G. Lopes, C. D. Cruz, F. C. M. Chaves, and C. S. Bezerra. 2018. Genetic diversity among accessions of sacha inchi (*Plukenetia volubilis*) by phenotypic characteristics analysis. *Acta Amazonica* 48: 93–97.

Ruales, J., and B. M. Nair. 1993. Saponins, phytic acid, tannins and protease inhibitors in quinoa (*Chenopodium quimoa*, Willd) seeds. *Food Chemistry* 48: 137–143.

Ruiz, C., C. Díaz, J. Anaya, and R. Rojas. 2013. Análisis proximal, antinutrientes, perfil de ácidos grasos y de aminoácidos de semillas y tortas de 2 especies de sacha inchi (*Plukenetia volubilis* y *Plukenetia huayllabambana*). *Revista de la Sociedad Química del Perú* 79(1): 29–36.

Sacilik, K., R. Öztürk, and R. Keskin. 2003. Some physical properties of hemp seed. *Biosystems Engineering* 86(2): 191–198.

Salas, L., D. Tapia, and F. Menegalli. 2015. Biofilms based on canihua flour (*Chenopodium pallidicaule*): Design and characterization. *Quimica Nova* 38(1): 14–21.

Saleem, Z. M., S. Ahmed, and M. Mohtasheemul Hasan. 2016. *Phaseolus vulgaris* Linn.: Botany, medicinal uses, phytochemistry and pharmacology. *World Journal of Pharmaceutical Research* 5(11): 1611–1616.

Salhuana, W. R. 2004. Evaluación de los recursos genéticos de maíz. In *Cincuenta años del Programa Cooperativo de Investigaciones en Maíz (PCIM)*, 252–253. Lima: *Programa Cooperativo de Investigaciones en Maíz*.

Salvador-Reyes, R., M. T. Pedrosa, and S. Clerici. 2020. Review Peruvian Andean maize: General characteristics, nutritional properties, bioactive compounds, and culinary uses. *Food Research International* 130: 108934.

Salvador-Reyes, R., A. P. Rebellato, J. A. Lima Pallone, R. A. Ferrari, M. T. Pedrosa, and S. Clerici. 2021. Kernel characterization and starch morphology in five varieties of Peruvian Andean maize. *Food Research International* 140: 110044.

Sandoval-Oliveros, M. R., and O. Paredes-López. 2013. Isolation and characterization of proteins from chia seeds (*Salvia hispanica* L.). *Journal of Agricultural and Food Chemistry* 1: 193–201.

Sanz-Penella, J. M., M. Wronkowska, M. Soral-Smietana, and M. Haros. 2013. Effect of whole amaranth flour on bread properties and nutritive value. *LWT-Food Science and Technology* 50: 679–685.

Sathe, S. K., B. R. Hamaker, K. W. Sze-Tao, and M. Venkatachalam. 2002. Isolation, purification, and biochemical characterization of a novel water soluble protein from Inca peanut (*Plukenetia volubilis* L.). *Journal of Agricultural and Food Chemistry* 50: 4906–4908.

Saturni, L., G. Ferretti, and T. Bacchetti. 2010. The gluten-free diet: safety and nutritional quality. *Nutrients* 2: 16–34.

Saunders, R. M., and R. Becker. 1984. *Amaranthus*: A potential food and feed resource. *Advance Journal of Food Science and Technology* 6: 357–396.

Schöeneberger, H., R. Gross, H. D. Cremer, and I. Elmadfa. 1982. Composition and protein quality of *Lupinus mutabilis*. *Journal of Nutrition* 112: 70–76.

Schöenlechner, R., S. Siebenhandl, and E. Berghofer. 2008. Pseudocereals. In *Gluten-Free Cereal Products and Beverages*, edited by E. K. Arendt, and F. Dal Bello, 149–190. Oxford: Elsevier.

Segura- Nieto, M., A. P. Barba de la Rosa, and O. Paredes-López. 1994. Biochemistry of amaranth proteins. In *Amaranth Biology, Chemistry, and Technology*, edited by O. Paredes-López, 75–106. London: CRC Press.

Serratos, J. A. 2012. El origen y la diversidad del maíz en el continente americano. Mexico: Universidad Nacional Autónoma de México, 4–12. http://m.greenpeace.org/mexico/Global/mexico/report/2012/9/GPORIGENMAIZ %20final%20web.pdf.

Shafaei, S. M., and S. Kamgar. 2017. A comprehensive investigation on static and dynamic friction coefficients of wheat grain with the adoption of statistical analysis. *Journal of Advanced Research* 8: 351–361.

Souci, S. W., W. Fachmann, and H. Kraut. 2000. *Food Composition and Nutrition Tables*. Stuttgart: Wissenschaft Verlags.

Steffolani, M. E., A. E. León, and G. T. Pérez. 2013. Study of the physicochemical and functional characterization of quinoa and kañiwa starches. *Starch/Staerke* 65(11–12): 976–983.

Stone, L. A., and K. Lorenz. 1984. The starch of amaranth – Physicochemical properties and functional characteristics. *Starch/Stärke* 36: 232–237.

Suleiman, R., K. Xie, and K. A. Rosentrater. 2019. Physical and thermal properties of chia, kañiwa, triticale, and farro seeds as a function of moisture content. *Applied Engineering in Agriculture* 35(3): 417–429.

Takeyama, E., and M. Fukushima. 2013. Physicochemical properties of *Plukenetia volubilis* L. seeds and oxidative stability of cold pressed oil (green nut oil). *Food Science Technology Research* 19(5): 875–882.

Tapia, M. 2015. El tarwi, lupino andino. *Fondo Ítalo Peruano* 15–16. https://docplayer.es/33372914-El-tarwi-lupino-andino.html.

Trugo, L. C., D. von Baer, and E. von Baer. 2003. Lupin. In *Encyclopedia of Food Sciences and Nutrition* (2nd ed.) edited by B. Caballero, 3623–3629. Oxford: Academic Press.

Ullah, R., M. Nadeem, A. Khalique, M. Imran, S. Mehmood, A. Javid, and J. Hussain. 2016. Nutritional and therapeutic perspectives of Chia (*Salvia hispanica* L.): A review. *Journal of Food Science and Technology* 53(4): 1750–1758.

UNALM/MINAGRI (Universidad Nacional Agraria de la Molina/Ministerio de Agricultura del Perú). 2014. Mapa de razas de maíz en el Perú. www.minam.gob.pe/diversidadbiologica/wp-content/uploads/sites/21/2014/02/razasmaizperu.pdf.

USDA (US Department of Agriculture), Agricultural Research Service. 2010. USDA National Nutrient Database for Standard Reference, Release 23. Nutrient Data Laboratory Home Page. www.ars.usda.gov/ba/bhnrc/ndl.

USDA (US Department of Agriculture), Agricultural Research Service. 2011. National Nutrient Database for Standard Reference, Release 24. Nutrient Data Laboratory Home Page. Washington, DC: US Department of Agriculture, Agricultural Research Service. www.ars.usda.gov/ba/bhnrc/ndl.

Valcárcel-Yamani, B., S. Caetano, and S. Lannes. 2012. Applications of quinoa (*Chenopodium quinoa* willd.) and amaranth (*Amaranth* spp.) and their influence in the nutritional value of cereal based foods. *Food and Public Health* 2(6): 265–275.

Valencia-Chamorro, S. A. 2003. Quinoa. In *Encyclopedia of Food Science and Nutrition*, edited by B. Caballero, 4895–4902. Oxford: Academic Press.

Vanier, N. L., C. D. Ferreira, I. da Silva Lindemann, J. Pozzada Santos, P. Zaczuk Bassinello, and M. E. Cardoso. 2019. Physicochemical and technological properties of common bean cultivars (*Phaseolus vulgaris* L.) grown in Brazil and their starch characteristics. *Revista Brasileira de Ciências Agrárias* 14(3): 56–75. ISSN 1981-0997.

Vega-Gálvez, A., M. Miranda, J. Vergara, E. Uribe, L. Puente, and E. A. Martínez. 2010. Nutrition facts and functional potential of quinoa (*Chenopodium quinoa* willd.), and ancient Andean grain: A review. *Journal of the Science of Food and Agriculture* 90: 2541–2547.

Vilche, C., M. Gely, and E. Santalla. 2003. Physical properties of quinoa seeds. *Biosystems Engineering* 86: 59–65.

Villacrés, E., C. Yánez, L. Armijos, M. B. Quelal, and M. Álvarez. 2016. *El Despertar Gastronómico del Maíz: Recetario; Estación Experimental Santa Catalina*. Quito, Ecuador.

Vrancheva, R., L. Krystev, A. Popova, and D. Mihaylova. 2019. Proximate nutritional composition and heat-induced changes of starch in selected grains and seeds. *Emirates Journal of Food and Agriculture* 31(9): 718–724.

Wang, B., and J. Wang. 2019. Mechanical properties of maize kernel horny endosperm, floury endosperm and germ. *International Journal of Food Properties* 22(1): 863–877.

Wang, S., F. Zhu, and Y. Kakuda. 2018. Review Sacha inchi (*Plukenetia volubilis* L.): Nutritional composition, biological activity, and uses. *Food Chemistry* 265: 316–328.

Yánez, C., J. L. Zambrano, C. Maicedo, V. H. Sánchez, and J. Heredia. 2003. *Catálogo de Recursos Genéticos de Maíces de Altura Ecuatorianos; INIAP, Estación Experimental Santa Catalina*. Quito, Ecuador. https://repositorio.iniap.gob.ec/jspui/bitstream/41000/43/1/iniapsc201.pdf.

Zambrano, J. L., C. F. Yánez, and C. A. Sangoquiza. 2021. Maize breeding in the highlands of Ecuador, Peru, and Bolivia: A review. *Agronomy* 11: 212–221.

Chapter 5

Latin-American Crops in Gluten-Free Applications

Silvia V. Melgarejo-Cabello[1], Jehannara Calle-Domínguez[2], María Alejandra Giménez[3], Claudia M. Haros[4], and Ritva Ann-Mari Repo-Carrasco-Valencia[1]
[1]Universidad Nacional Agraria La Molina, Peru
[2]Instituto de Investigaciones para la Industria Alimenticia (IIIA), La Habana, Cuba
[3]Universidad Nacional de Jujuy and Consejo Nacional de Investigaciones Científicas y Técnicas (CONICET), San Salvador de Jujuy, Argentina
[4]Universidad Nacional Agraria La Molina, Peru

CONTENTS

DOI: 10.1201/9781003088424-5

5.1   INTRODUCTION

Latin-American crops, such as quinoa, amaranth, kañiwa, tarwi, maize, chia, sacha inchi and black turtle beans have an excellent nutritional value and are widely cultivated and used in the region. Quinoa and tarwi contain antinutrients (saponins and alkaloids, respectively) which must be removed before consumption. Among these crops, kañiwa is the least known and has enormous potential as a very nutritious ingredient in the food industry. Amaranth is an ancient food crop traditionally consumed by people from Latin America. Maize is cultivated and used all over Latin America and has several traditional and modern food applications. Sacha inchi is a crop with an exceptionally high oil content with a very healthy fatty acid composition. Due to the physical–chemical and nutraceutical properties of chia seeds, they have interesting technological properties and potential uses for different food applications. Whole seed flours can be obtained from all of these crops, and they are promising ingredients for different kinds of products, including gluten-free applications.

5.2   ANDEAN GRAINS

The four native Andean grains, quinoa (*Chenopodium quinoa*), kañiwa (*Chenopodium pallidicaule*), kiwicha (*Amaranthus caudatus*) and tarwi (*Lupinus mutabilis*) have an exceptional nutritional composition. They can be used as ingredients in conventional and gluten-free bakery products to improve the nutritional value of these products. They have interesting techno-functional characteristics which could enhance the sensorial properties of the final products.

5.2.1 Processing of Quinoa

Quinoa contains macro- and micronutrients (starch, protein, fiber, fat, minerals and vitamins), which are found in different parts of the seed. The quinoa germ or embryo, represents approximately 30% of the whole seed (in the case of common cereals it represents only 5% of the seed and extends through the starchy endosperm), surrounds the perisperm like a ring covering the seed. It is part of the bran fraction, which is relatively rich in fats and proteins (Alonso-Miravalles and O'Mahony, 2018; García Solaesa et al., 2020; Mufari et al., 2018). The starch granules occupy the cells of the perisperm, which constitutes 60% of the grain, while most of the lipids and proteins are found in the germ (Ando et al., 2002; Prego et al., 1998).

Saponins, located in the outer layers of quinoa, generate a bitter taste and have anti-nutritional effects. The removal of saponins is easily accomplished by rinsing the seed in cold alkaline water (wet method) or mechanical abrasion (dry method). The wet method is generally used, generating a large amount of water waste and causing pollution in rivers and lakes. Abrasive de-hulling is an alternative method to eliminate the saponins. Han et al. (2019) investigated the effect of degree of milling (DOM) on the content of saponins and free and bound phenolics, and their antioxidant activity The saponin content decreased from 15.50 to 9.02 mg oleanolic acid equivalents/g dry weight (OAE/g DW) after using a disc mill device. These results indicate that the milling process could become an effective method for removing the saponins in quinoa.

Ando et al. (2002) tested abrasive grinding for the fractionation of quinoa and obtained a bran fraction (8.2%) and an embryo fraction (30.1%) that were significantly different in their composition; in addition, they were successfully separated from the perisperm (58.8%) and the proportion of protein fraction was not affected by abrasive grinding. Hemalatha et al. (2016) investigated sequential abrasive grinding and produced four fractions: husk (10–12%), de-hulled grain, bran (12–15%) and ground grain (mainly white perisperm). The most abundant fraction was grain flour at 75–77% of the total grain weight. Ruiz et al. (2016) explored combined dry and wet processing and obtained, through impact grinding, a coarse fraction rich in perisperm and a fine fraction mainly composed of the embryo, which had a different protein content.

D'Amico et al. (2019) investigated abrasive milling with commercial-quality seeds; 10 g of quinoa was sequentially ground for 8 min at 1-min intervals and sieved to separate it into the following components: de-hulled kernel, coarse (fraction), fine (fraction) and embryo fractions. They reported that in the first 2 min the coarse fraction was mainly separated (shell and bran), in 3 to 5 min the embryo was broken and after only 5 min the fine fraction was obtained, which contained parts of the perisperm. The approach of grinding the kernels by abrasive milling in layers allowed them to obtain very detailed information about the nutritional composition of the selected seed tissues. The outer layers of the kernels were rich in minerals and fiber (arabinose and galactose-rich polysaccharides). After 2 min of milling, there was no longer any coarse fraction obtained and removal of the embryo was boosted, which resulted in very high protein content of the abraded layers. After 8 min of milling, more than 65% of the original seed mass was removed and the polished kernels were low in protein, ash and fat, whereas the amount of glucose increased. Although quinoa has a completely different structure with respect to cereals, some similarities were found. The high concentrations of ash and dietary fiber were comparable to those of cereals, whereas the protein composition in the milling course was more affected in true cereals.

Comparison of fat localization showed a completely diverse picture, because in cereals the germ is located close to the bran, whereas in quinoa the embryo surrounds the perisperm. The ash content can be used to estimate the DOM yield. Thus, classification of quinoa flours can be performed as for cereals. To know the protein composition of the quinoa grain, extractions were carried out (based on the solubility of the proteins) with pure water and 2% sodium chloride, obtaining two fractions, albumins (43.3%) and globulins (27.9%) that correspond to Osbourne fractions. In a third extraction step, 1% sodium dodecyl sulfate (SDS) was used instead of 60–70% ethanol to extract the prolamines that represented 11.1%; glutelins (insoluble residues) are calculated as the mass difference of all soluble protein fractions with respect to total protein content and accounted for 21.8% (D'Amico et al., 2019).

The high proportion of albumins and globulins (more than 70% of the total proteins) and the lower abundance of prolamines and glutelins correspond to a typical dicotyledonous plant. The predominance of albumins was confirmed from the study by Ruiz et al. (2016), who found 40.3% water-soluble proteins in quinoa. These results demonstrate that the protein composition within quinoa grains is different from that of common cereals. Gluten proteins, composed of prolamines and glutelins, are predominantly found in the endosperm, while bran and especially the aleurone layer are rich in albumins and globulins. In summary, proportions of protein fractions are much more affected by husking in wheat than in quinoa.

Ballester-Sánchez et al. (2019) isolated starch from quinoa using a wet milling procedure. They investigated the effect of duration (1, 5 and 9 h) and temperature (30, 40 and 50°C) of maceration in an SO_2 solution with lactic acid on the recovery of starch and its quality. The effect of the steeping conditions on the starch was evaluated in terms of whiteness, proteins, lipids, amylase and the content of damaged starch. The highest level of recovery and the best quality of starch were obtained after 6.5 h of maceration at 30°C.

5.2.2   Processing of Kañiwa

Traditionally, kañiwa seeds are eaten roasted and ground into flour (*kañiwaco*). They are used to make soup, breakfast drinks, bread, cookies and cakes. Compared to quinoa, kañiwa has the advantage of not containing detectable amounts of saponins, so it does not require washing prior to consumption (Bustos et al., 2019). The grinding of grains can be carried out in various types of mills; flour improves the nutritional characteristics of bakery, baked or pasta products; however, the percentage used is limited by the absence of gluten, which affects the composition of the starch–gluten matrix, of utmost importance for textural properties.

Bustos et al. (2019) investigated the production of pasta prepared with wheat flour enriched with kañiwa wholemeal flour at replacement percentages of 10%, 20% and 30%. Wholemeal flour was prepared in a hammer mill, reaching granulometry of less than 250 μm. The partial replacement of wheat flour with kañiwa flour is a viable option for improving the nutritional value of breads, with acceptable dough yields. Rosell et al. (2009) evaluated the production of bread with 10% to 100% replacement of wheat flour with kañiwa flour obtained through a cyclone mill, concluding that up to 25% substitution could produce doughs with acceptable thermo-mechanical patterns and breads with good sensory acceptability, but variable in color.

Kañiwa flour is also a source of protein extraction. For example, Betalleluz-Pallardel et al. (2017) established optimal conditions for the extraction of kañiwa protein from flour with a granulometry of less than 500 μm; the optimal protein extraction conditions correspond to a temperature of 21°C, extraction time of 5 min and solvent (water)/flour ratio of 37:1 (v/w) at pH 10, resulting in a protein yield of 80.4%.

5.2.3   Processing of Kiwicha

Kiwicha is recognized for its excellent nutritional content, mainly for its good quality proteins. Kiwicha does not have detectable amounts of saponins, so it does not require pre-treatment washing, and it is gluten-free, which makes it an interesting ingredient for products aimed at celiac consumers (Álvarez-Jubete et al., 2009, 2010a; Calderón de la Barca et al., 2010).

Although numerous studies of milling kiwicha grains have been carried out at laboratory level, industrial use of these methodologies is still very complicated. This is due to factors such as poor development of grain cultivars suitable for mechanical harvesting and the present low yields associated with grain size, so that studies which allow optimization of these processes are necessary (Roa et al., 2014). Calzetta et al. (2006) developed a wet milling methodology with an optimal starch yield of 45.0%. This methodology begins with a previous soaking of seeds in an aqueous solution; subsequently, the wet materials are ground and then sieved to retain the fiber. The remaining suspension is centrifuged, the supernatant discarded, the mucilaginous layer separated and the precipitated starch suspended in distilled water. Finally, the starch is dried in a forced-air oven. Roa et al. (2014) used wet milling methodology, obtaining a starch with 98.9% carbohydrates and 0.6% proteins.

By dry milling, due to the structure and morphology of kiwicha grains, it is possible to separate the anatomical fractions to obtain flours of different composition. Tosi et al. (2000) developed a dry differential milling technique to produce three flour fractions: one rich in protein (40%), another rich in dietary fiber and a fraction containing approximately 79% starch. They determined that the grains should be dried before grinding to improve the protein yield in the first fraction; thus, the optimal conditions for grain drying (for an initial humidity of 11.5%) included treating with air at 90°C for 3 min, with a flour extraction yield of 36.9%; however, as a consequence of the drying effect, the available lysine decreased by 38.5%.

Roa et al. (2014) used abrasive milling, with a Suzuki MT mill that automatically separates pearl amaranth and bran, to obtain a starch-rich grinding fraction (88%) and a lipid-protein-rich fraction with 23% lipid and 46% protein contents, respectively. Next, they carried out a planetary ball mill grinding. They found that abrasive milling reduces the degree of crystallinity of the starch due to its amortization during these processes.

Sindhuja et al. (2005) developed biscuits with amaranth flour, finding that 25% substitution of wheat flour for amaranth flour was optimal. Also, due to its high starch content, amaranth is a promising material for edible film production (Roa et al., 2014). On the other hand, Basilio-Atencio et al. (2020) used whole kiwicha flour, obtained in a hammer mill and with a 0.1 mm mesh sieve, to obtain healthy extruded kiwicha snacks, managing to determine that the optimal extrusion conditions were at a temperature of 190°C and 14% initial moisture in the flour.

5.2.4 Processing of Tarwi

Tarwi seeds are recognized for their exceptionally high protein and oil content; their consumption is limited due to the high content of anti-nutrients such as bitter alkaloids, phytic acid, tannins, nitrates and trypsin inhibitors (Villacrés et al., 2020). For safe consumption, the seeds must be subjected to a treatment to considerably reduce the amount of anti-nutrients. This specific treatment, traditionally known as debittering, is therefore a critical step in tarwi processing.

Debittering consists of an aqueous thermal treatment, which is very efficient in reducing unwanted components of the seed, managing to reduce levels of alkaloids to average values below 0.2 g/kg of dry matter. Cortés-Avendaño et al. (2020) carried out a study to identify alkaloids after the debittering process in ten tarwi ecotypes from different regions of Peru. They managed to identify residue of only two types of alkaloid: lupanin (average 0.0012 g/100 g of dry matter) and sparteine (average 0.0014 g/100 g of dry matter). However, this methodology favors a reduction in the antioxidant capacity and of some of the main nutrients of the grain such as proteins, carbohydrates, ascorbic acid, phenolic compounds and carotenoids (Villacrés et al., 2020).

The debittering process of tarwi seeds requires large amounts of water and takes a long time. The water used for the elimination of alkaloids is traditionally used as a biocide for pests. Recent studies have been conducted with emerging procedures through biotechnological processes such as fungal fermentation, which allows a reduction of alkaloids and tannins, with the advantage of maintaining a higher content of other nutrients such as carotenoids and phenolic compounds (Romero-Espinoza et al., 2020; Villacrés et al., 2020).

Various types of mill can be used for grinding. Cortés-Avendaño et al. (2020) used a hammer mill to obtain flours with particle sizes between 100 and 500 μm. Wang et al. (2016) carried out two-stage milling: first by grinding with a pin mill and later in an impact mill, obtaining flours with 8 to 10% humidity and an average particle size of 10 to 100 μm. The particle size obtained from grinding is important for subsequent processes. They also managed to obtain flour with a high protein content using an electrostatic separation method, for flours with a smaller average particle size.

Rosell et al. (2009) used tarwi flour, obtained by milling in a laboratory cyclonic mill, in combination with wheat flour in bread formulations; however, they verified that breads with up to a minimum content of 12.5% of tarwi flour presented poor sensory acceptability.

5.2.5 Gluten-Free Products

The gluten-free products market has increased throughout Latin America, due to the rise in patients diagnosed with coeliac disease and consumers seeking healthier alternatives. However, these are not necessarily healthier, because the gluten-free bakery products on the market are made from starches or white rice flour, which are lacking in high-quality proteins, dietary fiber and important micronutrients. Andean grains such as quinoa, kiwicha, kañiwa and tarwi do not contain peptides like wheat gluten; therefore, they are appropriate raw materials for consumption by people with coeliac disease. Quinoa, kañiwa and kiwicha, the Andean amaranth, are sources of starches (more than 70% of their composition) (Ramos Diaz et al., 2013; Rosell et al., 2009; Schöenlechner, 2017a), which are necessary to create the structure of bread (Benavent-Gil and Rosell, 2018). On the other hand, tarwi is a legume with very high protein and oil content (almost 50% and around 20%, respectively). Tarwi oil could function as a natural emulsifier to retain the gas produced during fermentation of gluten-free breads (Vidaurre-Ruiz et al., 2019). Quinoa is appreciated because of its high protein quality, having a balanced essential amino acid composition, and it is also considered an excellent source of dietary fiber and minerals (Repo-Carrasco-Valencia et al., 2010). Kiwicha is a very good source of iron, calcium and zinc. It contains more zinc and iron than conventional corn and common beans (Burgos et al., 2018). The inclusion of these grains in the formulation of gluten-free products is promising.

The quality of wheat bread is the result of special properties of gluten proteins (gliadins and glutenins). They are responsible for the high water absorption capacity and give dough cohesiveness, viscosity, elasticity and gas-holding ability and a final product with a high volume and porous crumb. All other cereal flours produce breads of poor quality with low volume and a small-pored, inelastic crumb when baked under standardized conditions (Koehler et al., 2014). It is very difficult to replace gluten in bakery products and it requires the employment of a mix of permitted flours, proteins, hydrocolloids and special technologies in an attempt to replace the numerous functions of gluten (Arendt et al., 2008). Generally, basic materials in the production of gluten-free bread are starch-containing flours or starches from gluten-free sources (e.g., corn, rice, potatoes). To achieve an appropriate water absorption capacity and dough viscosity, several hydrocolloids can be used. Hydroxypropylmethyl-cellulose, carboxymethylcellulose, carrageenan, xanthan gum, guar gum and sodium alginate are examples of compounds used in the production of gluten-free bread (Koehler et al., 2014).

Dairy substitutes, like soy milk, can be produced from quinoa flour by a hot mashing process, or by drum drying (Schöenlechner, 2017b). These beverages provide novel alternatives to current dairy substitute products, having a relatively high protein content and low glycemic index (Pineli et al., 2015).

Quinoa can be used in bakery products such as cookies without the incorporation of wheat flours, despite the lack of gluten (Schöenlechner et al., 2006).

For biscuit and cookie making, a strong gluten network is not needed, making pseudo-cereals suitable for this purpose. Studies by Inglett et al. (2015) showed that amaranth-based cookies have good sensorial and textural attributes. Popped amaranth flour and wholegrain popped amaranth have been employed, resulting in cookies with good sensory properties (Calderón de la Barca et al., 2010). Malted amaranth flour cookies have shown superior properties to those of conventional wheat cookies (Chauhan et al., 2015).

The baking properties of amaranth, quinoa and buckwheat were studied by Álvarez-Jubete et al. (2010b). Their study demonstrated that pseudo-cereal flour can be introduced into gluten-free bread formulation with the aim of enhancing crumb softness and cohesiveness without adversely affecting the other sensory properties of the bread. The use of quinoa and lupin flour in the production of gluten-free and egg-free pasta was studied by Linares et al. (2019). They tested the effect of vegetable proteins and the oxidizing enzyme POx on the quality of noodles. After these trials, the final recipe containing lupin four, pea protein and POx had satisfactory gluten-free noodle quality and possessed a valuable nutritional composition with high protein and dietary fiber content.

Schöenlechner et al. (2010), in a study on the use of amaranth flour in bread making, found that water content had the greatest effect on bread properties, whereas variation in fat content did not have any significant influence. They also detected that the combined addition of fat and egg albumin resulted in the best overall sensory acceptance. Amaranth flour improves the freeze stability of gluten-free dough but reduces its shelf life (Leray et al., 2010). Calderón de la Barca et al. (2010) produced gluten-free bread with a uniform crumb and higher specific volume compared to other gluten-free breads using 60–70% popped and 30–40% raw amaranth flours. Sourdough fermentation could be a promising strategy in improving the quality of gluten-free bread. Jekle et al. (2010) demonstrated that amaranth sourdough properties could be enhanced by using different lactic acid bacteria. Coda et al. (2010) used *Lactobacillus plantarum* C48 and *L. lactis* subsp. *lactis* PU1 in sourdough with the aim of producing gluten-free bread using a blend of buckwheat, amaranth, chickpea and quinoa flours. Textural analysis showed that sourdough fermentation enhanced several characteristics of the gluten-free bread, thus approaching the features of wheat flour bread. Sensory analysis showed that sourdough fermentation allowed the production of breads with good palatability and overall taste appreciation.

Lupin protein concentrate has been used by Horstmann et al. (2017) and Ziobro et al. (2013) in gluten-free baking. Vidaurre-Ruiz (2020) reported that the inclusion of flour of the Andean lupin, tarwi, in a gluten-free bread formulation with quinoa flour and potato starch gives the final product the following nutritional content: 10.1% protein, 18.6% dietary fiber and 3.3% minerals. Similar results were obtained using tarwi flour in a kiwicha–potato starch formulation.

The protein content of these breads is between two to three times higher than commercial gluten-free breads.

Kañiwa is the least known of all Andean native grains. There is scarce information regarding the use of this nutritious grain in gluten-free products. Luna-Mercado and Repo-Carrasco-Valencia (2020) evaluated the techno-functional, thermo-mechanical and physico-chemical properties of two kañiwa varieties (Cupi and Illpa Inia), in order to assess their potential as ingredients in gluten-free bakery products. The characterization of these flours of two Peruvian kañiwa varieties provided new information on the suitability of kañiwa for use in gluten-free baking. Kañiwa is an interesting and novel ingredient for use in gluten-free products, enhancing their nutritional value.

Commercial products based on quinoa, amaranth, kañiwa and tarwi are shown in Figures 5.1–5.4.

5.3 MAIZE

Maize (*Zea mays* L.) is one of the cereals that celiac patients can consume, if suitable conditions are provided during production. Whole maize or its flours are widely used in the preparation of typical South America foods. Nevertheless, the absence of prolamins that form a viscoelastic network like gluten represents a challenge in the elaboration of pasta or baked products. Maize flour is not rich in essential amino acids like lysine and tryptophan; other gluten-free raw materials can be added to nutritionally enhance different gluten-free products (Figure 5.5).

5.3.1 Processing of Maize

5.3.1.1 Nixtamalization

Alkaline cooking of maize (nixtamalization) is a process used in the preparation of Mexican-style foods. During nixtamalization, maize is first cooked in water in the presence of lime, steeped overnight for several hours, and then washed to produce nixtamal, which is then stone-ground to form dough. Alternatively, nixtamalized maize dough can be dried and dry-milled to produce maize flour (Cravioto et al., 1945; Bressani, 1990). In this process, the pericarp is hydrolyzed mostly into arabinoxylans; the starch of the endosperm undergoes a partial gelatinization; some lipids are saponified, and proteins are partially solubilized and polymerized. Another important change is the calcium absorption including other minerals from lime (Santiago-Ramos et al., 2018). The extent of these changes and interactions among some components produces nixtamalized maize dough with different viscoelastic behavior (Vazquez-Carrillo et al., 2015). The nixtamalization process increases the calcium content, and other minerals,

Figure 5.1 Commercial gluten free foods with amaranth: **A.** Puffed amaranth compared to amaranth seeds; **B.** Amaranth bars with probiotics; **C.** Amaranth *tortillas*; **D.** "*Alegría*", **E.** Gluten free "*alfajores*" with amaranth. **Source: Authors' own work.**
Notes: A) Two small round plates next to each other, one of them with amaranth and the other with a puffier and therefore slightly bigger amaranth.

Figure 5.1 (continued)

B) Medium round plate with three different flavored popped amaranth (peanuts, chocolate and raisins), rectangular prism snacks (in their respective plastic bags) parallel to each other.

C) Medium round plate with a round amaranth tortilla; the seeds are embedded on the surface.

D) Five popped amaranth, short cylinder snacks in their respective plastic bags in a small round bowl.

E) Two alfajores next to each other. The alfajor is formed by two small, round amaranth cookies one at the top and the other at the bottom with a dairy, sweet ingredient called "dulce de leche" in the middle.

in maize flours, improves the bioavailability of niacin and iron and decreases the content of mycotoxins and phytic acid. However, during alkaline cooking, phenolic compounds and other important antioxidants can be lost (Cravioto et al., 1945; Bressani et al., 2002, de la Parra et al., 2007, Gutiérrez-Dorado et al., 2008).

In nixtamalization, the wet milling of nixtamal is an important operation. Two cylindrical stones are used to mill the maize limed samples; one of the stones is fixed, while the other rotates (Serna-Saldívar and Chuck-Hernández, 2019). The distance between the stones is adjusted to obtain a certain particle size distribution; at a higher compression force, the dough is soft, and is used to produce tortillas, but if the compression force decreases, the resulting dough turns into coarse particles, and is used in different products. Alternatively, steel plate dough mills are used. These mills are more efficient and produce dough at a lower temperature (Villada et al., 2017).

Nixtamalized maize flours are generally industrially obtained and have specific characteristics for the production of a particular food. Some companies offer more than 25 different nixtamalized flours formulated to meet certain requirements for color, pH, granulation, water absorption, viscosity and sensorial characteristics (Serna Saldívar and Chuck-Hernández, 2019). Interestingly, from a home-made limed maize in the Mayan and Aztec cultures, before the conquistadores arrived, it has now reached a wide international market in all continents of the world (Paredes-López et al., 2012).

5.3.1.2 Milling
According to Abarza and Schimpf (2009), different products are obtained from the grinding process of native maize from the Andean zone of Jujuy, Argentina, such as *frangollo* (broken maize grains that were peeled with lime), semolina and flour. The last two are mainly used to make soups (*calapi, caldo majado* and *tulpo*) or to elaborate bread known as *bollo* mixed with wheat flour

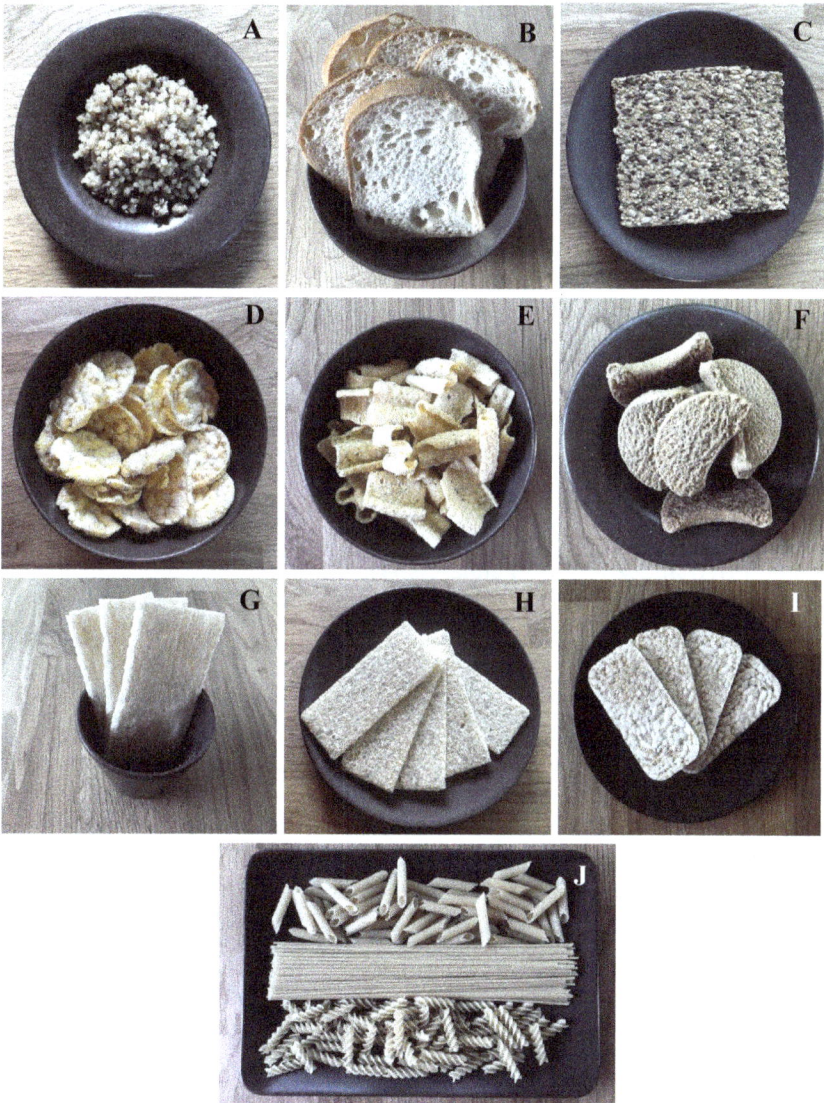

Figure 5.2 Commercial gluten free foods with quinoa: **A**. White and red quinoa ready-to-eat; **B**. Bread with quinoa and millet; **C**. Crispbread with sunflower seed and quinoa; **D**. Mini corn with quinoa cakes; **E**. Quinoa chips; **F**. Quinoa cookies; **G**. Crispy quinoa toast; **H**. Quinoa and rice crackbread; **I**. Whole rice and quinoa cakes; **J**. Penne, spaghetti or fusilli quinoa and rice pasta. **Source: Authors' own work.**
Notes: A) Cooked, ready-to-eat red and white quinoa on a small round plate.

Figure 5.2   (continued)

B) Five small slices of quinoa and millet bread in a small round bowl.

C) Three rectangular shaped crispbread with sunflower seed, quinoa and other grains on a medium size round plate.

D) Many circular mini corn crackers with quinoa in a small round bowl.

E) Many quinoa chips in a corrugated rectangular shape in a small round bowl.

F) Five half-moon shaped quinoa cookies, with a rough surface on a medium sized plate.

G) Three pieces of thin and rectangular crispy quinoa toast in a fan-like position in a cup.

H) Five thin and rectangular quinoa and rice crackerbreads in a fan-like position on a medium size plate.

I) Four rectangles with rounded corners, whole rice and quinoa cakes in a fan-like position on a medium size round plate.

J) Rectangular plate positioned horizontally with three types of pasta. Spaghetti is found horizontally across the middle of the plate, penne on the upper part of the plate and fusilli on the lower side of the plate.

(Cámara-Hernández and Arancibia-Cabezas, 2007). In this region, hydraulic mills brought by Spaniards were incorporated as a technological change; but this did not imply a cultural change since grain milling was a practice already known and used. On the contrary, grinding in stone mortars, domestic and community, is an important feature of many Andean societies. These mortars – *pecans, cutanas* – continued to be used together with hydraulic mills, generally differentiating uses of the flours that were produced in each. Currently, electric mills for community use have also been incorporated (Bugallo and Mamani, 2014). Semolina and flours obtained from maize from the Andean region of Argentina generally come from the whole grinding of the grain that, depending on culinary use, can be subjected to successive sieves to eliminate part of the bran.

5.3.1.3 Extrusion

Extrusion technology is considered one of the most economic processes for cereal and snack food industries. This process can produce dough with similar characteristics to that by using nixtamalization traditional processes (Corrales-Bañuelos et al., 2016). Extrusion is a suitable technology for obtaining gluten-free products such as pasta, even using whole grain maize flours. This process can also be applied to improve the stability of maize whole flour, reaching an adequate temperature to the inactivation of lipases and lipooxygenases responsible for rancidity, avoiding undesirable sensorial changes and increasing rancidity time (Paesani et al., 2020). According to Kljak et al. (2015), the different

Figure 5.3   Commercial products of kañiwa: **A.** Kañiwa and kiwicha bar; **B**. Kañiwa pop; **C**. Kañiwa flakes; **D**. Kañiwaco. **Source: Authors' own work.**
Notes: A) Five rectangular kañiwa and kiwicha bars on an oval shape plate. Three of them horizontally in the middle of the plate and two vertically at either side.
   B) Kañiwa popped seeds on a round medium size plate.
   C) Thin and small kañiwa flakes on a medium size plate.
   D) Toasted kañiwa flour on a medium size plate.

genotypes and different physical and chemical characteristics of maize lead to different nutritional values and different responses in the extrusion process. In this regard, Merayo et al. (2011) showed that the quality of maize spaghetti depended on extrusion conditions and endosperm hardness, finding that soft endosperm allowed for obtaining better quality spaghetti. Application of the

Figure 5.4   Commercial products of tarwi: **A**. Quinoa and tarwi cookies; **B**. Tarwi cheese; **C**. Tarwi and lucuma flour; **D**. Snack of tarwi (lupine) with spices.
Notes: A) Quinoa and tarwi heart-shaped cookies with a circle in the middle of each cookie, on a small round plate. **Source: Authors' own work.**
  B) Molded cylindrical tarwi cheese on a medium size plate.
  C) Tarwi and lucuma flour on a large plate.
  D) Plate with many white, oval-shaped tarwi seeds treated with spices.

extrusion process in different Andean maize varieties for spaghetti's manufacture allowed the obtention of products of adequate quality and sensory acceptance for celiac consumers (Giménez et al., 2014). The textural properties of maize-spaghetti and their behavior during overcooking depend on the degree of gelatinization reached during the extrusion-cooking process and the formation of the amylose-lipid complex (Merayo et al., 2011, Giménez et al., 2013).

Figure 5.5 Gluten free foods with Andean maize: **A.** Roasted yellow corn flavored with spices; **B.** Gluten free "*alfajores*" with capia maize; **C.** Spaghettis with culli, capia or yellow maize; **D.** Pasta made with Andean maize augmented with quinoa, beans or amaranth; **E.** Pudding made with Andean maize culli and Andean maize capia; **F.** Toasted yellow corn flour in preparation for an Andean drink "*Chilcan*". **Source: Authors' own work.**
Notes: A) Roasted yellow corn flavored with spices in a squared plastic bag.
 B) Two alfajores, formed by two small, round capia maize cookies, one at the top and the other at the bottom with a dairy, sweet ingredient called "dulce de leche" in the middle of them.

Figure 5.5   (continued)

C) Three different types of Andean maize spaghetti, front view (the top one is made with culli, middle one with capia and the bottom one with yellow maize).

D) Three plastic bags of Andean maize spaghetti positioned vertically, with quinoa, beans or amaranth.

E) Two large rectangular pudding cakes (one beige culli and the other dark brown capia), and beside each one there is a slide of them.

F) Toasted yellow corn flour in a rectangular plastic bag.

5.3.1.4   Toasting

Toasting is a poorly studied process of maize products, and its importance relies on the fact that they are widely consumed by the rural population of Mexico and South America in different typical preparations. Toasted grains exhibit improved digestibility, color, shelf life, flavor and texture, greater volume and crispiness and reduced anti-nutrient factors (Sandhu et al., 2017). The toasting process is carried out at dry high temperatures (250–270°C). The grain structure protects the starch granule during heat treatment, preventing thermal fractionation and a large melting degree. Therefore, it may be considered as an alternative heat treatment to obtain starch with some desirable properties, such as greater crystallinity (Carreras et al., 2015).

5.3.2   Gluten-Free Products

The health benefits of consuming whole grains and their flours are clear (Shepherd et al., 2012). However, a larger scale of production requires a thermal treatment for the destruction of lipolytic enzymes and lipoxygenases in order to prevent spoilage. On the other hand, the presence of bran is a critical factor in whole flours due to a negative effect on the quality of some products such as pastas and bread (Liu et al., 2016).

Despite the nutritional interest in the use of whole grain flours in gluten-free products, there are few studies on the use of whole-grain maize. Paesani et al. (2020) showed that the presence of whole-grain maize flours in gluten-free sugar-snap doughs and cookies increased the particle size and water absorption values while the elastic and viscoelastic modulus of the doughs decreased, allowing a greater expansion in the oven and a lower hardness. Segundo et al. (2019) formulated a gluten-free layer and sponge cakes with whole maize flour replaced by green banana flour, achieving nutritionally improved products with acceptable texture and sensory characteristics. According to Carmelo-Méndez et al. (2017) and Žilić et al. (2012) whole blue maize flour can be used to develop gluten-free pasta and cookies with high dietary fiber contents, increasing the total phenolic compound content and antioxidant capacity as well as a decreased starch hydrolysis due to the increase in resistant starch content. Red and blue maize flours were used for functional cookie preparation. The effect of the baking conditions, such as baking time and temperature, affected the content of phenolic compounds, antioxidant capacity and color of cookies. According to Žilić et al. (2016), the addition of citric acid to pigmented maize dough increased the phenolic compounds in maize cookies by anthocyanins stabilization and improved their accessibility. In addition, the Maillard reaction had a crucial influence on the increment of antioxidant capacity in cookies made from anthocyanin-rich maize flour. Flour derived from milled conventional maize (*Zea mays*) presents limitations in breadmaking due to its flavor and distinctive yellow color. Maize breads are characterized by a strong yellow color, a low specific volume, and a dense and firm crumb (Capriles and Areas, 2014). Because of its color, white maize flour is an interesting alternative for the production of gluten free bread. According to de la Hera et al. (2013), coarser particle size maize flours are best suited for making gluten free bread, since they give products with more volume and less firmness than bread made with finer flour.

Giménez et al. (2014) studied the acceptability of spaghetti of two races of Andean maize in celiac and non-celiac consumer groups. In this work, celiac consumers assigned high acceptability scores to the samples and described it as tasty, smooth, tender, novel, with a pleasant flavor and good quality. Also, they characterized maize spaghetti as a product that can be consumed every day and for the whole family. However, this product should be reformulated for non-celiac consumers. The combination of maize flour with other gluten-free flour of pseudo-cereals, legumes and vegetables among others provided an increase in the nutritional value of maize gluten free products such as pastas and bread (Padalino et al., 2013; Dib et al., 2018). The gluten free spaghetti and *alfajores* made with flours of maize, quinoa (*Chenopodium quinoa)*, amaranth (*Amaranthus caudatus*) and broad bean (*Vicia faba*) remarkably increased the content of dietary fiber, unsaturated fatty acids, iron and zinc. These nutritionally improved spaghettis had a higher "cooking loss" than spaghetti made with only maize, and adequate sensorial properties (Giménez et al., 2013, 2016).

Other gluten free products such as toasted maize grains with salt and other local condiments and *Chilcan* (toasted ground maize flour, with spices and sugar to prepare a drink) are commonly consumed in the Andean zone of Jujuy, Argentine as an energetic food.

5.4 SACHA INCHI AND BLACK TURTLE BEAN

Plukenetia volubilis (L.) and *Phaseolus vulgaris* (L.) species are plants from the oilseed and legume family. Around the world, they are known by their common name sacha inchi and black turtle bean, respectively; some cultivars of *Phaseolus vulgaris* (L.) are also known as common beans. However, in this chapter only black turtle beans are included. Their nutritional composition and the presence of bioactive compounds make these products a good alternative to extend their consumption, transforming them into raw ingredients such as flours and using them to develop new foods with added value. Sacha inchi and this legume seeds are novel foods that can be used as unconventional flours in some formulations. Their consumption could reduce malnutrition, guarantee food security and promote healthy eating, helping to ensure the longevity of the population by improving quality of life.

Lipids are the major compound present in sacha inchi seeds (35–60%), higher than values reported for chia and soybean seeds (Chirinos et al., 2013). Their lipid profile, rich in unsaturated (6.8–9.1%) and saturated fatty acids, is a good source of α-linolenic acid, linoleic acid and oleic acid (Wang et al., 2018) with a high content of tocopherols (Štěrbová et al., 2017). Furthermore, seeds present elevated protein content (27%) with relatively high levels of cysteine, tyrosine, threonine and tryptophan in their amino-acid profile. Castaño et al. (2019) and de la Rosa-Millán et al. (2019) reported that powder from these seeds contained more than 50% of protein.

Black turtle bean is a good source of carbohydrates (50–60%) being starch, the main component. These pulses present dietary fiber, ash (> 4%), high protein (20–25%) (Rocchetti et al., 2019) and low fat content (Ikezu et al., 2015). Bioactive compounds (i.e., saponins, flavonoids) and phenolics are also present (de la Rosa-Millán et al., 2019).

In this chapter various terms are used to indicate the particle size of solids like flour (particle size: 100–5000 μm) and powder (particle size: 50–200 μm) following the criteria of Bhandari (2013).

González et al. (2018) evaluated the nitrogen balance in blood samples of 30 healthy adults after consuming meals with 30g of sacha inchi and soy flour, respectively. They showed that nitrogen balance was similar in both cases and recommended these sources of vegetable protein as a good healthy option to be consumed. Some researchers highlighted their antioxidative, antihypertensive (Chirinos et al., 2020), antibacterial, anticancer and antidyslipidemic capacity (Chirinos et al., 2013; González-Aspajo et al., 2015; Štěrbová et al., 2017). Moreover, Alayón et al. (2018) evaluated the acute postprandial effect of sacha inchi oil on changes in carbohydrate metabolism in clinically healthy subjects. The authors showed that consumption of sacha inchi oil improved carbohydrate metabolism, increasing insulin sensitivity.

On the other hand, several authors have shown the anticancer activity (Kumar et al., 2017) of black turtle beans related to phytochemical compounds (Ikezu et al., 2015). However, there is scientific evidence regarding the low digestibility of pulse protein mainly due to anti-nutritional phytic acid, tannins and cell walls that are present in the seeds among other components (Sozer et al., 2017). Other authors have shown that a low-pH incubation reduces the allergenicity of black turtle bean lectin to enhance the safety of legume consumption (Audu et al., 2013).

5.4.1   Methods of Processing and Industrial Applications

Sacha inchi seeds are usually employed for oil extraction taking advantage of their high content (Wang et al., 2018) and their residues have been used for protein extraction (59%) (Chirinos et al., 2017). Black turtle beans are boiled in water to elaborate black bean stew or are blended with rice to prepare a typical dish consumed in Cuba named "*arroz congris*" (Figure 5.6).

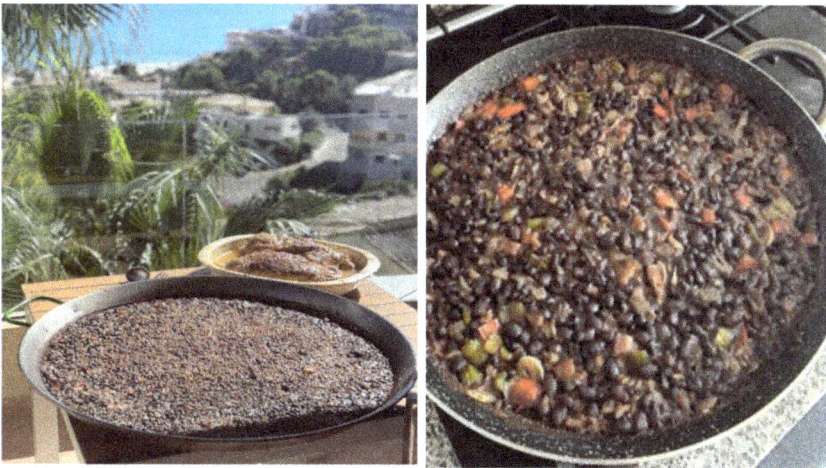

Figure 5.6   Traditional Cuban dish "*arroz congris* or *moros y cristianos*" rice with black turtle beans. **Source: Authors' own work.**
Note: The image on the left shows a large frying pan ("paellero") with rice and turtle black beans on a wooden table in a terrace with views of a city and vegetation at the sea coast. The image on the right shows an up-close view of the Cuban dish seen from the top, with whole turtle black beans, vegetables and rice.

In order to extend their industrial application, sacha inchi seed and black beans' potential must be explored. Analyzing technological characteristics shows that sacha inchi seeds have a strong flavor and contain heat-labile substances with a bitter taste (Bueno-Borges et al., 2018; Chirinos et al., 2013) that may negatively impact the overall sensory characteristics of processed foods.

Roasting, baking, steaming, boiling, extrusion, milling and drying are the conventional methods of processing seeds to obtain several raw materials and to process them into edible food. Roasted seeds had been employed to produce snacks (Wang et al., 2018). The authors recommend optimizing roasting conditions since this parameter may influence the flavor, antinutrients (Bueno-Borges et al., 2018), nutritional composition (Kim and Joo, 2019) and healthy properties. In order to obtain raw materials without anti-nutritional factors from sacha inchi seeds, several authors have employed different alternatives. Suwanangul et al. (2021) showed that the extrusion and autoclaving process causes the destruction of tannins, phytates, and trypsin inhibitors. A similar conclusion was reported by some authors who showed that an optimal roasting process (temperatures and time) influences the nutritional and phytochemical composition (Chirinos et al., 2013; Kim and Joo, 2019), increasing the antioxidant capacity and total phenolic compounds (Štěrbová et al., 2017). A similar study conducted by Bueno-Borges et al. (2018), demonstrated that roasting seeds for 15 min at 160°C decreased all assessed antinutrients except saponins.

In general, seeds can be aggregated into foods or be transformed into powder; they could also be added using the main by-product obtained after oil extraction, as pressed cake. Pressed cakes are the main by-product obtained from sacha-inchi oil extraction, contenting bioactive compounds with high antioxidant activities (Rawdkuen et al., 2016). Additionally, this product presents good physical properties (low solubility: 7.96%, high volume: 3.92 mL/g and low water adsorption capacity: 2.16 g/g) (Alcívar et al., 2020). Residual seed cake could be transformed into a powder in order to be used in the food industry. In the past, sacha inchi pressed-cake flour was mainly recommended in animal feeding (Alcívar et al., 2020; Viamonte-Garcés et al., 2020). Recently, several authors have proposed this material as nonconventional sources of isolated vegetable proteins from agro-industry since the protein content from the remaining cake was more than 50% (Chirinos et al., 2017; Mercado et al., 2015). In this context, some authors have shown the *in vitro* antioxidant and antihypertensive potential of bioactive peptides from protein hydrolysate (Chirinos et al., 2020).

Concerning flour from black turtle bean, some authors emphasized that different processing methods (boiling, cooking, roasting, sprouting and fermenting) significantly influenced their functional and technological parameters, likewise, their anti-nutritional factors (Audu and Aremu, 2015; Audu et al. 2013; Gobbetti et al., 2020). For example, during milling some critical parameters influence the quality of flour like particle size. In this sense,

Balandrán-Quintana et al. (1998) suggested grounding the seeds from pinto beans to pass through a 40-mesh screen in order to develop extruded whole pinto bean flour. While Gularte et al. (2011) employed powder from chickpeas, peas, lentils and beans with particle size lower than 210 μm. Chompoorat et al. (2018) sieved seeds to obtain powders with a particle size less than 250 μm in order to develop gluten-free cakes. Regarding the fermentation process, Gobbetti et al. (2020) explained that sourdough is a powerful process for exploiting the potential of legumes in the baking goods industry.

5.4.2   Baked Goods Application

Concerning food application, Lee (2019) elaborated a pound cake with 20% of sacha inchi powder and evaluated its technological properties, confirming good results in quality and sensory characteristics. Vázquez-Osorio et al. (2017) developed cookies including 25, 50 and 100% of sacha inchi flour. The best results were obtained at 50% inclusion and recommended evaluating the influence of 100% studying different strategies in order to obtain cookies with good technological and sensory properties. Aylas et al. (2015) substituted 10% of wheat flour with sacha inchi residual cake, obtaining a sweet bread with acceptable sensory parameters. Betancur et al. (2016) confirmed a high nutritional composition and sensory acceptance of the snack including sacha inchi seed flour as a novel ingredient.

Cappa et al. (2020) evaluated the behavior of 25 edible dry bean powder during the baking process in order to develop healthy cookies with higher protein content, lower rapid digestible starch and higher resistant starch. The authors concluded that bean powder with particle size similar to wheat flour (particle size ≤ 0.5 mm) showed better results in terms of cookie diameter and hardness.

5.4.3   Gluten-Free Products

Due to their high nutritional value and safety for human consumption sacha inchi and black turtle bean seeds are good alternatives for use in gluten free baked goods (Rawdkuen et al., 2016). Castaño et al. (2019) employed 100% residual cake seeds from sacha inchi to develop a gluten-free brownie; first, the powder was treated for 3h at 80°C to eliminate typical astringency.

To extend the use of the black turtle bean several authors have studied inclusion of its powder into formulations of gluten-free baked goods (Chompoorat et al., 2018; Gularte et al., 2011) showing good results in order to improve the technological properties and nutritional value. For example, Chompoorat et al. (2018) evaluated the influence of precooked red kidney beans on the viscoelastic properties of gluten-free cake batter and cupcake. The authors showed that different thermal processes improved batter viscoelasticity and sensory characteristics. They also concluded that boiling red kidney beans for 20 min with 3-h drying at 80°C is sufficient to obtain an acceptable legume powder for gluten free cupcake production. Therefore, Medina et al. (2018) obtained brownies from legume flour with added value, since this flour increases the antioxidant content and promotes inhibition of α-glucosidase.

Studies mentioned previously showed the powerful potential of seeds and common pulses to be included in baked goods. Processing methods and types of technologies employed to develop baked goods from sacha inchi and black turtle bean are the main parameters that must be controlled in order to develop value-added products. Some strategies like the addition of hydrocolloids, enzymes and starch blended with powders could improve the overall quality of products. Besides that, more scientific studies are needed in these fields because there is scarce information about the use of sacha inchi and black turtle bean powder in gluten-free products.

Commercial gluten-free products in Cuban markets can be seen in Figure 5.7. They can be added with sacha inchi flour or other nutritional ingredients.

Figure 5.7 Gluten free goods from Cuban bakeries with flour/starch from rice and/or potato; **A:** burger bread, **B:** loaf bread, **C:** muffins, **D:** brownie, **E:** brioche bread. **Source: Authors' own work.**

Notes: A) Four burger breads viewed from the front.

B) Four loaf breads, seen sideways, positioned horizontally and parallel to each other.

C) Muffins made in their paper muffin cups (of different sizes) viewed from above.

D) Square shaped brownies made viewed from above.

E) Many brioche breads viewed from above.

5.5 CHIA

Due to the technological importance of gluten, today the production of gluten-free food products is still a challenge for the industry. Indeed, large quantities of fat, sugars, structuring agents, and flavor enhancers are added to gluten free formulations to make textural and sensorial characteristics comparable to conventional products, resulting in nutritional and caloric intake imbalances (Montemurro et al., 2021). Chia seeds do not contain gluten (Ali et al., 2012) and due to their physical-chemical properties, these seeds have different technological capabilities and potential utilities for food application (Figure 5.8). Due to its nutraceutical and physicochemical properties, chia has been widely used as a whole seed, flour, seed mucilage, gel and oil to develop various enriched food products, such as bread, pasta, cakes, cookies, chips, cheese, yoghurt, meat, fish and poultry (Kaur and Bains, 2019).

One of the most important aspects of this seed is its high fiber content with a high proportion of mucilage, capable of absorbing large quantities of water (Reyes-Caudillo et al., 2008) (Figure 5.9). The addition of whole or ground chia had a favorable effect on bread crumb texture, as chia swells intensely with water and can be used as a hydrocolloid replacement in gluten free bread or as a fat replacer and thickener (Zettel and Hitzmann, 2018; De Lamo and Gómez, 2018; Ziemichod et al., 2019).

Another notable aspect of chia is its high content in omega-3-rich oil that is appropriately balanced with omega-6 (Ayerza, 2010). Therefore, chia seeds and co-products are excellent food ingredients with which to enrich foods by offering good technological benefit due to their mucilage properties, oil and proteins (Iglesias-Puig and Haros, 2013). Nonetheless, an in-depth study about baked foods is necessary because ingredients rich in omega-3 are oxidized while being baked (Miranda-Ramos et al., 2020), and those ingredients rich in proteins could favor acrylamide formation (Mesías et al., 2016) especially in foods with low water activity. Acrylamide is mainly produced at elevated temperatures and medium to low moisture content as a consequence of asparagine degradation produced by the Maillard reaction initiated by carbohydrates (Zamora and Hidalgo, 2011). However, lipid carbonyls can also play a role in these reactions. Asparagine degradation in the presence of lipid carbonyls is a two-step reaction in which the amino acid is first decarboxylated producing the corresponding biogenic amine (3-aminopro-pionamide) and then converted to acrylamide by elimination of ammonia by deamination (Zamora et al., 2009). The European Food Safety Authority (EFSA) has announced that the presence of acrylamide in food is a public health problem and that efforts must still be made to reduce its exposure (EFSA, 2015). Therefore, such aspects need to be thoroughly investigated. In this sense, chia seeds are considered a "novel food" in the European Union (EU) and have restrictions despite the

Figure 5.8   Gluten free chia foods: **A.** Lemon chia cookies; **B.** Soy lecithin with chia; **C.** *nachos* with chia; **D.** *tortillas* with chia. **Source: Authors' own work.**
Notes: A) Three round lemon cookies with whole chia seeds, with a rough surface, on a medium round plate.

B) Granules of soy lecithin with black chia seeds in a small cup, viewed from above.

C) "Nachos" with embedded chia seeds in a small bowl.

D) Medium round plate with a round amaranth chia; the seeds are embedded on the surface.

huge interest due to their technological, nutritional and functional potential. According to Regulation 258/97, new foods and new food ingredients are defined:

> as food that has not been consumed to a significant degree by humans in the EU prior to 1997, when the first Regulation on "novel food" came into force. Novel food can be newly developed innovative food or food produced using new technologies and production processes as well as food traditionally eaten outside of the EU.
>
> *OJEU, 1997*

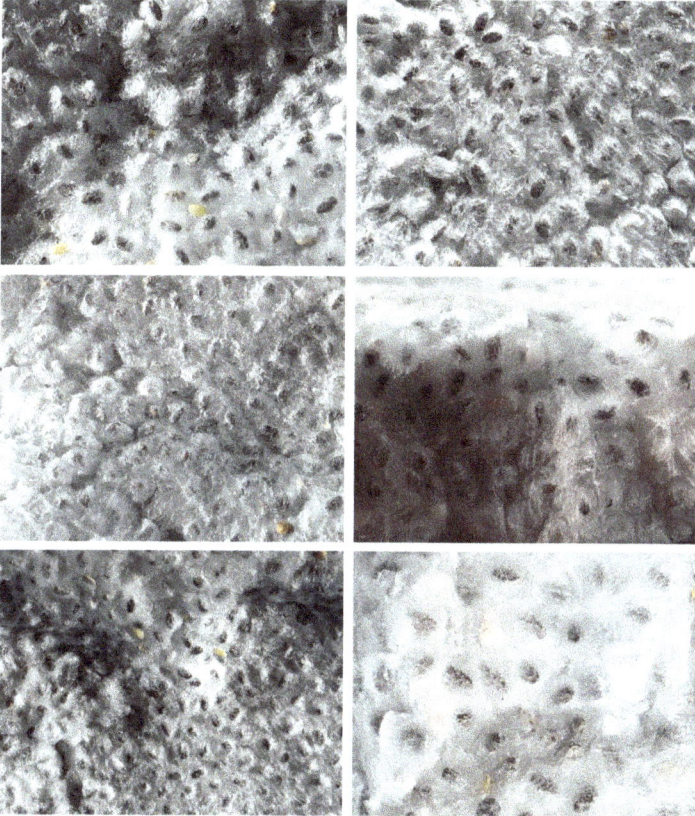

Figure 5.9   Lyophilized hydrated chia. **Source: Authors' own work.**

There are not many records about the intake of chia in Europe before 1997. So after considering extensive studies about the nutritional characteristics of chia, the EFSA issued a report in favor of its commercialization as "novel food" with restrictions (EFSA, 2009). Whole chia seeds could be used in baked products and breakfast cereals up to 10%; ground chia seeds up to 5% in bread; whole chia seeds up to 5% in sterilized ready-to-eat meals based on cereal/pseudo-cereal grains and/or pulses; pre-packaged chia seeds as such, and fruit/nut/seed mixes; and chia in confectionery products and chocolates; edible ices; fruit and vegetable products; non-alcoholic beverages and puddings (<120°C in their preparation) without limitations (EFSA, 2020). Recently, the use of two partially defatted powders of chia enriched with proteins or fibers was authorized as food supplements for the adult population (up to 7.5 and 12 g/day, respectively), or as nutritional ingredients in a variety of foods (yogurt, vegetable beverages, energy drinks, chocolate, fruit and pasta) at a level of 0.7–10% (Turck et al., 2019). Nowadays, technical and administrative formalities are accomplished by companies to allow the commercialization and use of other ingredients of interest in Europe. For this purpose, the nutritional potential of these ingredients will have to be validated and their innocuousness has to be guaranteed, as set out in the legislation currently in force. Outside the EU, whole chia and its seed fractions (whole chia flours, fractions rich in proteins, fractions rich in fibers, chia oil, chia partially defatted, etc.) have no restrictions regarding their utility in food preparation.

5.5.1 Processing Chia Seeds

Alternatives to the direct consumption of chia seeds could be recommended. Some processes such as soaking, sprouting or fermentation could increase its digestibility. From these alternatives, sprouting has gained popularity as it is a simple method that improves the nutritional value of the seeds, including increased phenols and mineral availability (Calvo-Lerma et al., 2020). Similarly, the milling of seeds has also been proven to enhance nutrient digestibility because of the matrix disruption and ease of nutrient release from the interior of plant cells (Grundy et al., 2016).

Individual chia components cannot be obtained by either wet or dry milling, as in the case of cereals/pseudo-cereals. The main chia components are extracted from whole seeds, or wholemeal flour, by milling or crushing whole seeds. The most popular procedure for particle size reduction and homogenization is cryogenic grinding (Gouveia et al., 2002). The temperature rise during grinding can be reduced with the use of a cryogenic fluid mainly to avoid thermo-labile compounds such as unsaturated fatty acids (Gouveia et al., 2002). Cryogenic fluids must be sufficiently inert not to give up components to the foods, without affecting the composition of the food or altering their sensorial characteristics. Besides, the low temperature keeps the material brittle during grinding and helps to achieve a finer particle size.

Chia oil, which is rich in omega-3, can be extracted by cold-pressing, solid-liquid hexane extraction or by supercritical CO_2 extraction (Ixtaína et al., 2010). The oil extraction residue has a high content of proteins (19–23%) and fiber (~ 37%); moreover, it has high potential to be applied to human food and animal feed. On the other hand, chia offers a higher fiber content than cereals/pseudo-cereals, and a high proportion of soluble fiber. Mucilage is a compound that exhibits high viscosity in water (Lin et al., 1994) and is extracted by seed hydration, followed by drying and sieving, and offers excellent techno-functional properties (Muñoz et al., 2009, 2012). The residual flour from the oil extraction process and the fiber fraction (mucilage) are rich in protein with a high biological value. All of these flours from chia processing are potential ingredients for developing gluten free products with a high nutritional/functional value and excellent techno-functional quality.

5.5.2 Gluten-Free Foods with Chia

5.5.2.1 Bread Products

The inclusion of chia or other oilseeds modifies the rheology of the dough, depending on the way in which it is made (flour or seeds, prehydration or not) and the percentage used (de Lamo and Gómez, 2018). To ensure commercial success of these inclusions, it is necessary to consider the acceptability of consumers, which may vary depending on the type of inclusion and its percentage (de Lamo and Gómez, 2018). In addition, as stated above, due to the mucilage release, chia and its by-products can also be regarded as a technological improvement for gluten-free breads (Zettel and Hitzmann, 2018).

Rice breads were elaborated with the addition of 15g of chia flour or seeds, either dry or pre-hydrated, per 100g of rice flour. In general, the addition of chia reduced bread specific volume, increased the crumb firmness with a darker crust/crumb color compared to the control sample, and the effect was more evident with the flour than with the seed (Steffolani et al., 2014). The previous hydration significantly modified the dough rheology but it seems did not produce significant differences in the final product formulations (Steffolani et al., 2014). On the other hand, Montemurro et al. (2021) developed a formulation of the novel "clean-label" gluten free bread including a commonly used mixture of maize and rice flour (ratio 1:1); it was fortified with quinoa and naturally hydrocolloid-containing flours, such as psyllium, flaxseed or chia, as structuring agents for their mucilage content. The use of chia at a 2.5% concentration could possibly replace gum in gluten-free bread formulations (Huerta et al., 2016).

In another study, rice flour bread was augmented with chia seeds and chia flour which resulted in products with good technological characteristics. The addition of chia seeds and flour in gluten-free bread provided an increase in the nutritional value of the products, mainly in the protein content; all the breads tested were well accepted by consumers, obtaining global acceptance values above 70%, with no significant difference between them (da Costa Borges et al., 2021). The development of premixes based on buckwheat and chia flour represents an alternative with high nutritional value that can be applied to the formulation of gluten-free breads, with adequate technological properties (Costantini et al., 2014; Coronel et al., 2021).

Some chia by-products such as partially defatted chia flour, hydroxypropylmethylcellulose gum (HPMC) and xanthan gum were used, in order to develop gluten-free breads with high nutritional and sensory quality (Ewerling et al., 2020). Even chia seeds' by-products of oil pressing can be regarded as valuable technological and functional food additives in gluten-free bakery products (Zdybel et al., 2019). On the other hand, rice breads with added chia protein hydrolysate resulted in good technological characteristics and anti-oxidant properties, and breads made with 3 mg of chia hydrolysate/g flour had good sensory characteristics, which were not lower than those of the control breads (Madruga et al., 2020).

The use of chia sourdoughs fermented with selected lactic acid bacteria strains as ingredients for gluten-free bread making, could have an impact on the functionality and healthiness of baked foods (Maidana et al., 2020a). The potential use of *Lactobacillus sanfranciscensis* W2 for the fermentation of chia flour to produce gluten-free maize/rice bread was investigated. The strain adapted to the environment of the chia flour, which decreased pH, specific volume, firmness and rate of bread staling and increased bread porosity compared to bread only made with chia seed flour (Jagelaviciute and Cizeikiene, 2021). Sorghum-based laboratory breads manufactured with different chia percentages or flaxseed sourdoughs fermented by *Weissella cibaria* CH28 together with *L. plantarum* FUA3171 and *L. fermentum* FUA3165 from fermented sorghum, significantly improved specific volume and visual appearance compared to 100% sorghum breads, with higher acceptability by panelists (Maidana et al., 2020a). The development of lactic acid bacteria during spontaneous chia flour fermentation (sourdough) was investigated by Maidana et al. (2020b). Lactic acid bacteria species from genus *Enterococcus* (*E. faecium*, *E. mundtii*) were the most abundant; they were isolated species from *Lactococcus* (*Lc. laths*), *Lactobacillus* (*L. rhamnosus*) and *Weissella* (*W. cibaria*). They were the dominant species in the final propagation stages while *Bacillus* and *Clostridium* were mostly present during the first fermentation stages.

5.5.2.2 Biscuits, Cookies, Cakes and Snacks

In addition to the development of breads, some formulations of gluten-free biscuits, cookies, cakes and snacks that include chia have been described in the literature. The substitution of buckwheat flour, millet flour, and chia seeds for wheat flour can be considered a suitable alternative for the preparation of gluten-free cookies. The optimum formulation was defined as that with the closest characteristics (diameter, expansion factor, thickness and hardness) to the control cookie elaborated with wheat, which was with 7.5% chia, 40% millet and 52.5% buckwheat (Ferreira Brites et al., 2019). In another investigation, an optimized gluten-free biscuit formulation with chia seeds and curcuma was developed with the same statically sensory characteristics as the control biscuits (Laczkowski et al., 2021).

The effect of using chia gel as a gluten substitute in cakes was also investigated which showed better acceptance by consumers than the control formulation made with rice flour (Hargreaves et al., 2018). Substituting rice flour with 10% prehydrated chia seed flour can achieve an acceptable piece center height and volume index of gluten-free layer cakes with a similar overall acceptability, texture, flavor and odor scores compared to those of gluten-free layer cake made with 100% rice flour and layer cake made with 100% wheat flour (Sung et al, 2020). Chia and azuki flour mixes were also investigated to develop a gluten-free chocolate cake with a higher amount of polyunsaturated fatty acids, mainly alpha-linolenic acid, which showed that the inclusion of chia can improve lipid composition (Gohara et al., 2016).

Chia was also utilized to prepare sweet snacks suitable for celiac people using popped amaranth, dried peaches, textured soy, corn flakes and whole sesame seeds agglutinated with glucose syrup, honey and soy lecithin and flavored with cinnamon in order to develop a highly nutritious product with a shelf life of around 4 months (Malka et al., 2020).

5.5.2.3   Pasta Products

Some investigators developed pasta products with chia. This seed is generally used in gluten-free pasta formulations for its nutritional value and thickening properties (Levent, 2017; Menga et al., 2017). Formulation of noodles with 20% of chia seed flour with DATEM (diacetyl tartaric acid ester of mono- and diglycerides) can be used with acceptable sensory attributes of raw and cooked samples, to improve the nutritional quality of noodles (Levent, 2017). On the other hand, partially defatted chia flour, high in protein content, dietary fiber and phenolic compounds was investigated as a gluten-free ingredient to develop pasta with an improved antioxidant capacity (Aranibar et al., 2018).

5.6   CONCLUSIONS

Quinoa, kañiwa, amaranth, tarwi, maize, chia, black turtle bean and sacha inchi are nutritious crops cultivated in Latin America. These grains are traditionally consumed by local inhabitants directly, without further processing and transformation. They also offer excellent ingredients for the modern food industry. They are all gluten free and can be used to produce bakery and pasta products without using wheat. The protein quality of quinoa, kiwicha and kañiwa is excellent; they can all substantially improve the nutritional value of traditional gluten-free products. Tarwi has an exceptionally high protein content, making it an interesting ingredient for the preparation of protein concentrates. The nixtamalization process used for maize in Mexico and Central America increases the calcium content in maize grains, improving the bioavailability of niacin and iron, and decreasing the content of mycotoxins and phytic acid. Sacha inchi has a very high oil content, being an excellent source of α-linolenic, linoleic and oleic acids and of tocopherols. It also has an elevated protein content. The flour of black turtle beans can be included in gluten-free baked goods, improving the technological properties and nutritional values of the final product. Due to the physical–chemical properties of chia seeds, they have interesting technological properties and potential uses for food applications. Chia has been widely used as a whole seed, flour, seed mucilage, gel and oil to develop various enriched food products, such as bread, pasta, cakes, cookies, chips, cheese, yoghurt, meat, fish and poultry.

These crops are staple foods for millions of people living in areas of Latin America where nutritional problems are common. They have been part of their traditional diets and are still important for food security. Additionally, these seeds are also a source of interesting ingredients for novel and specialty products beyond their traditional cultivation area, as consumers worldwide are increasingly conscious of their nutritional, healthy and sustainable foods. Unfortunately, there are important sectors of society in most geographical areas of the north and south, where excess weight and obesity have become endemic diseases. Thus, research is also needed to develop tasty and nutritious products but simultaneously with a low glycemic index for those consumers; crops like those analyzed in this chapter are called on to play a key role in the food strategy of the future which has already arrived.

ACKNOWLEDGMENTS

Authors would like to express their gratitude to the CYTED Program for its kind support through Project119RT0567, Food4ImNut Food4ImNut PID2019-107650RB-C21 funded by MCIN/AEI/10.13039/501100011033, Spain.

REFERENCES

Abarza, S. del V., and J. H. Schimpf. 2009. Los Maíces indígenas en la cultura alimentaria andina. *Revista Científica Ciencia* 4(9): 159–170.

Alayón, A. N., J. G Ortega Ávila, and I. Echeverri Jiménez. 2018. Carbohydrate metabolism and gene expression of sirtuin 1 in healthy subjects after Sacha inchi oil supplementation: A randomized trial. *Food and Function* 00: 1–9. doi:10.1039/C7FO01956D.

Alcívar, J. L., M. Martínez Pérez, P. Lezcano, I. Scull, and A. Valverde. 2020. Technical note on physical-chemical composition of Sacha inchi (*Plukenetia volubilis*) cake. *Cuban Journal of Agricultural Science* 54(1): 19–23.

Ali, Norlaily Mohd, Swee Keong Yeap, Wan Yong Ho, Boon Kee Beh, Sheau Wei Tan, and Soon Guan Tan. 2012. The promising future of chia, salvia hispanica L. *Journal of Biomedicine and Biotechnology*. https://doi.org/10.1155/2012/171956.

Alonso-Miravalles, L., and J. A. O'Mahony. 2018. Composition, protein profile and rheological properties of pseudocereal-based protein-rich ingredients. *Foods* 7(73): 1–17.

Álvarez-Jubete, L., M. Holse, A. Hansen, E. Arendt, and E. Gallagher. 2009. Impact of baking on vitamin E content of pseudocereals amaranth, quinoa, and buckwheat. *Cereal Chemistry* 86(5): 511–515.

Álvarez-Jubete, L., M. Auty, E. Arendt, and E. Gallagher. 2010a. Baking properties and microstructure of pseudocereal flours in gluten-free bread formulations. *European Food Research and Technology* 230: 437–445.

Álvarez-Jubete, L., H. Wijngaard, E. Arendt, and E. Gallagher. 2010b. Polyphenol composition and in-vitro antioxidant activity of amaranth, quinoa and buckwheat as affected by sprouting and bread baking. *Food Chemistry* 119: 770–778.

Ando, H., Y. C. Chen, H. Tang, M. Shimizu, K. Watanabe, and T. Mitsunaga. 2002. Food components in fractions of quinoa seed. *Food Science and Technology Research* 8(1): 80–84.

Aranibar, C., N. B. Pigni, M. Martinez, A. Aguirre, P. Ribotta, D. Wunderlin, and R. Borneo. 2018. Utilization of a partially-deoiled chia flour to improve the nutritional and antioxidant properties of wheat pasta. *LWT – Food Science and Technology* 89: 381–387.

Arendt, E. K., A. Morrissey, M. M. Moore, and F. dal Bello. 2008. Gluten-free breads. In *Gluten-Free Cereal Products and Beverages*, edited by E. K. Arendt and F. dal Bello, 289–319. New York: Academic Press.

Audu, S. S., and M. O. Aremu. 2015. Effect of domestic processing on the levels of some functional parameters in black turtle bean (*Phaseolus vulgaris* L.). *Food Science and Quality Management* 38: 55–59.

Audu, S. S., M. O. Aremu, and L. Lajide. 2013. Effects of processing on physicochemical and antinutritional properties of black turtle bean (Phaseolus vulgaris L.) seeds flour. *Oriental Journal of Chemistry* 29(03): 979–989.

Ayerza, R. 2010. Effects of seed colour and growing locations on fatty acid content and composition of two chia (Salvia hispanica L.) genotypes. *Journal of the American Oil Chemists' Society*, 87: 1161–1165. http://dx.doi.org/10.1007/s11746-010-1597-7.

Aylas, T., M. Pingus, and D. Alva. 2015. Effect of the partial replacement of wheat flour by cake of sacha inchi (*Plukenelia volubilis* L.) on the rheological properties of dough in sweet bread. *Revista Ciencia & Desarrollo* 2015: 16–21.

Balandrán-Quintana, R. R., G. V. Barbosa-Cánovas, J. J. Zazueta-Morales, A. Anzaldúa-Morales, and A. Quintero-Ramos. 1998. Functional and nutritional properties of extruded whole pinto bean meal (*Phaseolus Vulgaris* L.). *Journal of Food Science* 63(1): 113–116.

Ballester-Sánchez, J., J. V. Gil, M. T. Fernández-Espinar, and C. M. Haros. 2019. Quinoa wet-milling: Effect of steeping conditions on starch recovery and quality. *Food Hydrocolloids* 89: 837–843.

Bhandari, B. 2013. Introduction to food powders. In *Handbook of Food Powders. Processes and Properties*, edited by Bhesh Bhandari, Nidhi Bansal, Min Zhang and Pierre Schuck, 1–25. Cambridge, UK: Woodhead Publishing.

Basilio-Atencio, J., L. Condezo-Hoyos, and R. Repo-Carrasco-Valencia. 2020. Effect of extrusion cooking on the physical-chemical properties of whole kiwicha (*Amaranthus caudatus* L) flour variety Centenario: Process optimization. *LWT – Food Science and Technology* 128: 109426.

Benavent-Gil, Y., and C. M. Rosell. 2018. *Technological and Nutritional Applications of Starches in Gluten-Free Products. Starches for Food Application.* London: Elsevier.

Betalleluz-Pallardel, I., M. Inga, L. Mera, R. Pedreschi, D. Campos, and R. Chirinos. 2017. Optimization of extraction conditions and thermal properties of protein from the Andean pseudocereal canihua (*Chenopodium pallidicaule* Aellen). *International Journal of Food Science and Technology* 52: 1026–1034.

Betancur, Edwin, Amparo Urango, Marchena Luz, Luis Fernando Restrepo Betancur. 2016. Effect of adding sacha inchi (Plukenetia volubilis L.) seeds to a prototype of convenience food draft, on the nutritional composition and sensory acceptance. *Journal of Medicinal Plant Research* 10(29):435–441, DOI: 10.5897/JMPR2016.6064.

Bressani, R. 1990. Chemistry, technology, and nutritive value of maize tortillas. *Food Reviews International* 6(2): 225–264.

Bressani, R., J, C. Turcios, and A. S. C. de Ruiz. 2002. Nixtamalization effects on the contents of phytic acid, calcium, iron and zinc in the whole grain, endosperm and germ of maize. *Food Science and Technology International* 8(2): 81–86.

Bueno-Borges, Larissa Braga, Marco Aurélio Sartim, Claudia Carreño Gil, Suely Vilela Sampaio, Paulo Hercílio Viegas Rodrigues and Marisa Aparecida Bismara Regitano-d'Arce. 2018. Sacha inchi seeds from sub-tropical cultivation: Effects of roasting on antinutrients, antioxidant capacity and oxidative stability. *Journal of Food Science and Technology* 55: 4159–4166).

Bugallo, L. and L. Mamani. 2014. Molinos en la quebrada de Humahuaca: lugares de encuentro de gentes y caminos. La región molinera del norte jujeño, 1940-1980. In *Espacialidades Altoandinas. Avances de investigación desde el noroeste argentino,* edited by A. Benedetti and J. Tomasi, 63–118. Editorial de la Facultad de Filosofía y Letras, Universidad de Buenos Aires.

Burgos, V. E., M. J. Binaghi, P. A. Ronayne de Ferrer, and M. Armada. 2018. Effect of precooking on antinutritional factors and mineral bioaccessibility in kiwicha grains. *Journal of Cereal Science* 80: 9–15.

Bustos, M. C., M. I. Ramos, G. T. Pérez, and A. E. León. 2019. Utilization of kañawa (*Chenopodium pallidicaule* Aellen) flour in pasta making. *Journal of Chemistry* 2019: 4385045.

Calderón de la Barca, A. M., M. E. Rojas-Martínez, A. R. Islas-Rubio, and F. Cabrera-Chávez. 2010. Gluten-free breads and cookies of raw and popped amaranth flours with attractive technological and nutritional qualities. *Plant Foods for Human Nutrition* 65: 241–246.

Calvo-Lerma, J., C. P. Yépez, A. A. Grau, A. Heredia, and A. Andrés. 2020. Impact of processing and intestinal conditions on *in vitro* digestion of chia (*Salvia hispanica*) seeds and derivatives. *Foods* 9(3): 290.

Calzetta, A. N., R. J. Aguerre, and C. Suárez. 2006. Hydration kinetics of amaranth grain. *Journal of Food Engineering* 72: 247–253.

Cámara-Hernández, J., and D. Arancibia-Cabezas. 2007. *Maíces Andinos y sus Usos en la Quebrada de Humahuaca y Regiones Vecinas (Argentina).* Buenos Aires: Editorial Facultad de Agronomía, Universidad de Buenos Aires.

Camelo-Méndez, G., P. Flores, E. Agama-Acevedo, J. Tovar, and L. Bello-Pérez. 2017. Incorporation of whole blue maize flour increases antioxidant capacity and reduces *in vitro* starch digestibility of gluten-free pasta. *Starch – Stärke* 70: 1700126.

Cappa, C., J. D. Kelly, and P. K. W. Ng. 2020. Baking performance of 25 edible dry bean powders: Correlation between cookie quality and rapid test indices. *Food Chemistry* 302: 125338.

Capriles, V., and J. Areas. 2014. Novel approaches in gluten-free breadmaking: Interface between food science, nutrition, and health. *Comprehensive Reviews in Food Science and Food Safety* 13: 871–890.

Carrera, Y., R. Utrilla Coello, A. Bello-Pérez, J. Ávarez Ramírez, and E. Vernon Carter. 2015. In vitro digestibility, crystallinity, rheological, thermal, particle size and morphological characteristics of pinole, a traditional energy food obtained from toasted ground maize. *Carbohydrate Polymers* 123: 246–255.

Castaño, L. A. C., M. M. P. Valderrama, H. J. P. Díaz, and S. M. Montesino. 2019. Physicochemical and microbiological evaluation of flour obtained from the residual cake of sacha inchi (*Plukenetia volubilis* l.) for its potential use in the agri-food sector. *Italian Journal of Food Science* : 31(6SI): 69–78.

Chauhan, A., D. C. Saxena, and S. Singh. 2015. Total dietary fibre and antioxidant activity of gluten free cookies made from raw and germinated amaranth (*Amaranthus* spp.) flour. *LWT – Food Science and Technology* 63: 939–945.

Chirinos, R., G. Zuloeta, R. Pedreschi, E. Mignolet, Y. Larondelle, and D. Campos. 2013. Sacha inchi (Plukenetia volubilis): a seed source of polyunsaturated fatty acids, tocopherols, phytosterols, phenolic compounds and antioxidant capacity. *Food Chemistry* 141(3): 1732–1739.

Chirinos, R., M. Aquino, R. Pedreschi, and D. Campos. 2017. Optimized methodology for alkaline and enzyme-assisted extraction of protein from sacha inchi (*Plukenetia volubilis*) kernel cake. *Journal of Food Process Engineering* 40(2): e12412.

Chirinos, R., R. Pedreschi, and D. Campos. 2020. Enzyme-assisted hydrolysates from sacha inchi (*Plukenetia volubilis*) protein with in vitro antioxidant and antihypertensive properties. *Journal of Food Processing and Preservation* 44: e14969.

Chompoorat, P., P. Rayas-Duarte, Z. J. Hernández-Estrada, C. Phetcharat, and Y. Khamsee. 2018. Effect of heat treatment on rheological properties of red kidney bean gluten free cake batter and its relationship with cupcake quality. *Journal of Food Science and Technology* 55(12): 4937–4944.

Coda, R., C. G. Rizzello, and M. Gobbetti. 2010. Use of sourdough fermentation and pseudo-cereals and leguminous flours for the making of a functional bread enriched of γ-aminobutyric acid (GABA). *International Journal of Food Microbiology* 137(2–3): 236–245.

Coronel, E. B., E. N. Guiotto, M. C. Aspiroz, M. C. Tomás, S. M. Nolasco, and M. I. Capitani. 2021. Development of gluten-free premixes with buckwheat and chia flours: Application in a bread product. *LWT – Food Science and Technology* 141: 110916.

Corrales-Bañuelos, A. B., E. O. Cuevas-Rodríguez, J. A. Gutierrez-Uribe, E. M. Milán-Noris, C. Reyes-Moreno, J. Milán-Carrillo, and S. Mora-Rochín. 2016. Carotenoid composition and antioxidant activity of tortillas elaborated from pigmented maize landrace by traditional nixtamalization or lime cooking extrusion process. *Journal of Cereal Science* 69: 64–70.

Cortés-Avendaño, P., M. Tarvainen, J. P. Suomela, P. Glorio-Paulet, B. Yang, and R. Repo-Carrasco-Valencia. 2020. Profile and content of residual alkaloids in ten ecotypes of *Lupinus mutabilis* Sweet after aqueous debittering process. *Plant Foods for Human Nutrition* 75: 184–191.

Costantini, L., L. Luksic, R. Molinari, I. Kreft, G. Bonafaccia, L. Manzi, and N. Merendino. 2014. Development of gluten-free bread using tartary buckwheat and chia flour rich in flavonoids and omega-3 fatty acids as ingredients. *Food Chemistry* 165: 232–240.

Cravioto, R. O., R. K. Anderson, E. E. Lockhart, F. P. Miranda, and R. S. Harris. 1945. Nutritive value of the Mexican tortilla. *Science* 45(102): 91–93.

da Costa Borges, V., S. Santos Fernandes, E. da Rosa Zavareze, C. M. Haros, C. Prentice Hernandez, A. R. Guerra Días, and M. de las Mercedes Salas-Mellado. 2021. Production of gluten free bread with flour and chia seeds (*Salvia hispanica* L). *Food Bioscience* 43: 101294.

D'Amico, S., S. Jungkunz, G. Balasz, M. Foeste, M. Jekle, S. Tömösközi, and R. Schöenlechner. 2019. Abrasive milling of quinoa: Study on the distribution of selected nutrients and proteins within the quinoa seed kernel. *Journal of Cereal Science* 86: 132–138.

de la Hera E, M. Talegon, P. Caballero, and M. Gomez. 2013. Influence of maize flour particle size on gluten-free breadmaking. *Journal of the Science of Food and Agriculture* 93(4): 924–32.

de la Parra, C., S. O. Serna Saldivar and R. H Liu. 2007. Effect of processing on the phytochemical profiles and antioxidant activity of corn for production of masa, tortillas, and tortilla chips. *Journal of Agricultural and Food Chemistry* 55(10): 4177–4183.

de la Rosa-Millán, J., E. Heredia-Olea, E. Pérez-Carrillo, D. Guajardo-Flores, and S. R. O. Serna-Saldívar. 2019. Effect of decortication, germination and extrusion on physicochemical and *in vitro* protein and starch digestion characteristics of black beans (*Phaseolus vulgaris* L.). *LWT – Food Science and Technology* 102: 330–337.

de Lamo, B., and M. Gómez, M. 2018. Bread enrichment with oilseeds. A Review. *Foods* 7(11): 191.

Dib, A., A. Wójtowicz, L. Benatallah, A. Bouasla, and M. Zidoune. 2018. Effect of hydrothermal treated corn flour addition on the quality of corn-field bean gluten-free pasta. BIO Web of Conferences. *Contemporary Research Trends in Agricultural Engineering* 10: 02003.

EFSA (European Food Safety Authority). 2009. Scientific opinion of the panel on dietetic products nutrition and allergies on a request from the European Commission on the safety of 'Chia seed (*Salvia hispanica*) and ground whole Chia seed' as a food ingredient. *The EFSA Journal* 996: 1–26.

EFSA (European Food Safety Authority). 2015. Scientific opinion on acrylamide in food. EFSA Panel on Contaminants in the Food Chain (CONTAM). *EFSA Journal* 13(6):4104.

EFSA (European Food Safety Authority). 2020. Commission Implementing Regulation (EU) 2020/24 of 13 January 2020 authorising an extension of use of chia seeds (*Salvia hispanica*) as a novel food and the change of the conditions of use and the specific labelling requirements of chia seeds (Salvia hispanica) under Regulation (EU) 2015/2283 of the European Parliament and of the Council and amending Commission Implementing Regulation (EU) 2017/2470. *Official Journal of the European Union* L 8/12.

EU (European Union). 2020. Authorising an extension of use of chia seeds (*Salvia hispanica*) as a novel food and the change of the conditions of use and the specific labelling requirements of chia seeds (Salvia hispanica) under Regulation (EU) 2015/

2283 of the European Parliament and of the Council and Amending Commission Implementing Regulation (EU) 2017/2470. *Oficial Journal of the European Union*: Brussels: 12–16.

Ewerling, M., N. C. Steinmacher, M. R. dos Santos, D. L. Kalschne, N. E. de Souza, F. M. Arcanjo, A. H. P. de Souza, and A. C. Rodrigues. 2020. Defatted chia flour improves gluten-free bread nutritional aspects: A model approach. *Food Science and Technology* 40: 68–75.

Ferreira Brites L. T. G., F. Ortolan, D. W. da Silva, F. R. Bueno, T. de Souza Rocha, Y. K. Chang, and C. Joy Steel. 2019. Gluten-free cookies elaborated with buckwheat flour, millet flour and chia seeds. *Food Science and Technology* 39(2): 458–466.

García Solaesa, A., M. Villanueva, A.J. Vela, and F. Ronda. 2020. Protein and lipid enrichment of quinoa (cv.Titicaca) by dry fractionation. Techno-functional, thermal and rheological properties of milling fractions. *Food Hydrocolloids* 105: 1–9.

Giménez, M. A., R. J. González, J. Wagner, R.Torres, M. O. Lobo, and N. C. Sammán. 2013. Effect of extrusion conditions on physicochemical and sensorial properties of corn-broad beans (Vicia faba) spaghetti type pasta. *Food Chemistry* 136(2): 538–545.

Giménez, M. A., A. Gámbaro, M. Miraballes, A. Roascio, M. Amarillo, N. Sammán, and M. Lobo. 2014. Sensory evaluation and acceptability of gluten-free Andean corn spaghettis. *Journal of the Science of Food and Agriculture* 95: 186–192.

Giménez, M. A., S. Drago, M. Bassett, M. Lobo, and N. Samman. 2016. Nutritional improvement of corn pasta-like product with broad bean (*Vicia faba*) and quinoa (Chenopodium quinoa). *Food Chemistry* 199: 150–156.

Gobbetti, M., M. De Angelis, R. Di Cagno, A. Polo, and C. G. Rizzello. 2020. The sourdough fermentation is the powerful process to exploit the potential of legumes, pseudo-cereals and milling by-products in baking industry. *Critical Reviews in Food Science and Nutrition* 60(13): 2158–2173.

Gohara, A. K., A. H. P. Souza, E. M. Rotta, G. L. Stroher, S. T. M. Gomes, J. V. Visentainer, N. E. Souza, and Makoto Matsushita. 2016. Application of multivariate analysis to assess the incorporation of omega-3 fatty acid in gluten-free cakes. *Journal of the Brazilian Chemical Society* 27(1): 62–69.

Gonzales, G. F., J. Tello, A. Zevallos-Concha, L. Baquerizo, and L. Caballero, L. 2018. Nitrogen balance after a single oral consumption of sacha inchi (*Plukenetia volubilis* L.) protein compared to soy protein: A randomized study in humans. *Toxicology Mechanisms and Methods* 28(2): 140–147.

González-Aspajo, G., H. Belkhelfa, L. Haddioui-Hbabi, G. Bourdy, and E. Deharo. 2015. Sacha inchi oil (*Plukenetia volubilis* L.), effect on adherence of *Staphylococus aureus* to human skin explant and keratinocytes *in vitro*. *Journal of Ethnopharmacology* 171: 330–334.

Gouveia, S. T., G. S. Lopes, O. Fatibello-Filho, A. R. A. Nogueira, and J. A. Nóbrega. 2002. Homogenization of breakfast cereals using cryogenic grinding. *Journal of Food Engineering* 51: 59–63.

Grundy, M. M. L., K. Lapsley, and P. R. Ellis. 2016. A review of the impact of processing on nutrient bioaccessibility and digestion of almonds. *International Journal of Food Science & Technology* 51: 1937–1946.

Gularte, M. A., M. Gómez, and C. M. Rosell. 2011. Impact of legume flours on quality and in vitro digestibility of starch and protein from gluten-free cakes. *Food and Bioprocess Technology* 5(8): 3142–3150.

Gutiérrez-Dorado, R., A. E. Ayala-Rodríguez, J. Milán-Carrillo, J. López-Cervantes, J. A. Garzón-Tiznado, J. A. López-Valenzuela, O. Paredes-López, and C. Reyes-Moreno, 2008. Technological and nutritional properties of flours and tortillas from nixtamalized and extruded quality protein maize (*Zea mays* L.). *Cereal Chemistry* 85: 808–816.

Han, Y., J. Chi, M. Zhang, R. Zhang, S. Fan, L. Dong, F. Huang, and L. Liu. 2019. Changes in saponins, phenolics and antioxidant activity of quinoa (*Chenopodium quinoa* Willd) during milling process. *Food Science and Technology* 114: 1–7.

Hargreaves, S. M., R. P. Zandonadi, and R. P. Zandonadi. 2018. Flaxseed and chia seed gel on characteristics of gluten-free cake. *Journal of Culinary Science & Technology* 16(4): 378–388.

Hemalatha, P., D. P. Bomzan, B. V. S. Rao, and Y. N. Sreerama. 2016. Distribution of phenolic antioxidants in whole and milled fractions of quinoa and their inhibitory effects on α-amylase and α-glucosidase activities. *Food Chemistry* 199: 330–338.

Horstmann, S., M. Foschia, and E. Arendt. 2017. Correlation analysis of protein quality characteristics with gluten-free bread properties. *Food and Function* 8(7): 2465–2474.

Huerta, K. d M., J. d S. Alves, A. F. Coelho da Silva, E. H. Kubota, and C. Severo da Rosa. 2016. Sensory response and physical characteristics of gluten-free and gum-free bread with chia flour. *Food Science and Technology* 36: 15–18.

Iglesias-puig, E. and Haros, M. 2013. Evaluation of performance of dough and bread incorporating chia (*Salvia hispanica* L.). *European Food Research and Technology* 237: 865–874.

Ikezu, U. J. M., I. P. Udeozo, and D. E. Egbe. 2015. Phytochemical and proximate analysis of black turtle beans (Phaseolus vulgaris). *African Journal of Basic & Applied Sciences* 7(2): 88–90.

Inglett, G. E., D. Chen, and S. X. Liu. 2015. Physical properties of gluten-free sugar cookies made from amaranth–oat composites. *LWT – Food Science and Technology* 63: 214–220.

Ixtaína, V. Y., A. Vega, S. M. Nolasco et al. 2010. Supercritical carbon dioxide extraction of oil from mexican chia seed (*Salvia hispanica* L.): characterization and process optimization. *The Journal of Supercritical Fluids* 55(1): 192–199.

Jagelaviciute J., and D. Cizeikiene. 2021. The influence of non-traditional sourdough made with quinoa, hemp and chia flour on the characteristics of gluten-free maize/rice bread. *LWT – Food Science and Technology* 137: 110457.

Jekle, M., A. Houben, M. Mitzscherling, and T. Becker. 2010. Effects of selected lactic acid bacteria on the characteristics of amaranth sourdough. *Journal of the Science of Food and Agriculture* 90: 2326–2332.

Kaur, S. and K. Bains. 2019. Chia (*Salvia hispanica* L.) - A rediscovered ancient grain, from Aztecs to food laboratories. A review. *Nutrition & Food Science* 50(3): 463–479.

Kim, D.-S., and N. Joo. 2019. Nutritional composition of Sacha inchi (*Plukenetia volubilis* L.) as affected by different cooking methods. *International Journal of Food Properties* 22(1): 1235–1241.

Kljak, K., E. Šárka, P. Dostálek, P. Smrčková, and D. Grbeša. 2015. Influence of physico-chemical properties of Croatian maize hybrids on quality of extrusion cooking. *Food Science and Technology* 60(1): 472–477.

Koehler, P., H. Wieser, and K. Konitzer. 2014. Gluten-free products. In *Celiac Disease and Gluten. Multidisciplinary Challenges and Opportunities*, chap. 4, 174–214. London/Waltham, MA/San Diego, CA: Academic Press/Elsevier Science.

Kumar, S., V. K. Sharma, S. Yadav, and S. Dey. 2017. Antiproliferative and apoptotic effects of black turtle bean extracts on human breast cancer cell line through extrinsic and intrinsic pathway. *Chemistry Central Journal* 11(1): 56.

Laczkowski, M. S., M. R. Baqueta, V. M. A. T. de Oliveira, T. R.Goncalves, S. T. M. Gómes, P. H. Marco, M. Matsushita, and P. Valderrama. 2021. Application of chemometric tools in the development and sensory evaluation of gluten-free cracknel biscuits with the addition of chia seeds and turmeric powder. *Journal of Food Science and Technology – Mysore* DOI: 10.1007/s13197-020-04874-9.

Lee, M.-H. 2019. Study on the quality characteristics and functional analysis of pound cakes containing sacha inchi (*Plukenetia volubilis* L.) flour. *Culinary Science & Hospitality Research* 25(8): 28–40.

Leray, G., B. Oliete, S. Mezaize, S. Chevallier, and M. de Lamballerie. 2010. Effects of freezing and frozen storage conditions on the rheological properties of different formulations of non-yeasted wheat and gluten-free bread dough. *Journal of Food Engineering* 100: 70–76.

Levent, H. 2017. Effect of partial substitution of gluten-free flour mixtures with chia (*Salvia hispanica* L.) flour on quality of gluten-free noodles. *Journal of Food Science and Technology– Mysore* 54(7): 1971–1978.

Linares, L., R. Repo-Carrasco-Valencia, P. Glorio Paulet, and R. Schöenlechner. 2019. Development of gluten-free and egg-free pasta based on quinoa (*Chenopdium quinoa* Willd) with addition of lupine flour, vegetable proteins and the oxidizing enzyme POx. *European Food Research and Technology* 245: 2147–2156.

Liu, T., G. G. Hou, B. Lee, L. Marquart, and A. Dubat. 2016. Effects of particle size on the quality attributes of reconstituted whole-wheat flour and tortillas made from it. *Journal of Cereal Science* 71: 145–152.

Luna-Mercado, G. I., and R. Repo-Carrasco-Valencia. 2020. Gluten-free bread applications: Thermo-mechanical and techno-functional characterization of Kañiwa flour. *Cereal Chemistry* 98(3): 474–481.

Madruga, K., M. da Rocha, S. Santos Fernándes, M. de las M. Salas-Mellado. 2020. Properties of wheat and rice breads added with chia (*Salvia hispanica* L.) protein hydrolyzate. *Food Science and Technology* 40: 596–603.

Maidana, S. D., S. Finch, M. Garro, G. Savoy, M. Ganzle, and G. Vignolo. 2020a Development of gluten-free breads started with chia and flaxseed sourdoughs fermented by selected lactic acid bacteria. *LWT – Food Science and Technology* 125: 109189.

Maidana, S. D., C. A. Ficoseco, D. Bassi, P. S. Cocconcelli, E. Puglisi, G. Savoy, G. Vignolo, and C. Fontana. 2020b. Biodiversity and technological-functional potential of lactic acid bacteria isolated from spontaneously fermented chia sourdough. *International Journal of Food Microbiology* 316: 108425.

Malka, T., R. Bomben, L. Balmaceda, J. Leporatti, T. Batlle, and S. Zaniolo. 2020. Gluten-free amaranth-based sweet snack formulation. *Avances en Ciencias e Ingenieria* 11(2): 21–29.

Medina, J. J. R., K. Ramírez, J. Rangel-Peraza, and G. J. Aguayo-Rojas. 2018. Incremento del valor nutrimental, actividad antioxidante y potencial inhibitorio de α-glucosida en brownies a base de leguminosas cocidas. *Archivos Latinoamericanos de Nutrición* 68(2).

Menga, V., M. Amato, T. D. Phillips, D. Angelino, F. Morreale, and C. Fares. 2017. Gluten-free pasta incorporating chia (*Salvia hispanica* L.) as thickening agent: An approach to naturally improve the nutritional profile and the *in vitro* carbohydrate digestibility. *Food Chemistry* 221: 1954–1961.

Merayo, Y., R. González, S. Drago, R. Torres, and D. Greef. 2011. Extrusion conditions and Zea mays endosperm hardness affecting gluten-free spaghetti quality. *International Journal of Food Science & Technology* 46(11): 2321–2328.

Mercado, R. J. L., P. C. C. A. Elías, and C. G. L. Pascual. 2015. Obtención de un aislado proteico de torta de sacha inchi (*Plukenetia volubilis* L.) y evaluación de sus propiedades tecno-funcionales. *Anales Científicos* 76(1): 160–167.

Mesías, M., F. Holgado, G. Márquez-Ruiz, and F. J. Morales. 2016. Risk/benefit considerations of a new formulation of wheat-based biscuit supplemented with different amounts of chia flour. *LWT – Food Science and Technology* 7: 528–535.

Miranda-Ramos, K., M.C. Millán-Linares, and C.M. Haros. 2020. Effect of chia as breadmaking ingredient on nutritional quality, mineral availability, and glycemic index of bread. *Foods* 9(5): 663.https://doi.org/10.3390/foods9050663.

Montemurro, M., E. Pontonio, and C. G. Rizzello. 2021. Design of a "Clean-Label" gluten-free bread to meet consumer demand. *Foods* 10(2): 462.

Mufari, J. R., P. P. Miranda-Villa, and E. L. Calandri. 2018. Quinoa germ and starch separation by wet milling, performance and characterization of the fractions. *Lebensmittel-Wissenschaft & Technologie* 96: 527–534.

OJEU (Official Journal of the European Union). 1997. Regulation (EC) N° 258/97 of the European Parliament and of the Council of 27 January 1997, concerning novel foods and novel food ingredients. *Official Journal of the European Communities* No L 43/1, 14. 2. 97.

Padalino, L., M. Mastromatteo, L. Lecce, F. Cozzolino, and M. A. Nobile. 2013. Manufacture and characterization of gluten-free spaghetti enriched with vegetable flour. *Journal of Cereal Science* 57: 333–342.

Paesani C., A. Bravo-Núñez, and M. Gómez. 2020. Effect of extrusion of whole-grain maize flour on the characteristics of gluten-free cookies. *Food Science and Technology* 132: 109931.

Paredes López, O., F. Guevara Lara, and L. A. Bello Pérez. 2012. *Los Alimentos Mágicos de las Culturas Indígenas Mesoamericanas*. México D.F:Fondo de Cultura Económica. ISBN 978-968-16-7567-7.

Pineli, L. de L. de O., R. B. A. Botelho, R. P. Zandonadi, J.L. J. L. Solorzano, G. T. de Oliveira, C. E. G. Reis, and D. da S. Teixeira. 2015. Low glycemic index and increased protein content in a novel quinoa milk. *LWT – Food Science and Technology* 63: 1261–1267.

Prego, I., S. Maldonado, and M. Otegui. 1998. Seed structure and localization of reserves in *Chenopodium quinoa*. *Annals of Botany* 82(4): 481–488.

Ramos-Díaz, J.M., S. Kirjoranta, S. Tenitz, P. A. Penttilä, R. Serimaa, A.-M. Lampi, and K. Jouppila. 2013. Use of amaranth, quinoa and kañiwa in extruded corn-based snacks. *Journal of Cereal Science* 58: 59–67.

Rawdkuen, S., D. Murdayanti, S. Ketnawa, and S. Phongthai. 2016. Chemical properties and nutritional factors of pressed-cake from tea and sacha inchi seeds. *Food Bioscience* 15: 64–71.

Repo-Carrasco-Valencia, R. A. M., C. R. Encina, M. J. Binaghi, C. B. Greco, and P. A. Ronayne de Ferrer. 2010. Effects of roasting and boiling of quinoa, kiwicha and kañiwa on composition and availability of minerals in vitro. *Journal of the Science of Food and Agriculture* 90: 2068–2073.

Reyes-Caudillo E., A. Tecante, M.A. Valdivia-López. 2008. Dietary fibre content and antioxidant activity of phenolic compounds present in Mexican chia (*Salvia hispanica* L.) seeds. *Food Chemistry* 107(2): 656–663.

Roa, D. F., P. R. Santagapita, M. P. Buera, and M. P. Tolaba. 2014. Amaranth milling strategies and fraction characterization by FT-IR. *Food Bioprocess Technology* 7: 711–718.

Rocchetti, G., L. Lucini, J. M. L. Rodríguez, F. J. Barba, and G. Giuberti. 2019. Gluten-free flours from cereals, pseudocereals and legumes: Phenolic fingerprints and in vitro antioxidant properties. *Food Chemistry* 271: 157–164.

Romero-Espinoza, A. M., S. O. Serna-Saldívar, M. C. Vintimilla-Álvarez, M. Briones-García, and M. A. Lazo-Vélez. 2020. Effects of fermentation with probiotics on antinutritional factors and proximate composition of lupin (*Lupinus mutabilis* Sweet). *LWT – Food Science and Technology* 130: 109658.

Rosell, C. M., G. Cortez, and R. Repo-Carrasco. 2009. Breadmaking use of Andean crops quinoa, kañiwa, kiwicha, and tarwi. *Cereal Chemistry* 86(4): 386–392.

Ruiz, G.A., A. Arts, M. Minor, and M. Schutyser. 2016. A hybrid dry and aqueous fractionation method to obtain protein-rich fractions from quinoa (*Chenopodium quinoa* Willd). *Food Bioprocess Technology* 9: 1502–1510.

Sandhu, K. S., P. Godara, M. Kaur, and S. Punia. 2017. Effect of toasting on physical, functional and antioxidant properties of flour from oat (*Avena sativa* L.) cultivars. *Journal of the Saudi Society of Agricultural Sciences* 16: 197–203.

Santiago-Ramos, D., J. D. Figueroa-Cárdenas, and J. J. Véles-Medina. 2018. Viscoelastic behaviour of masa from corn flours obtained by nixtamalization with different calcium sources. *Food Chemistry* 248: 21–28.

Schöenlechner, R. 2017a. Pseudocereals in gluten-free products. In *Pseudocereals: Chemistry and Technology*, 193–216. Chichester, UK: John Wiley & Sons.

Schöenlechner, R. 2017b. Quinoa: Its unique nutritional and health-promoting attributes. In *Gluten-Free Ancient Grains. Cereals, Pseudocereals, and Legumes: Sustainable, Nutritious, and Health-Promoting Foods for the 21st Century*, edited by J. R. N. Taylor and J. M. Awika, 105–130. Woodhead Publishing Series in Food Science, Technology and Nutrition. Duxford, UK/Cambridge, MA/Kidlington, UK: Woodhead Publishing/Elsevier.

Schöenlechner, R., G. Linsberger, L. Kaczyk, and E. Berghofer. 2006. Production of gluten-free short dough biscuits from the pseudocereals amaranth, quinoa and buckwheat with common bean. Ernaehrung/Nutrition 30: 101–107.

Schöenlechner, R., I. Mandala, A. Kiskini, A. Kostaropoulos, and A. Berghofer. 2010. Effect of water, albumen and fat on the quality of gluten-free bread containing amaranth. *International Journal of Food Science and Technology* 45: 661–669.

Segundo, C., A. Giménez, M. Lobo, L. Iturriaga, and N. Sammán. 2019. Formulation and attributes of gluten-free cakes of Andean corn improved with green banana flour. *Food Science and Technology International* 26(2): 95–104.

Serna-Saldívar, S. O., and C. Chuck-Hernández. 2019. Food uses of lime-cooked corn with emphasis in tortillas and snacks. In Corn: Chemistry and Technology, edited by S. O. Serna-Saldívar, 469–500. Washington, DC: AACC International Press.

Shepherd, R., M. Dean, P. Lampila, A. Arvola, A. Saba, M. Vassallo, E. Claupein, M. Winkelmann, and L. Lähteenmäki. 2012. Communicating the benefits of wholegrain and functional grain products to European consumers. *Trends in Food Science & Technology* 25: 63–69.

Sindhuja, A., M. Sudha, and A. Rahim. 2005. Effect of incorporation of amaranth flour on the quality of cookies. *European Food Research and Technology* 221: 597–601.

Sozer, N., U. Holopainen-Mantila, and K. Poutanen. 2017. Traditional and new food uses of pulses. *Cereal Chemistry Journal* 94(1): 66–73.

Steffolani, E., E. de la Hera, G. Pérez, and M. Gómez. 2014. Effect of chia (*Salvia hispanica* L) addition on the quality of gluten-free bread. *Journal of Food Quality* 37(5): 309–317.

Štěrbová, L., P. Hlásná Čepková, I. Viehmannová, and D. C. Huansi. 2017. Effect of thermal processing on phenolic content, tocopherols and antioxidant activity of sacha inchi kernels. *Journal of Food Processing and Preservation* 41(2): e12848.

Sung, W. C., E. T. Chiu, A. Sun, and H. I. Hsiao. 2020. Incorporation of chia seed flour into gluten-free rice layer cake: Effects on nutritional quality and physicochemical properties. *Journal of Food Science* 85(3): 545–555.

Suwanangul, S., N. Jittrepotch, and K. Ruttarattanamongko. 2021. Effects of thermal treatments od physico-chemical properties and antinutritional factor of sacha inchi (*Plukenetia volubilis* L.). *Naresuan University Journal: Science and Technology* 29(3).

Tosi, E. A., E. D. Re, H. Lucero, and R. Masciarelli. 2000. Amaranth (*Amaranthus* spp.) grain conditioning to obtain hyperproteic flour by differential milling. *Food Science and Technology International* 6(6): 433–438.

Turck, D., J. Castenmiller, S. de Henauw, K. Hirsch-Ernst, J. Kearney, A. Maciuk, I. Mangelsdorf, H. J. McArdle, A. Naska, C. Peláez, K. Pentieva, A. Siani, F. Thies, S, Tsabouri, M. Vinceti, F. Cubadda, K.-H. Engel, T. Frenzel, M. Heinonen, R. Marchelli, M. Neuhäuser-Berthold, A. Pöting, M. Poulsen,Y. Sanz, J. R. Schlatter, H. van Loveren, L. Matijević, and H. K. Knutsen. EFSA Panel on Nutrition, Novel Foods and Food Allergens (NDA). 2019. Safety of chia seeds (*Salvia hispanica* L.) powders, as novel foods, pursuant to Regulation (EU) 2015/2283. *EFSA Journal* 17: e05716.

Vázquez-Carrillo, M. G., J. D. Santiago-Ramos, M. Gaytán-Martínez, E. Morales-Sánchez and M. Guerrero-Herrera. 2015. High oil content maize: Physical, thermal and rheological properties of grain, masa, and tortillas. *Food Science and Technology* 60(1): 156–161.

Vázquez-Osorio, D. C., J. D. Jaramillo Ramírez, G. A. Hincapié Llanos, and L. M. Vélez Acosta. 2017. Development of cookies with sacha inchi (*Plukenetia volubilis* L.) flour coming from residual cake. *UGCiencia* 23: 101–113.

Viamonte-Garcés, M. I., J. M. Sánchez-Campuzano, A. Ramírez-Sánchez, A. Tapuy Cabrera, and V. C. Andrade-Yucailla. 2020. Chemical characterization and fatty acid profile of sacha inchi flour (*Plukenetia volubilis*) as raw material, in the elaboration of diets for animal use. *Mol2Net* 2020: 6. DOI:10.3390/mol2net-06-xxxx.

Vidaurre-Ruiz, J. M. 2020. Desarrollo de panes libres de gluten con harinas de granos andinos. PhD Dissertation. Lima: Universidad Nacional Agraria La Molina.

Vidaurre-Ruiz, J. M., W. F. Salas-Valerio, and R. Repo-Carrasco-Valencia. 2019. Propiedades de pasta y texturales de las mezclas de harinas de quinua (*Chenopodium quinoa*), kiwicha (*Amaranthus caudatus*) y tarwi (*Lupinus mutabilis*) en un sistema acuoso. *Revista de Investigaciones Altoandinas* 21(1): 5–14.

Villacrés, E., M. Quelal, E. Fernández, G. García, G. Cueva, and C. Rosell. 2020. Impact of debittering and fermentation processes on the antinutritional and antioxidant compounds in *Lupinus mutabilis* Sweet. *LWT – Food Science and Technology* 131: 109745.

Villada, J. A., F. Sánchez-Sinencio, O. Zelaya-Ángel, E. Gutiérrez-Cortez, and M. E. Rodríguez-García. 2017. Study of the morphological, structural, thermal, and pasting corn transformation during the traditional nixtamalization process: From corn to tortilla. *Journal of Food Engineering* 212: 242–251.

Wang, J., J. Zhao, M. de Wit, R. M. Boom, and M. A. I. Schutyser. 2016. Lupine protein enrichment by milling and electrostatic separation. *Innovative Food Science and Emerging Technologies* 33: 596–602.

Wang, S., F. Zhu, and Y. Kakuda. 2018. Sacha inchi (*Plukenetia volubilis* L.): Nutritional composition, biological activity, and uses. *Food Chemistry* 265: 316–328.

Zamora, R., and F. J. Hidalgo. 2011. The Maillard reaction and lipid oxidation. *Lipid Technology* 23: 59–62.

Zamora, R., R. M. Delgado, and F. J. Hidalgo.2009. Conversion of 3-aminopropionamide and 3-alkylaminopropionamides into acrylamide in model systems. *Molecular Nutrition & Food Research* 53(12): 1512–1520.

Zdybel, B., R. Rozylo, and S. Sagan. 2019. Use of a waste product from the pressing of chia seed oil in wheat and gluten-free bread processing. *Journal of Food Processing and Preservation* 43(8): e14002.

Zettel, V., and B. Hitzmann. 2018. Applications of chia (*Salvia hispanica* L.) in food products. *Trends in Food Science & Technology* 80: 43–50.

Ziemichod, A., M. Wojcik, and R. Rozylo. 2019. *Ocimum tenuiflorum* seeds and *Salvia hispanica* seeds: Mineral and amino acid composition, physical properties, and use in gluten-free bread. CYTA-Journal of Food 17(1): 804–813.

Žilić, S., A. Serpen, G. Akıllıoğlu‖, V. Gökmen, and J. Vančetović. 2012. Phenolic compounds, carotenoids, anthocyanins, and antioxidant capacity of colored maize (*Zea mays* L.) kernels. *Journal of Agricultural and Food Chemistry* 60: 1224–1231.

Žilić, S., T. Tolgahan, J. Vančetović, and V. Gökmen. 2016. Effects of baking conditions and dough formulations on phenolic compound stability, antioxidant capacity and color of cookies made from anthocyanin-rich corn flour. *Lebensmittel-Wissenschaft und-Technologie* 65: 597–603.

Ziobro, R., L. Juszczak, M. Witczak, and J. Korus. 2016. Non-gluten proteins as structure forming agents in gluten free bread. *Journal of Food Science and Technology* 53(1): 571–580.

Chapter 6

Fractionation of Seeds or Grains

Marianela Capitani[1], Adriana Scilingo[2], Edgardo Calandri[3],
María Alejandra Giménez[4], Marcela Lilian Martínez[5],
Vanesa Ixtaína[2], Nancy Chasquibol Silva[6],
M. Carmen Pérez Camino[7], Natalia Bassett[4],
Víctor Delgado-Soriano[8], Ritva Ann-Mari Repo-Carrasco-
Valencia[8], Claudia M. Haros[9], and Mabel Tomás[2]
[1]TECSE, Olavarría, Buenos Aires, Argentina –
CONICET, CCT Tandil, Buenos Aires, Argentina
[2]Universidad Nacional de La Plata (UNLP),
La Plata, Buenos Aires, Argentina
[3]Instituto de Ciencias y Tecnología de los Alimentos - ICTA.
Córdoba, Argentina/Instituto de Ciencias y Tecnología de los
Alimentos Córdoba – ICyTAC (CONICET), Córdoba, Argentina
[4]Universidad Nacional de Jujuy, San Salvador de Jujuy, Argentina
[5]Instituto Multidisciplinario de Biología Vegetal (IMBIV),
(CCT Córdoba) – CONICET, Córdoba, Argentina
[6]Universidad de Lima, Peru
[7]Instituto de la Grasa (IG- CSIC), Seville, Spain
[8]Universidad Nacional Agraria La Molina (UNALM), Lima, Peru
[9]Instituto de Agroquímica y Tecnología de
Alimentos (IATA- CSIC), Valencia, Spain)

DOI: 10.1201/9781003088424-6

CONTENTS

6.1 QUINOA AND KAÑIWA

6.1.1 Milling

6.1.1.1 Dry Milling

Quinoa white flour was obtained at laboratory scale by roller milling; authors found enhanced bread quality, due to the reduction in bran content and the parallel increase in diastasic capacity of quinoa flour (Elgeti et al. 2015). The process, essentially the same as that (Chirinos et al. 2018) used in the wheat flour industry, opens up the possibility of scaling up production of quinoa white flour without the need of new machinery. In another study D'Amico et al. (2019) tried the abrasive milling of quinoa; the protein, fat and ash of the polished seeds decreased with processing time, while starch rose up strongly. At 8 min milling, 65% of seed mass is lost and with it, most of its nutritional quality. As abrasion was the main procedure used for saponin removal, the results should inspire a radical evaluation of nutritional loss during quinoa industrial processing. Kañiwa (*Chenopodium pallidicaule* Aellen) belongs to the same genus as quinoa, but the grain is smaller. A traditional way to ground kañiwa is the *kcona* milling, using stones as grinding device. Flour can be obtained by disc or hammer milling, from whole or toasted grains; this last option is known as *Cañihuaco*.

6.1.1.2 Wet Milling

Together with the well-known qualities of quinoa protein (Abugoch James 2009) its starch shows interesting functional properties (Li et al. 2016). Quinoa oil has also shown valuable characteristics (Mufari et al. 2020). While protein and oil are mainly found in the embryo, starch granules reside mostly in the perisperm (Prego et al. 1998). When wet enough, the embryo may easily be detached from the perisperm; this enables separation of the components. We showed that this is possible when seed humidity and roller milling conditions are carefully controlled (Mufari et al. 2018a). In a similar manner, Ballester-Sánchez (2019) obtained good yield and quality in quinoa starch separation, employing a plate mill. Dry milling and sieving of quinoa grain also yields enriched fractions, but it is not as effective as wet milling (Solaesa et al. 2020).

6.1.2 Oil Extraction

Quinoa and kañiwa are not considered oily grains; their content is around 6–8% (Bazile et al. 2013; Torrejón et al. 2016), so at the moment there is no industrial interest in it. Nevertheless, quinoa oil quality is well documented (Mufari et al. 2020) and future interest in it as a profitable by-product must not be discarded;

also kañiwa shows a good fatty acids profile (Torrejón et al. 2016). Traditionally, quinoa oil extraction consisted in boiling seeds in iron pots covered with soapstone lids. The oil that condensed against the lids was then collected; it was sold mainly in Baltimore (USA), hence it is known as "Baltimore oil" (Thoufeek Ahamed et al. 1998).

6.1.2.1   Solvent Extraction

This is the most common way to remove oil from different natural raw materials. Usually, non-polar petroleum derivatives are used, such as hexane or petroleum ether. Small scale processes involve soxhlet type devices (AOAC 1999). For industrial production, the whole flour or grain fractions could be extracted in a continuous process, similar to those employed in obtaining the oil from cereals (Willis and Marangoni 2017). In all cases, the raw material must be finely grounded, due to the extraction efficiency, which strongly depends on the particles' size. Supercritical extraction was also assayed for quinoa; the extracted oil showed an elevated presence of antioxidants and tocopherols but the yields were low (Benito-Román et al. 2018).

6.1.2.2   Cold Extraction

According to Eckey (1954) it is possible to extract quinoa oil with screw mills or twin screw extruders; nevertheless these processes are designed to maximize oil yield, so the high temperatures and pressure involved severely affect the nature of both proteins and oil. Today the solvent extraction method is more recommendable (Willis and Marangoni 2017). As we have shown before (Mufari et al. 2018a), it is possible to separate an oil enriched germ fraction from quinoa seeds, so that mild alternatives like those developed for cereal germs could also be applied to quinoa germ oil extraction enabling the recovery of a good quality protein (Stamenkovic et al. 2020). The process essentially involves a solid-liquid exchange where the oil moves from the germ to the continuous phase (solvent), on a countercurrent basis.

6.1.3   Protein and Fiber Isolation

6.1.3.1   Protein Isolation Methods

Quinoa is a well-valued food mainly due to the protein's nutritional quality; because of that it has been largely studied and characterized by several authors all round the world (Abugoch et al. 2008). The most common method for obtaining quinoa protein isolates is isoelectric precipitation, taking advantage of the fact that most proteins are insoluble at the isoelectric point (pI) (Mir et al. 2021). It requires two steps: solubilization at a proper pH, normally far from the pI value, and second, that the pH is gradually changed until it approaches the pI; at this moment the proteins get the lowest solubility and most of them aggregate and precipitate. Guerreo-Ochoa et al. (2015) found that optimal quinoa protein extraction conditions depend not only on pH but also on temperature, solvent/meal relationship and extraction time. On the other hand, Ruiz et al. (2016) found a correlation between isolated protein yield and the extraction pH. Kañiwa proteins concentrate was also obtained with a procedure that resembles the one used in quinoa (Betalleluz-Pallardel et al. 2017). Quinoa and kañiwa being both from *Chenopodium* genus it is not surprising that they show similar composition in their protein fractions, as can be seen in Table 6.1.

6.1.3.2   Fiber Extraction Methods

Quinoa seeds present an interesting content in fibers, around 3–4% (d.b.) by weight (Jancurová et al. 2009). In kañiwa, fiber content is widely variable but generally superior to quinoa (Repo-Carrasco-Valencia et al. 2010). Cereals fibers are in the outer shell, but in pseudo-cereals such as quinoa and kañiwa the shell is very thin; when an abrasive method is applied this narrow cover is immediately removed; if the action persists other components of the seed will become detached and nutritional quality can be affected (D'Amico et al. 2019). Ballester-Sánchez et al. (2019) found that wet milling of red quinoa produced a higher yield, recovery and purity than the dry milling procedure, but this last one gave breads with better technological qualities, attributed to the characteristics of its fibers. Quinoa soluble fiber showed a compact structure, higher reducing power and digestive enzyme inhibitory activity (Chen et al. 2021) in comparison with wheat.

6.1.3.3   Techno-functional Properties of Proteins

Mostly related with properties other than nutritional, techno-functional examples involve the capacity to modify the texture or sensory properties of food.

TABLE 6.1   PROTEIN COMPOSITION IN QUINOA AND KAÑIWA

Protein fraction	kañiwa (% d.b.)[a]	quinoa (% d.b.)[b]
Albumin	15.4–15.8	24.9 ± 8.2
Globulin	24.1–26.7	23.5 ± 5.6
Prolamin	9.6–9.9	5.1 ± 0.4
Glutelin	22.9–21.5	25.3 ± 2.1
Insoluble fraction	28.0–26.1	21.2 ± 13.0

Source: [a]Moscoso-Mujica et al. (2017); [b] Mufari et al. (2018b).
Notes: d.b. dry basis.

Also, they can influence the physical behavior of food or ingredients during their processing or storing. Hydration properties are related to the capacity of proteins to interact with a solvent, particularly water; protein–protein interaction mainly affects food structures and surface properties which are closely related to their ability to stabilize bubbles or emulsions. In this respect, emulsifying properties mainly depend on the ability of proteins to lower surface tension, and prevent coalescence of droplets and also to increase surface hydrophobicity. Quinoa protein isolates present a low emulsifying ability compared to bovine serum albumin (BSA) but they show good emulsion stability, close to BSA and higher than pearl millet, wheat and soybean. Enzymatic hydrolysis, saponins removal and protein-linked polysaccharides can modify these properties (Dakhili et al. 2019). Foaming capacity for quinoa proteins stay between soy protein and egg albumin and makes stable foam that could be used in food prototypes (Abugoch et al. 2008). The gelling properties of quinoa proteins depend on the pH of the extraction. If it is lower than 9 a semi-solid gel is obtained. When extraction pH overtakes 10 the gel is not formed; and according to Hermansson (1986), it could be used in liquid food. Quinoa proteins showed a water absorption capacity close to that of soy protein, but the oil absorption capacity was lower (Ashraf et al. 2012). Probably, the difference between oil and water absorption can reside in the conformational characteristics of quinoa albumins and globulins (Sathe and Salunkhe 1981). Quinoa proteins and soy proteins showed similar water holding capacities (Abugoch et al. 2008); additionally, this property of quinoa proteins is expected to confer improved texture on different food applications (Tang et al. 2015). It must be pointed out that the functional properties of quinoa protein can be affected by drying conditions (Mufari 2017). High temperatures may definitively affect functional properties of quinoa protein, as shown by Mir et al. (2021), who suggest that milder treatments would yield better results. Finally, interaction with other components in food can modify the quinoa protein behavior, as was observed by Roa-Acosta et al. (2020). These authors tried with quinoa flour which had received different abrasive milling treatments that changed protein conformation, thus affecting both pasting and water absorption behaviors.

6.1.4   Physicochemical, Thermal and Rheological Properties of Starch

The starch granules of quinoa are polygonal/irregular, of very small diameter (0.46–2.53 µm) and which have considerable variability in their amylose content (3.5–26.5%) (Table 6.2). In general, the high amount of amylopectin conducted its gelatinization at relatively lower temperatures (44.5–77.2°C) and exhibited a high pasting viscosity (4150.8 Cp) compared to normal maize starch (61.7–89.0°C and 3058.0 Cp, respectively). Due to these large differences in amylose content and physicochemical properties among quinoa starches, they have a wide variety of food and non-food applications such as in cosmetics, in biodegradable films and coatings, candy dusting, flavor carriers, in fat replacement and to produce creamy/smooth textures (Ahamed et al. 1996; Repo-Carrasco-Valencia and Valdez Arana 2017). Furthermore, the good freeze–thaw stability of quinoa starch suggests application as a thickener in frozen food products; it shows resistance to retrogradation, which also suggests possible uses in emulsions such as salad dressings, as well as in sauces, cream soups and pie fillings (Repo-Carrasco-Valencia and Valdez Arana 2017).

Kañiwa starches showed a similar amylose content to quinoa's (6.48–17.44%), both have lower contents than cereals such as maize (22.0–25.4%) or wheat (25–28%) starches (Table 6.2; JianYa-Qian 1999; Jian et al. 2020; Lindeboom 2005). Their granule size is very small with an average of around 1.45 μm. Additionally, their granule shape is irregular and polygonal with similar onset and peak gelatinization temperatures compared to quinoa starches, but presented higher values of firmness and lower syneresis than quinoa starches, which could be used in foods where consistency and firmness are required (Steffolani et al. 2013).

The high viscosity value obtained during the heating process of quinoa starches suggests a higher water absorption capacity, whereas the breakdown parameter (PV–HPV, where peak viscosity [PV] and hot paste viscosity [HPV]) provide information about stability during heating. This parameter is lower in quinoa, kañiwa and amaranth starches compared to conventional maize starches (Table 6.2), displaying a higher structural fragility during cooking than the last item in the table (Haros et al. 2006). During cooling, an important parameter to consider is the setback, which is the tendency to restructuration/retrogradation (Haros et al. 2006). High setback viscosities indicate low resistance to retrogradation mainly attributed to amylose, with a larger flexible structure than amylopectin. However, the lack of differences in setback values between maize and pseudo-cereals starches, between samples with different amounts of amylose, indicates that another phenomenon should be implicated as was suggested by Repo-Carrasco-Valencia and Valdez Arana (2017).

The gelatinization enthalpy (ΔH_G) varied from 1.66 to 14.9 J/g for quinoa starch and from 1.84 to 9.32 J/g for kañiwa starch, whereas the maize starch showed values between 10 and 19.0 J/g (with the exception of INIFAP-Mexico hybrids), demonstrating that, in general, higher energy was required to disrupt the crystalline structure of the last item in the table (maize starch) (Table 6.2). Low values in starch gelatinization and/or pasting processes of some varieties of quinoa and kañiwa might suggest a less crystalline structure, which could result in higher enzymatic/digestion susceptibility (Srichuwong et al. 2017). The relationship between the physicochemical and structural properties of both quinoa and kañiwa starches, is as yet uncertain (Jiang et al. 2020).

TABLE 6.2 PHYSICOCHEMICAL AND TECHNO-FUNCTIONAL PROPERTIES OF NATIVE STARCHES

Physicochemical characteristics	Units	Latin-American Maize	Quinoa	Kañiwa	Amaranth
Amylose	(%)	20[s]; 22.0[b][w]; 22.58[t]; 25.4; 27[y]	3.5–19.6[e]; 7[b]; 7.1[a]; 8.22–9.30[f]; 9.43–10.90[j]; 12.1[l]; 12.2[b]; 19.2–26.5[g]; 22.5[c]	6.48[l]; 10.70–17.44[f]	0–14[q]; 0.1–19.2[r]; 0.1–22.3[t]; 2.5[c]; 7.8[b]; 13.6[s]
Granule Size	(µm)	2.4–33.8[o][p]; 3-20[w][x][y]; 5–20[i]; 3.16–30.22[i]; 6-30>[u]	0.46–5.56[f]; 0.5–3.0[d]; 0.6–1.5[e]; 1[a]; 1–2[b]; 2.53[f]	1.45/< 2.0[f]	0.75–1.25[e]; 1–2[b][d]; 0.5–2[j]
Cristallinity	(%)	22[v]; 26[x]; 32[w]; 36.05[i]	21.00–29.67[i]; 35.0[a]; 35.4[b]; 36.3–39.6[f]	34.0–35.6[f]	24.5–45.5[r]; 27.9[q]
Shape	–	Irregular/ Polyhedric[l]; Irregular[u]; Spherical[w][x][y]	Polygonal[b]; Spherical[c]; Irregular/ Pol*nal[e][f][i]	Polygonal[l]; Irregular/ Polygonal[f]	Polygo*[b][q][r]; Spherical[e] Irregular/ Polygonal[i]
Pasting Properties (RVA)					
Pasting Temperature	(°C)	71.1–72.6[p]; 75.5[b]; 75.93[d]; 91.0[e]	62.7[f]; 63.0–64.0[e]; 64.50–68.44[g]; 66.5[b]; 66.8[b]; 72.6[i]	71.1–74.*)	70–75.7[r]; 71.7[b]; 73–78[q]
Peak time	(min)	4.47–4.73[c] 5.1[b]; 5.4[e]	4.97–5.17[f]; 5.1–6.9[e]; 6.72–7.00[g]; 7.0[h]	4.67–5.43[f]	N/A

(continued)

TABLE 6.2 (CONTINUED)

Physicochemical characteristics	Units	Latin-American Maize	Quinoa	Kañiwa	Amaranth
Peak Viscosity (PV)	(Cp)	1515–3058[p]; 367.2[e]; 2906[h]; 2669[i]	43.2–309.6[e]; 2983–3551[i]; 3322–4306[f]; 3064–3898 [g]; 3380[b]; 4150.8[b]	237.0–498.0[k]; 1907–2657[f]	216–2064[q]; 276–3456[r]; 1596–2556[f]; 1662.0[b]
Hot Paste Viscosity (HPV)	(Cp)	310–1196[p]; 1794[i]; 1852[h]	2231–3430[g]; 2250–3205[i]; 2653[b]; 3253.2[b]	N/A	792.0[b]
Breakdown (PV – HPV)	(Cp)	72.0[e]; 295.2[e]; 875[i]; 1052[h]; 1205–1882[p]	4.8–1056.0[e]; 34.8–214.8[e]; 313–733[i]; 897.6[b]; 439–1161[f]; 728[h]	5.04–28.0[k]; 234–673[f]	108–1020[q]; 120–1572[r]; 324–696[s]; 870.5[b]
Final viscosity (FV)	(Cp)	345.6[e]; 491–2133[p]; 2966(i); 3089[h]	50.4–346.8[e]; 2692–3705[i]; 3725–4109[f]; 3557–4340 [g]; 3867[b]; 4994.4[b]	269.04–810.00[k]; 2787–3332[f]	1296.0[b]
Setback (PV – TV)	(Cp)	64.8[e]; 181–957[p]; 1172[f]; 1237[h]	15.6–153.6[e]; 442–730[f]; 676–963[f]; 1215[h]; 1741.2[b]	60.0–316.8[k]; 1041–1332[f]	0–588[q]; 84–2292[r]; 368.4–1404[s]; 504.0[b]
			Thermal properties (DSC conditions: starch:water, 1:3 at 10°C/min)		
Onset Temperature (To)	(°C)	61.7[e]; 64.7[e]; 64.9[h]; 65.8[x]; 65.9[w]; 67.41[i]; 68.8–72.0[n]; 69.1–70.4[f]	44.6–50.6[e]; 50.7[h]; 51.2–55.4[g]; 54.25–55.72[f]; 54.5[a]; 57.89–61.76[i]; 59.9[b]	53.6[f]; 55.3–57.1[k]*; 58.3–58.9[m]*; 59.06–58.39[f]	62–66.1[q]; 62.3–69.3[r]; 66.3[b]; 66.8–68.0[f]; 69.5[s]

Peak Temperature (T_p)	(°C)	69.1[e]; 69.6[b]; 71.00[i]; 72.8[w]; 73.1[x]; 73.2–75.6[o]; 73.3[y]; 73.4–77.1[v]; 73.6–76.2[n];	50.5–61.7[e]; 58.6– 61.1[g]; 58.7[b]; 61.66– 63.01[f]; 62.6[a]; 63.77–67.44[i]; 64.5[b]	60.0[l]; 61.0–61.8[k]*; 62.8–64.8[m]*; 66.18–66.68[f]	68.8–73.9[q]; 69.2–78.0[r]; 73.4[s]; 73.4–78.8[t]; 74.5[b]
Final Temperature (T_f)	(°C)	76.0[b]; 78.47[i]; 79.7[w]; 79.9[x]; 80.2[y]; 82.0–84.3[o]; 85.1–89.0[n]	68.2–70.1[g]; 69.3[b]; 71.0[b]; 71.3[a]; 71.86–77.24[i]	68.1[l]; 68.8–69.0[k]*; 70.9–72.9[m]	77.7–89.6[r]; 78.8–88.8[q]; 81.2– 89.1[t]; 86.9[b]; 89[s]
Gelatinization Enthalphy(ΔH_G)	(J/g)	1.9–4.7[v]; 10.0– 13.7[o]; 10.5–13.9[w]; 12.5[b]; 13.1[e]; 13.30[i]; 18.5[y]; 18.9[w]; 19.0[x]	1.66[b]; 7.79–11.76[i]; 8.45– 9.00[f]; 10.3[f]; 10.4[b]; 12.4–13.5[g]; 12.8–14.9[e]	1.84–2.57[k]*; 3.5– 4.8[m]*; 7.49–9.32[f]; 8.00[l]	2.58[b]; 7.1–18.4[r]; 13–15.8[q]; 13.5[s]

Notes: N/A, not available; *flour; RVA: Rapid Visco-Analyzer; DSC: Differential Scanning Calorimetry; Cp: centipoise.

Source: [a]Tang et al. (2002); [b]Qian and Kuhn (1999), *Amaranthus cruentus;* [c]Tari et al. (2003), *Amaranthus paniculatus;* [d]Lindeboom et al. (2005), [e,z]2005; [f]Steffolani et al. (2013); [g]Ballester-Sanchez et al. (2019); [h]Selma-Gracia et al. (2020); [i]Jian et al. (2020); [j]Jane et al. (1994); [k]Bustos et al. (2019); [l]Luna-Mercado and Repo-Carrasco-Valencia (2020); [m]Salas-Valero et al. (2015); [n]Pérez et al. (2003), semi-dent maize, *Zea mays* c.v. Tilcara; [o]Haros et al. (2003), flint and dent maize; [p]Haros et al. (2004), semi-dent maize, *Zea mays* c.v. Tilcara: Zhu (2017); [q]*A. hypochondriacus,* [r]*A. cruentus,* [s]*A. caudatus,* [t]*A. hybridus;* [u]Méndez-Montealvo et al. (2008) INIFAP-Mexico (H515); [v]Méndez-Montealvo et al. (2005) INIFAP-Mexico; Agama-Acevedo et al. (2005) pigmented maize [w]black, [x]blue and [z]white from Mexico.

6.2 MAIZE

6.2.1 Grain Characteristics

The word "corn" has been used over time with different meanings; however, after the discovery of America, Europeans called it "Indian corn" and named it "mahis" in the native American language, which was later called "maize" (Salvador-Reyes and Pedrosa Silva Clerici 2020). In this chapter, the name maize will be used to indicate that it refers to the native races with production aimed at local markets.

Vitreousness- and hardness-associated properties are significantly correlated with the end usage of maize. The ratio of vitreous or hard to floury endosperm influences post-harvest resistance to insects and fungi, milling yield, dust formation during processing, rate of starch digestibility and textural properties of maize-based foods (Dombrink-Kurtzman and Bietz 1993; Córdova-Noboa et al. 2020; Osorio- Diaz et al. 2011; Cruz- Vázquez et al. 2019).

Currently there is great interest in the extraction of colorants from pigmented races (Jing and Giusti 2007). Purple and blue corn, characterized by high content of anthocyanin, are a good source for production of natural colorants for the food industry (Salinas et al. 1999). For a long time, its extracts were used as coloring agents for food and beverages, such as *chicha Morada* and *mazamorra Morada* (Saldaña et al. 2018). Most blue varieties of maize are planted in the highlands of Mexico and Peru and are still landraces inherited by past generations of local types. Major drawbacks of blue maize varieties are that plants yield lower amounts of grain, and that kernels are soft-textured. Due to the increased interest in colored maize, plant breeders have developed high-yielding hybrids adapted to other ecosystems such as subtropical regions. These hybrids with excellent yield potential produce harder kernels with higher concentrations of anthocyanins (Serna-Saldivar and Pérez-Carrillo 2019). Recently, several authors studied different techniques to obtain concentrated extracts of phenolic compounds (Monroy et al. 2020). However, there is little research on the milling properties of these maize varieties, so it is necessary to verify their suitability for conventional maize processing.

6.2.2 Starch and Lipid Extraction

The wet milling process is designed to efficiently take maize apart and purify its constituents (starch, oil, protein, and fiber), making them suitable for human and animal food ingredients use, industrial products, or as feedstocks for converting into other products with higher added value (Anderson and Watson 1982). Conventional wet milling is generally applied to hybrid maize as they have a high extractability of starch and oil.

Despite the great diversity of native maize in Mesoamerica and South America, little is known about their wet milling characteristics, so the separation of starch and oil from them is generally not practiced. In this regard, Uriarte-Aceves et al. (2015) showed that despite the wide variability found in the physical and chemical characteristics of different blue maize from northeast Mexico, they present great potential for wet milling with high yields and starch recovery and with an exceptional quality of whiteness.

Triglycerides from maize oil have a high level of linoleic acid (*c.*18:2), which is an essential polyunsaturated fatty acid for humans (Weber 1979). According to Giménez et al. (2020) and Ortíz-Prudencio (2006) different races of native maize from Mesoamerica and South America showed a kernel oil concentration of about 4–5% on dry weight basis. Development of maize cultivars with high oil without yield loss is a challenge in breeding programs (Rajendran et al. 2018). Specialty high oil maize contains 6% or more oil on dry weight basis.

6.2.3 Protein Fractions

The protein content in Latin-American maize races ranges from 5.7–12.5% of their dry weight (Giménez et al. 2020). Zeins are the most abundant in the grain; they are located in the endosperm. They have very low content of lysine and specially of tryptophan, which is why zeins are considered of very poor nutritional quality (Zarkadas et al. 1995; Rascón-Cruz et al., 2004).

6.2.4 Fiber

The most important sources of dietary fiber in maize are the bran and the tip tap (Boyer and Shannon 2003). The dietary fiber content in native maize races varies between 8.47 and 19.5% (Ortíz-Prudencio 2006). The high dietary fiber content of some Peruvian Andean maize can be related to the greater thickness and number of their pericarp layers compared with other maize cultivars (Salvador-Reyes et al. 2021). It is generally separated during milling in the shell and used for animal feed; however, it could be used in other formulated products.

6.2.5 Techno-Functional properties of Flours, Proteins, Starch and Fiber

Great variability in techno-functional properties has been observed in Andean and Mesoamerican maize races. Knowledge of the starch gelatinization process provides important information about functionality, energy requirements and end use. Some maize races from the north of Argentina were differentiated mainly due to the amylose content, thermal properties and firmness, and stability of their gels. In these races there were no significant differences in enthalpy of starch gelatinization; however, the floury kernels showed a lower gelatinization range than races with the highest endosperm hardness (Giménez et al. 2020). According to Narváez-González et al. (2007), the size of the starch granule and endosperm hardness greatly influenced the thermal and pasting properties of the flours of different races of Latin-American maize. The hard or vitreous grains presented small starch granules, while in the floury grains they were large. Kernels with the highest endosperm hardness tended to develop low viscosities, and enthalpies presented a long gelatinization range. The opposite was found for floury kernels. According to Salvador-Reyes et al. (2021) the differences in the chemical composition and endosperm structure between Peruvian Andean maize samples influenced their pasting properties. The floury races had a lower pasting temperature, high viscosity pastes and did not retrograde easily, making it an excellent option for improving the texture of soups, sauces, food mixes or ice creams where it is required to maintain the texture and avoid syneresis.

According to the broad techno-functional behaviors, the Mesoamerican and Andean maize races could be used in different technological processes to produce gluten-free foods. Their proteins, mainly zeins, provide a cohesive and extensible mass above its glass transition temperature, approximately 35°C, with moisture content greater than 20%, so they can be a suitable ingredient for this type of products (Taylor et al. 2018). Fiber of hybrid maize has been used to produce functional soluble dietary fiber. The product, known as *corn fiber gum*, is primarily composed of arabinoxylan and possesses a strong emulsifying property (Ai and Jane 2016). The majority of maize bran is almost completely insoluble dietary fiber (Boyer and Shannon 2003). Unfortunately, this can lead to undesirable changes in product quality (Rose et al., 2010).

6.2.5.1   Physico-Chemical, Thermal and Rheological Properties of Maize Starch

Unfortunately, little information was found with regard to Latin-American maize starches. The parameters described in Table 6.2 correspond to native maize from Argentina (types flint, semi-dent and dent), pigmented maize from Mexico (white, black and blue) or varieties/hybrids of maize (INIFAP [National Institute of Agricultural and Livestock Forestry Research]), Mexico).

Salvador-Reyes et al. (2021) have shown that some maize races from Peru presented starch granules with spherical shape at the center of their endosperm, while in other irregular and polyhedral shapes, most of the races studied by these authors presented only large starch granules (>10 μm), while a bimodal distribution presenting large granules (>10 μm) and small granules (<10 μm) was also observed. The presence of small granules could be related to the extreme conditions of growth (Tester 1997). In these races, the starch granules presented both smooth and porous surfaces. Although how pore formation occurs is still unknown; it has been shown that pore formation is directly related to the α-amylase activity of grain (Prompiputtanapon et al. 2020). The porous starches influence the mechanical and textural properties of food products (Sarifudin et al. 2020).

6.3 AMARANTH

6.3.1 Milling

Amaranth flour can be obtained by grinding the whole grain or by using strategies that enrich the product in any of its macro-components, starch, fiber, proteins and/or lipids (Roa et al. 2013; Roa-Acosta et al. 2017; Tosi et al. 2000). The grains directly ground in a planetary ball mill provided flour with 16.8, 7.7 and 73% weight for weight (w/w) for proteins (Nx6.25), lipids and carbohydrates (starch+ fiber) respectively (Roa et al. 2013). When an abrasive mill step prior to grinding was performed, two different products were obtained: one rich in starch and another rich in lipids and proteins (1.5, 5 and 88% w/w and 23, 47 and 10% w/w for lipids, proteins and carbohydrates, respectively). The products also differ in lipid oxidative stability during storage time, and in the starch's degree of crystallinity, which could affect its technological properties (Roa et al. 2014; Roa-Acosta et al. 2017). A successful differential milling technique to obtain *Amaranthus caudatus* high protein flours was described by Tosi et al. (2000): intense friction of specially conditioned dried grains in fluidized bed equipment was performed to maintain moisture within the seed and reduce it on the surface. Differential milling allowed a selective detachment of the different anatomical parts of the grain and minimized the loss of lysine during drying. After sieving and carrying out a pneumatic classification high protein and high starch flour was obtained. The protein percentage of this product was much higher than that of whole flour (near to 40% vs. 21% w/w respectively) with a yield close to 40%, while the differential grinding of unconditioned grains exhibited a lower yield (10%).

Amaranth wet grinding has been used as a way of obtaining starch. Seeds must be soaking in order to soften the grains. This phase, which is critical in the overall process, can be carried out in an aqueous medium (Pérez et al. 1993), in the presence of alkalis (Vázquez Batti et al. 2006), acids (Calzetta-Resio et al. 2006) or even detergents such as SDS (sodium dodecyl sulfate; Joaqui et al. 2020). Process conditions, use of additives and their concentration, soaking time, temperature, separation conditions, among others, have been studied in several publications (Pérez et al. 1993; Calzetta-Resio et al. 2006; Vázquez-Batti et al. 2006). Fractions rich in proteins and fiber can be separated out as by-products during starch production. Their characteristics and techno-functional properties strongly depend on the process conditions. It is possible to adjust the conditions to minimize the presence of proteins in the starch, or in the fraction rich in fiber depending on what it is desired (Calzetta-Resio et al. 2006). Joaqui et al. (2020) have recently studied the structural modifications of seed proteins when the milling process is carried out in the presence of sodium hydroxide (NaOH) and SDS using the Fourier-transform infrared spectroscopy (FT-IR). Important changes occur in the gelling capacity associated with modifications in the proteins' secondary and tertiary structure.

An alternative way implies using wet milling of the grains to prepare dispersion instead of obtaining starch. The grain in water grinding using a colloidal mill (model AD 35-R-ColMil) has allowed the development of a beverage amaranth like vegetable milk with a protein content similar to cowx' milk (Guzmán-Maldonado and Paredes-Lopez 1998; Manassero et al. 2020).

6.3.2   Oil Extraction

Amaranth lipids in the seed represent between 6 and 8.5% w/w of its weight with some variations among the different species (León-Camacho et al. 2001; Martirosyan et al. 2007; Gamel et al. 2007). Triacylglycerols contribute approximately 80% w/w, with various unsaturated and saturated fatty acids in a 3:1 ratio; linoleic acid is the most important fatty acid, followed by oleic and palmitic acids (Krist 2020). Phospholipids and squalene represent 10% and 5% of the oil, respectively (Gamel et al. 2007), making amaranth oil an interesting squalene source. Several sterols have been found, such as cholesterol and $\Delta 7$-campesterol, with the peculiarity that most of them are esterified. The main tocopherol is β-tocopherol, doubling that of α-tocopherol (Krist 2020). Given that β-tocopherol does not possess antioxidant activity, amaranth oil is much more resistant to oxidation than sunflower oil, preserving its characteristics for more than 9 months (Gamel et al. 2007). Amaranth oil with a high content of squalene and polyunsaturated fatty acids lowers total cholesterol, triglycerides, low density lipoproteins (LDL) and very LDL, diminishing overall risk of heart and circulatory diseases (Martyrosan et al. 2007). The presence of phytosterols, tocopherols and tocotrienols, along with squalene, provide antioxidant capacity. Although it is edible oil, its applications are not limited to the food area; they are extended to cosmetics and health. Squalene is a skin moisturizer, analgesic, healing and anti-inflammatory (Krist 2020).

6.3.3   Fiber Extractions

Amaranth seeds contain between 8 and 16% w/w of dietary fiber (DF) composed mainly of xyloglycans and pectic polysaccharides. This type of fiber is similar to fruits, vegetable and legume fiber, rather than to that of cereals (Lamothe et al. 2015). Insoluble DF (78% w/w) contains mainly galacturonic acid, arabinose, galactose, xylose and glucose, while soluble DF (22% w/w) contains glucose, galacturonic acid, and arabinose (Velarde Salcedo et al. 2019; Repo-Carrasco et al. 2009). Tosi et al. (2001) obtained a product rich in fiber by differential milling on conditioned grains and pneumatic classification, containing approximately 64 and 7% w/w of insoluble and soluble amaranth DF, respectively. Calzetta-Resio et al. (2006) have described obtaining a fraction very rich in fiber as a by-product of starch isolation. These authors used grain wet milling in the presence of sodium metabisulfite in order to obtain amaranth starch as the main product.

6.3.4 Protein Isolation

An amaranth protein isolate can be obtained by dispersing previously defatted flour in an aqueous medium, solubilizing the proteins at a pH>7 and then precipitating them at their isoelectric point, near to pH 4.5 (Martínez and Añón 1996; Salcedo Chávez et al. 2002; Cordero de los Santos et al. 2005). Amaranth seeds contain different types of protein: albumins, present in small amounts, globulins and glutelins which constitute the majority of reserve proteins. Amaranth globulins belong to the 7S and 11S families, along with a third type of globulin, P-glb, with great tendency to polymerization (Castellani et al. 2000), in which the precursor of 11S family A–B subunits is not cleaved (Martínez et al. 1997; Quiroga et al. 2007). The pH used for protein solubilization in the isolation process modifies the color and protein content of the isolate, in addition to the polypeptide composition and the conformation of the proteins, increasing the proportion of glutelins and the degree of denaturation when pH is close to 11 (Abugoch et al. 2010). Another technique for obtaining protein isolates is by the micellization process, in which flour proteins are solubilized at neutral pH in the presence of 0.8M NaCl, and then precipitated after ultrafiltration (membranes of 10 kDa nominal molecular weight limit) by diluting the solution in cold distilled water and separating them by centrifugation. Even though the proteins obtained by micellization showed a lower degree of denaturation and some improvement in functional behavior, the yield was lower than that achieved by isoelectric precipitation (Cordero de los Santos et al. 2005).

6.3.5 Techno-Functional Properties of Amaranth Flour and Proteins

The information available for evaluating the techno-functional quality of amaranth flours is scarce and is a field that needs to be further explored to promote the use of amaranth flours in diverse products. Tests designed for the study of weak wheat flours have been applied to characterize the functional behavior of amaranth flour (protein content 16% w/w, Nx5.85). Water imbibing capacity (WIC) and solvent retention capacity (SRC) determination showed that amaranth flour presented SRC values that were almost double those of wheat flour, similar to values found in a high percentage of damaged starch wheat. WIC values of amaranth and wheat flour were similar, while the initial hydration rate was four times lower in amaranth flour, showing that it is necessary to hydrate it before using it in cookies preparation to reach appropriate texture properties (Sabbione et al. 2019).

Amaranth seed major protein fractions are good candidates due to their variety in physicochemical and structural properties (Segura-Nieto et al. 1994; Martínez et al. 1997). Amaranth protein isolates provide important functional versatility due to the structural diversity in their composition, but exhibit a rather low solubility in aqueous solvents, limiting their use in the food industry (Segura-Nieto et al. 1994). Enzymatic hydrolysis is a chemical approach used to alter protein functionality improving their solubility, and the conditions can be selected according to the desired application in food formulation. Several works document this strategy by using plant proteases such as cucurbita or papain (Scilingo et al. 2002) or trypsin (Condés et al. 2009). The proteins obtained significantly improved their foaming properties and solubility even when heat treatments were applied (Condés et al. 2009).The pH strongly influences the functional behavior of proteins. Amaranth protein isolates showed the best foaming properties at acid pH (Bolontrade et al. 2013, 2016). Under these conditions, not only does denaturation occur due to pH effect (close to 2), but also the activity of an aspartic protease present in the isolate produces partial proteolysis modifying the size of the polypeptides, resulting in the improvement of foaming properties (Ventureira et al. 2012a). Recently, Malgor et al. (2020) described the application of this functional behavior in an amaranth lemon sorbet, using amaranth isolate as the protein source of the product. The best condition for emulsion formation and stabilization of amaranth protein isolates was also the acidic one (Ventureira et al. 2012b), even though isolates were able to form stable emulsions at pH close to neutrality. Under acidic pH conditions, there is higher soluble protein content in the system. In addition, when denaturation of these proteins occurs, the aspartic protease partially hydrolysates them, generating a greater exposure of hydrophobic sites capable of anchoring in the interface, hence forming a viscoelastic interfacial film more resistant to destabilization processes (Suárez and Añón 2018). Protein dispersion obtained at pH close to neutrality was also suitable for obtaining stable emulsions at high protein concentration due to the formation of flocs that considerably increased the viscosity of the continuous phase, indicating that the presence of insoluble protein in the continuous phase is not always detrimental to the stability of the emulsion. Amaranth protein isolate can also be used to form gels (Avanza et al. 2006) at concentrations higher than 7% weight in volume (w/v), increasing the water holding capacity of the matrices when protein concentration increases (up to 20% w/v). In addition, increasing the heating temperature (70 to 95°C) provides high luminosity and low water holding capacity gels. The main interactions involved in the gel network are disulfide bonds between 11S globulin and P-glb.

The search for foods that contribute to consumers' health discovers in amaranth a very interesting ingredient. Different authors have described biological activities derived from peptides encrypted in their proteins, which are released by digestion or by specifically designed enzymatic treatments. Among the properties studied it is worth mentioning their action on cholesterol metabolism, antioxidant, anticancer, anticoagulant and antihypertensive activities, among others (Caselato-Sousa and Amaya-Farfán 2012; Silva-Sánchez et al. 2008). Some of these bioactivities have been proven using *in vivo* studies as well as *in vitro*, *in sílico* and/or *ex vivo* tests (Nardo et al. 2020; Sabbione et al. 2016a, 2016b; Sisti et al. 2019).

6.3.6   Physico-Chemical, Thermal and Rheological Properties of Amaranth Starches

Amaranth starch shows outstanding properties from a techno-functional point of view (Table 6.2). In general, the small granule size (0.75-2 μm) and in some cases low amylose content provide unique techno-functional properties of amaranth starch for food and non-food applications such as fat replacers, development of biomaterials as carrier or encapsulation, food thickeners, paper coatings, laundry starch, dusting powers and cosmetics, among others (López et al. 1994; Repo-Carrasco-Valencia and Valdez Arana 2017; Zhu et al. 2017). The gelatinization temperature of the *Amaranth* spp. starch has a wide range, from 62.0–69.5 and 77.7–89.6°C. The high starch fraction from *A. cruentus* obtained by differential milling of the amaranth grain can be considered an interesting raw material for the production of precooked amaranth high-starch flours having a wide range of hydration properties (González et al. 2007). Amaranth starch had better stability after freezing and thawing cycles, in the presence or not of sugars or salt (Baker and Rayas-Duarte 1998a) and a slower retrogradation rate than maize, wheat and rice starches (Baker and Rayas-Duarte 1998b) – both properties important in food technology. In general, amaranth starches present lower viscosity and a higher temperature for gelatinizing than quinoa and kañiwa (Table 6.2).

6.3.6.1 Modification of Amaranth Starches

Amaranth native starch has been chemically and/or physically modified mainly into three species: *A. cruentus, A. hypocondriacus* and *A. paniculatus*. The chemical modifications were made by cross linking, acid treatment, oxidization, substitution of hydroxyl groups by hydroxypropylation, succinylation, octenyl succinylation, carboxymethylation or phosphorylation. Whereas physical modifications were done by heat-moisture treatment, annealing was performed by high hydrostatic pressure or γ-irradiation (Zhu 2017). The acid hydrolyzed amaranth starch could be used in the candy industry, as sausage or dressing stabilizer or in parenteral food and instant beverages (Repo-Carrasco-Valencia and Valdez Arana (2017). Starch derivatives (succinylated, acetylated and phosphorylated) have proved effective in encapsulating agents when used in the spray drying of flavors, pigments and probiotics (Repo-Carrasco-Valencia and Valdez Arana (2017). Besides, amaranth starch could be an excellent substrate for producing cyclodextrins because of its high dispersibility, high starch-granule susceptibility to amylases and its exceptionally high amylopectin content (Urban et al. 2012).

6.4 CHIA

6.4.1 Seed Conditioning and Oil Extraction

An essential step in processing oilseeds is preparation of the seeds, which includes seed cleaning, moisture conditioning and milling according to the process employed for oil extraction.

6.4.1.1 Conventional Processes: Cold Pressing and Solvent Extraction

Chia has been identified as a species with potential oil production because of its nutritional quality. The seed contains a high level of oil, 25–39%, rich in polyunsaturated fatty acids, omega-3 (α-linolenic acid 54 – 67%) and omega-6 (linoleic acid 12–21%) (Ayerza and Coates 2011; Ixtaína et al. 2011a; Martínez et al. 2012; Coorey et al. 2014). Many extraction techniques can be used to this purpose; however, variation in the methods and the conditions used produce differences in the yield and in the final chemical quality of the oil (Özcan et al. 2019; Fernandes et al. 2019). Conventional chia oil extraction processes include cold pressing and solid-liquid extraction with organic solvents. The oil yield, fatty acid profile and the physicochemical and quality characteristics of chia seed oils obtained by these methods have been studied by many researchers. Oil recovery under pressing is smaller than by solvent extraction (Ali et al. 2012; Dąbrowski et al. 2017). Martínez et al. (2012) and Dabrowski et al. (2017) could maximize chia oil seed extraction using a Komet screw press (Model CA 59 G, IBG Monforts, Germany). Martínez et al. (2012) hydrated chia seeds up to 10.1% moisture content and used a restriction die of 6 mm, a screw press speed of 20 rpm and a barrel temperature of 30°C to achieve an oil yield value of 26.1% (82.2% of the available oil). However, Dabrowski et al. (2017), who only reported chia seed moistening up to 9%, obtained oil yields, for both hot (110°C) and cold (room temperature) pressing conditions, of 26.3 and 24.1% (76.4 and 70.1% of the available oil), respectively. These conditions allowed obtention of the remaining oils with acceptable values for chemical quality parameters: peroxide (meq. O_2/kg) and acid (mg KOH/g) values below 2.5 and 1.6, respectively; specific extinction coefficients, k_{232} and k_{270}, below 1.43 and 0.22, respectively. Regarding minor compounds, total tocopherol content was between 498 and 616 mg/kg; phytosterol content of 4,800 mg/kg and phenolic compounds (mg D-catechin/kg), carotenoids (mg lutein/kg), squalene (mg/kg) around 6.5, 5.3 and 71.1, respectively. The oil oxidative stability was ~2 h with no significant differences observed in the fatty acids profile. Classical solvent extraction technique requires grinding the seed before extraction and enables the recovery of almost 95% of the available lipid fraction. Silva et al. (2016) observed that the increase in the amount of solvent had little influence on oil yield, and Kentish and Ashokkumar (2011) reported that pretreating milled seeds in an ultrasonic bath constituted an alternative to increasing the amount of extracted oil. The main solvents used in chia oil extraction have included hexane (Ixtaína et al. 2011a; Silva et al. 2016; de Souza et al. 2017; Dąbrowski et al. 2017), petroleum ether (Timilsena et al. 2017), acetone (Dąbrowski et al. 2017) and isopropanol and ethyl acetate (Silva et al. 2016). It is important to highlight that these solvents permit obtention of oils with acceptable values for chemical quality parameters, mainly acid and peroxide values; but great variability associated with minor compounds composition and oxidative stability was recorded. Dąbrowski et al. (2017) reported that the higher polarity of the organic solvent increases both the extraction efficiency of total lipids and bioactive components, especially phenolic compounds and carotenoids, producing oils with higher oxidative stability. Regarding the fatty acid profile of oils, the solvent used and seed to solvent ratio seem not to affect this characteristic. However, the main disadvantages in using these solvents are the damage caused to the environment and that the by-products, which may contain traces of solvent, cannot be used in the food industry due to a risk to human health.

6.4.1.2 Non-Conventional Processes: Supercritical Carbon Dioxide Extraction (CO_2-SE)

The supercritical CO_2 extraction (CO_2-SE) has emerged as an alternative to the processes conventionally used for oil extraction previously mentioned. CO_2-SE is an environmentally friendly method for obtaining high-quality products, free of toxic solvent residues (Jokić et al. 2014). CO_2 is used above its critical point (T_c 31.1°C, P_c 73.8 bar), presenting similar densities to those of the liquids and transport properties that most closely approximate those of gases. Little modification of the operating conditions close to the critical point affects either the yield of extractable compounds or their composition, promoting selectivity within the process (Ixtaína 2010).

Several authors carried out the extraction of chia seed oil with CO_2-SE using different operating conditions, both on a laboratory and pilot scale (Table 6.3). The extraction consists of two periods: an initial linear fast period and a second slow linear period. Most of the extraction of chia seed oil mainly occurred in the first extraction period (Ixtaína et al. 2010). Pressure and the time of extraction have a high positive impact on oil yield whereas the temperature presents a lesser influence (Ixtaína et al. 2010, 2011b; Rocha Uribe et al. 2011; Dąbrowski et al. 2018; Ishak et al. 2021). Thus, by optimizing the operating conditions it is possible to obtain yields and fatty acid compositions similar to those of conventional solvent extraction (Ixtaína et al. 2010, 2011a; Rocha Uribe 2011; Villanueva et al. 2019). As can be seen in Table 6.3, oil recovery achieved was >87% in all cases, and the maximum level was obtained at the highest pressures studied. Regarding fatty acid composition, α-linolenic and linoleic acids were the most abundant, about 65 and 20% of total fatty acids, respectively (Ixtaína et al. 2010, 2011b; Rocha Uribe et al. 2011; Villanueva et al. 2019; Ishak et al. 2021).

The addition of low levels of co-solvents such as water, ethanol, acetone, isopropanol and others, can modify the polarity of the CO_2-SE and consequently the solubility of different compounds. Dąbrowski et al. (2017, 2018) studied the CO_2-SE of chia seed oil using diverse conditions and acetone as a co-solvent. These authors obtained oils with similar fatty acid composition but with different total phytochemical content, highlighting the most abundant in squalene, sterols and tocopherols (1 h, pure CO_2-SE) and the richest in polyphenols and carotenoids (1 h, 10% acetone as co-solvent). These differences are due to the rise in polarity of solvent maximizing the content of amphiphilic compounds (polyphenols), whereas shorter extraction time increased concentration of the most apolar phytochemicals (tocopherols, squalene and phytosterols) (Dąbrowski et al. 2018).

A recent study of the CO_2-SE from edible and discarded chia seeds, which were rejected during post-harvest due to their low weight, small size and presence of broken pericarp, among others, reported that it is possible to use these seeds for the extraction of oil with similar characteristics to those of the edible seeds (Villanueva et al. 2020). Thus, this environmentally friendly technology has great potential to be used by the food industry to extract chia oil of high quality with bioactive compounds (antioxidants and phytosterols), depending on the operating conditions applied.

TABLE 6.3 SUPERCRITICAL CARBON DIOXIDE EXTRACTION (CO₂-SE) OF CHIA SEED OIL AS NON-CONVENTIONAL PROCESSES AT DIFFERING OPERATING CONDITIONS

Scale	Operating conditions					Maximum oil yield[1] (g oil/100 g seeds)	Maximum oil recovery[1,2] (%)	Reference
	P (bar)	T (°C)	t (min)	CO_2 flow rate	Co-solvents			
Lab	250/350/450	40/60/80	60/150/240	30 g/min	-	29.0 (450 bar, 80°C, 240 min)	88 (450 bar, 80°C, 240 min)	Ixtaina et al. (2010)
Pilot	250/450	40/60	135–423	133 g/min	-	33.0 (450 bar, 60°C, 138 min)	97 (450 bar, 60°C, 138 min)	Ixtaina et al. (2011)
Lab	136/272/408	40/60/80	40	1.8 g/min	-	7.2 (408 bar, 80°C, 40 min)	n.a.	Rocha-Uribe et al. (2011)
Lab	280	70/90	300	8 mL/min	-	29.9 (280 bar, 70°C, 300 min)	87 (280 bar, 70°C, 300 min)	Dąbrowski et al. (2017)
Lab	280	70	60/300	8 mL/min	2/6/10 % acetone	33.7 (280 bar, 70°C, 300 min, 10% acetone)	97 (280 bar, 70°C, 300 min, 10% acetone)	Dąbrowski et al. (2018)
Pilot	250/450	40/60	240	40 g/min	-	18.6 (450 bar, 60°C, 240min)	90.3 (450 bar, 60°C, 240min)	Villanueva Bermejo et al. (2019)
Pilot	250/450	40/60	240	40 g/min	-	18.6 (DCS) 25.2 (ECS) (450, 60°C, 240 min)	93.5 (DCS) 89.9 (ECS) (450, 60°C, 240 min)	Villanueva Bermejo et al. (2020)
Lab	220/280/340	40/60/80	300	4 mL/min	-	30.7 (335 bar, 45°C, 300 min)	87 (335 bar, 45°C, 300 min)	Ishak et al. (2021)

Notes: P, extraction pressure; T, extraction temperature; t, extraction time; DCS, discarded chia seeds; ECS, edible chia seeds.
1 Operating conditions for the maximum oil yield/recovery obtained are indicated in parentheses.
2 Values calculated with respect to the total oil content analyzed by Soxhlet or Folch procedures.
n.a. not available.

6.4.1.3 Residual Flours

6.4.1.3.1 Characterization of residual flours from different oil extraction processes

The residual flours from the oil-extracting processes of chia seeds are a good source of total dietary fiber (>40%, d.b.), proteins and minerals – mainly phosphorus, calcium and magnesium – polyphenolic compounds (chlorogenic and caffeic acids, quercetin, myricetin and kaempferol), and tocopherols with high antioxidant activity (Reyes-Caudillo et al. 2008; Capitani et al. 2012). Chia flour from the pressing process exhibits residual oil content 7.06–11.39% (d.b.), with high omega-3 and omega-6 fatty acids concentration (6.85 and 2.16%, respectively) (Capitani et al. 2012; Aranibar et al. 2018).

6.4.1.3.2 Characterization of residual flour from seeds after removal of mucilage

Capitani et al. (2013a) performed a comparative evaluation of the physicochemical and functional properties of chia flours obtained from seeds with and without mucilage. The flour obtained from seeds with previous mucilage extraction exhibited a higher content of insoluble dietary fiber (45.62% d.b.) than the others (41.13% d.b.), with a significant decrease in their soluble dietary fiber content (1.51% d.b.). The flour without mucilage exhibited a lower absorption and water holding capacity than those of the whole flour. This behavior can be associated with the mucilage, soluble dietary fiber, capable of holding water inside its matrix. Both types of chia flours presented a low absorption of organic molecules and oil-holding capacity, being significantly higher in chia flour without mucilage. These properties turn those flours into important ingredients in the manufacture of fried products due to their low fatty mouth-feel contribution. Moreover, they show the potential use of a by-product of the mucilage extraction with favorable nutritional characteristics.

6.4.2 Protein and Fiber Isolation

6.4.2.1 Protein Rich Fraction

The fractionation procedure by dry sieving to obtain protein rich fraction (PRF) from chia defatted flour consists of sifting the flour using a mesh through which the high protein content fraction passes, and the fiber-enriched fraction is retained (Vázquez-Ovando et al. (2010). A high globulin proportion (64.86%) and a minor glutenin level (20.21%) were present, containing the PRF 44.62% of crude protein. Capitani et al. (2013b) obtained PRF from chia flours from different oil extraction processes. After sieving both meals, the fractions that passed through the sieve mesh exhibited an increase in protein content (45.6 and 63.5% fraction cold-pressing and solvent, respectively). Julio et al. (2019) determined the amino acid composition of the chia PRF, indicating an important content of essential sulfur amino acids but limited lysine and tryptophan. These authors obtained differing protein fractions (albumins, globulins, glutelins and prolamins) from PRF by solubility gradient, with globulins being the predominant fraction (64.86%). Also, oil in water (O/W) emulsions with PRF and their fractions at different pH levels in their native and denatured state were obtained. Diverse authors studied the physicochemical, functional and biological properties of the PRF (Segura-Campos et al. 2013a, 2013b; Vázquez-Ovando et al. 2013).

6.4.2.2 Protein Isolates

Chia seed is a source of vegetable protein with biological activities. The extraction of protein fractions from seeds with previous mucilage removal was carried out by Olivos-Lugo et al. (2010), Sandoval-Oliveros and Paredes-López (2013), Timilsena et al. (2016a), López et al. (2018), and Urbizo-Reyes et al. (2019), followed by lipid extraction using solvents or a cold screw-press. Other authors generated protein isolates from defatted seeds but without mucilage removal (Cárdenas et al. 2018; Julio et al. 2019). Chia protein isolates showed good water-holding capacity (4.06 g/g) and excellent oil-retention capacities (4.04 g/g), making it an interesting additive in bakery products and food emulsions (Olivos-Lugo et al. 2010). The denaturation temperatures of crude albumins, globulins, prolamins and glutelins were 103, 105, 86, and 91°C, respectively, indicating excellent thermal stability for albumins and globulins and suggesting that they are suitable for food products undergoing heat treatment (Sandoval-Oliveros and Paredes-López 2013; López et al. 2018). Chia protein hydrolysates obtained by the sequential hydrolysis with microwave treatment showed high *in vitro* antioxidant activity (Urbizo-Reyes et al. 2019).

6.4.3 Fiber Rich Fraction

Chia fiber rich fraction (FRF) was produced by dry processing of defatted flour (Vázquez-Ovando et al. 2009; Capitani et al. 2012), and pressing extraction (Capitani et al. 2012; Oliveira-Alves et al. 2017). The generated fraction presented a high content of total dietary fiber (51.98–56.46% d.b.) consisting mostly of insoluble dietary fiber (47.65–53.45% d.b.) with low soluble dietary fiber content (3.01–4.33% d.b.). Moreover, the chia FRF exhibited high antioxidant activity associated with the polyphenolic compounds, mainly caffeic acid and rosmarinic and salvianolic acids, and the presence of tocopherols in the pressing fraction (Vázquez-Ovando et al. 2009; Capitani et al. 2012; Oliveira-Alves et al. 2017).

6.4.3.1 Mucilage: Extraction and Characterization

Chia mucilage (CM) is an anionic heteropolysaccharide consisting mainly of the sugars xylose and glucose in a 2:1 ratio, with significant amounts of glucuronic and galacturonic acids (Timilsena "al. 2016b"). The complex monosaccharide composition in terms of xylose as building backbone, is similar to other mucilage, such as those extracted from *Plantago psyllium* seed, basil seed and Chinese quince (Liu et al. 2021). CM is localized in the fruit exocarp and exuded from the polygonal cells of the epidermis coat of chia seeds when they are placed in an aqueous solution, being composed mainly of carbohydrates fiber. Upon full hydration, filaments became apparent and conformed to a transparent "capsule" attached to the seed (Muñoz et al. 2012; Capitani et al. 2013c; Salgado-Cruz et al. 2013). Different methods have been developed for CM extraction obtaining a wide range of yields, influenced by the applied operating conditions (Table 6.4). In a recent study by Muñoz-Tebar et al. (2021), CM was extracted from defatted flour chia obtaining a yield of 15.3%, relating this high yield with the compounds, such as proteins, released after grinding the seeds. CM contains different protein levels after extraction (2.6–11.95%), which could be related to the methods used to separate the seeds from the liquid mucilage, soaking conditions, and proteins covalently linked to the polysaccharide (Capitani et al. 2015; Timilsena et al. 2016b); Ferreira Ignácio Câmara et al. 2020). The CM fatty acid profile includes significant levels of linolenic acid (56.16%) and linoleic acid (20.64%) (Ferreira Ignácio Câmara et al. 2020). CM solubility increases with temperature, with pH being maximized at 60°C (87%) (Capitani et al. 2013c) and alkaline pH (97%) (Timilsena et al. 2016b). Timilsena et al. (2016b) indicate that CM resists pyrolytic decomposition (T >250°C). CM dispersed in water presents irregular shape, fibrous microgel particles and average size $D_{4,3} \sim 700$ μm (Goh et al. 2016). García-Salcedo et al. (2018) studied its pasting properties indicating that its relative viscosity did not exhibit dependency on temperature. Rheological measurements reported a weak viscoelastic gel with shear-thinning behavior dependent properties even at low levels (Capitani et al. 2015; Timilsena et al. 2015; Goh et al. 2016; García-Salcedo et al. 2018; Punia et al. 2019). Velázquez-Gutiérrez et al. (2015) determined the sorption isotherms (25 and 40°C), and some thermodynamic properties indicating a sigmoidal shape for the sorption isotherms of freeze-dried CM, characteristic of glassy biopolymer isotherms. The pore radius of CM varied from 0.87 to 6.44 nm and maximum stability 7.56–7.63 kg H_2O/100 kg of dry solids (a_w 0.34–0.53) in the studied range. The glass transition Tg of the mucilage was found to be between 42.93 and 57.93°C.

6.4.4   Techno-Functional Properties of Residual Flours, Proteins and Soluble Fiber (Mucilage)

6.4.4.1   Water-holding Capacity, Absorption and Adsorption Capacity

By-products from oil chia seed extraction (flours, proteins and fiber fraction) present high WHC for the mucilaginous fraction which has excellent

TABLE 6.4 CHIA MUCILAGE EXTRACTION YIELD AT DIFFERING OPERATING CONDITIONS

T (°C)	seed: water ratio	t (h)	pH	Operating conditions		Mucilage yield (% d.b.)	Reference
				solubilization	separation		
25	1:20	1	-	magnetic stirring	vacuum-assisted filtration/drying	7.78	Marin Flores et al. (2008)
25	1:20	2	-	sonication	vacuum-assisted filtration/drying	10.79	Marin Flores et al. (2008)
25	1:20	2	-	magnetic stirring and sonication	vacuum-assisted filtration/drying	15.1	Marin Flores et al. (2008)
20/40/80	1:20/1:30/1:40	2	4/6/8	magnetic stirring	drying oven/rubbing	6.97	Muñoz et al. (2012)
25	1:10		-	without stirring	freeze-drying/rubbing	3.8	Capitani et al. (2013c)
50	1:20	0.45	-	magnetic stirring	centrifugation/freeze-drying	10.9	Segura-Campos et al. (2014)
30-80	1:10–1:30	2-4	-	magnetic stirring	drying oven with forced air circulation/rubbing	4.95	Campos et al. (2016)
25	1:20	4	-	magnetic stirring	centrifugation/ethanol precipitation/ freeze-drying	1.20	Goh et al. (2016)
	1:20	2	-	magnetic stirring	freeze-drying/rubbing	5.6	Timilsena et al. (2016b)
15-85	1:12–1:40.8	½/3	-	magnetic stirring	freeze-drying/rubbing	11.6	Orifici et al. (2018)
27	1:10–1:40	2	8	magnetic stirring	cold pressing/freeze-drying	8.46	Tavares et al. (2018)
refrigeration	1:20	24	-	pre-heat (55°C) and sonication	vacuum-assisted filtration/ drying 40°C, 12h	7.65	Urbizo-Reyes et al. (2019)

Notes: T, extraction temperature; t, extraction time; d.b.: dry basis.

water-holding properties (Vázquez-Ovando et al. 2009, Capitani et al. 2013b). By-products from solvent extraction exhibit better functional properties than others, by their lower level of residual lipids and the higher protein content (Capitani et al. 2013b). In contrast, chia by-products present low water adsorption capacity(WAdC).

6.4.4.2 Oil Holding, Absorption and Adsorption Capacities, and Absorption Capacity of Organic Molecules

Vázquez-Ovando et al. (2009) and Capitani et al. (2013b) reported low oil-holding capacity (OHC) (1.09–2.06 g/g) in chia flours, fiber and protein fraction, suggesting that these by-products are potential ingredients in fried products since they would provide a non-greasy sensation. Chia by-products could efficiently interact with fats, biliary acids, cholesterol, drugs and toxic compounds at the intestinal level, due to the absorption capacity of organic molecules (0.79–1.94 g/g). The FRF exhibits the highest capacity to link to organic molecules (Vázquez-Ovando et al. 2009; Capitani et al. 2013b).

6.4.4.3 Emulsifying Properties, Foaming Capacity and Foam Stability

Chia by-products present high emulsifying activity and emulsion stability (44.33–56.00, 34.33–94.84 mL/100 mL, respectively), being at the higher levels of these parameters in the solvent extracted by-products than others because of their low level of residual lipids and higher protein content (Vázquez-Ovando et al. 2009; Capitani et al. 2013b). In chia PRF, Vázquez-Ovando et al. (2013) reported emulsion activity of 50–56%, independent of pH and emulsion stability of 92% at pH 8 and 10. Capitani et al. (2012) reported fairly stable emulsions with chia flours and fibrous fraction from solvent by-products due to their high proportion of proteins and fibers as emulsifying and stabilizing agents. Capitani et al. (2016) and Guiotto et al. (2016) obtained stable O/W emulsions with ≥0.75% of chia mucilage (CHM) during refrigerated storage by a reduction in the mobility of oil particles in a tridimensional network. Foaming capacity (FC) and foam stability (FS) of chia PRF indicated that it is not a good foaming agent but presents a foam stability of 80% (Vázquez-Ovando et al. 2013).

6.5   SACHA INCHI

6.5.1   Decapsulation and Seed Conditioning

Sacha inchi seeds, and the most widespread species, *Plukenetia volubilis* and *Plukenetia huayllabambana* are relatively large almonds (1.3–2.1cm) with a high amount of oil content (49–54 wt%), followed by the proteins (24–25 wt%). The seeds also contain minor quantities of fiber (11–15 wt%), water (5–7 wt%), carbohydrates (3–7 wt%) and ashes (3–4 wt%). The oils are formed in high proportion by essential fatty acids, $\omega3$-α-linolenic (>48%) and linoleic (>26%). Before extracting the oil, one of the post-harvest essential stages of decapsulation are implemented – drying (artificially, or naturally under the sun) – as well as the subsequent stage, dehulling. For all these stages, there are currently different technological possibilities for releasing the sacha inchi seeds from the capsule and shell that contain them, with the least possible damage. To facilitate oil extraction using any of the chosen methods, it is sometimes necessary to reduce the seed size by grinding.

6.5.2 Oil Extraction Processes

In general terms, four kinds of extraction methods are implemented for sacha inchi oils, namely, pressing extraction, aqueous extraction, solvent extraction and supercritical fluid extraction.

6.5.2.1 Conventional Processes

Extraction of sacha inchi oil from seeds by pressing is cost-efficient and the most extended method in the industry for obtaining genuine extra virgin oil, and it is the shortest process. The pressure extraction equipment in most cases consists of a perforated cylinder subjected to high pressure. It is quite easy to control, needs low investment and has a low operating cost. Furthermore, few parameters besides the speed of rotation and the geometry of the press cylinder need to be controlled. In some cases, to guarantee a steady oil temperature, a cooling system is installed at the outlet. With the pressing process, usually 75–95 wt% of the oil is recovered from the seeds. The effects of the drying (60, 75, 90°C) and press head (60, 75, 90°C), temperature on oil yields from sacha inchi (*P. volubilis* L.) seeds using a single screw press machine have been studied (Muangrat et al. 2018). As no temperature or solvent is used, most of the health-beneficial components remain in the oil, such as tocopherols and phytosterols. It is the eco-friendliest extracting system, together with extraction by steam drag. Steam dragging is a technique that uses a stream of water vapor for the extraction; after some maceration time, the plant tissue is broken, releasing oil. Oil extraction is facilitated by incorporating several classes of proteases and cellulase enzymes. Nevertheless, even using optimal conditions of enzyme loading, water-to-sample ratio, extraction time and temperature, oil yield is lower than 30 wt%. Contrary to the solvent-free processes mentioned above, there are procedures which use solvents for oil extraction. Soxhlet (AOAC 1990), with petroleum ether, is the most widespread and preferred solvent extraction method for many seed oils. However, in the case of sacha inchi, this method is preferred for extracting the oil remaining after the first pressure extraction, which is usually in the range of 5–25 wt%.

6.5.2.2 Non-conventional Processes

The supercritical extraction using carbon dioxide (CO_2) is the most widespread non-conventional procedure for extracting sacha inchi oils. Studies of the influence of temperature on the system, pressure ranges and CO_2 flow were tracked in lab scale and at pilot plant scale (Triana-Maldonado et al. 2017). Maximum recovery was 60 wt%, achieved at the CO_2-SE pilot scale. Other different solvents such as *n*-propane were also assayed (Zanqui et al. 2016).

6.5.2.3 Residual Flour

After the oil is obtained, the sacha inchi cake is ground in a hammer mill to obtain ~500μm particles. Then it is exhaustively defatted, usually by an organic solvent method. The flour obtained from the sacha inchi cake has a very high percentage of protein (>50 wt%) and can be used for human consumption in the meat, dairy, bakery and confectionery industries. Sacha inchi protein has been reported to have high content of cysteine, tyrosine, threonine and tryptophan and low content of phenylalanine (Ruiz et al. 2013). In order to eliminate its astringent taste, an extrusion process is necessary, and this is conducted at a moderate temperature of 60–90°C for a very short time, 5–7 sec. This process improves the biological quality and inactivates tannins, improving digestibility of protein and other nutrients.

6.5.3 Isolates and Protein Hydrolysates

The production of sacha inchi protein isolates is of growing interest to the food industry because of its increasing application in food markets, nutraceutical products, and functional foods. A water-soluble albumin fraction comprising ~25 wt% of sacha inchi cake has been obtained (Sathe et al. 2002; Quinteros et al. 2016). This is composed of two glycosylated polypeptides, with estimated molecular weights (MW) of 32.8 and 34.8 kDa, respectively. It contains all the essential amino acids in adequate amounts when compared to the FAO/WHO (FAO 2007) recommended pattern for a human adult. Protein from the sacha inchi cake can also be removed by optimized enzyme-assisted extraction (Chirinos et al. 2016), or by an alkaline process using isoelectric precipitation.

6.5.4   Techno-Functional Properties of Sacha Inchi Cake Flour

The values of solubility, oil absorption capacity, foaming capacity, foam stability and emulsifying capacity are techno-functional properties that influence sensorial characteristics, and are essential in different food systems. In the case of sacha inchi cake flour, those properties show values of higher magnitudes compared to other protein isolates and can be particularly useful for the production of sausages, mayonnaise and ice cream, among other food products (Mercado et al. 2015).

6.6   TARWI

Tarwi or Andean lupin (*Lupinus mutabilis*) is a leguminous crop with a remarkable chemical composition and an elevated content of both protein and oil. Tarwi thus has a twofold advantage: it is a valuable food crop in the fight against malnutrition, as well as a promising cash crop for edible oil production. It has more protein than soybean and even more than other lupine species. According to Carvajal-Larenas et al. (2016), the average seed protein content (g/100 g db) reported in the literature for different *Lupinus* species is: *L. albus*, 38.2; *L. angustifolius*, 33.9; *L. luteus*, 42.2; and *L. mutabilis*, 43.3. Tarwi oil is of nutritionally good quality, its main fatty acids being oleic and linoleic acids (Repo-Carrasco-Valencia 2020). The low linolenic acid content gives tarwi oil good stability against oxidation. The oil does not contain any erucic acid, which is known to be present in the oil of other lupines such as *L. angustifolius* (Hatzold et al. 1983). The main carbohydrates in tarwi are oligosaccharides. The stachyose content in tarwi is relatively high. Starch is present only in very small amounts, therefore products of high viscosity cannot be produced, although this is compensated by the relatively high pentosan content (Beirão da Costa 1989). The main hindrance to the wider use of tarwi as food and animal feed has been its content of toxic, bitter quinolizidine alkaloids. Lupanin, spartein, 4-hydroxylupanin and 13-hydroxylupanin are the principal alkaloids found in *L. mutabilis*. The alkaloid content in most common varieties of tarwi is in average 5 g/100 g (Cortés-Avendaño et al. (2020).

6.6.1   Alkaloid Extraction

The debittering of tarwi seeds is necessary for the removal of toxic alkaloids. This process is traditionally carried out as follows: The lupin grains are hydrated for 12 h at room temperature, followed by cooking for 1 h. The cooking water is discarded and the seeds are washed with running water for 5 days. Simultaneous extraction of fat and alkaloids may be applicable for tarwi. The efficiency of the extraction procedure was studied using different solvents by Beirão da Costa (1989). The solvent which produced the best results was ethanol 35%. Concerning alkaloid extraction by water, some trials showed that this may be improved 100% if the lupin material is primarily subjected to enzymatic treatment with cellulases.

Debittered tarwi can be consumed directly as a snack (Villacrés et al. 2010) but it can also be used as an ingredient in many different products such as salads, traditional dishes, bread, cookies, noodles, extruded drinks, protein isolate and concentrate, and refined oil (Hatzold et al. 1983; Córdova-Ramos et al. 2020) as shown in Figure 6.1.

Lupine flour can be obtained after the oil extraction process either ground through a roller mill (Koivunen et al. 2016) or hammer mill or grinder (Zaworska et al. 2017). The flour can be used in different products, such as breads, muffins and pasta (Wandersleben et al. 2018; Nasar-Abbas and Jayasema 2011; Martínez-Villaluenga et al. 2010)

6.6.2   Oil Extraction

Lupine oil can be obtained by pressing or by the conventional method using a solvent. To perform the extraction of the oil, Hatzold et al. (1983) carried out a process which included the following steps: cleaning, peeling and cooking at 80°C to finally extract the oil using hexane as a solvent. The crude oil was degummed and debittered by washing with diluted acids, then neutralized, blanched and deodorized. The alkaloids are concentrated in the defatted oil cake. Crude oil, however, contains only small amounts of alkaloids. At the end of the refining process the alkaloid content of the edible oil had decreased to the very low level of 5 ppm.

6.6.2.1   Characterization of Residual Flour

Residual lupin flour and its extracted components, such as protein or fiber, can be added to foods to enhance their nutritional quality (Villarino et al. 2016). Lupin protein isolates are extracted by solubilization of protein from wet-milled kernels or lupin flour (defatted or non-defatted) at high alkaline pH (e.g., 9), removal of the insoluble portion (dietary fiber) through centrifugation, followed by acid precipitation of the major globulin proteins (i.e., α- and β-conglutins) at pH 4.5 (Johnson et al. 2017). The insoluble fiber residue, acid-precipitated protein, and acid-soluble "whey fraction" can then be dried for use as staple food ingredients against deteriorative processes.

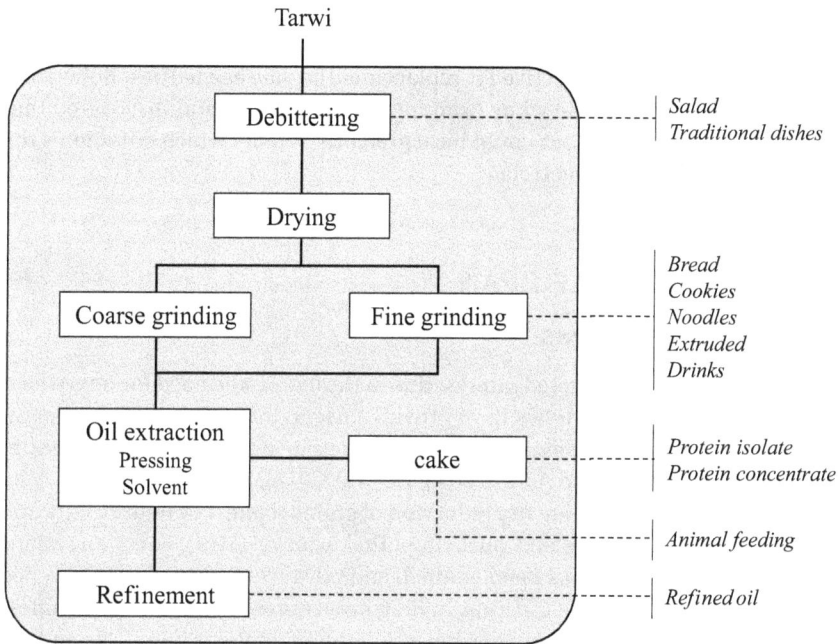

Figure 6.1 Products obtained from tarwi.
Notes: Tarwi can undergo successive processes in which different products/ingredients can be obtained at each step. The processes are: debittering, in which the seeds could be used to prepare salad and traditional dishes; followed by drying and then coarse grinding or fine grinding (in which the flour can be used to prepare: bread, cookies, noodles, extruded products and drinks); after that oil obtention by pressing and/or by solvent extraction, followed by the refinement step to obtain the refined oil; with the remaining cake (after extraction) the isolate or concentrate protein is obtained, or the cake can be used for animal nutrition. **Source: Authors' own work.**

Sironi et al. (2005) studied the techno-functional properties of lupin protein isolate obtained by isoelectric precipitation. They found that this fraction has excellent emulsifying properties but low viscosity and reduced gel-forming properties. The acid-soluble whey fraction obtained in isoelectric precipitation also contains valuable protein components, such as the bioactive peptide γ-conglutin; it has stable foam-forming properties (Wong et al. 2013). Due to their functional properties, which are important in food manufacture, lupin protein fractions have been categorized as emulsifying fraction (α-, β-conglutin) and foaming fraction (γ-conglutin-rich) (Sironi et al. 2005).

Only limited studies have been reported on the use of lupin fiber fractions as ingredients for the development of high-fiber foods. Archer et al. (2004) used lupin kernel fiber as an effective fat replacement in sausage patties. Roberfroid (2007) studied acid-soluble whey from lupin fractionation and discovered that the oligosaccharides present could have prebiotic activity which stimulates the growth of beneficial gut bacteria.

6.7   BLACK TURTLE BEAN

6.7.1   Milling Process

Dry bean milling has attracted interest due to the increasing need for non-wheat food ingredients. Some studies have provided useful information on the potential of black turtle bean (*Phaseolus vulgaris*) flour as a functional ingredient in the food system (Audu and Aremu 2015; He et al. 2018).

Milling is a process where the reduction of grains to meal or flour occurs and includes grinding, sieving and purifying (Thakur et al. 2019). For many years, wheat and corn milling has been studied, and advances in knowledge have led to an understanding of the different milling procedures used today. Little information is available concerning dry bean milling on different flours. Nowadays, study of the milling of dry beans is increasing due to the beans' rich nutritional profile.

In general, for the production of black turtle bean flour, seeds that were previously (or not) subjected to some treatment processes, are now dried in air forced ovens or in sun-dryers and finally they are ground. The ground seeds are sieved through a single mesh to obtain flour (Figure 6.2).

To obtain defatted black bean flour, the full-fat flour is soaked in a solvent; after being allowed to stand, the mixture is removed, dried and finally pulverized (Figure 6.3). The particle size of milled dry beans is an important variable on flour quality. For size reduction, a combination of forces (e.g., abrasion, shearing, compression and impact) might be used. Dry beans are often milled using a one-pass system that utilizes a plate mill (disk mill, burr mill) or a hammer mill (Raigar and Mishra 2018). The hammer mill is an impact-type mill where particle size reduction depends on hammer design. A disadvantage of hammer milling is that it is less uniform in the particle size distribution and less energy efficient compared to a roller mill. A burr mill, also known as a plate mill, consists of a set of knives, which use cutting, shearing and crushing actions for particle size reduction. It has two circular plates and the material is fed between them. It utilizes a multiple-stage approach in a series of rolls for fine particle size reduction. Carter and Manthey 2019 have demonstrated the effectiveness of using a

...

(Fix)

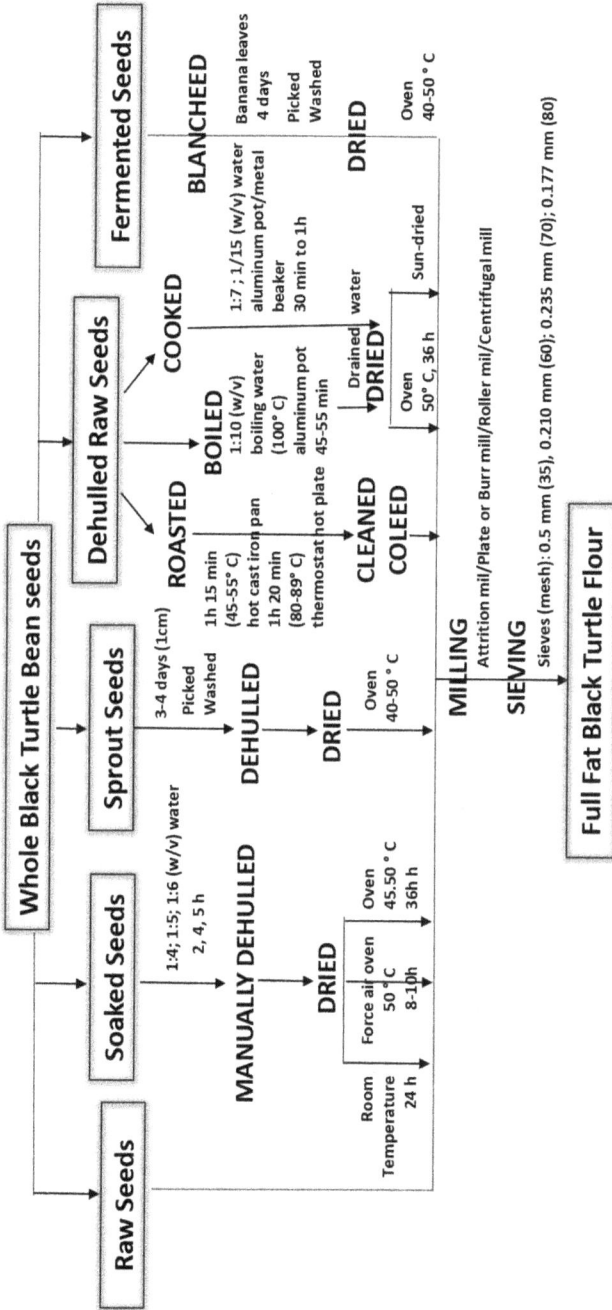

Figure 6.2 Flow chart for the production of fatted black turtle bean flours.

Notes: Flow chart of whole black turtle bean seeds, describing the six different types of processes for obtaining the full black turtle flour. These processes start with the raw seeds or soaked seeds (which are manually de-hulled and then dried) or sprout seeds (which are de-hulled and dried) or de-hulled raw seeds (which can be roasted and then cleaned, cooked or boiled and dried or cooked and dried) or fermented seeds (which are blanched and dried); all of the above are then followed by milling and sieving. **Source: Authors' own work.**

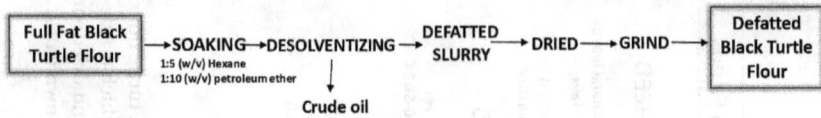

Figure 6.3 Flow chart for the production of defatted black turtle bean flours. Source: Authors' own work.
Notes: Flow chart of the production of defatted black turtle bean flours, which undergo the following processes: soaking the flours, solvent extraction (by hexane or petroleum ether) of crude oil followed by cake desolventizing, defatted slurry, dried and ground to obtain defatted black turtle flour.

centrifugal mill for milling black beans. A centrifugal mill consists of an impeller that spins at high velocity and is located inside a cutting or an abrading screen.

6.7.2 Protein Concentrates and Isolates

Black turtle bean flour can be further processed by fractionation to isolate its components (e.g., protein, starch and dietary fiber) for using in food applications. Protein isolates and concentrates have been extensively studied as food and dietary supplements for several oilseeds, but there is very little information on their production procedure and corresponding functionality. The easiest method of separating protein from the cellular structure of the seed is by extracting with an aqueous solvent for successful utilization of food products. Figure 6.4 is a diagram showing how to obtain black turtle bean protein isolate and concentrate.

In general, the defatted flour with alkaline solution is mixed by stirring for few minutes. The slurry is centrifuged and the supernatants are treated with 95% (v/v) ethanol; the pH is adjusted by stirring with an acid solution and the precipitated proteins are recovered by filtration. The protein concentrate is dried or lyophilized in a forced-air oven. To obtain the isolate, the defatted bean flour is suspended in water and the pH adjusted with an alkaline solution to extract the protein while stirring. Then, the supernatants' pH is adjusted to 4 with an acid solution and the precipitate is obtained by centrifugation. The precipitate is washed with distilled water and after the pH has been readjusted, it is finally dried in an oven. Recently, more protein researchers were concerned about the conformational changes of the food protein molecules by adjusting the pH to neutral after a low acidic condition treatment (i.e., pH-shifting) (Jiang et al. 2017). According to He et al. (2020), low pH changes in treatment during the protein extraction process could have great potential to produce improvements in emulsifying, foaming and fat-holding properties, and lead to the reduced

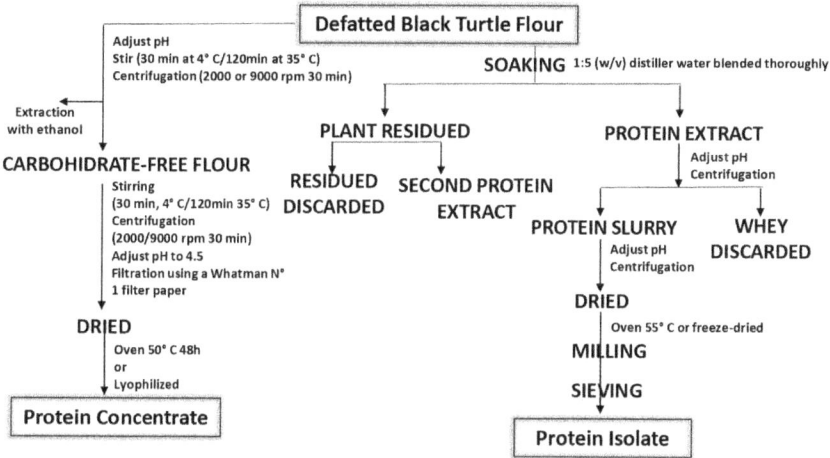

Figure 6.4   Flow chart for the production of protein concentrates and protein isolates of black turtle bean flours. **Source: Authors' own work.**
Notes: Flow chart for the production of protein concentrate and isolate from defatted black turtle beans flour. To produce the protein concentrate, defatted black turtle flour has to be processed to carbohydrate-free flour (by adjusting the pH and the extraction of ethanol) and then has to be dried. On the other hand, to produce a protein isolate, the defatted black turtle flour has to be soaked in water to produce a protein extract, then the obtention of protein slurry (by adjusting the pH and centrifugation) and finally it undergoes drying, milling and sieving. The whey is discarded. After the first soaking, the plant residues undergo a second protein extraction and the residue is discarded.

solubility and water-holding capacity of the isolate of black turtle bean protein with a high level of consumer safety.

6.7.3 Oil Extraction

Edible oils from plant sources are receiving growing interest in various food applications and industries, due to their high concentration of bioactive lipid components, such as polyunsaturated fatty acids and phytosterols that have shown various health benefits (Aremu et al. 2015). Oil can be extracted from the black turtle bean seeds using traditional methods (small-scale), mechanical (hydraulic and screw) presses that can be manual, semi-automated or automated, and extraction with solvents (for example, hexane, fluid carbon dioxide), or a combination of two of these methods (Aremu et al. 2015). Values of the physicochemical parameters reported by Audu et al. (2013) have shown that black turtle bean oils might be useful as edible oils due to their stability as frying oils. Unsaturated fatty acids predominate in raw and processed black turtle bean flours with an adequate amount of essential fatty acids (Padhi et al. 2017; Audu et al. 2013).

6.7.4   Techno-Functional Properties of Flours, Protein Concentrate, Oil and Fiber

Several studies have reported that the type of mill used, as well as conditions pertaining during pretreatment and milling, can influence the physical, chemical and pasting properties of the flour (Audu et al. 2013; Audu and Aremu 2015; Carter and Manthey 2019; Okafor et al. 2015). These changes in the external and internal structure of black bean seeds affected their milling and flour properties. Pretreatment processing and milling significantly affected the bulk density of black turtle bean flours by increasing it. Both soaking and cooking affected the final flour color by decreasing the lightness due to the high leaching of seed coat pigments into the cooking water. All the processing methods are shown in Figure 6.2, which enhanced water absorption (WAC) and foaming capacities of these flours. Boiling, cooking, roasting, sprouting and fermentation affected the oil absorption capacities (OAC) and lowest gelation concentration or least gelatinous capacity (LGC) by increasing them. All the processing methods reduced foaming stability. Emulsion capacity for the raw black turtle bean flour was lower (Audu and Aremu 2015). Cooking, roasting and sprouting increased emulsion capacity while boiling and fermenting reduced it. The relatively low emulsion capacity of the fatted and defatted black turtle bean flours could be due to the nature and type of protein. This indicates that black turtle bean flour might be useful in the production of sausages, soups and cakes. Boiling, cooking and sprouting increased emulsion stability. The swelling index of full-fat and defatted flour was lower, probably due to presence of fiber content. The swelling index would have affected the texture of food prepared from such flours (Okafor et al. 2015). Wang et al. (2010) reported that cooking black turtle beans in water significantly increased the content of protein, starch, soluble dietary fiber (SDF), insoluble dietary fiber (IDF), total dietary fiber (TDF), Mn and P (in dry weight), while it reduced the ash content, K, Mg, trypsin inhibitory activity (TIA), tannin content, sucrose and oligosaccharides (raffinose, stachyose and verbascose) content.

6.8   GENERAL CONCLUSIONS

Processing is key to the fractionation of seeds and grains and has been applied on diverse Latin-American crops like quinoa, kañiwa, maize, amaranth, chia, sacha inchi and legumes. An overview of the different types of milling and their impact on the characteristics of their by-products was provided, mainly for quinoa, kañiwa, amaranth and black turtle bean.

The protein isolation and dietary fiber extraction methods were described characterizing the main components of each fraction from the different crops studied. Also, the techno-functional properties of proteins and the physico-chemical, thermal and rheological properties of the starches, flours and mucilage were extensively summarized.

For oil seeds such as chia, sacha inchi and others, the corresponding extraction through conventional (cold pressing and solvent extraction) or non-conventional processes (supercritical carbon dioxide extraction) were also described, in terms of the influence of the operating conditions on the oil yield, the fatty acid composition and minor compounds (tocopherols, phenolic compounds).

This chapter brings to the fore knowledge and useful information about the fractions of these ancient crops and their valuable potential for food and other outstanding applications

ACKNOWLEDGMENTS

Authors would like to express their gratitude to the CYTED (Spain) for its kind support through Project la ValSe-Food 119RT0567, Food4ImNut Food4ImNut PID2019-107650RB-C21 funded by MCIN/AEI/10.13039/501100011033, Spain.

REFERENCES

Abugoch James, L. E. 2009. Quinoa (*Chenopodium quinoa* Willd.): Composition, chemistry, nutritional, and functional properties. *Advances in Food and Nutrition Research* 58: 1–31.

Abugoch James, L. E., N. Romero, C. A. Tapia, J. Silva, and M. Rivera. 2008. Study of some physicochemical and functional properties of quinoa (*Chenopodium quinoa* Willd) protein isolates. *Journal of Agricultural and Food Chemistry* 56(12): 4745–4750.

Abugoch James, L. E., E. N. Martínez, and M. C. Añón. 2010. Influence of pH on structure and function of amaranth (*A. hypochondriacus*) protein isolates. *Cereal Chemistry* 87: 448–453.

Agama-Acevedo, E., M.-A. Ottenhof, I. A. Farhat, O. Paredes-López, J. Ortíz-Cereceres, and L. A. Bello-Pérez. 2005. Isolation and characterization of starch from pigmented maizes. *Agrociencia* 39(4): 419–429.

Ahamed, N., R. Singhal, P. Kulkarni, and M. Pal. 1996. Physicochemical and functional properties of *Chenopodium quinoa* starch. *Carbohydrate Polymers* 31: 99–103.

Ai, Y. and J. Jane. 2016. Macronutrients in corn and human nutrition. *Comprehensive Reviews in Food Science and Food Safety* 15(3): 581–598.

Anderson, R.A., and S. A. Watson. 1982. The corn milling industry. In *CRC Handbook of Processing and Utilization in Agriculture*, edited by I. A. Wolf, Vol. 2, Part 1, 31–61. Boca Ratón, FL CRC Press.

AOAC (Association of Official Agricultural Chemists). 1990. *Official Methods of Analysis of AOAC 925.10:1990*, 15th ed.. Washington, DC: AOAC International.

AOAC (Association of Official Agricultural Chemists). 1999. *Official Methods of Analysis of AOAC*. Washington, DC: AOAC. Ce 2–66.

Aranibar, C., N. B. Pigni, M. Martínez et al. 2018. Utilization of a partially-deoiled chia flour to improve the nutritional and antioxidant properties of wheat pasta. *Journal of Food Science and Technology* 89: 381–387.

Archer, B. J., S. K. Johnson, H. M. Devereux, and A. L. Baxter. 2004. Effect of fat replacement by inulin or lupin-kernel fibre on sausage patty acceptability, post-meal perceptions of satiety and food intake in men. *British Journal of Nutrition* 91: 591–599.

Aremu, M., H. Ibrahim, and T. Bamidele. 2015. Physicochemical characteristics of the oils extracted from some Nigerian plant foods – A review. *Chemical and Process Engineering Research* 32: 36–52.

Ashraf, S., S. M. G. Saeed, S. A. Sayeed, and R. Ali. 2012. Impact of microwave treatment on the functionality of cereals and legumes. *International Journal of Agriculture & Biology* 14: 356–370.

Audu, S., and M. Aremu. 2015. Effect of domestic processing on the levels of some functional parameters in black turtle bean (*Phaseolus vulgaris* L.). *Food Science and Quality Management* 38: 55–59.

Audu, S., M. Aremu, and L. Lajide. 2013. Effects of processing on physicochemical and antinutritional properties of black turtle bean (*Phaseolus vulgaris* L.) seeds flour. *Oriental Journal of Chemistry* 29(3): 979–989.

Avanza, M. V., M. C. Puppo, and M. C. Añón. 2006. Structural characterization of amaranth protein gels. *Journal of Food Science* 70: E223–E229.

Ayerza, R., and W. Coates. 2011. Protein content, oil content and fatty acid profiles as potential criteria to determine the origin of commercially grown chia (*Salvia hispanica* L.). *Industrial Crops and Products* 34: 1366–1371.

Baker, L., and P. Rayas-Duarte 1998a. Freeze-thaw stability of amaranth starch and the effects of salt and sugars. *Cereal Chemistry* 75(3): 301–307.

Baker, L., and P. Rayas-Duarte 1998b. Retrogradation of amaranth starch at different storage temperatures and the effects of salt and sugars. *Cereal Chemistry* 75(3): 308–314.

Ballester-Sánchez, J., J. V. Gil, M. T. Fernández-Espinar, and C. M. Haros. 2019. Quinoa wet-milling: Effect of steeping conditions on starch recovery and quality. *Food Hydrocolloids* 89: 837–843.

Bazile, D., D. Bertero, and C. Nieto. 2013. State of the art report on quinoa around the world in 2013. Retrieved from www.fao.org/3/contents/ca682370-10f8-40c2-b084-95a8f704f44d/i4042e00.htm.

Beirão da Costa, M.L. 1989. Aspects of lupin composition as food. In *Proceedings of the Joint CEC–NCRD Workshop held in Israel* (Ginozar Kibbutz). "Lupin production and bio-processing for feed, food and other by-products": 94–105.

Benito-Román, O., M. Rodríguez-Perrino, M. T. Sanz, R. Melgosa, and S. Beltrán. 2018. Supercritical carbon dioxide extraction of quinoa oil: Study of the influence of process parameters on the extraction yield and oil quality. *Journal of Supercritical Fluids* 139: 62–71.

Betalleluz-Pallardel, I., M. Inga, L. Mera et al. 2017. Optimisation of extraction conditions and thermal properties of protein from the Andean pseudocereal cañihua (*Chenopodium pallidicaule* Aellen). *International Journal of Food Science Technology* 52(4): 1026–1034.

Bolontrade, A. J., A. A. Scilingo, and M. C. Añón. 2013. Amaranth proteins foaming properties: Adsorption kinetics and foam formation-Part 1. *Colloids and Surfaces B: Biointerfaces* 105: 319–327.

Bolontrade, A. J., A. A. Scilingo, and M. C. Añón. 2016. Amaranth proteins foaming properties: Film rheology and foam stability-Part 2. *Colloids and Surfaces B: Biointerfaces* 141: 643–650.

Boyer, C. D. and C. J. Shannon. 2003. Carbohydrates of the kernel. In *Corn: Chemistry and Technology*, 2nd ed., edited by P. J. White and L. A. Johnson, 289–311. St Paul, MN: AACC International.

Bustos, M. C., M. I. Ramos, G.T. Pérez, and A. E. León. 2019. Utilization of kañiwa (*Chenopodium pallidicaule* Aellen) flour in pasta making. *Journal of Chemistry* 2019: 4385045.

Calzetta-Resio, A. N., M. P. Tolaba, and C. Suárez. 2006. Effects of steeping conditions on wet-milling attributes of amaranth. *International Journal of Food Science and Technology* 41: 70–76.

Campos, B. E., T. Dias Ruivo, M. R. da Silva Scapim, G. Scaramal Madrona, and R. Bergamasco. 2016. Optimization of the mucilage extraction process from chia seeds and application in ice cream as a stabilizer and emulsifier. *LWT – Food Science and Technology* 65: 874–883.

Capitani, M. I., V. Spotorno, S. M. Nolasco, and M. C. Tomás. 2012. Physicochemical and functional characterization of by-products from chia (*Salvia hispanica* L.) seeds of Argentina. *LWT – Food Science and Technology* 45 (1): 94–102.

Capitani, M. I., S. M. Nolasco, and M. C. Tomás. 2013a. Effect of mucilage extraction on the functional properties of chia meals. In *Food Industry*, edited by I. Muzzalupo, Chap. 19, 421–437. Rijeka: InTech.

Capitani, M. I., S. M. Nolasco, and M. C. Tomás. 2013b. Characterization and functionality of by-products from chia (*Salvia hispanica* L.) seeds of Argentina. In *Dietary Fibre: Sources, Properties and their Relationship to Health*, edited by D. Betancur-Ancona, L. Chel-Guerrero, and M. Segura-Campos, 141–158. New York: Nova Science Publishers.

Capitani, M. I., V. Y Ixtaína, S. M Nolasco, and M. C. Tomás. 2013c. Microstructure, chemical composition and mucilage exudation of chia (*Salvia hispanica* L.) nutlets from Argentina. *Journal of the Science of Food and Agriculture* 93 (15): 3856–3862.

Capitani, M. I., L. J. Corzo-Ríos, L. A Chel-Guerrero et al. 2015. Rheological properties of aqueous dispersions of chia (*Salvia hispanica* L.) mucilage. *Journal of Food Engineering* 149: 70–77.

Capitani, M. I., S.M Nolasco, and M. C Tomás. 2016. Stability of oil-in-water (O/W) emulsions with chia (*Salvia hispanica* L.) mucilage. *Food Hydrocolloids* 61: 537–546.

Cárdenas, M., C. Carpio, J. Welbaum, E. Vilcacundo, and W. Carrillo. 2018. Chia protein concentrate (*Salvia hispanica* L.) anti-inflammatory and antioxidant activity. *Asian Journal of Pharmaceutical and Clinical Research* 11 (2): 382–386.

Carter, C. and F. Manthey. 2019. Seed treatments affect milling properties and flour quality of black beans (*Phaseolus vulgaris L.*). *Cereal Chemistry* 96(4): 689–697.

Carvajal-Larenas, F. E., A. R. Linnemann, M. J. R. Nout, M. Koziol, and M. A. J. S. van Boekel. 2016. *Lupinus mutabilis*: composition, uses, toxicology, and debittering. *Critical Reviews in Food Science and Nutrition* 56: 1454–1487.

Caselato-Sousa, V. M. and J. Amaya-Farfán. 2012. State of knowledge on amaranth grain: A comprehensive review. *Journal of Food Science* 77: 93–10.

Castellani, O. F., E. N. Martínez, and M. C. Añón. 2000. Globulin-P structure modifications induced by enzymatic proteolysis. *Journal of Agricultural and Food Chemistry* 48: 5624–5629.

Chen H., M. Xiong, T. Bai, D. Chen, Q, Zhang, D. Lin, Y. Liu, A. Liu, Z. Huang, and W. Qin. 2021. Comparative study on the structure, physicochemical, and functional properties of dietary fiber extracts from quinoa and wheat. *LWT – Food Science and Technology* 149: 111816.

Chirinos, R., M. Aquino, R. Pedreschi, and D. Campos. 2016. Optimized methodology for alkaline and enzyme-assisted extraction of protein from sacha inchi (*Plukenetia volubilis*) kernel cake. *Journal of Food Process Engineering* 40(2): ee12412.

Chirinos, R., K. Ochoa, A. Aguilar-Galvez et al. 2018. Obtaining of peptides with in vitro antioxidant and angiotensin I converting enzyme inhibitory activities from cañihua protein (*Chenopodium pallidicaule* Aellen). *Journal of Cereal Sci*ence 83: 139–146.

Condés, M. C., A. Scilingo, and M. C. Añón. 2009. Characterization of amaranth proteins modified by trypsin proteolysis. Structural and functional changes. *LWT – Food Science and Technology* 42: 963–970.

Coorey, R., A. Tjoe, and V. Jayasena. 2014. Gelling properties of chia seed and flour. *Journal of Food Science* 79: 859–866.

Cordero-de-los-Santos, M. Y., J. A. Osuna-Castro, A. Borodanenko, and O. Paredes-López. 2005. Physicochemical and functional characterisation of amaranth (*Amaranthus hypochondriacus*) protein isolates obtained by isoelectric precipitation and micellisation. *Food Science and Technology International* 11(4): 269–280.

Córdova-Noboa, H. A., E. O. Oviedo-Rondon, A. Ortíz et al. 2020. Effects of corn kernel hardness and grain drying temperature on particle size and pellet durability when grinding using a roller mill or hammermill. *Animal Feed Science and Technology* 271: 114715.

Córdova-Ramos, J., P. Glorio-Paulet, A. Hidalgo, and F. Camarena. 2020. Efecto del proceso tecnológico sobre la capacidad antioxidante y compuestos fenólicos totales del lupino (*Lupinus mutabilis* Sweet) andino. *Scientia Agropecuaria* 11(2): 157–165.

Cortés-Avendaño, P., M. Tarvainen, J, P. Suomela et al. 2020. Profile and content of residual alkaloids in ten ecotypes of *Lupinus mutabilis* Sweet after aqueous debittering process. *Plant Foods for Human Nutrition* 75: 184–191.

Cruz-Vázquez, C., A. Villanueva-Carvajal, G. Estrada-Campuzan, and A. Dominguez-Lopez. 2019. Tamales texture properties as a function of corn endosperm type. *International Journal of Gastronomy and Food Science* 16: 100153.

Curti, C., R. Curti, N. Bonini, and A. Ramón. 2018. Changes in the fatty acid composition in bitter *Lupinus* species depend on the debittering process. *Food Chemistry* 263: 151–154.

D'Amico, S., S. Jungkunz, G. Balasz et al. 2019. Abrasive milling of quinoa: Study on the distribution of selected nutrients and proteins within the quinoa seed kernel. *Journal of Cereal Science* 86: 132–138.

Dąbrowski, G., I. Konopka, S. Czaplicki, and M. Tańska, M. (2017). Composition and oxidative stability of oil from Salvia hispanica L. seeds in relation to extraction method. *European Journal of Lipid Science and Technology* 119(5): 1600209.

Dąbrowski, G., I. Konopka, and S. Czaplicki. 2018. Supercritical CO_2 extraction in chia oils production: impact of process duration and co-solvent addition. *Food Science and Biotechnology* 27(3): 677–686.

Dakhili, S., L. Abdolalizadeh, S. M. Hosseini, S. Shojaee-Aliabadi, and L. Mirmoghtadaie. 2019. Quinoa protein: Composition, structure and functional properties. *Food Chemistry* 299: 125–161.

de Souza, A. L., F. P. Martínez, S. B. Ferreira, and C. R. Kaiser. 2017. A complete evaluation of thermal and oxidative stability of chia oil. *Journal of Thermal Analysis and Calorimetry* 130: 1307–1315.

Dombrink-Kurtzman, M.A., and J. A Bietz. 1993. Zein composition in hard and soft endosperm of maize. *Cereal Chemistry* 70: 105–108.

Eckey, E. W. 1954. *Vegetable Fats and Oils*. New York: Reinhold, 75–79.

Elgeti, D., M. Jekle, and T. Becker. 2015. Strategies for the aeration of gluten-free bread - A review. *Trends in Food Science and Technology* 46(1): 75–84.

FAO (Food and Agriculture Organization). 2007. *Protein and Amino Acid Requirements in Human Nutrition*. Report of a joint WHO/FAO/UNU expert consultation. WHO Technical Report Series No. 935. Geneva: FAO.

Fernandes, S. S., D. Tonato, M. A. Mazutti et al. 2019. Yield and quality of chia oil extracted via different methods. *Journal of Food Engineering*, 262: 200–208.

Ferreira Ignácio Câmara A. K., P. Kiyomi Okurob, R. Lopes da Cunhab, A. M. Herreroc, C. Ruiz-Capillasc, and M. A. Rodrigues Pollonio. 2020. Chia (*Salvia hispanica* L.) mucilage as a new fat substitute in emulsified meat products: Technological, physico-chemical, and rheological characterization. *LWT – Food Science and Technology* 125: 109193.

Gamel, T. H., A. S. Mesallam, A. A. Damir, L. A. Shekib, and J. P. Linssen. 2007. Characterization of amaranth seed oils. *Journal of Food Lipids* 14: 323–334.

García-Salcedo, A. J., O. L. Torres-Vargas, A. del Real, B. Contreras-Jiménez, and M. E Rodríguez-García. 2018. Pasting, viscoelastic, and physicochemical properties of chia (*Salvia hispanica L.*) flour and mucilage. *Food Structure* 16: 59–66.

Giménez, M. A., C. N. Segundo, M. O. Lobo, and N. C. Samman. 2020. Physicochemical and techno-functional characterization of native corn reintroduced in the Andean Zone of Jujuy, Argentina. *Proccedings II Congreso Internacional de Ia ValSe-Food Network* 53(1): 7.

Goh, K. K. T., L. Matia-Merino, J. Hong Chiang, R. Quek, S. Jun Bing Soh, and R. G. Lentle. 2016. The physico-chemical properties of chia seed polysaccharide and its microgel dispersion rheology. *Carbohydrate Polymer* 149: 297–307.

González, R., E. Tosi, E. Re, M. C. Añon, A. Pilosof, and K. Martinez K 2007. Amaranth starch-rich fraction properties modified by high-temperature heating. *Food Chemistry* 103: 927–934.

Guerreo-Ochoa, M. R., R. Pedreschi, and R. Chirinos. 2015. Optimised methodology for the extraction of protein from quinoa (*Chenopodium quinoa* Willd). *International Journal of Food Science and Technology* 50(8): 1815–1822.

Guiotto, E. N., M. I. Capitani, S. M. Nolasco, and M. C. Tomás 2016. Stability of oil-in-water (O/W) emulsions with sunflower (*Helianthus annuus* L.) and chia (*Salvia hispanica* L.) by-products. *Journal of the American Oil Chemists' Society* 93: 133–143.

Guzmán-Maldonado, S. H. and O. Paredes-López. 1998. Production of high-protein flours as milk substitutes. In *Functional Properties of Food Components*, edited by J. R. Whitaker and A. Lopez, 66–79. Washington, DC: American Chemical Society.

Haros, M., M. P. Tolaba, and C. Suarez. 2003. Influence of corn drying on its quality for the wet-milling process. *Journal of Food Engineering* 60: 177–184.

Haros, M., O. P. Pérez, and C. M. Rosell. 2004. Effect of steeping corn with lactic acid on starch properties. *Cereal Chemistry* 81: 10–14.

Haros, M., W. Blaszczak, O. E. Pérez, J. Sadowska, and C. M. Rosell. 2006. Effect of ground corn steeping on starch properties. *European Food Research and Technology* 222: 194–200.

Hatzold, T., I. Elmadfa, and R. Gross. 1983. Edible oil and protein concentrate from *Lupinus mutabilis*. *Plant Foods for Human Nutrition* 32: 125–132.

He, S., B. K. Simpson, H. Sun et al. 2018. *Phaseolus vulgaris* lectins: A systematic review of characteristics and health implications. *Critical Reviews in Food Science and Nutrition* 58 (1): 70–83.

He, S., J. Zhao, X. Cao et al. 2020. Low pH-shifting treatment would improve functional properties of black turtle bean (*Phaseolus vulgaris* L.) protein isolate with immunoreactivity reduction. *Food Chemistry* 330: 127217.

Hermansson, A. M. 1986. Soy protein gelation. *Journal of the American Oil Chemists' Society* 63(5): 658–666.

Ishak, I., N. Hussain, R. Coorey, and M. Abd Ghani. 2021. Optimization and characterization of chia seed (*Salvia hispanica* L.) oil extraction using supercritical carbon dioxide. *Journal of CO_2 Utilization* 45: 101430.

Ixtaína, V. Y. 2010. *Caracterización de la semilla y el aceite de chía (*Salvia hispanica *L.). Aplicaciones en tecnología de alimentos.* Doctoral Thesis, Facultad de Ciencias Exactas, Universidad Nacional de La Plata (FCE-UNLP).

Ixtaína, V. Y., A. Vega, S. M. Nolasco et al. 2010. Supercritical carbon dioxide extraction of oil from mexican chia seed (*Salvia hispanica* L.): characterization and process optimization. *The Journal of Supercritical Fluids* 55(1): 192–199.

Ixtaína, V.Y., M. L. Martínez, V. Spotorno et al. 2011a. Characterization of chia seed oils obtained by pressing and solvent extraction. *Journal of Food Composition and Analysis* 24: 166–174.

Ixtaína, V. Y., F. Mattea, D. A. Cardarelli et al. 2011b. Supercritical carbon dioxide extraction and characterization of Argentinean chia seed oil. *Journal of the American Oil Chemists' Society* 88(2): 289–298.

Jancurová, M., L. Minarovičová, and A. Dandár. 2009. Quinoa – a review. *Czech Journal of Food Science* 27(2): 71–79.

Jane, J. L., T. Kasemsuwan, S. Leas et al. 1994. Anthology of starch granule morphology by scanning electron microscopy. *Starch/Stärke* 46: 121–129.

Jiang, F., Ch. Du, Y. Guo et al. 2020. Physicochemical and structural properties of starches isolated from quinoa varieties. *Food Hydrocolloids* 101: 105515.

Jiang, S., J. Ding, J. Andrade et al. 2017. Modifying the physicochemical properties of pea protein by pH-shifting and ultrasound combined treatments. *Ultrasonics Sonochemistry* 38: 835–842.

Jing, P., M. M. Giusti. 2007. Effects of extraction conditions on improving the yield and quality of an anthocyanin-rich purple corn (*Zea mays* L.) color extract. *Journal of Food Science* 72: C363–367.

Joaqui, B. A., A. Bolaños-Monilla, J. E. Bravo-Gomez, J. F. Solanilla-Duque, and D. F. Roa-Acosta. 2020. Wet milling of the amaranth grain: Relationship between the secondary structure of the protein and its ability to form gel. *Sylwan* 164: 31–48.

Johnson, S., J. Clements, C. Villarino, and R. Coorey. 2017. Lupins: Their unique nutritional and health-promoting attributes. In *Gluten-Free Ancient Grains Cereals, Pseudocereals, and Legumes: Sustainable, Nutritious, and Health-Promoting Foods for the 21st Century*, edited by J. R. N. Taylor and J. M. Awika, Chap. 8, 179–222. Cambridge: Woodhead Publishing Series in Food Science, Technology and Nutrition.

Jokić, J., P. Vidović, and K. Aladić. 2014. Supercritical fluid extraction of edible oils. In *Handbook on Supercritical Fluids: Fundamentals, Properties and Applications*, edited by J. Osborne, 205–228. New York: Nova Science Publishers.

Julio, L. M., J. C. Ruiz-Ruiz, M. C. Tomás, and M. R. Segura-Campos. 2019. Chia (*Salvia hispanica*) protein fractions: Characterization and emulsifying properties. *Journal of Food Measurement and Characterization* 13: 3318–3328.

Kentish, S. and M. Ashokkumar. 2011. The physical and chemical effects of ultrasound. In *Ultrasound Technologies for Food and Bioprocessing*, edited by H. Feng, G. Barbosa-Cánovas, and J. Weiss, 1–12. New York: Food Engineering Series Springer.

Koivunen, E., K. Partanen, S. Perttilä et al. 2016. Digestibility and energy value of pea (*Pisum sativum* L.), faba bean (*Vicia faba* L.) and blue lupin (narrow-leaf) (*Lupinus angustifolius*) seeds in broilers. *Animal Feed Science and Technology* 218: 120–127.

Krist, S. 2020. *Vegetable Fats and Oils*. Cham: Springer.

Lamothe, L. M., S. Srichuwong, B. L. Reuhs, and B. R. Hamaker. 2015. Quinoa (*Chenopodium quinoa* W.) and amaranth (*Amaranthus caudatus* L.) provide dietary fibres high in pectic substances and xyloglucans. *Food Chemistry* 167: 490–496.

León-Camacho, M., D. L. García-González, and R. Aparicio. 2001. A detailed and comprehensive study of amaranth (*Amaranthus cruentus* L.) oil fatty profile. *European Food Research & Technology* 213: 349–355.

Li, G., S. Wang, and F. Zhu. 2016. Physicochemical properties of quinoa starch. *Carbohydrate Polymers* 137: 328–338.

Lindeboom, N., P. Chang, K. Falk, and R. Tyler . 2005. Characteristics of starch from eight quinoa lines. *Cereal Chemistry* 82(2): 216–222.

Liu, Y., Z. Liu, X. Zhu, X. Hu, H. Zhang, Q. Guo, R. Y. Yada and S. W. Cui. 2021. Seed coat mucilages: Structural, functional/bioactive properties, and genetic information. *Comprehensive Reviews in Food Science and Food Safety* 20(3): 2534–2559.

López, D. N., R. Ingrassia, P. Busti et al. 2018. Structural characterization of protein isolates obtained from chia (*Salvia hispanica* L.) seeds. *LWT – Food Science and Technology* 90: 396–402.

López, M. G., L.A. Bello-Pérez, and O. Paredes-López, O. 1994. Amaranth carbohydrates. In *Amaranth – Biology, Chemistry and Technology*, edited by O. Paredes-Lopez, 107– 131. Boca Raton, FL:CRC Press.

Luna-Mercado, G. I. and R. Repo-Carrasco-Valencia. 2020. Gluten-free bread applications: Thermo-mechanical and techno-functional characterization of Kañiwa flour. *Cereal Chemistry* 00: 1–8.

Malgor, M., A. C. Sabbione, and A. Scilingo. 2020. Amaranth lemon sorbet, elaboration of a potential functional food. *Plant Foods for Human Nutrition* 75: 404–412.

Manassero, C. A., M. C. Añón, and F. Speroni. 2020. Development of a high protein beverage based on amaranth. *Plant Foods for Human Nutrition* 75: 599–607.

Marín Flores, F. M., M. J. Acevedo, R. M. Tamez, M. Nevero, and A. L. Garay. 2008. WO/ 2008/0044908 Method for obtaining mucilage from *Salvia hispanica* L. Paris: UN Word Internacional Property Organization.

Martínez, M. L., M. A. Marín, C. M. Salgado Faller et al. 2012. Chia (*Salvia hispanica* L.) oil extraction: Study of processing parameters. *LWT – Food Science and Technology* 47: 78–82.

Martínez, N. E. and M. C. Añón. 1996. Composition and structural characterization of amaranth protein isolates: an electrophoretic and calorimetric study. *Journal of Agricultural and Food Chemistry* 44: 2523–2530.

Martínez, N. E., O. F. Castellani, and M. C. Añón. 1997. Common molecular features among amaranth storage proteins. *Journal of Agricultural and Food Chemistry* 45: 3832–3839.

Martínez-Villaluenga, C., A. Torres, J. Frias, and C. Vidal-Valverde. 2010. Semolina supplementation with processed lupin and pigeon pea flours improve protein quality of pasta. *LWT – Food Science and Technology* 43: 617–622.

Martirosyan, D. M., L. A. Miroshnichenko, S. N. Kulakova, A. V. Pogojeva, and V. I. Zoloedov. 2007. Amaranth oil application for coronary heart disease and hypertension. *Lipids in Health and Disease* 6: 1–12.

Méndez-Montealvo, G., J. Solorza-Feria, M. Velázquez del Valle, N. Gómez-Montiel, O. Paredes-López, and L. A. Bello-Pérez. 2005. Chemical composition and calorimetric characterization of hybrids and varieties of maize cultivated in Mexico. *Agrociencia* 39(3): 267–274.

Méndez-Montealvo, G., F. J. García-Suárez, O. Paredes-López, and L. A. Bello-Pérez. 2008. Effect of nixtamalization on morphological and rheological characteristics of maize starch. *Journal of Cereal Science* 48: 420–425.

Mercado, J., C. Elías, and G. Pascual. 2015. Protein isolated from cake of sacha inchi (*Plukenetia volubilis* L.) and evaluation of its techno-functional properties. *Anales Científicos* 76 (1): 160–167.

Mir N. A., S. Riar Ch, and S. Singh. 2021. Improvement in the functional properties of quinoa (*Chenopodium quinoa*) protein isolates after the application of controlled heat-treatment: Effect on structural properties. *Food Structure* 28: 100189.

Monroy, Y. M., R. A. F. Rodrigues, A. Sartoratto, and F. Cabral. 2020. Purple corn (*Zea mays* L.) pericarp hydroalcoholic extracts obtained by conventional processes at atmospheric pressure and by processes at high pressure. *Brazilian Journal of Chemical Engineering* 37: 237–248.

Moscoso-Mujica, G., A. Zavaleta, Á. Mujica, M. Santos, and R. Calixto. 2017. Fraccionamiento y caracterización electroforética de las proteínas de la semilla de kañihua (*Chenopodium pallidicaule* Aellen) Fractionation and electrophoretic characterization of (*Chenopodium pallidicaule* Aellen) kanihua seed proteins. *Revista Chilena de Nutrición* 44(2): 144–152.

Muangrat, R., P. Veeraphong, and N. Chantee. 2018. Screw press extraction of Sacha inchi seeds: Oil yield and its chemical composition and antioxidant properties. *Journal of Food Processing and Preservation* 42(6): ee13635.

Mufari, J. R., P. P. Miranda-Villa, and E. L. Calandri. 2018a. Quinoa germ and starch separation by wet milling, performance and characterization of the fractions. *LWT – Food Science and Technology* (96): 527–534.

Mufari, J., H. A. Gorostegui, P. P. Miranda-Villa, A. E. Bergesse, and E. L. Calandri. 2020. Oxidative stability and characterization of quinoa oil extracted from whole meal and germ flours. *Journal of the American Oil Chemists' Society* 97(1): 57–66.

Mufari, J. R. 2017. *Elaboración de subproductos derivados de la quinoa (Chenopodium quinoa Willd.) a partir de distintos procesos extractivos*. Doctoral Thesis, Universidad Nacional de Córdoba.

Mufari, J., P. Miranda-Villa, A. Bergesse, N. Cervilla, E. Calandri. 2018b. Physico-chemical analysis and protein fraction compositions of different quinoa cultivars. *Acta Alimentaria* 47(4): 462–469.

Muñoz, L., A. Cobos, O. Díaz, and J. M. Aguilera. 2012. Chia seeds: Microstructure, mucilage extraction and A hydration. *Journal of Food Engineering* 108: 216–224.

Muñoz-Tebar, N., A. Molina, M. Carmona, and M. I. Berruga. 2021. Use of chia by-products obtained from the extraction of seeds oil for the development of new biodegradable films for the agri-food industry. *Foods* 10: 620–629.

Nardo, A. E., S. Suárez, A. V. Quiroga, and M. C. Añón. 2020. Amaranth as a source of antihypertensive peptides. *Frontiers in Plant Science* 11: 1–15.

Narváez-González, E. D., J. de Dios Figueroa-Cárdenas, S. Taba, E. Castaño Tostado, and R. Martínez Peniche. 2007. Efecto del tamaño del gránulo de almidón de maíz en sus propiedades térmicas y de pastificado. *Revista Fitotecnia Mexicana* 30(3): 269–277.

Nasar-Abbas, S. and V. Jayasena. 2011. Effect of lupin flour incorporation on the physical and sensory properties of muffins. *Quality Assurance and Safety of Crops and Foods* 4: 41–49.

Okafor, D., R. Enwereuzoh, J. Ibeabuchi et al. 2015. Production of flour types from black bean (*Phaseolus vulgaris*) and effect of pH and temperature on functional physicochemical properties of the flours. *European Journal of Food Science and Technology* 3(2): 64–84.

Oliveira-Alves, S. C., D. Barbosa Vendramini-Costa, C. Bau Betim Cazarin et al. 2017. Characterization of phenolic compounds in chia (*Salvia hispanica* L.) seeds, fiber flour and oil. *Food Chemistry* 232: 295–305.

Olivos-Lugo, B. L., M. A. Valdivia-López, and A. Tecante. 2010. Thermal and physico-chemical properties and nutritional value of the protein fraction of Mexican chia seed (*Salvia hispanica* L.). *Food Science and Technology International* 16 (1): 89–96.

Orifici, S. C., M. I. Capitani, M. C. Tomás, and S. M. Nolasco. 2018. Optimization of muci-lage extraction from chia seeds (*Salvia hispanica* L) using response surface method-ology. *Journal of the Science of Food and Agriculture* 98(12): 4495–4500.

Ortíz-Prudencio, S. A. 2006. *Determinación de la composición química proximal y fibra dietaria de 43 variedades criollas de maíz de 7 municipios del Sureste del Estado de Hidalgo*. Tesis de Licenciado en Nutrición. México. Universidad Autónoma, Pachuca de Soto, México.

Osorio-Díaz, P., E. Agama-Acevedo, L. A. Bello-Pérez et al. 2011. Effect of endosperm type on texture and in vitro starch digestibility of maize tortillas. *LWT – Food Science and Technology* 44: 611–615.

Özcan, M. M., F. Y. Al-Juhaimi, I. A. Mohamed Ahmed, M. A. Osman, and M. A. Gassem. 2019. Effect of different microwave power setting on quality of chia seed oil obtained in a cold press. *Food Chemistry* 278(25): 190–196.

Padhi, E., R. Liu, M. Hernandez, R. Tsao, and D. Ramdath. 2017. Total polyphenol con-tent, carotenoid, tocopherol and fatty acid composition of commonly consumed Canadian pulses and their contribution to antioxidant activity. *Journal of Functional Foods* 38: 602–611.

Pérez, E., Y. A. Bahnassey, and W. M. Breene. 1993. A simple laboratory scale method for isolation of amaranth starch. *Starch/Stärke* 45: 211–214.

Pérez, O. E., M. Haros, C. Suárez, and C. M. Rosell. 2003. Effect of steeping time on the starch properties from ground whole corn. *Journal of Food Engineering* 60: 281–287.

Prego, I., S. Maldonado, M. Otegui. 1998. Seed structure and localization of reserves in *Chenopodium quinoa*. *Annals of Botany* 82: 481–488.

Prompiputtanapon, K., W. Sorndech, and S. Tongta. 2020. Surface modification of tapioca starch by using the chemical and enzymatic method. *Starch/Stärke* 72: 1900133.

Punia, S. and S. B. Dhull. 2019. Chia seed (*Salvia hispanica* L.) mucilage (a heteropolysaccharide): Functional, thermal, rheological behaviour and its utiliza-tion. *International Journal of Biological Macromolecules* 140(1): 1084–1090.

Qian, J. Y. and M. Kuhn. 1999. Characterization of *Amaranthus cruentus* and *Chenopodium quinoa* starch. *Starch/Stärke* 51: 116–120.

Quinteros, M. F., R. Vilcacundo, C. Carpio, and W. Carrillo. 2016. Isolation of proteins from sacha inchi *Plukenetia volubilis* L. in presence of water and salt. *Asian Journal of Pharmaceutical and Clinical Research* 9(3): 193–196.

Quiroga, A., E. N. Martínez, and M. C. Añón. 2007. Amaranth globulin polypeptide hetero-geneity. *Protein Journal* 26: 327–333.

Raigar, R. and H. Mishra. 2018. Grinding characteristics, physical, and flow specific properties of roasted maize and soybean flour. *Journal of Food Processing and Preservation* 42(1): 1–9.

Rajendran, A., D. Chaudhary, and N. Singh. 2018. Prospecting high oil in corn (*Zea mays* L.) germplasm for better quality breeding. *Maydica* 63(1): 1–5.

Rascón Cruz, Q., S. Singawa García, J. Osuna Castro, N. Bohorova, and O. Paredes López, O. 2004. Accumulation, assembly, and digestibility of amarantin expressed in trans-genic tropical maize. *Theoretical and Applied Genetics* 108(2): 335–342.

Repo-Carrasco-Valencia, R. 2020. Nutritional value and bioactive compounds in andean ancient grains. *Proceedings II International Conference of La ValSe- Food Network "Development of Food Ingredients from Iberoamerican Ancestral Crops"* 53: 1–5.

Repo-Carrasco-Valencia, R., J. Peña, H. Callio, and S. Salminen. 2009. Dietary fiber and other functional components of two varieties of crude and extruded kiwicha (*Amaranthus caudatus*). *Journal of Cereal Science* 49: 219–224.

Repo-Carrasco-Valencia, R., J. K. Hellström, J. M. Pihlava, and P. H. Mattila. 2010. Flavonoids and other phenolic compounds in Andean indigenous grains: Quinoa (*Chenopodium quinoa*), kañiwa (*Chenopodium pallidicaule*) and kiwicha (*Amaranthus caudatus*). *Food Chemistry* 120(1): 128–133.

Repo-Carrasco-Valencia, R. and J. Valdez Arana. 2017. Chapter 3: Carbohydrates of kernels. In *Pseudocereals: Chemistry and Technology*, edited by C. M. Haros and R. Schönlechner, chap. 3, 49–70. Chichester: John Wiley and Sons.

Reyes-Caudillo, E., A. Tecante, and M. A. Valdivia-López. 2008. Dietary fibre content and antioxidant activity of phenolic compounds present in Mexican chia (*Salvia hispanica* L.) seeds. *Food Chemistry* 107(2): 656–663.

Roa, D. F., P. R. Santagapita, M. P. Buera, and M. P. Tolaba. 2013. Amaranth milling strategies and fraction characterization by FT-IR. *Food and Bioprocess Technology* 7: 711–718.

Roa, D., P. Santagapita, M. Buera, and M. Tolaba. 2014. Ball milling of amaranth starch-enriched fraction. Changes on particle size, starch crystallinity, and functionality as a function of milling energy. *Food and Bioprocess Technology* 7: 2723–2731.

Roa-Acosta, D., C. González-Callejas, and Y. Calderón-Yonda. 2017. Control of abrasive grinding of amaranth grain to obtain two fractions with industrial potential. *Biotecnología en el sector agropecuario y agroindustrial Special edition* 1: 59–66.

Roa-Acosta, D. F., J. E. Bravo-Gómez, M. A. García-Parra, R. Rodríguez-Herrera, and J. F. Solanilla-Duque. 2020. Hyper-protein quinoa flour (*Chenopodium quinoa* Wild): Monitoring and study of structural and rheological properties. *LWT – Food Science and Technology* 121(4): 108952.

Roberfroid, M. 2007. Prebiotics: The concept revisited. *Journal of Nutrition* 137: 830S–837S.

Rocha Uribe, J. A., J. Y. Novelo Pérez, H. Castillo Kauil, G. Rosado Rubio, and C. G. Alcocer. 2011. Extraction of oil from chia seeds with supercritical CO_2. *The Journal of Supercritical Fluids* 56(2): 174–178.

Rose, D.J., G. E. Inglett, and S. X. Liu. 2010. Utilization of corn (*Zea mays*) bran and corn fiber in the production of food components. *Journal of the Science of Food and Agriculture* 90: 915–924.

Ruiz, C., C. Dîaz, J. Anaya, and R. Rojas. 2013. Proximate analysis, antinutrients, fatty acids and amino acids profiles of seeds and cakes from 2 species of sacha inchi: *Plukenetia volubilis* and *Plukenetia huayllabambana*. *Revista de la Sociedad Química del Perú* 79(1): 29–36.

Ruiz, G. A., W. Xiao, M. Van Boekel, M. Mino, and M. Stieger. 2016. Effect of extraction pH on heat-induced aggregation, gelation and microstructure of protein isolate from quinoa (*Chenopodium quinoa* Willd). *Food Chemistry* 209: 203–210.

Sabbione, A. C., A. E. Nardo, M. C. Añón, and A. A. Scilingo. 2016a. Amaranth peptides with antithrombotic activity released by simulated gastrointestinal digestion. *Journal of Functional Foods* 20: 204–214.

Sabbione, A. C., G. Rinaldi, M. C. Añón, and A. A. Scilingo. 2016b. Antithrombotic effects of *Amaranthus hypochondriacus* proteins in rats. *Plant Foods for Human Nutrition* 71: 19–27.

Sabbione, A. C., S. E. Suárez, M. C. Añón, and A. A. Scilingo. 2019. Amaranth functional cookies exert potential antithrombotic and antihypertensive activities.
International Journal of Food Science and Technology 54: 1506–1513.

Salas-Valero, L. M., D. R. Tapia-Blácido, and F. C. Menegalli. 2015. Biofilms based on
canihua flour (*Chenopodium pallidicaule*): design and characterization. *Química
Nova* 38(1): 14–21.

Salcedo-Chávez, B., J. A. Osuna-Castro, F. Guevara-Lara, J. Domínguez-Domínguez, and
O. Paredes-López. 2002. Optimization of the isoelectric precipitation method to
obtain protein isolates from amaranth (*Amaranthus cruentus*) seeds. *Journal of
Agricultural and Food Chemistry* 50: 6515–6520.

Saldaña, E., J. Ríos-Mera, H. Arteaga et al. 2018. How does starch affect the sensory
characteristics of mazamorra morada? A study with a dessert widely consumed by
Peruvians. *International Journal of Gastronomy and Food Science* 12: 22–30.

Salgado-Cruz, M. P., G. Calderón-Domínguez, J. Chanona-Pérez et al. 2013. Chia (*Salvia
hispanica* L.) seed mucilage release characterisation. A microstructural and image
analysis study. *Industrial Crops and Products* 51: 453–462.

Salinas, M. Y., M. Soto, F. Martínez, V. González, and R. Ortega. 1999. Análisis de
antocianinas en maíces de grano azul y rojo provenientes de cuatro razas. *Revista
Fitotecnica. Mexicana* 22: 161–174.

Salvador-Reyes, R. and M. T. Silva Pedrosa Clerici. 2020. Peruvian Andean maize: General
characteristics, nutritional properties, bioactive compounds, and culinary uses.
Food Research International 130: 108934.

Salvador-Reyes, R., A. P. Rebellato, J. Azevedo Lima Pallone, R. A. Ferrari, and M. T. Pedrosa
Silva Clerici. 2021. Kernel characterization and starch morphology in five varieties
of Peruvian Andean maize. *Food Research International* 140: 110044.

Sandoval-Oliveros, M. R. and O. Paredes-López. 2013. Isolation and characterization
of proteins from chia seeds (*Salvia hispanica*). *Journal of Agricultural and Food
Chemistry* 61: 193–201.

Sarifudin, A., T. Keeratiburana, S. Soontaranon, C. Tangsathitkulchai, and S. Tongta.
2020. Pore characteristics and structural properties of ethanol-treated starch
in relation to water absorption capacity. *LWT – Food Science and Technology*
129: 109555.

Sathe, S. K., and D. K. Salunkhe. 1981. Functional properties of the great northern bean
(*Phaseolus vulgaris* L.) proteins: Emulsion, foaming, viscosity, and gelation properties. *Journal of Food Science* 46(1): 71–81.

Sathe, S. K., B. R. Hamaker, K. Wai, C. Sze-Tao, and M. Venkatachalam. 2002. Isolation,
purification and biochemical characterization of a novel water-soluble protein
from Inca peanut (*Plukenetia volubilis*). *Journal of Agricultural and Food Chemistry*
50(17): 4906–4908.

Scilingo, A., S. Molina Ortíz, E. N. Martínez, and M. C. Añón. 2002. Amaranth protein
isolates modified by hydrolytic and thermal treatments. Relationship between
structure and solubility. *Food Research International* 35: 855–865.

Segura-Campos, M. R., I. M. Salazar-Vega, L. A. Chel-Guerrero, and D. A. Betancur-Ancona.
2013a. Biological potential of chia (*Salvia hispanica* L.) protein hydrolysates and
their incorporation into functional foods. *LWT – Food Science and Technology*
50(2): 723–731.

Segura-Campos, M. R., F. Peralta-González, L. A. Chel-Guerrero, and D. A. Betancur-Ancona. 2013b. Angiotensin I-converting enzyme inhibitory peptides of chia (*Salvia hispanica*) produced by enzymatic hydrolysis. *International Journal of Food Science* 2013: 158482.

Segura-Campos, M. R., Ciau-Solís N., Rosado-Rubio G., Chel-Guerrero L. A. and Betancur-Ancona D. A. 2014. Chemical and functional properties of chia seed (*Salvia hispanica* L.) gum. *International Journal of Food Science* 2014: 241053.

Segura-Nieto, M., A. P. Barba de la Rosa, and O. Paredes-López. 1994. Biochemistry of amaranth proteins. In *Amaranth: Biology, Chemistry and Technology*, edited by O. Paredes-López, 75–106. Boca Ratón, FL: CRC Press.

Selma-Gracia, R., J. M. Laparra, and C. M. Haros. 2020. Potential beneficial effect of hydro-thermal treatment of starches from various sources on *in vitro* digestion. *Food Hydrocolloids* 103: 105687.

Serna-Saldivar, S. O. and C. Chuck-Hernández. 2019. Food uses of lime-cooked corn with emphasis in tortillas and snacks. In *Corn: Chemistry and Technology*, edited by S. O. Serna-Saldivar, 469–500. Washington, DC: AACC International Press.

Serna-Saldivar, S. O., and C. E. Pérez Carrillo. 2019. Food uses of whole corn and dry-milled fractions. In *Corn: Chemistry and Technology*, edited by S. O. Serna-Saldivar, 435–467. Washington, DC: AACC International Press.

Silva, C., V. A. dos Santos García, and C. M. Zanette. 2016. Chia (*Salvia hispanica* L.) oil extraction using different organic solvents: oil yield, fatty acids profile and techno-logical analysis of defatted meal. *International Food Research Journal* 23(3): 998–1004.

Silva-Sánchez, C., A. P. Barba de la Rosa, M. F. León-Galván et al. 2008. Bioactive peptides in amaranth (*Amaranthus hypochondriacus*) seed. *Journal of Agricultural and Food Chemistry* 56: 1233–1240.

Singh, B., J. P. Singh, A. Kaur, and N. Singh. 2017. Phenolic composition and antioxidant potential of grain legume seeds: A review. *Food Research International* 101: 1–16.

Sironi, E., F. Sessa, and M. Duranti. 2005. A simple procedure of lupin seed protein frac-tionation for selective food applications. *European Food Research and Technology* 221: 145–150.

Sisti, M. S., A. Scilingo, and M. C. Añón. 2019. Effect of the incorporation of amaranth (*Amaranthus mantegazzianus*) into fat- and cholesterol-rich diets for Wistar rats. *Journal of Food Science* 82: 3075–3082.

Solaesa, Á. G., M. Villanueva, A. J. Vela, and F. Ronda. 2020. Protein and lipid enrichment of quinoa (cv.Titicaca) by dry fractionation. Techno-functional, thermal and rheo-logical properties of milling fractions. *Food Hydrocolloids* 105: 1–9.

Srichuwong, S., D. Curti, S. Austin et al. 2017. Physicochemical properties and starch digestibility of whole grain sorghums, millet, quinoa and amaranth flours, as affected by starch and non-starch constituents. *Food Chemistry* 233: 1–10.

Stamenković, O. S., M. D. Kostić, M. B.Tasić et al. 2020. Kinetic, thermodynamic and optimization study of the corn germ oil extraction process. *Food and Bioproducts Processing* 120: 91–103.

Steffolani, M. E, A. E. León, and G. A. Pérez. 2013. Study of the physicochemical and func-tional characterization of quinoa and kañiwa starches. *Starch/Stärke* 65: 976–983.

Suárez, S. E. and M. C. Añón. 2018. Comparative behaviour of solutions and dispersions of amaranth proteins on their emulsifying properties. *Food Hydrocolloids* 74: 115–123.

Tang, H., K. Watanabe, and T. Mitsunaga, T. 2002. Characterization of storage starches from quinoa, barley and adzuki seeds. *Carbohydrate Polymers* 49: 13–22.

Tang, Y., X. Li, B. Zhang, P. Chen, R. Liu, and R. Tsao. 2015. Characterisation of phenolics, betanins and antioxidant activities in seeds of three *Chenopodium quinoa* Willd. genotypes. *Food Chemistry* 166: 380–388.

Tari, T., U. Annapure, R. Singhal, and P. Kulkarni. 2003. Starch-based spherical aggregates: Screening of small granule sized starches for entrapment of a model flavouring compound, vanillin. *Carbohydrate Polymers* 53: 45–51.

Tavares, L.S., L.A. Junqueira, I. C. de Oliveira Guimarães, and J. Vilela de Resende. 2018. Cold extraction method of chia seed mucilage (*Salvia hispanica* L.): effect on yield and rheological behavior. *Journal of Food Science and Technology* 55: 457–466.

Taylor, J., J. O. Anyango, P. J. Muhiwa, S. L. Oguntoyinbo, and J.R. Taylor. 2018. Comparison of formation of visco-elastic masses and their properties between zeins and kafirins. *Food Chemistry* 245: 178–188.

Tester, R.F. 1997. Influence of growth conditions on barley starch properties. *International Journal of Biological Macromolecules* 21: 37–45.

Thakur, S., M. G. Scanlon, R. T. Tyler, A. Milani, and J. Paliwal. 2019. Pulse flour characteristics from a wheat flour miller's perspective: A comprehensive review. *Comprehensive Reviews in Food Science and Food Safety* 18 (3): 775–97.

Thoufeek Ahamed, N., R. S. P. R. Singhai, and M. Kulkarni Pal. 1998. A lesser-known grain, Chenopodium quinoa: Review of the chemical composition of its edible parts. *Food and Nutrition Bulletin* 19(1): 61–70.

Timilsena, Y. P., R. Adhikari, S. Kasapis, and B. Adhikari. 2015. Rheological and microstructural properties of the chia seed polysaccharide. *International Journal of Biological Macromolecules* 81: 991–999.

Timilsena, Y. P., B. Wang, R. Adhikari, and B. Adhikari. 2016a. Preparation and characterization of chia seed protein isolate–chia seed gum complex coacervates. *Food Hydrocolloids* 52: 554–563.

Timilsena, Y. P., R. Adhikari, S. Kasapis, and B. Adhikari. 2016b. Molecular and functional characteristics of purified gum from Australian chia seeds. *Carbohydrate Polymers* 136: 128–136.

Timilsena, Y. P., J. Vongsvivut, R. Adhikari, and B. Adhikari. 2017. Physicochemical and thermal characteristics of Australian chia seed oil. *Food Chemistry* 228: 394–402.

Torrejón, I., B. L. Martín, T. B. de la Puente, J. R. Nasser, and R. Rizzi. 2016. La Kañiwa: nueva alternativa alimentaria para la prevencion de la desnutricion y las enfermedades cardiovasculares. *Revista de Salud Pública* 20(2): 17.

Tosi, E., E. Re, H. Lucero, and R. Masciarelli. 2000. Amaranth (*Amaranthus* spp.) grain conditioning to obtain hyperproteic flour by differential milling. *Food Science and Technology International* 5: 60–63.

Tosi, E. A., E. Ré, H. Lucero, and R. Masciarelli. 2001. Dietary fiber obtained from amaranth (*Amaranthus cruentus*) grain by differential milling. *Food Chemistry* 73: 441–443.

Triana-Maldonado, D. M., S. Torijano-Gutiérrez, and C. Giraldo-Estrada. 2017. Supercritical CO_2 extraction of oil and omega-3 concentrate from Sacha inchi (*Plukenetia volubilis* L.) from Antioquia, Colombia. *Grasas y Aceites* 68(1): e172.

Urban, M., M. Beran, L. Adamek et al. 2012. Cyclodextrin production from amaranth starch by cyclodextrin glycosyltransferase produced by *Paenibacillus macerans* CCM. *Czech Journal of Food Science* 30: 15–20.

Urbizo-Reyes, U., M. F. San Martin-González, J. García-Bravo, A. López Malo Vigil, and A.M. Liceaga. 2019. Physicochemical characteristics of chia seed (*Salvia hispanica*) protein hydrolysates produced using ultrasonication followed by microwave-assisted hydrolysis. *Food Hydrocolloids* 97: 105–187.

Uriarte-Aceves, P., E. Cuevas-Rodriguez, R. Gutiérrez Dorado et al. 2015. Physical, compositional, and wet-milling characteristics of Mexican blue maize (*Zea mays* L.) landrace. *Cereal Chemistry Journal* 92: 491–496.

Vázquez Batti, A. I., A. C. Resio, and M. P.Tolaba. 2006. Optimization of amaranth wet milling in alkaline medium. Proceedings XXII Interamerican Congress of Chemical Engineering, CIIQ 2006 and V Argentinian Congress of Chemical Engineering, CAIQ 2006 – Innovation and Management for Sustainable Development. October, Buenos Aires, pp. 811–816.

Vázquez-Ovando, A., G. Rosado-Rubio, L. A. Chel-Guerrero, and D. A. Betancur-Ancona. 2009. Physicochemical properties of a fibrous fraction from chia (*Salvia hispanica* L.). *LWT – Journal of Food Science and Technology* 42: 168–173.

Vázquez-Ovando, A., G. Rosado-Rubio, L. A. Chel-Guerrero, and D. A. Betancur-Ancona. 2010. Dry processing of chia (*Salvia hispanica* L.) flour: Chemical characterization of fiber and protein. *CyTA Journal of Food* 8(2): 117–127.

Vázquez-Ovando, A., D. A. Betancur-Ancona, and L. A. Chel-Guerrero. 2013. Physicochemical and functional properties of a protein-rich fraction produced by dry fractionation of chia seeds (*Salvia hispanica* L.). *CyTA Journal of Food* 11(1): 75–80.

Velarde-Salcedo, A. J., E. Bojórquez-Velázquez, and A. P. B. de la Rosa. 2019. Amaranth. In *Whole Grains and their Bioactives: Composition and Health*, edited by J. Johnson and T. C. Wallace, 209–250. Hoboken, NJ: John Wiley & Sons.

Velázquez-Gutiérrez, S. K., A. C. Figueira, M. E. Rodríguez-Huezo, A. Román-Guerrero, H. Carrillo-Navas, C. Pérez-Alonso. 2015. Sorption isotherms, thermodynamic properties and glass transition temperature of mucilage extracted from chia seeds (*Salvia hispanica* L.). *Carbohydrate Polymers* 121: 411–419.

Ventureira, J., E. N. Martínez, and M. C. Añón. 2012a. Effect of acid treatment on structural and foaming properties of soy amaranth protein mixtures. *Food Hydrocolloids* 29: 272–279.

Ventureira, J. L., A. J. Bolontrade, F. Speroni et al. 2012b. Interfacial and emulsifying properties of amaranth (*Amaranthus hypochondriacus*) protein isolates under different conditions of pH. *LWT – Food Science and Technology* 45: 1–7.

Villacrés, E., M. Navarrete, O. Lucero, S. Espín, and E. Peralta. 2010. Evaluación del rendimiento, características físico-químicas y nutracéuticas del aceite de chocho (*Lupinus mutabilis* sweet). *Revista Tecnológica ESPOL – RTE* 23: 57–62.

Villanueva-Bermejo, D., M. V. Calvo, P. Castro-Gómez, T. Fornari, and J. Fontecha. 2019. Production of omega 3-rich oils from underutilized chia seeds. Comparison between supercritical fluid and pressurized liquid extraction methods. *Food Research International* 115: 400–407.

Villanueva-Bermejo, D., T. Fornari, M. V. Calvo et al. 2020. Application of a novel approach to modelling the supercritical extraction kinetics of oil from two sets of chia seeds. *Journal of Industrial and Engineering Chemistry* 82: 317–323.

Villarino, C. B. J., V. Jayasena, R. Coorey, S. Chakrabarti-Bell, and S. K. Johnson. 2016. Nutritional, health and technological functionality of lupin flour addition to bread and other baked products: Benefits and challenges. *Critical Reviews in Food Science and Nutrition* 26: 835–857.

Wandersleben, T., E. Morales, C. Burgos-Díaz et al. 2018. Enhancement of functional and nutritional properties of bread using a mix of natural ingredients from novel varieties of flaxseed and lupine. *LWT – Food Science and Technology* 91: 48–54.

Wang, N., D. Hatcher, R. Tyler, R. Toews, and E. Gawalko. 2010. Effect of cooking on the composition of beans (*Phaseolus vulgaris* L.) and chickpeas (*Cicer arietinum* L.). *Food Research International* 43(2): 589–594.

Weber, E. J. 1979. The lipids of corn germ and endosperm. *Journal of the American Oil Chemists' Society* 56: 637–641.

Willis, W. M. and A. G. Marangoni. 2017. Food lipids. Chemistry, nutrition, and biotechnology. In *Food Lipids: Chemistry, Nutrition, and Biotechnology*, edited by C. Akoh, 899–939. Boca Ratón, FL: CRC Press.

Wong, A., K. Pitts, V. Jayasena, and S. Johnson. 2013. Isolation and foaming functionality of acidsoluble protein from lupin (*Lupinus angustifolius*) kernels. *Journal of Science of Food and Agriculture* 93: 3755–3762.

Zanqui, A. B., C. Marques da Silva, D. Rodrigues de Morais et al. 2016. Sacha inchi (*Plukenetia volubilis* L.) oil composition varies with changes in temperature and pressure in subcritical extraction with *n*-propane. *Industrial Crops and Products* 87: 64–70.

Zarkadas, G. G., Z. R. Yu, R. I. Hamilton, P. L. Pattison, and N. G. W. Rose. 1995. Comparison between the protein-quality of northern adapted cultivars of common maize and quality protein maize. *Journal of Agricultural and Food Chemistry* 43: 84–93.

Zaworska, A., M. Kasprowicz-Potocka, A. Frankiewicz, Z. Zduńczyk, and J. Juśkiewicz. 2017. Effects of fermentation of narrow-leafed lupine (*L. angustifolius*) seeds on their chemical composition and physiological parameters in rats. *Journal of Animal and Feed Science* 25: 326–334.

Zhu, F. 2017. Structures, physicochemical properties, and applications of amaranth starch. *Critical Reviews in Food Science and Nutrition* 57(2): 313–325.

Chapter 7

Food Uses of Selected Ancient Grains

Claudia M. Haros[1], Marcela Lilian Martínez[2],
Bernabé Vázquez Agostini[3], and Loreto A. Muñoz[4]
[1]Instituto de Agroquímica y Tecnología de
Alimentos (IATA), Paterna, Valencia, Spain
[2]Instituto Multidisciplinario de Biología Vegetal (IMBIV,
CONICET-UNC), Córdoba, Argentina
[3]Universidad Nacional de Jujuy and Consejo
Nacional de Investigaciones Científicas y Técnicas
(CONICET), San Salvador de Jujuy, Argentina
[4]Universidad Central de Chile, Santiago de Chile

CONTENTS

DOI: 10.1201/9781003088424-7

7.1 INTRODUCTION

The consumption of grains, cereals and pseudo-cereals such as Andean maize (*Zea mays*), amaranth (*Amaranthus* spp), quinoa (*Chenopodium quinoa* Willd), kañiwa (*Chenopodium pallidicaule* Aellen), chia (*Salvia hispanica* L.), sacha inchi (*Plukenetia volubilis*) and legumes as black turtle bean (*Phaseolus vulgaris*) and tarwi (*Lupinus mutabilis*) date backs to ancient times. Andean crops have been cultivated for thousands of years (since pre-Colombian times) including pseudo-cereals such as quinoa, kañiwa and amaranth, as well as the Andean lupin or tarwi and several pigmented varieties/races of corns. Purple corn may well be the most representative and consumed among this type of cereal (Ayala, 1998; Ranilla et al.; Apostolidis et al., 2009). On the other hand, chia is originally from the central valley of Mexico and northern Guatemala (Peláez et al., 2019); it began to be used in human food around 3500 BC and acquired importance as a staple crop in Central Mexico between 1500 and 900 BC (Ayerza and Coates, 2005; Cahill, 2003). The seed was one of the main crops of pre-Columbian societies, consumed by Aztecs and Mayans in many food preparations; it was also used in medicine and paintings – surpassed only by corn and beans (Muñoz et al., 2012). Finally, black beans are native to some regions of the American continent and were domesticated from the wild legume distributed from northern Mexico to northeastern Argentina; nowadays black beans are popular around the world and consumed in various preparations. An example would be black bean-chili in Bolivia, "feijoada" in Brazil, "gallo pinto" in Costa Rica, mixed with rice called "Moros y Cristianos" in Cuba, in soup called "sopa de frijoles negros" in Puerto Rico and in a thick corn tortilla (pupusa) stuffed with black beans or in many dishes in Mexico like intact or refried (boiled, mashed) with rice and tortillas, in stews, in soups, in mixed dishes or in casseroles (The Bean Institute, 2014; Wright, 2008).

All these crops have played an important role in human nutrition as sources of nutrients. There are multiple and diverse food uses of these grains, depending on the region or country where they are consumed; for example, in the quotidian Andean diet, plant foods were usually boiled into soups or stews in large pots and commonly consumed or in bowls or on plates (Hastorf, 2015). In ancient times, grains were consumed whole and/or ground in millstone or small stone mortar, to later be incorporated into different foods such as bread, stews and diverse traditional dishes. In the last two decades, these ancestral grains have been transformed from denigrated foods (produced and consumed almost exclusively by people from rural locations) into new ingredients thanks to the knowledge we have about them today, but also thanks to the chefs who have played a leading role in the revalorization and rediscovering of these ancestral Latin-American crops (Ayerza and Coates, 2005; Izquierdo and Roca, 1998; McDonell, 2019).

This chapter is focused on the food uses of ancient Latin-American grains and how they have been used in food consumption throughout history, and how they are an important component in our daily or common diets.

7.2 BAKERY AND BREAD PRODUCTS

A growing number of investigations have studied the use of alternative ingredients to wheat in the production of nutrient-rich bakery products. For the inclusion of high levels of whole flours/seeds of alternative cereals/pseudo-cereals/oilseeds/legumes in bread, it would be necessary to make modifications to traditional technological breadmaking procedures, which could enable the development of a range of new baking products with enhanced nutritive and sensory value. The addition of these unconventional flours, particularly in high amounts, results in technological challenges such as: increases in dough yield, resulting in moist and shorter dough; decreases in fermentation tolerance; lower volume, tense and non-elastic crumb; and various flavor changes, depending on ingredients and bread type, due to the higher amount of fiber and dilution of gluten (Sanz-Penella et al., 2013; Sanz-Penella and Haros, 2017). In this sense, pseudo-cereals are an excellent alternative, being more frequently used due to consumer demand. In particular, amaranth is the pseudo-cereal that has been mostly studied for breadmaking purposes (Table 7.1).

The use of amaranth in the formulation gradually and significantly increases protein, lipid, minerals and dietary fiber content compared to white bread (Table 7.1). The inclusion of whole amaranth flour also shows a slight tendency to decrease specific volume and increases crumb firmness, whereas crumb structure sometimes exhibits no significant changes (Sanz-Penella et al., 2013; Miranda-Ramos et al., 2019). Color tristimulus values were significantly affected when whole amaranth flour was used in both crumb and crust. Despite bread made with amaranth not achieving greater acceptability than wheat bread, particularly with a high percentage of substitution, consumers concluded they preferred amaranth bread due to its more nutritious condition even though its taste and aroma were different from traditional bread.

Popped amaranth was also evaluated as a bread supplementation (Bodroza-Solarov et al., 2008), which had a similar effect on bread quality than whole amaranth flours. If the purpose of pseudo-cereal inclusion in bread making is to increase protein amount and quality, protein isolates could be used in order to avoid quality deterioration of the end product (Tömösközi et al., 2011).

TABLE 7.1   CHARACTERISTICS OF BREAD MADE WITH AMARANTH AND OTHER INGREDIENTS

Amaranthus	Ingredient	Level of Substitution in flour basis	Formulation	Benefits	Loss of quality	References
A. caudatus	Whole Flour	up to 100%	Wheat	↑Nutritional value	↑Crumb Firmness, ↓Specific Volume, ↓Acceptability	Rosell et al., 2009
	Whole Flour	up to 40%	Wheat, Chia, and/or Quinoa	↑Nutritional value, ↑Minerals (Ca, Fe, Zn)	↑Crumb Firmness, ↓Specific Volume, ↑Phytates	Miranda-Ramos, 2020
	Whole Flour	up to 40%	Wheat	↑Nutritional value, ↓Microbial load	↑Antinutritional Components, ↓Acceptability	Zula et al., 2020
A. cruentus	Popped Grains		Wheat	↑Crumb quality, taste, odour and freshness	N.D	Simurina et al., 2005
	Popped grains	up to 20%	Wheat	↑Nutritional value, ↑Squalene, ↑color and flavor, ↑Uniform porosity	↑Crumb Firmness, ↓Crumb Elasticity, ↑Crumb Density	Bodroza-Solarov et al., 2008
	Whole flour	up to 40%	Wheat, phytase	↑Nutritional Value, ↑Fe Availability, ↓Phytates	↓Specific Volume, ↑Crumb Firmness	Sanz-Penella et al., 2012
	Whole flour	up to 50%	Wheat	↑Nutritional Value	↓Specific Volume, ↑Crumb Firmness, ↑Colour	Sanz-Penella et al., 2013

	Form	Amount	Substrate	Effects	Effects	Reference
	Whole flour	up to 50%	Wheat, *Bifidobacterium* phytases	↑Nutritional Value, ↓Phytates, ↑Mineral Availability (Ca, Fe, Zn)	↓Specific Volume, ↑Crumb Firmness	Garcia-Mantrana et al., 2014
	Whole flour	25%	Wheat	↑improved the Hb concentrations, ↓GI	N.D	Laparra & Haros, 2016
A. hypochondriacus	Whole flour	up to 50%	Chapatti Indean Flat Bread	↑Nutritional Value, ↑Fe, Ca, Mg, ↑Lysine, ↑Protein Digestibility	N.D.	Banerji et al., 2018
	Whole flour, Flakes and croats	up to 25%	Wheat	↑Specific Volume, ↑Porosity, ↑Nutritional Value	↓Whitness, ↓Shape stability ↓Consumer characteristics	Mykolenko et al., 2020
	Whole flour	up to 50%	Wheat	↑Nutritional Value	↓Crumb Staling, ↑Phytates, ↓Acceptability	Miranda-Ramos et al., 2019
	Whole flour	25%	Wheat	↓GI *in vivo*	N.D.	Laparra & Haros, 2018
Amaranthus dubius Mart. ex Thell	Whole flour	up to 20%	Wheat	↑Nutritional Value, ↑Digestibility	N.D.	Montero-Quintero et al., 2015
A. spinosus	Whole flour	up to 50%	Wheat	↑Nutritional Value	↓Crumb Staling, ↑Phytates, ↓Acceptability	Miranda-Ramos et al., 2019

Source: Authors' own work.

Another option would be to use defatted hyperproteic amaranth flours as an alternative ingredient to supplement breads (Tosi et al., 2002). Traditional Brazilian bread made with cheese was also studied with the addition of amaranth flour (Lemos et al., 2012) or as a raw material for sourdough fermentation (Jekle et al., 2010; Houben et al., 2010). Amaranth is a suitable ingredient for fermentation by various species of *Lactobacillus*, and sourdough fermentation was able to produce doughs with viscosity and elasticity similar to those found in pure wheat flours (Houben et al., 2010). However, more balanced fermentation quotient values should encourage improvement in bread flavor (Jekle et al., 2010).

Quinoa (*Chenopodium quinoa*) was also investigated for the development of lactic ferment as an alternative to biopreservation of packed bread. It is an optimal substrate for growth and production of improved amounts of antifungal compounds by *Lactobacillus plantarum* isolated from sourdough (phenyllactic and hydroxyphenyllactic acids), allowing a quantity reduction of the chemical preservative calcium propionate commonly added to bread (Dallagnol et al., 2015). Strains of *Lactobacillus plantarum* and *Lactococcus lactis* subsp. *lactis*, previously selected for biosynthesis of γ-aminobutyric acid (GABA), were used for sourdough fermentation of cereal, pseudo-cereal and leguminous flours, where pseudo-cereals and chickpea flours were the most suitable for enrichment with GABA (Coda et al., 2010). In general, whole quinoa flour could be a good replacement for wheat flour (Figure 7.1.A) in bread formulations, increasing the product's nutritional value in terms of dietary fiber, minerals, proteins and healthy fats, with only a small depreciation in bread quality (Stikic et al., 2012; Iglesias-Puig et al., 2015; Ballester et al., 2019a, 2019b, 2019c) taking into account that a high level of substitution did not show the small depreciation in bread quality (Rosell et al., 2009; Wang et al., 2015). The same trend was observed through the replacement of wheat flour by 25% kañiwa (*Chenopodium pallidicaule*) with a good sensory acceptability but variable in color (Rosell et al., 2009).

Compared to wheat breads, the products resulting from composed flours such as quinoa, amaranth or kañiwa/wheat, affect their quality in terms of reduction of specific volume, increment of density, firmness and chewiness of crumb bread, its darkness, redness and/or yellowness (depending on the color of the raw material) with acceptable marks by consumers (Rosell et al., 2009; Iglesias-Puig et al., 2015; Ballester et al., 2019a, 2019b; Miranda-Ramos et al., 2019). Despite the change of color, pseudo-cereal breads were well accepted, showing good commercial potential among consumers, with most of those interviewed stating that they would buy the product mainly for the health benefits it could bring (Calderelli et al., 2010; Bilgiçli and İbanoğlu, 2015; Iglesias-Puig et al., 2015; Miranda-Ramos et al., 2019; Ballester et al., 2019b; Miranda-Ramos and Haros,

Figure 7.1   Foods with quinoa: A. Bread with whole black quinoa flour, 25%; B. Quinoa balls with cocoa; C. Chocolate with quinoa; D. Quinoa and buckwheat cookies; E. Quinoa, cinnamon and lemon cookies; F. Quinoa cookies with cocoa; G. Crispy chocolate quinoa toast; H. Puffy quinoa, compared with quinoa seeds. **Source: Authors' own work.**
Notes: A) A slice of bread with 25% of whole black quinoa flour bread.

B) Four, thin rectangular quinoa and buckwheat crackers, in a cup; one side of the cracker is covered with chocolate.

C) Spherical, popped cereals with quinoa on a small round plate.

D) Three squashed, ball-like quinoa and buckwheat cookies, on a small round plate.

E) Three rectangular quinoa, cinnamon and lemon cookies; on the surface there is a cookie mold drawing of four parallel lines down the middle. Placed on a small round plate.

F) Three circular quinoa cookies with cocoa, with a rough surface, on a small round plate.

G) Three squared chocolate with quinoa, on a small round plate.

H) Puffier white quinoa compared to raw white quinoa seeds, each on a small round plate.

2020). Ballester et al. (2019a, 2019b, 2019c) developed bakery products with 25% replacement of wheat by Real quinoa of different colors in order to evaluate its functionality as a bread-making ingredient, with excellent results in sensory evaluation, nutritional/functional value but lower technological bread quality than the counterpart without quinoa. In this sense, the fiber fraction with high amounts of antioxidant compounds was isolated, by wet and dry milling, and included in bread formulation at 5% to get the same functional value as the samples with 25% of quinoa but with equal bread quality than the control sample (wheat) (Ballester et al., 2020).

In general, the pseudo-cereals improved mineral content but also the phytate content of breads (Bilgiçli and İbanoğlu, 2015; García-Mantrana et al., 2014; Iglesias-Puig et al., 2015; Miranda-Ramos et al., 2019; Miranda-Ramos and Haros, 2020). High phytate contents were easily avoided by the use of exogenous phytases, the use of flours from germinated seeds or the use of sourdough or acidified sponges (Park and Morita, 2005; García-Mantrana et al., 2015; Dallagnol et al., 2013; Iglesias-Puig et al., 2015).

In regard to chia, the first bread products described in the literature with this whole grain were developed for women, rich in fibers and proteins, fortified with iron and folic acid, with or without soy and/or linseed, which showed high acceptability by consumers (Bautista Justo et al., 2007). Subsequently, Iglesias-Puig and Haros (2013), following the regulations in force in Europe at that time (EFSA, 2009; OJEU, 2009), bread products with up to 5% substitution levels in flour bases were developed (Figure 7.2.A). The objective of this investigation was to develop new cereal-based products with increased nutritional quality by using chia ingredients such as its seeds, whole flour, semi-defatted flour and low-fat flour, in order to evaluate its potential as a bread-making ingredient. Later, because of its high nutritional properties, consumption of chia has spread widely within the European Community (EC); and in the EU list of novel foods, the European Food Safety Authority (EFSA) has authorized the use of chia seeds and/or other chia ingredients to higher levels and/or their inclusion in other foods different than bread (Turck et al., 2019a).

Whole chia seeds may be marketed in the EC as a food ingredient and used in baked products and breakfast cereals up to 10%; ground chia seeds up to 5% in bread; whole chia seeds up to 5% in sterilized ready-to-eat meals based in cereal/pseudo-cereal grains and/or pulses; prepackaged chia seed as such and fruit/nut/seed mixes; and chia in confectionery products and chocolates; edible ices; fruit and vegetable products; non-alcoholic beverages and puddings (<120°C in their preparation) without limit, according to the EC (EU, 2020). Recently, the use of two partially defatted powders of chia enriched

Figure 7.2 Foods with chia: A. Bread with chia seeds, 5%; B. Fresh pasta with chia seeds, 10%; C. Whole rice with chia, quinoa, spelt and flaxseeds; D. Digestive chia cookies; E. Biscuits with curcuma and chia; F. Chia and spelt *roscos*; G. Crackers of chia and quinoa; H. Chia and quinoa sticks; I. Sea biscuits with chia. **Source: Authors' own work.**

Notes: A) Five slices of bread with chia seeds on a medium-size round plate.

B) Close-up of white tagliatelle spaghettis with black chia seeds.

C) Whole rice with ready-to-eat chia, quinoa, spelt and flax seeds on a small round plate.

D) Seven circular digestive cookies with chia and a rough surface, in a small round bowl.

E) Three yellow rectangular cookies with curcuma and chia; on the surface there is a cookie mold drawing of four parallel lines down the middle; placed on a small round plate.

F) Seven chia and spelt doughnut-shaped cookies in a small bowl.

G) Many crackers with chia and quinoa of irregular quadrilateral shape in a small bowl.

H) Many popped sticks with chia and quinoa in a small bowl.

I) Many circular sea biscuits with chia in a small bowl.

with proteins or fibers was authorized as food supplements for the adult population (up to 7.5 and 12 g/day, respectively), or as nutritional ingredients in a variety of foods (yogurt, vegetable beverages, energy drinks, chocolate, fruit and pasta) at a level of 0.7–10% (Turck et al., 2019b). Restrictions in the use of chia/chia ingredients – principally in bakery products – are only valid in the EU; the rest of the world does not restrict their uses. The main reason for these restrictions are the high amount of asparagine (and consequent formation of acrylamide in baked products) in chia and its ingredients (Galluzzo et al., 2021), especially these that have higher protein concentrations than the original seeds. The partial replacement of wheat by chia seeds, whole chia flour and defatted chia flour in bread (up to a level of 5–6%) obtained high consumer acceptance in earlier studies, and it could extend the shelf-life of bread, since it inhibits the kinetics of retrogradation of amylopectin during storage (Iglesias-Puig and Haros, 2013; Sayed-Ahmad et al., 2018). However, the bread-making process may affect the stability and/or bioavailability of nutrients/bioactive compounds owing to various chemical and enzymatic reactions during kneading, fermentation and baking, and accordingly this enriched bread would provide a greater or lower health benefit (Benitez et al., 2018; Miranda-Ramos et al., 2020; Miranda-Ramos and Haros, 2020). In general, formulations with chia ingredients significantly increased the levels of proteins with high biological value, healthy lipids, ashes/minerals and soluble dietary fiber, with similar or better technological quality to the control bread. This is mainly due to the increase in specific bread volume, decrease in crumb firmness, with change in crumb color and higher overall acceptability by consumers (Miranda-Ramos et al., 2020; Miranda-Ramos and Haros, 2020; Guiotto et al., 2020; Ahmed et al., 2020).

Only a few scientific investigations have been found that make use of Andean maize (*Zea mays*), black turtle bean (*Phaseolus vulgaris*), tarwi (*Lupinus mutabilis*) or sacha inchi (*Plukenetia volubilis*) in bakery product formulations. There is more information on gluten-free products, which were discussed in Chapter 6. Regarding the case of maize flours from crops grown in the Andean region, which some authors have termed as Andean maize, these can be used in bread, pasta, cookies and cakes, providing color and flavor to the processed product (Salvador-Reyes and Pedrosa Silva Clerici, 2020). This type of maize represents a potential ingredient for the development of new products: it can provide a variety of color, flavor and texture, contributing to the reduction of the use of additives in the food industry, besides being ancient grains, which have survived naturally against extreme adverse growing conditions (Salvador-Reyes and Pedrosa Silva Clerici, 2020). In the case of black turtle bean, there are no scientific publications in terms of its use in the bread-making process. In general, legumes and in particular lupins can be used to fortify the protein content of pasta, bread, biscuits, salads, hamburgers and sausages, and can substitute for milk and soya beans (Carvajal-Larenas et al., 2016). Replacement of wheat flour by <12.5% of tarwi produced breads with good sensory acceptability but variable color and doughs with acceptable thermomechanical patterns (Rosell et al., 2009). Bread with 10% of tarwi flour had a protein efficiency ratio (76%) higher than the control (28%) and similar acceptability (Carvajal-Larenas, 2019). However, it was an effort to evaluate the decrease or elimination of non-nutritional compounds, and changes in color in tarwi derivatives (protein concentrated or isolated) to apply in three types of bread products (loaf, bun and sweet) (Güemes-Vera et al., 2008).

In the case of sacha inchi as a bread ingredient, a bread loaf enriched with extruded sacha inchi cake has been developed (Rodríguez et al., 2018). The incorporation of this bakery ingredient significantly increased the levels of ash, fiber, fat and protein, and decreased the carbohydrate content. The best rheo-logical properties were presented at 6.3% incorporation level, which highlights a retention of 3.33% of alfa-linolenic fatty acid (omega-3), increase of texture and crumb darkening. There was no significant difference between the control and the bread with sacha inchi in terms of the attributes of color, appearance, aroma, flavor and texture. The elaboration of bread with the incorporation of coconut pulp flour and sacha inchi nibs was also investigated (Ordoñez et al., 2019). The best results in the sensory evaluation (aroma, texture and volume-symmetry), physicochemical and farinographic tests were found in breads made with 7.5% sacha inchi nibs with coconut flour (12.5%). The effect of partial replacement of wheat flour by cake of sacha inchi on the rheological properties of dough in sweet bread was also studied (Toralva Aylas et al., 2015).

The Latin-American crops are known to be rich in some bioactive compounds, such as flavonoids, phenolic acids, trace elements, dietary fibers, essential fatty acids and vitamins with known effects on human health. In this sense, partial substitution of wheat flour by whole flours from these seeds/fruit or a fraction of them (single or combined) constitutes a viable option to improve the nutritional value of bakery products, with acceptable technological performance of dough blends and composite products which provides a wide gamma of products with a positive perception by consumers.

7.3 BISCUITS, COOKIES AND CAKES

Cookies/biscuits/cakes are other cereal products in the bakery industry in which efforts have been made to improve the composition of the product by the inclusion of ancient grains. They typically have: low fiber and water levels, high fat and sugar and air cells varying in size embedded within the protein–starch–lipid matrix (Brennan and Samyue, 2004; Poutanen et al., 2014). Some examples are shown in Figures 7.1–7.4.

The incorporation of amaranth flour in sugar snap cookies improved the cookie's surface cracking as a consequence of the reduction in their breaking strength. As a result of increasing the percentage of amaranth (from 5 to 35%) in the formulation, an increase in thickness was also observed (Sindhuja et al., 2005). The sensory evaluation revealed that cookies containing 25% of amaranth scored the best marks due to their golden-brown color, larger size and uniformity, malty-sweet flavor and tenderness.

In another study, sensory evaluation of biscuits prepared with wholemeal barley or wheat flour supplemented with amaranth showed that addition at 20% and 30% levels to barley and wheat flour, respectively, gave the best color, and at levels of 10–20% gave the best biscuit taste (Sandak and El-Hofi, 2000).

Wheat flour and corn starch could be successfully replaced by up to 20% amaranth flour in conventional and up to 30% in reduced fat pound cakes without negatively affecting sensory quality in fresh products (Dias-Capriles et al., 2008).

Biscuits were developed with amaranth – and other foods – to increase their nutritional value with high acceptance by consumers (Nidhi and Indira, 2012; Zanwar and Pawar, 2015). With the help of sucralose and soluble fibers, it is possible to develop a sweet biscuit with amaranth flour partially reduced in sugars and fats with low caloric content and high nutritional value (Torres Palacios et al., 2019). However, the diameter of the cookies proportionally decreased with the increasing levels of supplementation with amaranth/oat/sorghum, whereas hardness of cookies increased, obtaining highest overall acceptability scores when the level of substitution is at 10% (Raihan et al., 2017).

The performance of quinoa–wheat flour blends was also evaluated in cakes biscuits and cookies, providing products with various technological and sensory characteristics (Lorenz and Coulter, 1991; Wang et al., 2015; Watanabe et al., 2014). Cake quality was acceptable with 5–10% of quinoa flour, taste improved and cake grain became more open and the texture less silky as the level of quinoa substitution increased. On the other hand, cookie spread and top grain scores decreased with increasing levels of quinoa flour blended with high-spread cookie flour. Flavor improved with an incorporation of up to 20% quinoa flour in the blend, whereas cookie spread and appearance improved with a quinoa/low-spread flour blend by using 2% lecithin (Lorenz and Coulter, 1991). The cookies containing quinoa flour had greater oxidative stability and nutritional quality, and were rich in dietary fiber, essential amino acids, linolenic acid and minerals, with good sensory acceptability (Wang et al., 2015; Watanabe et al., 2014). Quinoa-tempeh – fermented with *Rhizopus oligosporus* – powder is more suitable than quinoa powder as an ingredient for biscuits, and it may be added to flour in amounts of up to 20% according to the

Figure 7.3 Foods with amaranth: A. Amaranth chocolate chip cookies; B. Amaranth multigrain crispbread; C. Amaranth with chocolate; D. Amaranth snacks: moka flavor; with raisins, peanuts and pumpkin seeds; with cranberries; chocolate flavor; with peanuts and probiotics; with chocolate and probiotics; with cranberries and probiotics. Probiotic: *Bacillus coagulans* GBI-30. **Source: Authors' own work.**
Notes: A) Eight round amaranth chip cookies with a rough surface created by the seeds, on a medium size plate.

B) Five rectangular amaranth multigrain crispbread with cracks and holes on the surface, in a small bowl.

C) Seven cubes, of three different types and colors (white, pink and brown), of amaranth with chocolate on a small plate.

D) Seven different snacks of popped amaranth, in their respective transparent plastic bags, on a large rectangular plate. Distribution is two cubes at the top (moka and raisins, peanuts and pumpkin seeds), followed by two other cubes (with cranberries and one with chocolate flavor), and then three rectangular shaped snacks (all with probiotics) positioned horizontally, parallel to one another (with peanuts, chocolate and cranberries, respectively).

biscuit quality and sensory analysis (Matsuo, 2006). Quinoa affected the scores of sensory properties of cookie samples, in terms of color, taste, crispness and overall acceptability, with the exception of odor scores (Demir et al., 2017). Even wheat flour substitution with quinoa flour severely decreased biscuit acrylamide, and almost maintained biscuit browning index in comparison to the control sample (Sazesh et al., 2020). There is no information in the literature concerning kañiwa as a raw material to develop cookies/biscuits or cakes, but there are many recipes and commercial products with kañiwa, with and without other crops.

Concerning chia, there are many investigations which describe the use of its seeds and whole flour to develop cookies, biscuit or cake formulations with the objective of improving their nutritional/functional quality, and also to take advantage of by-products of the chia industry such as defatted chia flour, mucilage fractions, its oils and/or enriched protein flours. Usually, the addition of chia seeds and chia flour in cookies/biscuits/cakes improves their nutritional value and sensory acceptance of the products (Goyat et al., 2018; Alcántara Brandao et al., 2019). Supplementation of cookies with 10% defatted chia flour could be recommended to improve antioxidant quality without reducing technological or sensorial properties (Mas et al., 2020). In another study, it was found that the mixture of canola-chia oils and gelatin enhances the nutritional and sensory quality of the pound cake reduced in margarine without changing final cake quality (Sánchez-Paz et al., 2020). In this sense, to avoid lipid oxidation, partial substitution of margarine by microencapsulated chia oil in carnauba wax solid lipid microparticles was investigated, to obtain cookies containing omega-3 and omega-6 (Carvalho de Almeida et al., 2018; Hernándes Venturini et al., 2019). On the other hand, chia mucilage also proved to be a new alternative for replacing fats or eggs in cake products, preserving quality attributes while making them healthier foods (Borneo et al., 2010; Ferrari Felisberto et al., 2015; Santos Fernándes and Salas Mellado, 2017). Sensory analysis showed that chia gel stored in different conditions could replace egg in chocolate cake without impairing acceptability. It is possible that cold storage improves chia gel functionality, especially when it is stored frozen (Rodrígues dos Reis Gallo et al., 2020). In cookie formulations with roasted chia, the baking loss, hardness and brightness were inversely proportional to roasting time, affecting the texture and sweetness scores in a consumer preference test (Song et al., 2019). It is important to remark that although nutritional properties are improved by the incorporation of chia/chia by-products into biscuit/cookie formulations, the increase in content of process contaminants – such as acrylamide, hydroxymethyl-furfural and/or furfural, furan and/or methyl furans – and the extent of the lipid oxidation should be carefully considered (Mesías et al., 2016; Myrisis et al., 2022).

As was previously mentioned, Andean maize flour can be used in cookies and cakes, as a potential ingredient for the development of new and innovative products in terms of natural colors and new textures and flavors (Salvador-Reyes and Pedrosa Silva Clerici, 2020). On the other hand, fruits and extracts from sacha inchi were studied such as integral sacha inchi with linseed sticks (Yovera Méndez and Ruddy Gianny, 2018); cookies and cakes with residual sacha inchi cake flour after oil extraction (Toralva Aylas and Rodas Pingus, 2015; Vásquez

Osorio et al., 2017). On websites several recipes are described with this fruit or its derivates.

In general, lupin flour is widely considered an excellent raw material for supplementing different food products owing to its high protein content (Kohajdová et al., 2011) and is largely used as an egg and/or butter substitute; for example, in cakes, pancakes, biscuits, brioche, and/or croissants (Tronc, 1999). The optimal formulation of cookies made with tarwi oil and chia was at 12% and 3% level, respectively, with an acceptability of 6.58 on a 7-point scale (Salvatierra-Pajuelo et al., 2019).

7.4   PASTA PRODUCTS

The production of pasta has been favored with the inclusion of ancient grains and their by-products. Fresh and dry gluten-free spaghetti were developed varying the proportions of maize, quinoa and soy flours between 20.00 and 31.25%, 7.50 and 30.00% and 2.50 and 10.00%, respectively (Mastromatteo et al., 2011). These authors evidenced that the spaghetti samples with highest maize and quinoa content presented the best extensibility values, and the sensory quality did not significantly vary among formulations. Moreover, Giménez et al. (2015) made spaghetti from Andean corn, capia and cully varieties from northern Argentina and compared the sensory profile of these types of spaghetti with those made with rice and wheat flours. These authors observed that celiac consumers assigned high acceptability scores to the corn products. Then, Giménez et al. (2016) determined the nutritional quality of spaghetti made with maize flour enriched with 30% broad bean flour and 20% quinoa flour. These gluten-free spaghetti showed a significant increase in net protein utilization and decreased digestibility compared with a non-enriched control sample. One portion of this product supplied the 10–20% of recommended daily intake of fiber. Addition of quinoa flour had a positive effect on the Fe supply as did broad bean flour on Zn. Meanwhile, Deepa et al. (2017) made three types of pasta: unmicronized maize flour pasta with guar gum, micronized maize flour pasta with guar gum and micronized maize flour pasta with guar and xanthan gum. Pasta samples prepared with micronized flour showed increased firmness, improved color index and overall acceptability, compared with those of unmicronized flour.

Many pasta formulations have been developed based on amaranth flour and in general, the results demonstrated that this raw material affects texture firmness, stickiness and cooking time (Rayas-Duarte et al., 1996; Schöenlechner et al., 2010; Fiorda et al., 2013). Rayas-Duarte et al. (1996) showed that an ideal level of substituted grain would optimize nutritional quality without destroying functional properties. These authors reasoned that a multigrain pasta with better nutritional quality, acceptable cooking properties and good sensory attributes can be produced, using buckwheat, amaranth or lupin flour without exceeding the following substitution levels: 30% for light buckwheat, 15% for dark buckwheat, 25% for amaranth, and 15% for lupin. Cárdenas-Hernández et al. (2016) reported that a pasta based on amaranth flour (25.5–50.0%) and semolina (51.0–21.0%) with 15.0% of egg presented acceptable technological quality compared to a control pasta (85.0% semolina and 15.0% egg). Even though the incorporation of amaranth produced a reduction in pasta luminosity values and sensory acceptance, all the formulations exhibited a significantly higher content of protein (17.47%), crude fiber (2.64%), ash (2.03%), fat (4.40%), total phenol content (2.25 mg Ferulic acid equivalents/g) compared to the control pasta (15.13; 0.72; 0.90; 2.12; 0.98 respectively).

Pantoja Tirador and Prieto Rosales (2014) developed pasta made with 80% wheat, 10% tarwi and 10% quinoa flours with an acceptable sensorial quality and high nutritional value providing 19.06% protein; 3.23% fat; 0.65% ash and 2.16% fiber. The incorporation of 20% (w/w) of quinoa flour to semolina improved the nutritional profile without affecting the technological and sensory quality of pasta (Lorusso et al., 2017). Furthermore, these authors demonstrated that the quinoa flour fermentation with lactic acid bacteria can also enhance the favorable outcomes of quinoa. Moreover, Demir and Bilgicli (2020) found that the replacement of wheat semolina with 10, 20 and 30% of raw and germinated quinoa flours in pasta formulations significantly improved the nutritional quality of the product, keeping technological and sensory qualities within acceptable limits. These authors communicated that as the raw and germinated quinoa flour ratios increased in pasta, ash (0.81 to 1.35%), crude fat (0.49 to 1.39%) and protein (12.56 to 16.65%), total phenol content (0.54 to 1.22 mg GAE/g), antioxidant activity (12.39 to 27.69%) and mineral matter (Ca, Fe, K, Mg, P and Zn) amounts significantly increased.

Kañiwa flour can be used to replace partial wheat flour in pasta to increase its nutritional value. Bustos et al. (2019) developed enriched pasta with wheat and kañiwa blends. These authors studied three levels of kañiwa flour substitutions: 10, 20 and 30% and observed that this flour replaced negatively affected pasta cooking properties, such as water absorption and cooking loss. With respect to texture parameters, Kañiwa pasta firmness and chewiness generally decreased when kañiwa content increased. A replacement of 20% of kañiwa flour improved the nutritional quality of pasta, increasing the dietary fiber (57.0 to 97.2 g/kg pasta) and protein quality (130.5 to 136.4 g/kg pasta), obtaining a functional pasta with satisfactory cooking quality.

Ground chia seed and defatted chia flour have been widely used for the development of these types of products. Home style noodles were made with 10% substitution of durum flour with ground chia seeds. The final product changed from "yellow" to "yellowish-gray" preserving acceptable taste properties with an increase in macroelements (calcium by 2.2, magnesium by 2 and phosphor by 1.4, times respectively) and of microelements' concentration (copper by 2, zinc by 1.5 and iron by 3.5, times respectively). Moreover, the microstructure of the product, vitamin value, and physicochemical indicators of quality of pasta products were not affected during their lifetime (Naumova et al., 2017). Levent et al. (2017), using chia flour in pasta based on rice and corn flour, obtained a healthier product with acceptable sensory results, with 20% of chia flour incorporation in combination with diacetyl tartaric esters of mono (and di-)glycerides. In addition, Aranibar et al. (2018) evaluated the nutritional, sensory and technological quality of pasta supplemented with partially de-oiled chia flour (PDCF) at different proportions (2.5%, 5.0% and 10.0%).These authors concluded that incorporation of PDCF into wheat pasta allows significant improvement to nutritional properties (total dietary fiber, 2.86 to 9.08%, d.b; ω-3/ω-6 ratio, 0 to 2.14; total phenolic content, 10 to 35 mg GAE/100g pasta and antioxidant capacity) preserving in general the color, sensory, texture and cooking characteristics of pasta samples compared to non-supplemented pasta.

Legume flours were used in combination with precooked rice in order to produce gluten free spaghetti (Bouasla et al. 2017). The precooked rice pasta was enriched with different levels (10, 20 and 30%) of legume flours (yellow pea, chickpea and lentil). The products obtained presented an improved chemical profile compared with control pasta (rice pasta). In fact, the content of proteins, fat, ash, insoluble and soluble fiber was among 9.68–13.95, 0.08–0.38, 0.69–1.29, 2.66–5.15 and 1.09–1.70% d.b., respectively; while the control sample presented the following composition 8.25, 0.09, 0.50, 2.52, 0.69% d.b., respectively. These authors maintained that incorporation of legume flour reduces lightness and expansion ratios, and increases yellowness, adhesiveness and firmness, without affecting cooking time. On the other hand, Sudha et al. (2012), Nielsen et al. (1980), and Bahnassey and Khan (1986) who studied the fortification of pasta with legume flours recognized significant structural changes in the protein network, which were reflected in greater cooking loss. Similar trends were observed by Zhao et al. (2005) who maintained that the cooked weight of pasta was not significantly affected, but cooking loss increased with the rise in content of legume flours. In general, the addition of legume flours decreased pasta firmness, so the incorporation of additives like gluten, glycerol mono-stearate or sodium stearoyl lactylate, in combination, improved the quality characteristics of pasta.

In the literature there are several applications of tarwi flour for the development of pasta. Pepe Guato (2011) made wheat flour pasta with a degree of substitution with tarwi flour of 0, 15, 20, 25 and 30%. The best treatment was pasta with 20% replacement tarwi flour, as it presented a higher than average acceptability, lower stickiness and greater firmness; and shorter cooking time. Regarding the chemical composition, it presented 9.41% humidity; carbohydrates 69.94%; protein 22.56%; fiber 2.81%. Ponce et al. (2018) who studied a partial substitution of wheat flour with tarwi flour in the production of long pasta concluded that substitution of 25% tarwi flour and 18% egg in the mixture was the best formulation to obtain a more nutritious and good quality pasta. In addition, Albuja and Yépez (2017) developed gluten-free rice pasta with partial substitution of tarwi flour, egg and guar gum addition. The best pasta was achieved with 20% tarwi flour, 30% egg and 0.15% guar gum. This pasta compared with a control of 100% rice presented an increase in protein 12.06 to 20.64% and fiber 0.32 to 0.75% d.b. Finally, Pinares Huamani (2019) formulated pasta with a partial substitution of tarwi flour (between 8 and 22%) and adding powdered egg shell (between 0.8 and 2.2%). The replacement of 10% tarwi and 1% egg shell powder was the best alternative regarding general acceptability and cooking properties. The pasta showed a high nutritional value: 12.33% protein; 0.27% fat; 73.58% carbohydrates; 139.9 mg / 100 g of calcium and 7.25 μg / 100 g of vitamin D.

7.5   SNACKS AND BREAKFAST MIXTURES

Ritva Repo-Carrasco-Valencia et al. (2011) designed a nutritious extruded snack product using corn and three varieties of quinoa. The optimal mix was 70% quinoa and 30% corn. This product offers a nutritious alternative to traditional snacks. In addition, Figueiredo de Sousa et al. (2019) developed snacks bars (SB) with different corn bran (CB) proportions (0, 10, 25, 40 and 55%). The SB with the highest proportions of CB had the highest levels of dietary fiber. Formulations with up to 40% CB were well-accepted by the judges, compared with the formulation without any CB. On the other hand, Carvajal et al. (2019) developed a snack based on corn, beans and sweet potato with the following proportion (w/w): 80/10/10 and 70/15/15. This last formulation presented the higher nutritional content: protein (12.48%), ash (1.36%) and fiber (1.28%) in relation to the commercial control (100% corn) that presented protein (8.32%), ash (0.57%) and fiber (0.64%) content.

Chávez-Jauregui et al. (2000) prepared an extrudate based on amaranth flour with a fat, protein and carbohydrate content of 0.18%, 15.82% and 80.77%, respectively. Moreover, Gearhart and Rosentrater (2014) made an extrudate snack with 100% of amaranth containing 6 to 8%, 13 to 18% and 63% of fat, protein and carbohydrate, respectively. In addition, Ramos Díaz (2015) produced expanded CB snacks containing up to 50% amaranth, quinoa and kañiwa and at most 20% lupine maintaining textural and sensorial properties as well as added nutritional value.

The development of cereal bar formulations has been investigated by authors such as Delgado and Barraza (2014) who produced bars based on quinoa, sacha inchi and amaranth with a maximum inclusion of 24.9% of quinoa in the formulation. The general acceptance of the bar with the highest quinoa content presented the lowest score, while the formulation containing 21.8% was the most accepted. Regarding the proximal analysis, it was reported that the bars have a protein content between 2.67 and 2.25 g/g d.b. Moreover, Calisto-Guzmán (2009) formulated quinoa bars with red beans and honey that obtained good sensory acceptance. Another alternative are the quinoa flakes which have a protein content of 8.5%, fiber 3.8% and minerals such as calcium 114 mg, phosphorus 60 mg and iron 4.7 mg /100 g (González Rojas and Moya Valenzuela, 2004; Rojas et al., 2010). According to a report by Bergesse et al. (2015), the nutritional difference between quinoa flakes and corn flakes is their essential amino acid content.

Repo-Carrasco-Valencia et al. (2003) formulated two dietary mixtures: quinoa-kañiwa-beans and quinoa-amaranth-beans, with high nutritional value. At the same time, Repo-Carrasco-Valencia et al. (2009) produced expanded snacks based only on kañiwa with two different varieties. The products presented on average the following composition: proteins, 14.13%; fat, 5.63%; crude fiber, 4.56%; ash, 4.23%; carbohydrates, 71.45%; total dietary fiber, 19.52%; phenolic compounds, 2.48 mg gallic acid equivalent/g d.b.; antioxidant activity, 4125 µg trolox eq./g d.b.

Coorey et al. (2012) produced chips with rice, potato and chia flours. The chips elaborated with 0, 5, 10, 12 and 15% of chia flour substitution presented the following nutritional parameters: protein, 3.87, 5.10, 5.97, 6.59, 7.14%; fat, 3.88, 4.46, 5.30, 3.78, 3.32%; dietary fiber, 2.79, 2.91, 8.20, 10.43, 11.72%; ash, 5.89, 6.29, 7.11, 6.54, 7.33%; calcium, 0.00, 39.75, 94.85, 110.41, 126.21%; and antioxidant activity, 1.32, 5.55, 25.66, 35.01, 44.04%, respectively. There were no significant differences in sensory acceptability between a commercial chip sample and the 5% chia chips. Chemical analysis indicated that all four trial chips were excellent sources of omega-3.

Delgado and Jáuregui (2014) evaluated the effect of the proportion of quinoa (21.8%, 24.9% and 15.7%), amaranthus (21.8%, 16.9% and 22%) and sacha inchi (5.3%, 7.1% and 11.2%) on the acceptability and nutritional profile of an energy bar. It was determined that the proportion of quinoa, amaranth and sacha inchi did not influence the general acceptability, flavor and texture of an energy bar unlike the proximate composition, where significant influence was observed, reporting that the formulation with 21.8% of quinoa, 21.8% amaranth and 5.3% saccha inchi presented higher protein (2.85%) and fiber (1.21%) and lower fat (4.37%) content than the rest of the combinations. Moreover, Verduga Verdezoto (2019) designed an energy bar with 40.00% sacha inchi, 30.00% honey, 6.66% of amaranth, quinoa, oat and 10.00% dehydrated pineapple with good sensorial acceptance. The product contained 9.74% humidity, 23.93% fat, 14.21% protein, 1.82% ashes, 13.74% fiber and 50.30% carbohydrates.

Tiwari et al. (2011) developed deep-fried snacks with nutritional benefits and an acceptable quality of cereal (rice) and legumes. The protein and fat content ranged from 13.5% to 17.5% and 23.2% to 30.0%, respectively. Patil et al (2016) prepared wheat-based extrudates using four different legume flours: lentil, chickpea, green pea, and yellow pea flour added at different levels (0%, 5%, 10%,

Figure 7.4   A. Natural sachi inchi snack, B. Sacha inchi snack with salt, C. Wasabi-flavored snack with sacha inchi, pistachio nuts, raisins and maize, D: Wheat pancakes with sacha inchi (10%), E: Vanilla flavored hydrolyzed collagen with sacha inchi. **Source: Authors' own work.**
Notes: A) Oval shaped, toasted sacha inchi seed snacks covered in salt on a small plate.

B) Toasted, salty wasabi-flavored snacks with sacha inchi seeds, pistachio nuts, raisins and maize grains, on a small plate.

C) Two large, rectangular wheat pudding cakes with sacha inchi (10%), positioned on a plate and parallel to each other.

D) Dust-like vanilla flavored hydrolyzed collagen with sacha inchi in a small bowl.

and 15%). The snack with legumes caused a slight increase in protein content but the extrusion technique increased protein digestibility by 37%–62% w/v.

Tarazona and Bocanegra Reyes (2018) developed an extruded snack based on rice (90%) and tarwi (10%) flours. The product presented great sensory acceptability and nutritional properties (13.11% of protein and 4.03% of fiber).

Some commercial snacks made with quinoa, chia, amaranth and sacha inchi are shown in Figures 7.1–7.4.

7.6   DRINKS AND FERMENTED BEVERAGES

One of the most important traditional beverages in Andean cultures is the so-called "chicha", which is a fermented or unfermented drink made from maize (Nicholson, 1960). This beverage has been produced since pre-Hispanic times in the north-west regions of Argentina, and Andean regions of Bolivia, Colombia, Ecuador and Peru (Elizaquível et al., 2015). "Chicha" has played a very important role in Andean culture as an essential part of political, social, ceremonial, and religious transactions. In addition, an ethnohistorical information report declared that fermented beverages were highly esteemed and critical for ceremonial gatherings (Hastorf, 2015; Hastorf and Johannessen, 1993). Cereal fermented products, those derived from maize, have been important in Latin America; in this case chicha is produced by handmade ancestral procedures. Chicha has been defined by Lorence-Quiñones et al. (1999) as a clear, yellowish, effervescent, alcoholic beverage prepared from maize with flavor similar to that of cider; the alcoholic content varies from 2 to 12% (v/v). When pigmented maize varieties are used, their color varies from red to purple, for example, purple maize is used to produce the traditional Peruvian drink and dessert called *chicha morada* and *mazamorra morada* (Salvador-Reyes and Clerici, 2020).

On the other hand, amaranth has also been used to produce different kinds of beverages due to its characteristics as a raw material with nutritional and therapeutic properties in the prevention and treatment of different diseases (Aderibigbe et al., 2020). In recent years, there has been an interest in functional food and beverages with beneficial physiological properties; in this sense, extruded amaranth grains have been used to develop an instant nutraceutical beverage with high antioxidant capacity, rich in dietary fiber and proteins. These can contribute to the prevention of cardiovascular disease, type-2 diabetes, among others (Milán-Carrillo et al., 2012); amaranth proteins have been used to formulate a functional high protein beverage suitable for vegans, celiac patients, and lactose intolerants (Manassero et al., 2020). An amaranth instant beverage was developed by mixing amaranth, rice and corn flour with milk whey and powder milk by spray dryer to obtain an instant beverage with nutritional and high protein properties (Arcila and Mendoza, 2006).

In addition, some mix of malted cereals and pseudo-cereals such as sorghum, buckwheat, quinoa and amaranth were used to produce gluten-free beer; the results showed high fermentability of malted amaranth suggesting that it is possible to produce gluten-free beer as an alternative for celiac people (de Meo et al., 2011).

Quinoa has also been used to produce different beverages, fermented and unfermented, gluten free and with functional properties. A beverage rich in phenolic components, high protein and high antioxidant capacity with antidiabetic and antihypertensive potential was developed by Kaur and Tanwar (2016) with germinated and malted quinoa seeds. The use of different techniques such as soaking, germination and malting can improve bio-accessibility by enzyme action, improve nutrient content and decrease the antinutrients in grains (Owusu-Mensah et al., 2011). In the same context, quinoa has also been used to produce a fermented beverage with probiotic cultures including *Bifidobacterium* sp., *Lactobacillus acidophilus*, and *Streptococcus thermophilus*, obtaining a significant increase in proteins and phenolic compounds, and producing an end-product which can increase health benefits due to the content and bio-accessibility of its bioactive compounds (Karovičová et al., 2020). A fermented quinoa-based beverage was developed by Ludena Urquizo et al. (2017) who used two quinoa varieties (Rosada de Huancayo and Pasankalla) fermented by three strains of *Lactobacillus* to produce a high protein, fiber, vitamins and mineral content and low saponins beverage. Results showed that both varieties can be used for this kind of beverage.

Tarwi has also been used to produce beverages; its proteins exhibit excellent techno-functional properties such as solubility and emulsification, which make them an attractive raw material producing plant-based beverages (Nawaz et al., 2020). In this context, tarwi has been used to produce spray-dried milk as an alternative to cows' milk for the institutional and vegetarian market in Chile; this vegetable milk showed a protein, a similar fat and ash content to cows' milk, while minerals such as calcium and phosphorus were slightly lower (Camacho et al., 1988). In addition, lupin-based milk alternatives and different exo-polysaccharides which produce lactic acid bacteria were investigated by Hickisch et al. (2016) to produce a yogurt substitute for vegetarians or vegan consumers. For this purpose, protein isolate was used to produce the milk which was thermally treated with two temperatures (80° and 140°C) and two strains of *Lactobacillus,* and one *Pediococcus* were used to produce the exo-polysaccharide. The more intensive thermal treatment resulted in better rheological and textural properties and lower syneresis tendency, which revealed that lupin proteins are appropriate for producing plant-based yogurt.

On the other hand, Sacha inchi is recognized as a good source of unsaturated lipids and bioactive compounds; in this sense, the seed has been used to produce an enriched yogurt with a significant increase in protein content, n-3 and n-6 fatty acids, tocopherols, phytosterols, phenolic compounds, dietary fiber and essential amino acid, making it a valuable resource for high value compounds to be used in food and beverages (Wang et al., 2018). Chia seed has been used to elaborate beverages since the pre-1600 period, for example, an Aztec beverage called "Chinatoles" was made with roast and ground seeds; the whole seeds were also used to prepare a refreshing drink "chia fresca" which has recently gained great popularity in Mexico (Cahill, 2003). Today, the whole chia seed and protein fractions are used in many commercial beverages mixed with fruit juices. In this context, the Company "Mamma Chia" has developed a great variety of beverages using organic seeds in the development of organic chia beverages, milks and energy beverages, as well as different kinds of squeeze pouches as protein smoothies which have been produced using protein fractions. A known company developed chia beverages with functional properties and rich in nutrients. Moreover, another company produced fruit juices with chia seeds highlighting omega-3 content. In addition, in 2009 a patent was created to develop a chia seed beverage with enhanced gastrointestinal regularity and heart health effects (Minatelli et al., 2009).

Finally, Andean and ancestral grains offer a significant nutritional composition which allows development of fermented and non-fermented beverages with distinctive sensorial properties and providing a significant contribution to the prevention of different diseases.

7.7 INFANT FOODS

According to UNICEF (2019), at least one in three children under the age of five shows signs of chronic malnutrition – undernourished or overweight – and one in two suffers from hidden hunger, impairing their ability to develop their full potential. Proteins are especially important in childhood growth and their main source is in food of animal origin. Moreover, meat and derivatives are not economically accessible for the entire population, hence the importance of looking for other sources of cheaper proteins with a good amino acid profile. In this context, native crops from Latin America play a key role as an economic, sustainable, nutritious and versatile opportunity to be used in various infant dishes and preparations.

In 2011, the Food and Agriculture Organization of the United Nations (FAO, 2011) considered quinoa as a strategic crop for food security due to its characteristics: nutritional quality – represented by its essential amino acid composition; wide genetic variability to develop varieties that improve production performance; and its adaptation to adverse climate and soil conditions (Cerezal Mezquita et al., 2012). In the same sense, kiwicha and kañiwa are foods also recognized for their high protein quality and content of calcium, phosphorus, iron, potassium, zinc, vitamin E and B complex (Ayala, 2004; Mujica and Jacobsen, 2006). Therefore, some combination of these could be an effective substitute for animal protein (Bhat and Karim, 2009).

A research work regarding the frequency of use of Andean grains in Lima – the Peruvian capital – showed that the majority of mothers (44%) stated that they used quinoa once or twice per month, 91% of them never used kiwicha and 100% of them never used kañiwa in their children's preschool diet. These interesting results show that in this case local crops are scarcely consumed; despite the necessity of improving children's diets and to provide a wide variety of dishes (Baltazar-Ñahui, 2015).

Repo-Carrasco-Valencia and Hoyos (1993) developed an infant dish for children from 5 to 24 months of age using grains of quinoa, kañiwa, kiwicha, castilla bean (*Vigna sinensis*), broad bean (*Vicia faba*) and rice (*Oriza sativa*). The protein content ranged from 11.35 to 15.46% with a good balance of essential amino acids and good acceptability. Cerezal Mezquita et al. (2011) used quinoa and lupine (*Lupinus albus* L.) mixtures with traditional cereals such as corn and rice, obtaining gluten-free products with higher protein quality for children under 24 months of age who suffer from celiac disease. In Peru, an instant protein mixture was developed with yellow maize, red quinoa, kiwicha and lentils, as an easy-to-prepare option to include in infant diets (Huamaní-Condori, 2019). Cerezal Mezquita et al. (2007) developed a similar food, where quinoa and tarwi were combined with the aim of obtaining a high protein formulation with additives and chicken flavoring. In this case, a powder is reconstituted by adding water to obtain porridge for children from 2 to 5 years.

In Argentina, Jiménez et al. (2020) studied quinoa and amaranth flours from germinated and non-germinated grains to develop baby purees, obtaining products with good sensory acceptance. Furthermore, the nutritional and rheological properties achieved improved protein digestibility and an appropriate texture for swallowing, both characteristics which may be taken into consideration during infant food development.

Biscuits are snacks massively consumed by the school population, usually made with refined wheat flour. In Ecuador, Gaibor-Monar al. (2016) developed biscuits based on quinoa and amaranth flours – to replace wheat flour. The utilization of these two regional crops improved their nutritional quality, since these Andean grains have around 16% of protein with an excellent amino acids balance and some minerals such as iron and calcium. In addition, due to their versatility, they can be used in products such as desserts, popcorn and other sweets with minimal loss of nutrients during cooking.

Improving food quality in vulnerable sectors, such as childhood, involves a great challenge because of the high cost of healthy food, lack of correct nutritional information and constant exposure of many children, adolescents and families to ultra-processed food advertising.

7.8   TRADITIONAL FOODS

Food consumption is one of the most important acts in human life. Beyond its nutritional message, food is framed by social, symbolic and material aspects of human culture. According to UNESCO (2003), traditional gastronomy constitutes an element of identity, social cohesion and cultural distinction. Regional traditional cuisines constitute part of the intangible heritage of communities, and set the background where those foods and culinary knowledge and practices are transmitted to following generations, reinforcing their cultural identity (Meléndez Torres and Cañez de la Fuente, 2009).

Besides the cultural role of Latin-American crops, their contribution in daily nutrition and the possible economic impact in the region are fundamental factors. There is enough scientific evidence that favors the total or partial replacement of crops such as rice and wheat or its complement with these crops (Campos et al., 2018). Therefore, talking about and vindicating dishes with native crops is of the utmost importance to maintaining traditional culture and improving nutrition and local economies.

Maize – with its wide spectrum of colors and shapes – is one of the crops with the greatest gastronomic and cultural preponderance, occupying a relevant place in food, as well as in ancestral ritual (Abarza and Schimpf, 2009). Probably one of the most important cultural roles of dishes prepared with maize is as an offering, presented in front of the divinity (Mazzetto, 2013). Maize is a staple food in Mexican and Central-American diets, and has moved throughout South America, where it is consumed in a wide range of traditional dishes such as "tamales" and "arepas" in Colombia and Venezuela, "indio Viejo" and "guirilas" in Nicaragua, "humitas" in Bolivia and Chile, and "locro" in Argentina (Tanumihardjo et al., 2020).

In pre-Hispanic times, amaranth was used in religious ceremonies. According to Mazzetto (2013), to celebrate the feast of Xiuhtecuhtli – God of Fire – "tamales" were prepared with amaranth leaves, since it was conceived of as the body of the gods. Its cultivation and consumption were almost eradicated during the viceroyalty due to the evangelization process (Guerrero-Jacinto and Viesca-González, 2015).

Quinoa was cultivated and used by pre-Inca civilizations and its expansion was consolidated with the Inca Empire, extending from Colombia to Chile and Argentina, and was replaced by cereals upon the arrival of the Spanish, despite constituting a staple food of the population at that time. In more recent decades, quinoa has achieved worldwide renaming, but its regional production is mainly exported. Local populations – especially autochthonous – usually prefer to sell their quinoa production and buy cheaper commodities, such as wheat or rice. Therefore, its local consumption is reserved for special occasions and dishes, such as those mentioned in Table 7.2.

TABLE 7.2   LATIN-AMERICAN CROPS AND THEIR TYPICAL DISHES

Crop	Typical dish	History/origin	Consumption/ preparation
Maize	Tortilla	Mexico. Its origin is debated, probably from the Teotihuacan valley 1500 BC–AQ 500.	It is consumed filled with meats, vegetables, with cheese. Multiple variants.
Maize flour	Tamal o hallaca	Mesoamerican origin. From Mexico to northern Argentina. Used in rituals and tombs as an offering.	Stuffed with vegetables, meat, seasonings, It is wrapped in the same corn or banana leaves.
Maize dough	Tlacoyos	Pre-Hispanic origin, from Nahuatl tlaoyo Mexico	Thick oval dough filled with beans, cheese, chili and onion.
Whole maize	Pozole	Mexico. Pre-Hispanic origin, prepared by the Mexican to venerate the god Xipe Tótec.	In ancient times it was made with human flesh and blood.
Maize dough	Humitas	From south of Colombia to Argentina From quechua: humint'a	Dough made from crushed maize kernels, with goat's cheese and seasonings, wrapped in its green husk
Maize dough	Maize cake	From south of Colombia to Argentina	Similar to humita, with egg, sugar and cooked in oven
Maize flour	Tulpo	Argentinean northwest and Bolivia	Thick soup of cornmeal, sheep fat, vegetables and jerky (ground in a mortar).

TABLE 7.2 (CONTINUED)

Crop	Typical dish	History/origin	Consumption/ preparation
Whole maize	Locro	Argentinean, Chilean and Bolivian norwest	Stew with grated corn, meat, vegetables and seasonings.
Kernels of different colors	Tijtincha	Argentinean norwest and Bolivia. Consumed especially on August 1st, *Pachamama Day*	Cob boiled overnight until grains are soft, then added to a beef, lamb, or chicken broth, with vegetables
Maize	Calapurca	Argentina, Chile and Bolivia From Aymara qala phurk'a, qala, 'stone', and phurk'a, 'roast', 'on the embers.	Corn stew with mote and beef or llama that is cooked on hot stones
White or purple maize (also with quinoa/ cañihua)	Mazamorra	Colombia, Argentina, Peru. Widely consumed in Argentina during the Viceroyalty period, until the beginning of the 20th century. In the Caribbean region, bananas are added.	Dessert with milk, sugar and vanilla or cinnamon, cloves, raisins, plums and fruits such as pineapple and peach.
Amaranth	Alegría (Figure 7.3.D)	Mexico. Dish declared intangible heritage of the city of Santiago de Tulyehualco.	Honey, peanuts and raisins.
Red and black amaranth	Olli (cakes)	Mexico. They were made to venerate Xochipilli God of the game.	They were made with different types of amaranth that gave it a different color.
Amaranth	Gachas	Mexico, Peru	With cinnamon, yogurt or milk for breakfast.
Quinoa	Quínoa puree with potatoes	Peru.	It is a traditional garnish
Quinoa	Chaulafan	Ecuador. Local variation of Chinese fried rice. The Peruvian equivalent is known as *chaufa*.	Rice is replaced by quinoa.

(continued)

TABLE 7.2   (CONTINUED)

Crop	Typical dish	History/origin	Consumption/preparation
Kañihua flour	Pito de kañihua	Bolivia, Peru	Drink consumed for breakfast. Sweetened with honey, cinnamon, and cloves.
Tarwi	Ají de tarwi	Ecuador	Spicy preparation to dress *locro*, stews, empanadas, meats.
Tarwi	Ceviche de tarwi	Ecuador. Typical dish of the coastal area	It is mixed with fish and garnished with citrus fruits such as lime, orange.
Chía seeds	Chianpinolli	Mexico, Guatemala, El Salvador	Dense flour that is later used as a raw material for breads, dough, etc.
Chía flour	Chapatas	Mexico, Guatemala, El Salvador	Thin bread
Chía flour	Pinole	Mexico, Guatemala, El Salvador	To prepare energy drinks, combined with cinnamon and honey
Toasted chía seeds	Atole	Mexico	Hot and thick drink made from corn and chia seeds. In some places, fruits and cinnamon were added.

Source: Authors' own work.

Kañiwa is a less extended crop – not spread beyond Bolivia and Peru – with qualities that make it a nutritional support when others fail. On the other hand, tarwi is used in a wide variety of dishes ranging from purees, sauces and soups to main dishes such as stews and a version of ceviche, as well as in desserts, and tarwi flour can even be used in popular drinks (Mikuy and Mikuy, 2013).

Chia crops have been cultivated in Mexico and Guatemala since pre-Columbian times as part of the basic diet along with corn and beans. Aztec warriors used to be rewarded with bags of chia seeds for their military exploits and for their widows for their loss on the battlefields. Traditionally, they were added to drinks, as a thickener and due to their fiber content (McQuown, 1958).

7.9 HAUTE CUISINE

Not all ancestral crops have earned a position in haute cuisine restaurants, only some of them have achieved that recognition, but that is exclusively due the "celebrity" chefs who have seen the great potential that these grains have. Since ancestral seeds have been recognized for their contribution to nutrition, in addition to their important characteristics in terms of disease prevention, many of them have begun to be used in culinary preparations and are beginning to be desired by many guests in the most prestigious restaurants around the world. An example would be Ferran Adriá, the Spanish chef with a three-star restaurant "elBulli" who has used quinoa in his preparations called Q*uinoa gelada de foiegras d´anec amb consomé* which is frozen duck foie gras quinoa with consommé (Svejenova et al., 2007). On the other hand, elite chefs from Perú have constructed the *haute* Peruvian *cuisine* redefining the *cuisine* and positioning themselves as national and global culinary leaders; examples of these are Gaston Acurio and Virgilio Martinez who use different grains from Andean highlands in their dishes (McDonell, 2019).

Haute cuisine is also used in hospital food services which is significantly correlated with overall patient satisfaction (Cox, 2006). In this context, chefs are using ancestral seeds to create interesting dishes, for example bulgur wheat salad with berries, quinoa with roasted chicken and multigrain pancake. On the other hand, the magazine *Time Out* from Mexico highlights some *haute cuisine* restaurants that are using ancestral seeds to produce elaborate dishes such as "amaranth tamal coated in salsa verde and quelites" (Time Out Magazine, 2016). In Seville, Spain, the recognized restaurant Abantal (one Michelin star), uses chia seed, amarant and quinoa for its preparations (Figure 7.5).

7.10 CONCLUSIONS

Latin-American crops and especially grains, such as cereals and pseudo-cereals are known to be a source of nutrients and bioactive compounds that contribute to the prevention of many diseases; these outstanding food sources had been part of the diet of Latin-American cultures throughout history. In recent decades and thanks to their recognized nutritional and functional contribution, they have been reintroduced and today are cultivated and used as a source of food and ingredients for the elaboration of countless products not only in the Latin-American region but around the world. This chapter describes food components, ingredients, and products with the aim of opening, and reopening, additional alternatives for these crops, which may be classified as superfoods.

Figure 7.5   A. Quinoa dish; B. Amaranth-based preparation with *haute cuisine*.
Notes: A) Quinoa dish composed of a round, soft orangey, flat-disk base with cooked quinoa inside the disk. On top of the quinoa, larger (than quinoa) orange jelly-like balls are found and then on top of them a white foam-like larger ball. Decorated with green leaves. **Source: Courtesy of Julio Fernández, ABANTAL Restaurant, Sevilla, Spain.**

   B) Amaranth dish with a foam-like, flat-disk base, acting as a bowl, where cooked amaranth is inside. On top of the quinoa rests a small square shaped type of fish.

ACKNOWLEDGMENTS

Authors would like to express their gratitude to the CYTED Program through Project119RT0567, FONDECYT Project 1201489 from the National Agency for Research and Development (ANID, Chile) and Food4ImNut Food4ImNut PID2019-107650RB-C21 funded by MCIN/AEI/10.13039/501100011033, Spain for its kind support.

REFERENCES

Abarza, S., and J.H. Schimpf. 2009. Los maíces indígenas en la Cultura alimenticia Andina. *Ciencia, Universidad de Jujuy* 4(9): 159–170.

Aderibigbe, O. R., O.O. Ezekiel, S.O. Owolade, J.K. Korese, B. Sturm, and O. Hensel. 2020. Exploring the potentials of underutilized grain amaranth (*Amaranthus* spp.) along the value chain for food and nutrition security: A review. *Critical Reviews in Food Science and Nutrition* 62(3): 656–669.

Ahmed, I.B.H., A. Hannachi, and C.M. Haros. 2020. Combined effect of Chia flour and soya lecithin incorporation on nutritional and technological quality of fresh bread and during staling. *Foods* 9(4): 446.

Albuja Vaca, D. E., and C.A. Yépez Vizuete. 2017. Desarrollo de pasta de arroz libre de gluten con sustitución parcial por harina de chocho (*Lupinus Mutabilis* Sweet) a través de un diseño experimental de proceso-mezclas. Bachelor's thesis, Quito: USFQ. http://repositorio.usfq.edu.ec/handle/23000/6408.

Alcantara Brandao N., M. Boges de Lima Dutra, A.L. Andrade Gaspardi, and M.R. Segura-Campos. 2019. Chia (*Salvia hispanica* L.) cookies: Physicochemical/microbiological attributes, nutrimental value and sensory analysis. *Journal of Food Measurement and Characterization* 13(2): 1100–1110.

Aranda Tarazona, J. J., and G.I. Bocanegra Reyes. 2018. Evaluación de parámetros durante la extrusión de una mezcla de harinas de tarwi (*Lupinus mutabilis*) y arroz (*Orysa sativa*) para la producción de un snack. http://repositorio.uns.edu.pe/handle/UNS/3052.

Aranibar, C., N. B. Pigni, M. Martínez, A. Aguirre, P. Ribotta, D. Wunderlin, and R. Borneo. 2018. Utilization of a partially-deoiled chia flour to improve the nutritional and anti-oxidant properties of wheat pasta. *LWT – Journal of Food Science and Technology* 89: 381–387.

Arcila, N., and Y. Mendoza. 2006. Elaboration of an instant beverage of amaranth seeds (*Amaranthus cruentus*) and its potential use in the human diet. *Revista de la Facultad de Agronomía* 23(1): 114–124.

Ayala, G. 1998. Aporte de los cultivos andinos a la nutrición humana. In *Raíces Andinas: Contribuciones al conocimiento y a la capacitación*, 101–112). Lima: Universidad Nacional Mayor de San Marcos.

Ayala, G. 2004. Aporte de los cultivos andinos a la nutrición humana. In *Raíces Andinas: Contribuciones al conocimiento y a la capacitación*, 2nd edn., 101–112). Lima: Universidad Nacional Mayor de San Marcos.

Ayerza, R., and W. Coates. 2005. *Chia: Rediscovering a Forgotten Crop of the Aztecs*. Tucson, AZ: University of Arizona Press.

Bahnassey, Y., and K. Khan. 1986. Fortification of spaghetti with edible legumes. II. Rheological, processing, and quality evaluation studies. *Cereal Chemistry* 63(3): 216–219.

Ballester-Sánchez, J., J.V. Gil, C.M. Haros, and M.T. Fernández-Espinar. 2019a. Effect of incorporating white, red or black quinoa flours on the total polyphenol content, antioxidant activity and colour of bread. *Plant Food for Human Nutrition* 89: 837–843.

Ballester-Sánchez, J., E. Yalcin, M.T. Fernández-Espinar, and C.M. Haros. 2019b. Rheological and thermal properties of royal quinoa and wheat flour blends for breadmaking. *European Food Research and Technology* 254: 1571–1582.

Ballester-Sánchez J., M.C. Millán-Linares, M.T. Fernández-Espinar, and C.M. Haros. 2019c. Development of healthy, nutritional bakery products by incorporation of quinoa. *Foods* 8(9): pii: E379.

Ballester-Sánchez J., M.T. Fernández-Espinar, and C.M. Haros. 2020. Isolation of red quinoa fiber by wet- and dry-milling and application as a potential functional bakery ingredient. *Food Hydrocolloids* 101: 105513.

Baltazar-Ñahui, R. 2015. *Conocimientos, actitudes y prácticas sobre uso de granos andinos en la alimentación del preescolar de madres en una institución educativa*. Thesis, Universidad Nacional Mayor de San Marcos, Lima, Perú. Retrieved from https://cyb ertesis.unmsm.edu.pe/bitstream/handle/20.500.12672/5443/Baltazar_%c3%b1r. pdf?sequence=1&isAllowed=y.

Banerji, A., L. Ananthanarayan, and S. Lele. 2018. Rheological and nutritional studies of amaranth enriched wheat chapatti (Indian flat bread). *Journal of Food Processing and Preservation* 42(1): e13361.

Bautista-Justo M., A.D. Castro Alfaro, E. Camarena Aguilar, K. Wrobel, K. Wrobel, G.A. Guzmán, Z. Gamiño Sierra, and V. da Mota Zanella. 2007. Integral bread development with soybean, chia, linseed, and folic acid as a functional food for women. *Archivos Latinoamericanos de Nutrición* 57(1): 78–84.

The Bean Institute. 2014. Dry beans commonly consumed (by nationality). Retrieved from https://beaninstitute.com/.

Benítez V., R.M. Esteban, E. Moniza, N. Casado, Y. Aguilera, and E. Molla. 2018. Breads fortified with wholegrain cereals and seeds as source of antioxidant dietary fibre and other bioactive compounds. *Journal of Cereal Science* 82: 113–120.

Bergesse, A. E., P.N. Boiocchi, E.L. Calandri, N.S. Cervilla, V. Gianna, C.A. Guzmán, P. Miranda, P. Montoya, and J.R. Mufari. 2015. Aprovechamiento integral del grano de Quinoa. Aspectos tecnológicos, fisioquímicos, nutricionales y sensoriales. http://hdl.handle.net/11086/1846.

Bhat, R., and A.A. Karim. 2009. Exploring the nutritional potential of wild and underutilized legumes. *Comprehensive Reviews in Food Science and Food Safety* 8(4): 305–331.

Bilgiçli N., and S. İbanoğlu. 2015. Effect of pseudo cereal flours on some physical, chemical and sensory properties of bread. *Journal of Food Science and Technology* 52: 7525–7529.

Bodroza-Solarov M., B. Filipcev, Z. Kevresan, A. Mandic, and O. Simurina. 2008. Quality of bread supplemented with popped *Amaranthus cruentus* grain. *Journal of Food Process Engineering* 31: 602–618.

Borneo R., A. Aguirre, and A.E. León. 2010. Chia (*Salvia hispanica* L) gel can be used as egg or oil replacer in cake formulations. *Journal of the American Dietetic Association* 110(6): 946–949.

Bouasla, A., A. Wójtowicz, and M.N. Zidoune. 2017. Gluten-free precooked rice pasta enriched with legumes flours: Physical properties, texture, sensory attributes and microstructure. *LWT – Journal of Food Science and Technology* 75: 569–577.

Brennan C. S., and E. Samyue. 2004. Evaluation of starch degradation and textural characteristics of dietary fibre enriched biscuits. *International Journal of Food Properties* 7: 647–657.

Bustos, M. C., M.I. Ramos, G.T. Pérez, and A.E. Leon. 2019. Utilization of Kañawa (*Chenopodium pallidicaule* Aellen) flour in pasta making. *Journal of Chemistry* 2019: 4385045, 1–8. https://doi.org/10.1155/2019/4385045.

Cahill, J. P. 2003. Ethnobotany of chia, *Salvia hispanica* L. (Lamiaceae). *Economic Botany* 57(4): 604–618.

Calderelli A. V. S., M. de Toledo Benassi, J.V. Visentainer, and G. Matioli. 2010. Quinoa and flaxseed: Potential ingredients in the production of bread with functional quality. *Brazilian Archives of Biology and Technology – An International Journal* 53: 981–986.

Calisto Guzmán, L. A. 2009. Desarrollo de producto snack a base de materias primias no convencionales: poroto (*Phaseolus vulgaris* L.) y quinua (*Chenopodium quinoa* Wild). www.repositorio.uchile.cl/handle/2250/105325.

Camacho, L., M. Vasquez, M. Leiva, and E. Vargas. 1988. Effect of processing and methionine addition on the sensory quality and nutritive value of spray-dried lupin milk. *International Journal of Food Science & Technology* 23(3): 233–240.

Campos, D., R. Chirinos, L. Gálvez Ranilla, and R. Pedreschi. 2018. Chapter Eight - Bioactive potential of andean fruits, seeds, and tubers. In F. Toldrá (Ed.), *Advances in Food and Nutrition Research*, edited by F. Toldrá (Vol. 84, 287–343). New York: Academic Press.

Cárdenas-Hernández, A., T. Beta, G. Loarca-Piña, E. Castaño-Tostado, J.O. Nieto-Barrera, and S. Mendoza. 2016. Improved functional properties of pasta: Enrichment with amaranth seed flour and dried amaranth leaves. *Journal of Cereal Science* 72: 84–90.

Carvajal-Larenas, F.E. 2019. Nutritional, rheological and sensory evaluation of *Lupinus mutabilis* food products - A review. *Czech Journal of Food Sciences* 37(5): 301–331.

Carvajal-Larenas F.E., A.R. Linnemann, M.J.R. Nout, M. Koziol, and M.A.J.S. Van Boekel. 2016. *Lupinus mutabilis*: Composition, uses, toxicology, and debittering. *Critical Reviews in Food Science and Nutrition* 56:1454–1487.

Carvalho de Almeida M. M., C.R.F. Lopes, A. de Oliveira, S.S. de Campos, A.P. Bilck, R. Hernandez Barros Fuchs, O. Hess Gonçalves, P. Velderrama, A. Kamal Genena, and F.V. Leimann. 2018. Textural, color, hygroscopic, lipid oxidation, and sensory properties of cookies containing free and microencapsulated chia oil. *Food and Bioprocess Technology* 11(5): 926–939.

Cerezal Mezquita, P., A. Carrasco Verdejo, K. Pinto Tapia, N. Romero Palacios, and R.Arcos Zavala. 2007. Suplemento alimenticio de alto contenido proteico para niños de 2- 5 años: Desarrollo de la formulación y aceptabilidad. *Interciencia* 32: 857–864. Retrieved from http://ve.scielo.org/scielo.php?script=sci_arttext&pid=S0378-18442007001200013&nrm=iso.

Cerezal Mezquita, P., V. Urtuvia Gatica, V. Ramírez Quintanilla, and R. Arcos Zavala. 2011. Desarrollo de producto sobre la base de harinas de cereales y leguminosa para niños celíacos entre 6 y 24 meses; II: Propiedades de las mezclas. *Nutrición Hospitalaria* 26: 61–169. Retrieved from http://scielo.isciii.es/scielo.php?script= sci_arttext&pid=S0212-16112011000100019&nrm=iso.

Cerezal Mezquita, P., E. Acosta Barrientos, G. Rojas Valdivia, N. Romero Palacios, and R. Arcos Zavala. 2012. Desarrollo de una bebida de alto contenido proteico a partir de algarrobo, lupino y quinoa para la dieta de preescolares. *Nutrición Hospitalaria* 27: 232–243. Retrieved from http://scielo.isciii.es/scielo.php?script=sci_artt ext&pid=S0212-16112012000100030&nrm=iso.

Chávez-Jáuregui, R. N., M.E.M.P. Silva, and J.A. Arêas. 2000. Extrusion cooking process for amaranth (*Amaranthus caudatus* L.). *Journal of Food Science* 65(6): 1009–1015.

Coda R., C.G. Rizzello, and M. Gobbetti. 2010. Use of sourdough fermentation and pseudo-cereals and leguminous flours for the making of a functional bread enriched of γ-aminobutyric acid (GABA). *International Journal of Food Microbiology* 137: 236–245.

Coorey, R., A. Grant, and V. Jayasena. 2012. Effect of chia flour incorporation on the nutritive quality and consumer acceptance of chips. *Journal of Food Research* 1: 85–95.

Cox, S.A. 2006. Improving hospital foodservice. *Food Technology Magazine* 60(6). www. ift.org/news-and-publications/food-technology-magazine/issues/2006/june/featu res/improving-hospital-foodservice.

Dallagnol A.M., M. Pescuma, G.F. de Valdez, and G. Rollán. 2013. Fermentation of quinoa and wheat slurries by Lactobacillus plantarum CRL 778: proteolytic activity. *Applied Microbiology and Biotechnology* 97: 3129–40.

Dallagnol A.M., M. Pescuma, G. Rollán, M.I. Torino, and G.F. de Valdéz. 2015. Optimization of lactic ferment with quinoa flour as bio-preservative alternative for packed bread. *Applied Microbiology and Biotechnology* 99: 3839–3849.

de Meo, B., G. Freeman, O. Marconi, C. Booer, G. Perretti, and P. Fantozzi. 2011. Behaviour of malted cereals and pseudo-cereals for gluten-free beer production. *Journal of the Institute of Brewing* 117(4): 541–546.

de Sousa, M. F., R.M. Guimarães, M. de Oliveira Araújo, K.R. Barcelos, N.S. Carneiro, D.S. Lima, ... and M.B. Egea. 2019. Characterization of corn (*Zea mays* L.) bran as a new food ingredient for snack bars. *Food Science and Technology* 101: 812–818.

Deepa, C., S. Sarabhai, P. Prabhasankar, and H.U. Hebbar. 2017. Effect of micronization of maize on quality characteristics of pasta. *Cereal Chemistry* 94(5): 840–846.

Delgado, L., and G. Barraza. 2014. Efecto de la proporción de *Chenopodium quinoa* (quinua), *Amaranthus caudatus* (kiwicha) y *Plukenetia volubilis* l. (sacha inchi) en la aceptabilidad general y el análisis proximal de una barra energética. *Cientifi-k* 2(2): 56–70.

Demir, B., and N. Bilgiçli. 2020. Changes in chemical and anti-nutritional properties of pasta enriched with raw and germinated quinoa (*Chenopodium quinoa* Willd.) flours. *Journal of Food Science and Technology* 57: 3884–3892.

Demir, M.K., and M. Kilinc. 2017. Utilization of quinoa flour in cookie production. *International Food Research Journal* 24(6): 2394–2401.

Dias-Capriles V., E. Lopes-Almeida, R.E. Ferreira, J.A. Gomes-Arêas, C. Joy-Steel, and Y. Kil Chang. 2008. Physical and sensory properties of regular and reduced-fat pound cakes with added amaranth flour. *Cereal Chemistry* 85: 614–618.

EFSA (European Food Safety Authority). 2009. Opinion on the safety of 'Chia seeds (*Salvia hispanica* L.) and ground whole Chia seeds' as a food ingredient. *EFSA Journal* 996: 1–26.

Elizaquível, P., A. Pérez-Cataluña, A. Yépez, C. Aristimuño, E. Jiménez, P. S. Cocconcelli, ... and R. Aznar. 2015. Pyrosequencing vs. culture-dependent approaches to analyze lactic acid bacteria associated to chicha, a traditional maize-based fermented beverage from Northwestern Argentina. *International Journal of Food Microbiology* 198: 9–18.

EU (European Union). 2020. Authorising an Extension of Use of Chia Seeds (*Salvia hispanica*) as a Novel Food and the Change of the Conditions of Use and the Specific Labelling Requirements of Chia Seeds (*Salvia hispanica*) under Regulation (EU) 2015/2283 of the European Parliament and of the Council and Amending Commission Implementing Regulation (EU) 2017/2470; *Official Journal of the European Union* 12–16, Brussels.

FAO (Food and Agriculture Organization/Oficina Regional para America Latina y el Caribe). 2011. *La quinoa: cultivo milenario para contribuir a la seguirdad alimentaria mundial.* Rome: FAO.

Ferrari Felisberto M. H., A.L. Wahanik, C. Rodrigues Gomes-Ruffi, M.T. Pedrosa Silva Clerici, Y. Kil Chang, and C. Joy Steel. 2015. Use of chia (*Salvia hispanica* L.) mucilage gel to reduce fat in pound cakes. *LWT – Food Science and Technology* 63(2): 1049–1055.

Fiorda, F. A., M.S. Soares Jr, F.A. da Silva, M.V. Grosmann, and L.R. Souto. 2013. Microestructure, texture and colour of gluten-free pasta made with amaranth flour, cassava starch and cassava bagasse. *LWT – Food Science and Technology* 54(1): 132–138.

Gaibor-Monar, F. M., J.P. Torres-Cadena, and L.V. Yépez-Martínez. 2016. Valor nutricional de las galletas a base de amaranto y quinua asociado a la aceptabilidad microbiológica. *Revista Caribeña de Ciencias Sociales* 12(8): 1–20.

Galluzzo F.G., G. Cammilleri, L. Pantano, G. Lo Cascio, A. Pulvirenti, A. Macaluso, A. Vella, and V. Ferrantelli. 2021. Acrylamide assessment of wheat bread incorporating chia seeds (*Salvia hispanica* L.) by LC-LM/MS. Food additives and contaminants Part A - Chemistry analysis control exposure & risk assessment. DOI: 10.1080/19440049.2020.1853823.

García-Mantrana I., V. Monedero, and M. Haros. 2014. Application of phytases isolated from bifidobacteria in the development of cereal-based products with amaranth. *European Food Research and Technology* 238: 853–862.

García-Mantrana I., V. Monedero, and M. Haros. 2015. Myo-inositol hexakisphosphate degradation by *Bifidobacterium pseudocatenulatum* ATCC 27919 improves mineral availability of high fibre rye-wheat sour bread. *Food Chemistry* 178: 267–275. doi: 10.1016/j.foodchem.2015.01.099.

Gearhart, C., and K.A. Rosentrater. 2014. Extrusion processing of amaranth and quinoa. Paper No. 1 41912019, presented at ASABE and CSBE/SCGAB Annual International Meeting, Montreal, PQ, Canada July 13–July 16, 2014, 1–36. St. Joseph, MI: American Society of Agricultural and Biological Engineers.

Giménez, M. A., A. Gámbaro, M. Miraballes, A. Roascio, M. Amarillo, N. Sammán, and M. Lobo. 2015. Sensory evaluation and acceptability of gluten-free Andean corn spaghetti. *Journal of the Science of Food and Agriculture* 95(1): 186–192.

Giménez, M. A., S.R. Drago, M.N. Bassett, M.O. Lobo, and N.C. Sammán. 2016. Nutritional improvement of corn pasta-like product with broad bean (*Vicia faba*) and quinoa (*Chenopodium quinoa*). *Food Chemistry* 199: 150–156.

González Rojas, M. J., and A.C. Moya Valenzuela. 2004. Producción y comercialización en Bogotá de hojuelas de quinua empacadas. http://hdl.handle.net/10554/7089.

Goyat Y., S.J. Passi, S. Suri, and H. Dutta 2018. Development of chia (*Salvia hispanica*, L.) and quinoa (*Chenopodium quinoa*, L.) seed flour substituted cookies - physico-chemical, nutritional and storage studies. *Current Research in Nutrition and Food Science* 6(3): 757–769.

Güemes-Vera N., R.J. Peña-Bautista, C. Jiménez-Martínez, G. Dávila-Ortiz, and G. Calderón-Domínguez. 2008. Effective detoxification and decoloration of *Lupinus mutabilis* seed derivatives, and effect of these derivatives on bread quality and acceptance. *Journal of the Science of Food and Agriculture* 88: 1135–1143.

Guerrero Jacinto, M. and F. Viesca González. 2016. *Recetario de Santiago Tulyehualco, El Amaranto Como Patrimonio Gastronómico*. Benito Juárez, Mexico: Editorial Época.

Guiotto E.N., M.C. Tomás, and C.M. Haros. 2020. Development of high nutritional breads with by-products of chia (*Salvia hispanica* L.) seeds. *Foods* 9(6): 819.

Hastorf, C. A. 2015. Andean luxury foods: Special food for the ancestors, deities and the élite. *Antiquity* 77(297): 545–554.

Hastorf, C. A., and S. Johannessen. 1993. Pre-Hispanic political change and the role of maize in the Central Andes of Peru. *American Anthropologist* 95(1): 115–138.

Herndandes Venturini, L., T. Fernades Moya Moreira, T. Barlati Vieira da Silva, M.M. Carvalho de Almeida, F.C.R. Lopes, A. de Oliveira, S. Silva de Campos, A.P. Bilck, R. de Souza Leone, A.A. Coelho Tanamati, O. Hess Gonçalves, F.V. Leimann 2019. Partial substitution of margarine by microencapsulated chia seeds oil in the formulation of cookies. *Food and Bioprocess Technology* 12(1): 77–87.

Hickisch, A., R. Beer, R.F. Vogel, and S. Toelstede. 2016. Influence of lupin-based milk alternative heat treatment and exopolysaccharide-producing lactic acid bacteria on the physical characteristics of lupin-based yogurt alternatives. *Food Research International* 84: 180–188.

Houben A., H. Götz, M. Mitzscherling, and T. Becker. 2010. Modification of the rheological behavior of amaranth (*Amaranthus hypochondriacus*) dough. *Journal of Cereal Science* 51: 350–356.

Huamaní-Condori, J. 2019. *Desarrollo de una mezcla instantánea proteica a partir de maíz amarillo (Zea mays indurata St.), quinua roja (Chenopodium quinoa Willd), kiwicha (Amaranthus caudatus) y lenteja (Lens culinaris)*. Engineering Thesis. Universidad Peruana Unión, Lima, Perú. Retrieved from http://repositorio.upeu.edu.pe/handle/UPEU/2034.

Iglesias-Puig, E., and Haros, M. 2013. Evaluation of performance of dough and bread incorporating chia (*Salvia hispanica* L.). *European Food Research and Technology* 237(6): 865–874.

Iglesias-Puig E., V. Monedero, and M. Haros. 2015. Bread with whole quinoa flour and bifidobacterial phytases improve contribution to dietary mineral intake and their bioavailability without substantial loss of bread quality. *LWT – Food Science and Technology* 60: 71–77.

Izquierdo, J., and W. Roca. 1998. *Under-Utilized Andean Food Crops: Status and Prospects of Plant Biotechnology for the Conservation and Sustainable Agricultural Use of Genetic Resources*. Paper presented at the Symposium on Plant Biotechnology as A Tool for the Exploitation of Mountain Lands, Leuven, Belgium. www.fao.org/tempref/GI/Reserved/FTP_FaoRlc/old/prior/recnat/recursos/biodiv/andinos.pdf.

Jekle M., A. Houben, M. Mitzscherling, and T. Becker. 2010. Effects of selected lactic acid bacteria on the characteristics of amaranth sourdough. *Journal of the Science of Food and Agriculture* 90: 2326–2332.

Jiménez, D., M. Miraballes, A. Gámbaro, M. Lobo, and N. Sammán. 2020. Baby purees elaborated with andean crops. Influence of germination and oils in physico-chemical and sensory characteristics. *LWT – Food Science and Technology* 124: 108901.

Karovičová, J., Z. Kohajdová, M. Lauková, L. Minarovičová, M. Greifová, J. Hojerová, and G. Greif. 2020. Utilisation of quinoa for development of fermented beverages. *Potravinarstvo Slovak Journal of Food Sciences* 14: 465–472.

Kaur, I., and B. Tanwar. 2016. Quinoa beverages: Formulation, processing and potential health benefits. *Romanian Journal of Diabetes Nutrition and Metabolic Diseases* 23(2): 215–225.

Kohajdová Z., J. arovičováand, and Š. Schmidt. 2011. Lupin composition and possible use in bakery – A review. *Czech Journal of Food Science* 29(3): 203–211.

Laparra J. M., and M. Haros. 2016. Inclusion of ancient Latin-American crops in bread formulation improves intestinal iron absorption and modulates inflammatory markers. *Food & Function* 7(2): 1096–1102.

Laparra J. M., and M. Haros. 2018. Inclusion of whole flour from Latin-American crops into bread formulations as substitute of wheat delays glucose release and uptake. *Plant Foods for Human Nutrition* 73(1): 13–17.

Lemos A. R., V.D. Capriles, M.E. Machado Pinto e Silva, and J.A. Gomes Arêas. 2012. Effect of incorporation of amaranth on the physical properties and nutritional value of cheese bread. *Ciencia e Tecnología de Alimentos* 32: 427–431.

Levent, H. 2017. Effect of partial substitution of gluten-free flour mixtures with chia (*Salvia hispanica* L.) flour on quality of gluten-free noodles. *Journal of Food Science and Technology* 54(7): 1971–1978.

Lorence-Quiñones, A., C. Wacher-Rodarte, and R. Quintero-Ramírez. 1999. Cereal fermentations in Latin American countries. In *Fermented Cereals. A Global Perspective*, 99–114). Rome: FAO Agricultural Services Bulletin.

Lorenz K., and L. Coulter. 1991. Quinoa flour in baked products. *Plant Foods Human Nutrition* 41: 213–224.

Lorusso, A., M. Verni, M. Montemurro, R. Coda, M. Gobbetti, and C.G. Rizzello. 2017. Use of fermented quinoa flour for pasta making and evaluation of the technological and nutritional features. *Food Science and Technology* 78: 215–221.

Ludena Urquizo, F. E., S.M. García Torres, T. Tolonen, M. Jaakkola, M.G. Pena-Niebuhr, A. von Wright, … and C. Plumed-Ferrer. 2017. Development of a fermented quinoa-based beverage. *Food Science & Nutrition* 5(3): 602–608.

McDonell, E. 2019. *Reproducing "Indian Food": Race, Value, and Development in Peru's Quinoa Boom-Bust*. Bloomington, IN: Indiana University Press.

McQuown, N. A. 1958. The general history of the things of New Spain, by Bernardino de Sahagún. *The Hispanic American Historical Review* 38(2): 235–238.

Manassero, C. A., M.C. Añón, and F. Speroni. 2020. Development of a high protein beverage based on amaranth. *Plant Foods for Human Nutrition* 75(4): 599–607.

Mas A. L., F.I. Brigante, E. Salvucci, N.B. Pigni, M.L. Martinez, P. Ribotta, D.A. Wunderlin, and M.V. Baroni. 2020. Defatted chia flour as functional ingredient in sweet cookies. How do processing, simulated gastrointestinal digestion and colonic fermentation affect its antioxidant properties? *Food Chemistry* 316: 126279.

Mastromatteo, M., S. Chillo, M. Iannetti, V. Civica, and M.A. Del Nobile. 2011. Formulation optimisation of gluten-free functional spaghetti based on quinoa, maize and soy flours. *International Journal of Food Science & Technology* 46(6): 1201–1208.

Matsuo M. 2006. Suitability of quinoa fermented with Rhizopus oligosporus as an ingredient of biscuit. *Journal of the Japanese Society for Food Science and Technology* 53(1): 62–69.

Mazzetto, E. 2013. La comida ritual en las fiestas de las veintenas mexicas: un acercamiento a su tipología y simbolismo. *Amérique Latine Histoire et Mémoire* 25. doi:https://doi.org/10.4000/alhim.4461

Meléndez Torres, J. M., and G.M. Cañez de la Fuente. 2009. La cocina tradicional regional como un elemento de identidad y desarrollo local: el caso de San Pedro El Saucito, Sonora, México. *Estudios sociales (Hermosillo, Son.)* 17: 181–204. Retrieved from www.scielo.org.mx/scielo.php?script=sci_arttext&pid=S0188-45572009000300008&nrm=iso.

Mesías M., F. Holgado, G. Márquez Ruiz, and F.J. Morales. 2016. Risk/benefit considerations of a new formulation of wheat-based biscuit supplemented with different amounts of chia flour. *LWT – Food Science and Technology* 73: 528–535.

Mikuy, A., and S. Mikuy. 2013. *Traditional High Andean Cuisine*. Santiago: Food and Agriculture Organization of the United Nations. Regional Office for Latin America and the Caribbean (FAO/RLC).

Milán-Carrillo, J., A. Montoya-Rodríguez, R. Gutiérrez-Dorado, X. Perales-Sánchez, and C. Reyes-Moreno. 2012. Optimization of extrusion process for producing high antioxidant instant amaranth (*Amaranthus hypochondriacus* L.) flour using response surface methodology. *Applied Mathematics* 3(104): 1516–1525.

Minatelli, J. A., H. Stephen, R. Moerck, and U. Nguyen. 2009. United States Patent No. US20090181114A1. U. Nutraceuticals.

Miranda-Ramos K., and C.M. Haros. 2020. Combined effect of chia, quinoa and amaranth incorporation on the physico-chemical, nutritional and functional quality of fresh bread. *Foods* 9(1859): 1–22.

Miranda-Ramos, K., N. Sanz-Ponce, and C.M Haros. 2019. Evaluation of technological and nutritional quality of bread enriched with amaranth flour. *LWT – Food Science and Technology* 114: 108418.

Miranda-Ramos K., M.C. Millán-Linares, and C.M. Haros. 2020. Effect of chia by-products as breadmaking ingredients on nutritional quality, mineral availability and glycaemic index of bread. *Foods* 9(5): 663.

Montero-Quintero, Keyla Carolina, Rafael Moreno-Rojas, Edgar Ali Molina, Máximo Segundo Colina, and Adriana Beatriz Sánchez-Urdaneta. 2015. Evaluation of amaranth enriched bread for dietary regimes. *Interciencia* 40(7): 473–478.

Mujica, A., and S. Jacobsen. 2006. La quinua (*Chenopodium quinoa* Willd.) y sus parientes silvestres. Botánica Económica de los Andes Centrales. In *Botánica Económica de los Andes Centrales*, edited by M. Moraes, B. Øllgaard, L.P. Kvist, F. Borchsenius, and H. Balslev, 449–457. La Paz, Bolivia: Universidad Mayor de San Andrés.

Muñoz, L. A., A. Cobos, O. Díaz, and J.M. Aguilera. 2012. Chia seeds: Microstructure, mucilage extraction and hydration. *Journal of Food Engineering* 108(1): 216–224.

Mykolenko S., D. Zhygunov, and T. Rudenko. 2020. Baking properties of different amaranth flours as wheat bread ingredients. *Journal of Food Science and Technology-Ukraine* 14(4). DOI:10.15673/fst.v14i4.1896

Myrisis, G., S. Aja and C. M. Haros. 2022. Substitution of critical ingredients of cookie products to increase nutritional value. *Biology and Life Sciences Forum* 17(1): 15. https://doi.org/10.3390/blsf2022017015.

Naumova, N., A. Lukin, and V. Erlikh. 2017. Quality and nutritional value of pasta products with added ground chia seeds. *Bulgarian Journal of Agricultural Science* 23: 860–865.

Nawaz, M. A., M. Tan, S. Øiseth, and R. Buckow. 2020. An emerging segment of functional legume-based beverages: A review. *Food Reviews International* 38(5): 1064–1101.

Nicholson, G. E. 1960. Chicha maize types and chicha manufacture in Peru. *Economic Botany* 14(4): 290–299.

Nidhi B., and V. Indira. 2012. Organoleptic and nutritive scoring of value added cereal products from grain amaranth (*Amaranthus* spp.). *Indian Journal of Nutrition and Dietetics* 49: 31–37.

Nielsen, M. A., A.K. Sumner, and L.L. Whalley. 1980. Fortification of pasta with pea flour and air-classified pea protein concentrate. *Cereal Chemistry* 57(3): 203–206.

OJEU (Official Journal of the European Union). 2009. Authorizing the placing on the market of Chia seed (*Salvia hispanica*) as novel food ingredient under regulation (EC) No 258/97 of the European Parliament and of the Council 294: 14–15.

Ordoñez E. S., K.A. Castillo, D. Reátegui, and V.E. Condori. 2019. Elaboration of bread with the incorporation of coconut pulp flour and sacha inchi (*Plukenetia volubilis* L.) nibs. *Agroindustrial Science* 9(2): 189–198.

Owusu-Mensah, E., I. Oduro, and K. J. Sarfo. 2011. Steeping: A way of improving the malting of rice grain. *Journal of Food Biochemistry* 35(1): 80–91.

Pantoja Tirado, L. R., and G.P. Prieto Rosales. 2014. *Evaluación tecnológica y sensorial de pastas alimenticias enriquecidas con harina de quinua (Chenopodium quinua wild.) y tarwi (Lupinus mutabilis sweet)*. Masters Thesis. Nuevo Chimbote – Perú: Universidad Nacional del Santa.

Park S.H., and N. Morita. 2005. Dough and breadmaking properties of wheat flour substituted by 10% with germinated quinoa flour. *Food Science and Technology International* 11: 471–476.

Patil, S. S., M.A. Brennan, S.L. Mason, and C.S. Brennan. 2016. The effects of fortification of legumes and extrusion on the protein digestibility of wheat based snack. *Foods* 5(2): 26.

Peláez, P., D. Orona-Tamayo, S. Montes-Hernandez, M.E. Valverde, O. Paredes-López, and A. Cibrian-Jaramillo. 2019. Comparative transcriptome analysis of cultivated and wild seeds of *Salvia hispanica* (chia). *Scientific Reports* 9: 9761.

Pepe Guato, M. F. 2011. *Comparación de las Mezclas de Harina de Trigo (Triticum spp) y Chocho (Lupinus mutabilis) en la Evaluación Sensorial de Pastas.* Bachelor's Thesis). http://repositorio.uta.edu.ec/handle/123456789/3055.

Pinares Huamaní, C. 2019. Formulación de pasta alimenticia (Tallarín de Casa) con sustitución parcial de Tarwi (*Lupinus mutabilis* Sweet) y adición de cáscara de huevo en polvo. http://repositorio.unamba.edu.pe/bitstream/handle/UNAMBA/818/T_0509.pdf?sequence=1&isAllowed=y.

Poleth Carvajal Basantes, S. 2019. Efecto de los parámetros de extrusión sobre la composición nutricional de un snack a base de maíz, fréjol y camote. http://repositorio.utn.edu.ec/bitstream/123456789/8827/2/ARTICULO%20.pdf

Ponce, M., D. Navarrete, and M.G. Vernaza. 2018. Sustitución parcial de harina de trigo por harina de lupino (*lupinus mutabilis* sweet) en la producción de pasta larga. *Información Tecnológica* 29(2): 195–204.

Poutanen K., N. Sozer, and G. Della Valle. 2014. How can technology help to deliver more of grain in cereal foods for a healthy diet? *Journal of Cereal Science* 59: 327–336.

Raihan M., and C.S. Saini. 2017. Evaluation of various properties of composite flour from oats, sorghum, amaranth and wheat flour and production of cookies thereof. *International Food Research Journal* 24(6): 2278–2284.

Ramos Diaz, J. M. 2015. Use of amaranth, quinoa, kañiwa and lupine for the development of gluten-free extruded snacks. http://urn.fi/URN:ISBN:978-951-51-1657-4, http://hdl.handle.net/10138/157282.

Ranilla, L. G., E. Apostolidis, M.I. Genovese, F.M. Lajolo, and K. Shetty. 2009. Evaluation of indigenous grains from the Peruvian Andean Region for antidiabetes and antihypertension potential using in vitro methods. *Journal of Medicinal Food* 12(4): 704–713.

Rayas-Duarte, P., C.M. Mock, and L.D. Satterlee. 1996. Quality of spaghetti containing buckwheat, amaranth, and lupin flours. *Cereal Chemistry* 73(3): 381–387.

Repo-Carrasco-Valencia, R., and N.L. Hoyos. 1993. Elaboración y evaluación de alimentos infantiles con base en cultivos andinos. *Archivos Latinoamericanos de Nutrición* 43(2): 168–175.

Repo-Carrasco-Valencia, R., C. Espinoza, and S.E. Jacobsen. 2003. Nutritional value and use of the Andean crops quinoa (*Chenopodium quinoa*) and kañiwa (*Chenopodium pallidicaule*). *Food Reviews International* 19(1–2): 179–189.

Repo-Carrasco-Valencia, R., A.A. de la Cruz, J.C.I. Alvarez, and H. Kallio. 2009. Chemical and functional characterization of kañiwa (*Chenopodium pallidicaule*) grain, extrudate and bran. *Plant Foods for Human Nutrition* 64(2): 94–101.

Repo-Carrasco-Valencia, R., J.J. Pilco, and C.R. Encina-Zelada. 2011. Desarrollo y elaboración de un snack extruido a partir de quinua (*Chenopodium quinoa* Willd.) y maíz (*Zea mays* L.). *Ingeniería Industrial* 29: 209–224.

Rodrigues dos Reis Gallo, L., R. Braz Assunção Botelho, V. Cortez Ginani, L. de Lacerda de Oliveira, R. Figueiredo Resende Riquette, and E. dos Santos Leandro. 2020. Chia (*Salvia hispanica* L.) Gel as egg replacer in chocolate cakes: Applicability and microbial and sensory qualities after storage. *Journal of Culinary Science & Technology* 18(1): 29–39.

Rodríguez G., S. Avellaneda, R. Pardo, E. Villanueva, and E. Aguirre. 2018. Bread leaf enriched with extruded cake from sacha inchi (*Plukenetia volubilis* L.): Chemistry, rheology, texture and acceptability. *Scientia Agropecuaria* 9(2): 199–208.

Rojas, W., J.L. Soto, M. Pinto, M. Jäger, and S. Padulosi. 2010. Granos andinos: avances, logros y experiencias desarrolladas en quinua, cañahua y amaranto en Bolivia. www.bioversityinternational.org/fileadmin/_migrated/uploads/tx_news/ Granos_andinos__avances__logros_y_experiencias_desarrolladas_en_quinua__ ca%c3%b1ahua_y_amaranto_en_Bolivia_1413.pdf.

Rosell, C. M., G. Cortez, and R. Repo-Carrasco. 2009. Breadmaking use of Andean crops quinoa, kañiwa, kiwicha, and tarwi. *Cereal Chemistry* 86: 386–392.

Salvador-Reyes, R., and M.T. Pedrosa Silva Clerici. 2020. Peruvian Andean maize: General characteristics, nutritional properties, bioactive compounds, and culinary uses. *Food Research International* 130: 108934.

Salvatierra-Pajuelo Y.M., M.E. Azorza-Richarte, and L.M. Paucar-Menacho. 2019. Optimization of the nutritional, textural and sensorial characteristics of cookies enriched with chia (*Salvia hispanica*) and oil extracted from tarwi (*Lupinus Mutabilis*). *Scientia Agropecuaria* 10(1): 7–17.

Sánchez-Paz L.A., C. Pérez-Alonso, O. Dublan-García, J.C. Arteaga-Arcos, M. Mayorga-Rojas, L. Romero-Salazar, and M. Díaz-Ramírez 2020. Effect of a mixture of canola-chia oils and gelatin addition on a pound cake reduced in margarine. *Journal of Food Processing and Preservation* 44(1): e14298.

Sandak R.N., and A.A.A. El-Hofi. 2000. Utilization of barley and *Amaranthus* seeds in biscuit processing. *Egyptian Journal of Agricultural Research* 78: 1241–1252.

Santos Fernandes S., and M.M. Salas-Mellado. 2017. Addition of chia seed mucilage for reduction of fat content in bread and cakes. *Food Chemistry* 227: 237–244.

Sanz-Penella J.M., and C.M. Haros. 2017. Food uses of whole pseudocereals. Chapter 8 In: *Pseudocereals: Chemistry and Technology*, edited by C.M. Haros and R. Schönlechner, chap. 8, 163–192. Chichester: John Wiley & Sons. ISBN: 978-1-118-93828-7.

Sanz-Penella J.M., J.M. Laparra, Y. Sanz, and M. Haros. 2012. Bread supplemented with amaranth (*Amaranthus cruentus*): effect of phytates on in vitro iron absorption. *Plant Foods for Human Nutrition* 67: 50–56.

Sanz-Penella J.M., M. Wronkowska, M. Soral-Śmietana, and M. Haros. 2013. Effect of whole amaranth flour on bread properties and nutritive value. *LWT – Food Science and Technology* 50: 679–685.

Sayed-Ahmad B., T. Talou, E. Straumite, M. Sabovics, Z. Kruma, Z. Saad, A. Hijazi, and O. Merah. 2018. Evaluation of nutritional and technological attributes of whole wheat based bread fortified with chia flour. *Foods* 7: 35.

Sazesh, B. and M. Goli. 2020. Quinoa as a wheat substitute to improve the textural properties and minimize the carcinogenic acrylamide content of the biscuit. *Journal of Food Processing and Preservation* 44(1): 14563. DOI:10.1111/jfpp.14563.

Schöenlechner, R., J. Drausinger, V. Ottenschlaeger, K. Jurackova and E. Berghofer. 2010. Functional properties of gluten-free pasta produced from amaranth, quinoa and buckwheat. *Plant Foods for Human Nutrition* 65(4): 339–349.

Simurina O., M. Bodrom-Solarov, and B. Filipcev. 2005. *Expanded Seed of Amaranthus Cruentus as a Component in Speciality Bread Formulations*. Paper presemted at the 2nd Central European Meeting/5th Croatian Congress of the Food Technologists, Biotechnologists and Nutritionists, Opatija, Croatia, October 17–20, 2004.

Sindhuja A., M.L. Sudha, and A. Rahim. 2005. Effect of incorporation of amaranth flour on the quality of cookies. *European Food Research and Technology* 221: 597–601.

Song, K.Y., K.Y. Joung, S.Y. Shin, and Y.S. Kim. 2019. Effects of chia (*Salvia hispanica* L.) seed roasting conditions on quality of cookies. *Italian Journal of Food Science* 31(1): 54–66.

Stikic R., D. Glamoclija, M. Demin, B. Vucelic-Radovic, Z. Jovanovic, D. Milojkovic-Opsenica, S.E. Jacobsen, and M. Milovanovic. 2012. Agronomical and nutritional evaluation of quinoa seeds (*Chenopodium quinoa* Willd.) as an ingredient in bread formulations. *Journal of Cereal Science* 55: 132–138.

Sudha, M. L., and K. Leelavathi. 2012. Effect of blends of dehydrated green pea flour and amaranth seed flour on the rheological, microstructure and pasta making quality. *Journal of Food Science and Technology* 49(6): 713–720.

Svejenova, S., C. Mazza, and M. Planellas. 2007. Cooking up change in haute cuisine: Ferran Adrià as an institutional entrepreneur. *Journal of Organizational Behavior* 28(5): 539–561.

Tanumihardjo, S. A., L. McCulley, R. Roh, S. López-Ridaura, N. Palacios-Rojas, and N.S. Gunaratna. 2020. Maize agro-food systems to ensure food and nutrition security in reference to the Sustainable Development Goals. *Global Food Security* 25: 100327.

Time Out Magazine. 2016. Haute cuisine in Mexico City. *Time Out*. www.timeoutmexico. mx/ciudad-de-mexico/restaurantes/haute-cuisine-in-mexico-city.

Tiwari, U., M. Gunasekaran, R. Jaganmohan, K. Alagusundaram, and B.K. Tiwari. 2011. Quality characteristic and shelf life studies of deep-fried snack prepared from rice brokens and legumes by-product. *Food and Bioprocess Technology* 4(7): 1172–1178.

Tömösközi S., L. Gyenge, Á. Pelcéder, T. Abonyi, R. Schönlechner, and R. Lásztity. 2011. Effects of flour and protein preparations from amaranth and quinoa seeds on the rheological properties of wheat-flour dough and bread crumb. *Czech Journal of Food Science* 29: 109–116.

Toralva Aylas, A. D., and M. Rodas Pingus. 2015. *Efecto de la sustitución parcial de la harina de trigo por torta de sacha inchi (Plukenetia volubilis L) sobre las propiedades reológicas y sensoriales en el bizcocho*. Peru: Universidad Nacional de Callao, Facultad de Ingeniería Pesquera y de Alimentos, Escuela Profesional de Ingeniaría de Alimentos.

Toralva Aylas A. D., M. Rodas Pingas, and D.M. Guerrero Alva. 2015. Effect of the partial replacement of wheat flour by cake of sacha inchi (*Plukenelia volubilis* L.) on the rheological properties of dough in sweet bread. *Revista Ciencia & Desarrollo* 10: 16–21.

Torres Palacios, L. M., I.F. Pallares, and M.P. Tarazona Diaz. 2019. Development of a sweet biscuit reduced in fat and sugar, fortified with amaranth flour. *Nutricion Clinica y Dietetica Hospitalaria* 39(2): 135–140.

Tosi, E. A., E.D. Re, R. Masciarelli, H. Sanchez, C. Osella, and M.A. de la Torre. 2002. Whole and defatted hyperproteic amaranth flours tested as wheat flour supplementation in mold breads. *LWT – Food Science and Technology* 35: 472–475.

Tronc, E. 1999. Lupin flour: A new ingredient for human food. *Grains Legumes* 25: 3, 24.

Turck D., J. Castenmiller, S. de Henauw, K. Hirsch-Ernst, J. Kearney, A. Maciuk, I. Mangelsdorf, H. McArdle, A. Naska. C. Pelaez et al. 2019a. Safety of chia seeds (*Salvia hispanica* L.) as a novel food for extended uses pursuant to Regulation (EU) 2015/2283. *European Food Safety Authority Journal* 17: e05657.

Turck D., J. Castenmiller, S. de Henauw, K. Hirsch-Ernst, J. Kearney, A. Maciuk, I. Mangelsdorf, H. McArdle, A. Naska. C. Pelaez et al. 2019b. Safety of chia seeds (*Salvia hispanica* L.) powders, as novel foods, pursuant to Regulation (EU) 2015/ 2283. *European Food Safety Authority Journal* 17: e05716.

UNESCO (Producer). (2003). Convención para la Salvaguardia del Patrimonio Cultural Inmaterial. *Instrumentos Normativoswe.* https://es.unesco.org/about-us/legal-affa irs/convencion-salvaguardia-del-patrimonio-cultural-inmaterial.

UNICEF. (2019). Estado Mundial de la Infancia 2019 – Niños, alimentos y nutrición: Crecer bien en un mundo en transformación. Retrieved from www.unicef.org/es/infor mes/estado-mundial-de-la-infancia-2019.

Vásquez Osorio D. C., J.D. Jaramillo Ramírez, G.A. Hincapié Llanos, and L.M. Vélez Acosta. 2017. Development of cookies with sacha inchi (*Plukenetia volubilis* l.) flour coming from residual cake. UGCiencia, 23: 101–113.

Verduga Verdezoto, K. A. 2019. *Elaboración de una barra energética a partir de Sacha Inchi.* Bachelor's Thesis. Quito: UCE). www.dspace.uce.edu.ec/handle/25000/19677.

Wang S., A. Opassathovorn, and F. Zhu. 2015. Influence of quinoa flour on quality characteristics of cookie, bread and Chinese steamed bread. *Journal of Texture Studies* 46: 281–292.

Wang, S., F. Zhu, and Y. Kakuda. 2018. Sacha inchi (*Plukenetia volubilis* L.): Nutritional composition, biological activity, and uses. *Food Chemistry* 265: 316–328.

Watanabe, K., M. Kawanishi-Asaoa, C. Myojin, S. Awata, K. Ofusa, and K. Kodama. 2014. Amino acid composition, oxidative stability, and consumer acceptance of cookies made with quinoa flour. *Food Science and Technology Research* 20: 687–691.

Wright, E. M. 2008. *Mapping QTL for agronomic and canning quality traits in black bean (Phaseulus vulagaris L.).* Masters Thesis. East Lansing, MI: Michigan State University.

Yovera Méndez S. L., and L.M. Ruddy Gianny. 2018. *Palitos integrales de semilla de sacha inchi (Plukenetia volubilis) y linaza (Linum usitatissimum) y efecto dietético en escolares con sobrepeso.* Masters Thesis. Peru: Universidad Nacional Jose Faustino Sánchez Carrión Facultad de Bromatología y Nutrición Escuela Profesional de Bromatologia y Nutriciónhuacho.

Zanwar S.R., and V.S. Pawar. 2015. Preparation of biscuit from amaranth, sago and Echinochloa colonum (Bhagar). *Bioinfolet - A Quarterly Journal of Life Sciences* 12: 254–255.

Zhao, Y. H., F.A. Manthey, S.K. Chang, H.J. Hou, and S.H. Yuan. 2005. Quality characteristics of spaghetti as affected by green and yellow pea, lentil, and chickpea flours. *Journal of Food Science* 70(6): 371–376.

Zula A. T., D.A. Ayele, and W.A. Egigayhu. 2020. Proximate, antinutritional, microbial, and sensory acceptability of bread formulated from wheat (*Triticum aestivum*) and amaranth (*Amaranthus caudatus*). *International Journal of Food Science* 9429584.

Chapter 8

Nutritional Composition, Bioactive and Anti-Nutritional Compounds of Latin-American Crop Grains

Norma Sammán[1], María C. Rossi[1], Sonia Calliope[1],
and Ritva Ann-Mari Repo-Carrasco-Valencia[2]
[1]Universidad Nacional de Jujuy and Consejo
Nacional de Investigaciones Científicas y Técnicas
(CONICET), San Salvador de Jujuy, Argentina
[2]Universidad Nacional Agraria La Molina, Peru

CONTENTS

DOI: 10.1201/9781003088424-8

8.1   NUTRITIONAL COMPOSITION OF LATIN-AMERICAN GRAINS OF INTEREST

Latin America produces food crops which have great importance worldwide. Among them are maize (*Zea maize*), quinoa (*Chenopodium quinoa*), amaranth (*Amaranthus spp.*), chia (*Salvia hispanica L.*), beans such as black turtle bean (*Phaseolus vulgaris*), tarwi (*Lupinus mutabilis*), and also some lesser-known ones, such as kañiwa (*Chenopodium pallidicaule*) and sacha inchi (*Plukenetia spp.*). Latin-American grains are an important source of macronutrients, vitamins, minerals, and bioactive compounds. Consumption of these grains, especially quinoa has increased mainly due to their nutritional profile, in particular, their high protein value and mineral content (Motta et al., 2016a, 2016b).

Suitable nutrition is a critical element in the prevention of numerous diseases related to current lifestyle conditions. The leading chronic diseases due to unhealthy food habits nowadays are cancer, diabetes, and other non-communicable diseases. With growing global health awareness due to the increasing number of non-communicable diseases, demand for healthier food options that prevent and control these diseases has increased. Furthermore, the main challenge today, and even in future, will be providing sufficient safe and healthy foods – including the crops already mentioned – in populations' diets to help meet these challenges.

The word corn has been used over time and it has been given different meanings; after the discovery of America, Europeans called it Indian corn and named it "mahis" in native American language. Later it was called maize (Galinat, 1971; Salvador-Reyes and Pedrosa Silva Clerici, 2020). So in this chapter, the name "maize" will refer to the native races, with production aimed mainly at local markets. The terms "grains" and "seeds" are used along the text indiscriminately.

8.2   MACRONUTRIENT CONTENT

Maize is a very old crop and originated through the domestication process carried out by ancient inhabitants of Mesoamerica, from *teosinte*, which used to grow naturally in Mexico. Due to its great genetic diversity as well as the influence of different agroecological conditions where it is grown, its nutritional composition presents great variability. Maize has been used in all areas since ancient times as a food (Bañuelos Pineda et al., 2018).

The protein content in maize races varies between 5.7–12.5 g/100 g dry weight (db) (Bressani, 1991; Bello Perez et al., 2016), and is mainly distributed in the germ and endosperm. Carbohydrates are their major component, specifically starch, which represents 63–73 g/100 g db; it is found mainly in the endosperm and is an energy reserve for the embryo. Digestibility of maize starch decreases with the increasing amylose content and the length of the branched-chain of amylopectin (Rincon Londoño et al., 2016).

Maize also contains, although to a lesser extent, other important components for the diet such as lipids and fiber (Gimenez et al., 2020). The grains of native maize have lipid content from 2.70–6.14 g/100 g. According to Navarro-Cortez et al. (2016), it depends on the races as well as climatic factors, such as drought. Lipids are found mainly in the germ (76–83%), then in the aleuronic layer (13–15%) and pericarp and endosperm (1–2%). Regarding fiber content, several studies have shown lower postprandial glycemic and insulin responses after the ingestion of high-fiber corn-based foods (Lv et al., 2017; Colín-Chávez, 2020). The macronutrient content of maize reported in several studies are shown in Table 8.1.

Amaranth is considered to be one of the oldest crops in Latin America; there is evidence that it was used by the Mayas, Aztecs, and Incas approximately 10,000 years BC. Amaranth grains have significant amounts of proteins and starch; they are also an excellent source of insoluble fiber, principally lignin and cellulose. According to Bhosale et al. (2021) and Bressani (2018), the total fiber content in amaranth is higher compared to widespread cereals. The lipid fraction of amaranth grain represents from 6–10%; it is composed of triacylglycerides, fatty acids, phospholipids, squalene, phytosterols, and tocopherols. Amaranth, unlike quinoa, contains few bitter substances such as saponins that must be eliminated prior to consumption (Coelho, 2018). The nutritional composition of the grains described are presented in Table 8.1.

Quinoa (*Chenopodium quinoa*) is a native species of the South American Andes and its center of origin would be the surroundings of Lake Titicaca. This species has been cultivated in the Andes for the past 5,000 to 7,000 years, and during that time has spread to altitudes around 4,000 meters above sea level and to sea level in southern Chile. This distribution reveals its high adaptability, which, together with the nutritional quality of its protein, has led to its reevaluation as a new alternative crop that has extended its cultivation to other regions outside Latin America (Abugoch-James, 2009). The protein content ranges from 12.1 to 20.8 g/100 g) and in particular its soluble protein value is remarkable; albumins and globulins (31 and 37 g/100 g protein respectively) were identified as the major fractions. Another important aspect is that quinoa grains do not contain gluten, which allows their inclusion in foods for celiac or gluten intolerant patients. In addition, quinoa contains 49 g/100 g of carbohydrates; quinoa, like other Andean cereals, has higher sugar content compared to common cereals (6.2 g/100 g db). Starch represents most of the carbohydrate content (approximately 55 g/100 g

TABLE 8.1   PROXIMATE COMPOSITION OF LATIN-AMERICAN GRAINS*

Grain	Moisture	Proteins*	Lipids*	CH*	Fiber*	Ashes*	References
Maize	8–19	5.7–12.5	2.70–6.14	63–73	8.47–19.5	1.13–1.84	Cazares Sánchez et al. (2015)
Amaranth	10.50	13.4	6.43	55.30	11.30	2.89	Coelho et al. (2018)
Quinoa	11.30	12.10	6.31	57.20	10.40	2.01	Abugoch-James (2009); Gargiulo et al. 2019
Chia	5.82	25.32	30.22	34.57	37.50	4.07	Ullah et al. (2016)
Kaniwa	11.1	15.4	7.2	66.5	24	4.4	Repo-Carrasco-Valencia et al. (2009, 2019)
Tarwi	7.7	44.3	16.5	28.2	7.1	3.3	Huaccho-Human and Lope-Surichaqui (2007)
Turtle bean	5–8.2	25.66–25.93	1.52–1.59	67.83–68.09	5.4–7.6	4.71–4.65	Hayat et al. (2014)
Sacha Inchi	3.3–8.32	24.2–27.0	33.4–54.3	13.4–30.9	72.4	2.7–6.46	Wang et al. (2018)

Notes: CH: Carbohydrates; *Values expressed as g/100 g dry basis.

db) and is located in the perisperm. The starches are rich in amylopectin that gelatinizes at low temperatures (Haros and Schöenlechner, 2017). The lipid content of quinoa is two to three times higher compared to common cereals such as buckwheat and wheat, reaching 4 g/100 g db (Vilcacundo and Hernández-Ledesma, 2017).

Kañiwa (*Chenopodium pallidicaule*) is a little-known Andean native grain. It is very closely related to quinoa (*Chenopodium quinoa* Willd) and was considered as a variety of quinoa until it was classified as a species of its own (Gade, 1970). Kañiwa does not contain detectable quantities of saponins. It is one of the most resilient crops cultivated in the Andes. For highland farmers, kañiwa is very important because it is the only crop that can resist frost. The most intensive production of kañiwa occurs in the southern Andes of Peru and Bolivia in the surroundings of Lake Titicaca. The grains of kañiwa are much smaller than quinoa´s as can be seen in Figure 8.1.

The protein content of kañiwa seeds is between 14.4 and 16.9 g/100 g. Bioactive peptides in Andean grains have been studied by Chirinos et al. (2018), and these peptides show antidiabetic, antioxidant, and antihypertensive properties *in vitro*. The fat content of kañiwa is between 5.7 and 8.9 g/100 g and is mainly located in the embryo. Its oil content is considerably higher compared to common cereal grains. Kañiwa is an important source of dietary fiber, which is mainly insoluble, as is common in all grains.

All these crops, but specifically maize, quinoa, amaranth, and kañiwa are included in recipes that are part of the diet of native people in America. After the conquest process, foods were mixed to form new dishes that up until today continue to be part of gastronomic cultural identity (García, 2015). Maize is used in foods such as tortillas, arepas, snacks, and breakfast cereals, as well as being a source of starch to use as main products or ingredients in the food industry. Amaranth can be consumed in the form of sprouts, young leaves in salad, or ground to be served in soups, among many other preparations. In the form of flour, quinoa and amaranth are used as raw materials to make tortillas, bread, cookies, cakes, smoothies, strains, energy bars, cakes, and countless other products.

Chia (*Salvia hispanica L.*) seeds are brown with dark brown and white irregular spots (Alcocer Villacís, 2018); they provide significant amounts of lipids, proteins, carbohydrates, and dietary fiber. On the other hand, chia's high fiber content makes its inclusion in the diet meet the recommended daily intake of this component (Ullah et al., 2016). The human health benefits of chia seeds have drawn much attention from researchers and consumers because of the essential fatty acids, high protein levels, and other components such as phenolic compounds (Zettel and Hitzmann, 2018; Katunzi-Kilewela et al., 2021).

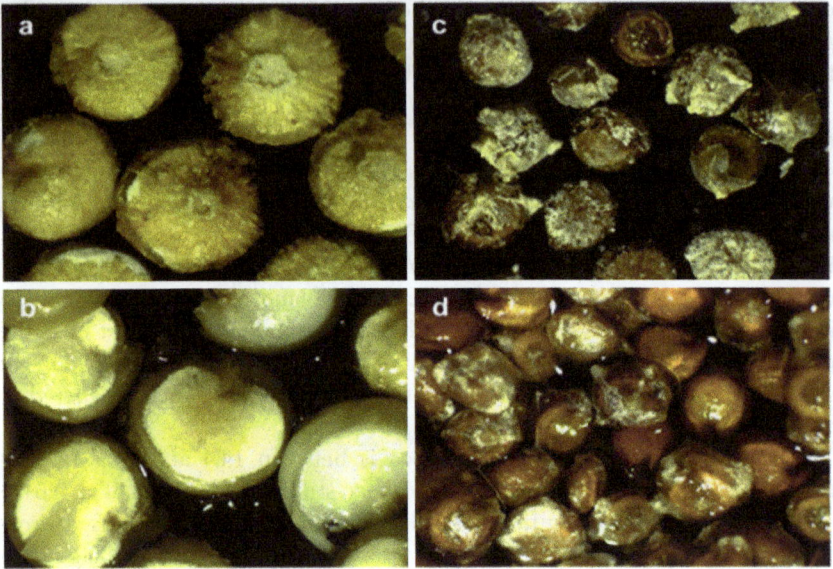

Figure 8.1 Quinoa grain *Amarilla Sacaca* variety: (a) raw; (b) washed. Kañiwa grain *Cupi* variety: (c) raw; (d) washed **Source: Repo-Carrasco-Valencia (unpublished).**

Tarwi or chocho (*Lupinus mutabilis*) is a legume of high nutritional value that is distinguished by its protein content and agronomic characteristics such as rusticity, its ability in the uptake and fixation of atmospheric nitrogen in the soil benefiting crops, and its adaptability to dry ecological environments, located between 2,800 and 3,600 meters above sea level. Also, tarwi has a high lipid content rich in mono- and polyunsaturated fatty acids. Regarding starch, it is usually absent, and the carbohydrates are mainly oligosaccharides (e.g., stachyose and raffinose) and the cell wall stores polysaccharides. The macronutrient content of tarwi grains are shown in Table 8.1. Since chronic diseases such as obesity and diabetes are increasing, sugar and certain carbohydrates are considered the main culprits of bad health, while unsaturated fats and proteins are greatly needed as part of a balanced diet. Tarwi's nutritional composition makes it a very suitable ingredient of healthier foods; it is used mainly as flour to enrich bread, cereal bars, and specific foods for different age groups (Suca, 2015; Salvatierra-Pajuelo et al., 2019).

Sacha inchi (*Plukenetia spp.*), also known as Inca peanut, wild peanut, sacha peanut, or mountain peanut, is an oilseed plant that grows in the lowlands of the Peruvian Amazon. It has been cultivated and consumed for centuries, as demonstrated by the clay pots in the shape of its fruit, which were found in pre-Inca cultures, traditionally part of various indigenous dishes in the Peruvian Amazon. The proximal composition shows that the seeds have remarkable high-fat content, between 34.23 and 49.28% in different varieties (Arévalo et al., 2019). Sacha inchi seeds spark attention as an important and alternative source of omega-3, which shows health benefits in lifestyle-related diseases (Chirinos et al., 2013).

Common beans (*Phaseolus vulgaris* L.) are economically important edible seeds cultivated and consumed worldwide. *P. vulgaris* stands out as a source of protein, starch, and dietary fiber. There is information on the slower digestibility of bean starch compared to the same cereal macromolecules. This desirable behavior of bean starch is associated with its amylose content, the length of the amylopectin branched chain, and its tendency to retrograde (Los *et al*, 2018). Starches that are digested slowly show a low glycemic index which has shown health benefits like improving glucose tolerance in both healthy people and diabetics. Other studies have shown that a diet characterized by a low glycemic index reduces risk factors for diabetes and dyslipidemia (Bello-Pérez et al., 2020). The proximate composition of this legume can be observed in Table 8.1.

8.3   FATTY ACIDS PROFILE

The plant triacylglycerol provides essential fatty acids (EFA); humans cannot synthesize them and they are required for good health. There are two families of EFA, and the fatty acids that give rise to them are linoleic (ω-3) and α-linolenic (ω-6). They must be provided by foods.

The total content of unsaturated fatty acids in maize oils is greater than 80% which could provide positive health effects (Serna-Saldivar et al. 2017). Palmitic (11.4–15.0%), stearic (2.2–3.5%), oleic (31.4–46.6%), and linoleic (39.0-53.0%) fatty acids are mainly found in native maize oils (Guzman et al. 2015). The fatty acid composition of this grain is shown in Table 8.2.

The lipids of amaranth grains are mainly found in the germ, most of which are polyunsaturated; the fatty acid composition varies according to the cultivar, growing conditions, and genetic traits of the seed (D'Amico et al., 2017). Ranges of the main fatty acids in amaranth are shown in Table 8.2.

Regarding quinoa, the predominant fatty acid groups are polyunsaturated and monounsaturated. Fatty acid values in the grain are 8.1, 52.3, and 23 g/100 g of linolenic, linoleic, and oleic fatty acids, respectively.

Chia oil contains mainly α-linolenic, linoleic, oleic, palmitic, and stearic acids, with a predominant amount of α-linolenic acid. Chia oil shows higher quantities of α-linolenic acid compared to flaxseed oil (Khattab and Zeitoun, 2013).

TABLE 8.2   FATTY ACIDS (FA) CONTENT IN LATIN-AMERICAN GRAINS (G/100 G FA)*

Grain/ seed/fruit	16:0	18:0	18:1	18:2	18:3	Reference
Maize	0.56	0.07	1.25	2.1	0.06	Sánchez-Ortega (2014)
Amaranth	17.0–21.35	2.75–4.11	20.2–32.1	33.5–43.9	NA	Cuevas-Espinal et al. (2017)
Quinoa	9.6–10	0.84–0.94	25.8–29.12	46.6–49.5	7.34–8.51	Vera (2014)
Chia	6.39–7.72	2.36–3.74	6.59–9.12	16.9–22.5	56.9–64.7	Coates (2011)
Kaniwa	17.9	0.4	23.5	42.6	6	Repo-Carrasco-Valencia (2011)
Tarwi	9.48	6.66	36.8	39.6	3.91	Ejigui et al. (2005)
Turtle bean	8.37–10	4.36–5.14	25–32.6	41.2–45.7	8.7–10.4	Ejigui et al. (2005)
Sacha Inchi	1.7–2.1	1.1–1.3	3.5–4.7	12.4–14.1	12.8–16	Chirinos (2013)

Notes: NA= Not available. *Values expressed as g/100 g dry basis.

The ω6/ω3fatty acid ratio of Chilean chia oil (0.29) is similar to the chia seed oil of Argentina (Ayerza, 2010) and markedly lower compared to most vegetable oils, such as maize (76.57), rapeseed (2.26), soybean (6.68), sunflower (30.77), and olive (17.86). The use of chia seed oil in the human daily diet is beneficial since the use of vegetable oils with high content of polyunsaturated fatty acids (PUFAs) have been shown to provide several health benefits (Villanueva et al., 2017). Low ω6/ω3 (normal ratio ranges are from 1:1 to 4:1) fatty acid ratio has also been associated with the reduction of cardiovascular disease risk (Ganesan et al., 2014). Costantini et al. (2014) substituted 10% of wheat flour with chia flour to make nutritionally improved bread; authors found an increase in the content of fat, dietary fiber, and total phenolic compounds, and a decrease in the content of carbohydrates.

Tarwi seeds are rich in lipids, with a range from 12 to 24 g/100 g (dry basis). In decreasing order, fatty acids per 100 grams of seed are oleic, linoleic, palmitic, and stearic (Elena, 2011; Suca, 2015).

Sacha inchi seeds show large amounts of unsaturated fatty acids, with high concentrations of linolenic and linoleic acids. Ranges in different varieties are between 39 and 47% and 34 to 41 g/100 g (dry basis) respectively (Maurer et al., 2012; Chirinos, 2013; Chirinos et al. 2016). The oil also contains other fatty acids such as oleic, palmitic, and stearic to a lower extent. Despite its composition which is rich in fatty acids susceptible to oxidation, sacha inchi oil presented better oxidative stability than linseed and sesame oils, allowing it to preserve its sensorial characteristics. This fact could be attributed to the rich presence of natural antioxidant substances such as tocopherols, carotenoids, and polyphenols (Sánchez-Sánchez, 2012).

Black turtle beans are also considered a source of unsaturated fatty acids. This bean has a predominance of PUFAs (0.30 g/100 g wb), followed by saturated fatty acids (0.2 g/100 g wb), and finally monounsaturated (0.07 g/100 g wb). Among the individual fatty acids are mainly found linoleic, alpha-linolenic, and palmitic acid (Wang et al., 2010).

8.4 AMINO ACIDS PROFILE

Proteins provide essential (eaa) and non-essential (neaa) amino acids for human metabolism, not only for the growth of infants and children but also for the constant replacement and turnover of body proteins in adults. Of the total amino acids that make up food proteins, eight are considered eaa for adults and nine for children (histidine, isoleucine, leucine, lysine, methionine, phenylalanine, threonine, tryptophan, and valine), because the human body cannot synthesize them and they must be provided through diet. High-quality protein is one that contains all the eaa in the right proportions. A useful method to evaluate the protein quality of foods is based on the proportional content of eaa with the age group requirements (amino acid scoring system) (FAO FINUT, 2017).

Maize protein is deficient in lysine and tryptophan (limiting amino acids) with a score of approximately 0.6 and 0.84 respectively according to the FAO/WHO (FAO, 1985; WHO, 2007) standard requirement.

One of the most important characteristics of amaranth grain is that its protein content is higher and better balanced in essential amino acids than most cereals. Amaranth proteins are relatively rich in tryptophan, histidine, valine, phenylalanine, lysine, and threonine and poor in methionine and leucine with respect to values reported by the FAO/WHO amino acid pattern requirements for 2–5 year-old children (FAO, 1985; WHO, 2007). Lysine content of A. *hypochondriacus* (5.95 g/16 g N) is higher compared to 2.90 g/16 g N in wheat flour. A suitable content of both lysine and tryptophan with a low content of leucine makes it a high-quality supplement for maize, which is rich in leucine but poor in lysine and tryptophan. The amino acid profile and the chemical score confirm the suitability of amaranth to mix with other cereals and achieve proteins of higher nutritional quality (Motta et al., 2019).

Quinoa provides all the eaa; and compared to similar commodities, it has high protein content and a balanced eaa profile with the exception of tryptophan (Filho et al., 2017).

The amino acid profile of quinoa shows significant differences between cultivars and geographical origins due to the different agroecological conditions that are cultivated. Essential amino acids, i.e., methionine, leucine, lysine, phenylalanine, tyrosine, isoleucine, threonine, tryptophan, and valine were present in several cultivars of quinoa, but their distribution patterns were different between them. Table 8.3 shows the amino acids content of different crops (Valenzuela Zamudio and Segura Campos, 2020). Compared to the requirement for children WHO (2007), the quinoa amino acid is adequately balanced, with a chemical score of 0.82 for tryptophan.

The kañiwa protein has an excellent biological quality because of the balanced eaa composition, especially because of its high lysine content. The lysine content of kañiwa is twice that of wheat (Repo-Carrasco-Valencia et al., 2003). Kañiwa proteins meet the FAO/WHO/UNU (UN University) protein reference pattern for children (FAO, 1985; WHO, 2007).

Coates (2011) found that chia seeds from lower altitude ecosystems showed 61% more protein content than those from higher cultivation areas. Nevertheless, differences in the profile of individual eaa among the seeds were not statistically significant (P <0.05). Percentages of amino acids in chia are shown in Table 8.3.

The amino acid composition of tarwi seed proteins indicates that it is deficient in sulfur amino acids and rich in lysine (Laurente-Flores, 2016).

Like other legumes, black turtle bean proteins also contain greater amounts of eaa including lysine that is deficient in cereal grains. Therefore, beans and cereal proteins are nutritionally complementary in eaa and the combined consumption of beans and cereals may alleviate these mutual deficiencies ensuring a balanced diet (Audu et al., 2013).

The main amino acids present in sacha inchi are arginine, valine, threonine, methionine, histidine, isoleusine, leucine, phenylalanine, and tryptophan and low amounts of lysine (Gutiérrez et al., 2011; Ruiz et al., 2013). The amino acid content for different grains is shown in Table 8.3.

TABLE 8.3   AMINO ACIDS PROFILE OF LATIN-AMERICAN GRAINS (MG/G PROTEIN)

		Histidine	Isoleucine	Leucine	Lysine	Threonine	Tryptophan	Valine	Methionine	References
Maize		30.7	37.6	125.2	34	38.4	5.9	53.9	ND	Nuss and Tanumihrdjo 2010
Amaranth	Olpir	20	36	62	80	49	ND	48	20	Pisaříková 2005
	Koniz	17	38	69	80	45	ND	53	23	
	Elbrus	17	34	59	59	47	ND	47	20	
Quinoa	Amilda	34.8	29.5	73.7	64.1	44.3	7.5	41.1	18.6	González et al. 2012
	Robura	24.4	23.2	58.2	50.9	35.4	9	31.1	15	
	Sayana	26	23.7	55.3	48.4	33.8	8.3	30.7	30.7	
Chia		22.1	28.1	51.2	40.3	29.2	0.1	46.1	ND	Ayerza 2013
Kaniwa		27	34	61	56	34	11	42	31	Repo-Carrasco-Valencia et al. 2009
Tarwi		ND	48	79	59	36	7	45	4	Suca 2015
Turtle bean		32	41	77	65	29	12	47	ND	Audu et al.,2013
Sacha Inchi		19	30	52	31	40	12	52	13	Ruiz et al. 2013

Note: ND = Not determined.

8.5 VITAMINS AND MINERALS

In maize, phosphorus and potassium are the most abundant minerals. It also contains two fat-soluble vitamins, vitamin E and provitamin A (carotenoids). Most of the carotenoids are located in the hard endosperm of the grain and in small quantities in the germ. Maize β-carotene and cryptoxanthin content are 20% and 51% respectively of the total carotenoids present in the grain; vitamin E is mainly located in the germ (Shah et al., 2016). It is important to mention that among the different varieties of maize, several of the carotenes present in each one of them and their proportions vary. The water-soluble vitamins are more abundant in the aleurone layer of the grain; part is lost when these grains are subject to the wet cooking process (Siyuan et al., 2018). Table 8.4 shows the vitamin content of these Latin-American grains.

Amaranth's high content of calcium, iron, and magnesium is very noteworthy; the contribution of riboflavin and folates is higher compared to other cereals of widespread consumption such as maize and rice. The niacin content is the lowest. The mineral contents are of great importance throughout all biological stages of humans making this Mesoamerican and Andean grain an important ingredient in the preparation of different foods like bread, cookies, drinks, and infant formulas, among others (Suárez et al., 2013; Hernández, 2018).

Regarding quinoa, it can be observed that both in minerals and in the content of other nutrients, there is great variability between the mean values reported by different authors; this variability is due to environmental and genetic factors (Bravo et al., 2013; Miranda et al., 2012) (Table 8.4). It is worth mentioning that amaranth and quinoa are rich in iron, copper, manganese, and zinc. Kañiwa is a particularly good source of tocopherols (Repo-Carrasco-Valencia et al., 2003). Tocopherols are compounds with high antioxidant capacity and other important physiological functions; some have the function of vitamin E. Tocopherols exist as four different isomers with antioxidant power, that is, in decreasing order: δ> γ> β> α. Kañiwa has α-tocopherols and γ-tocopherols, γ-tocopherols being the main compounds. The tocopherol content in kañiwa is superior to that in common cereals. Results from previous *in vitro* and *in vivo* studies suggest that γ- and δ-tocopherols could be additional potential anticancer agents compared to α-tocopherol (Das Gupta and Suh, 2016; Yang and Suh, 2013). Due to its broad cancer-preventive activity and availability, the naturally occurring tocopherol mixture, rich in γ-tocopherol or similar tocopherol mixtures may have a significant role in practical application (Das Gupta and Suh, 2016). In this respect, kañiwa is an outstanding source of these valuable phytochemicals.

TABLE 8.4   VITAMINS AND MINERALS IN IN LATIN-AMERICAN GRAINS (MG/100 G)**

Nutrient	Maize	Amaranth	Quinoa	Chia	Kaniwa	Tarwi	Turtle Bean	Sacha Inchi
Pro Vitamin A	17.93–266.2	-	-	13.2*	-	-	-	-
Vitamin C	-	-	16.4	-	-	-	-	-
Vitamin E	0.07	-	-	8.23	-	57.4	-	78.6–137
Niacin	2.6	-	-	-	1.34	1.17	-	-
Thiamin (B1)	0.16	-	0.4	0.18	-	0.47	0.34	-
Riboflavin (B2)	0.23	0.21	-	0.04	0.55	0.24	0.19	-
Pyridoxine (B6)	0.47	0.22	-	-	-	-	-	-
Manganese	-	-	-	2.3	-	-	-	-
Iron	2.3	5.3	4.4	16.4	15	1.08	6.22	10.3
Zinc	-	-	-	3.7	-	-	2.68	-
Magnesium	147	344	-	390	-	-	-	-
Calcium	158	303	47	714	110	72.8	179.1	240
References	Nuss and Tanumihardjo (2010)	Coelho et al. (2018)	Nascimento et al. (2014)	Ayerza (2013)	Repo-Carrasco-Valencia (2011)	Suca (2015)	Ejigui et al. (2005)	Chirinos et al. (2013)

Notes: * µg; **Values expressed as g/100g db.

Regarding minerals, kañiwa has a relatively high iron, calcium, and zinc content. Compared to unenriched wheat flour (iron, 0.68 mg/100 g; zinc, 0.98 mg/100 g; and calcium, 18.46 mg/100 g db; concentrations of these minerals are considerably higher in Andean grains (Dyner et al., 2007; Repo-Carrasco-Valencia et al., 2010). Iron content in kañiwa is higher than in rice (1.32 mg/100 g) and finger millet (2.13 mg/100 g db) (Hemalatha et al., 2007). Pachon et al. (2009) analyzed iron and zinc content in conventional and nutritionally enhanced beans and maize. Andean grains contain more zinc and iron than conventional maize and beans.

Chia seeds have been characterized as a good source of B vitamins such as riboflavin, thiamine, vitamin E, and carotenes with provitamin A activity and minerals like calcium, phosphorus, magnesium, potassium, iron, zinc, and copper. It is also noted that chia's sodium content is low. All of these minerals and vitamins make chia a great addition to foods for all ages (Muñoz et al., 2013; da Silva, 2017). They are used in different foods from cookies, cereal bars, and more elaborate products (Zettel and Hitzmann, 2018; Coelho et al., 2014).

Tarwi has been designated by the FAO (2016) as an *underutilized* food during the *Year of Legumes* since tarwi is native to the Andes and its cultivation is maintained in different production systems, from Ecuador to Chile and northeastern Argentina. After the Spanish conquest its production declined, despite providing not only proteins, but also vitamins and minerals. The main mineral contents of tarwi are phosphorus, calcium, and iron. Vitamins present in tarwi are C, niacin, thiamine, and riboflavin in decreasing order according to their amount per 100 grams of this food (Elena et al., 2011; Fernández-Mejía and Guivar-Delgado, 2020).

As with other Latin-American crops, information on the vitamin and mineral composition of sacha inchi is very scarce. For this reason, the data is very focused on its macronutrients. Some authors refer to Vitamin E as the most important vitamin in this crop, in the form of tocopherols and tocotrienols (de Souza et al., 2013). This, along with its richness in essential fatty acids, makes sacha inchi a great potential for food preparation and inclusion in the usual diet (Clavijo et al., 2015; Paucar-Menacho et al., 2015).

In all the crops mentioned, an important contribution of both vitamins and minerals is observed. This shows the relevance of highlighting their production and promoting consumption by all population groups across the world.

8.6 BIOACTIVE COMPOUNDS

Current nutritional approaches are starting to reflect a fundamental change in the understanding of the diet–health relationship. Increasing knowledge about the impact of diet on regulation at a genetic and molecular level is changing the way the role of nutrition is considered, resulting in new dietary strategies.

Nowadays it is considered that diet must not only provide adequate nutrients to meet metabolic requirements, but consequently can also contribute to improving human health through these and other components. These compounds with health-promoting properties are called bioactives. They are essential and nonessential compounds (e.g., vitamins or polyphenols) that occur in nature and are part of the food chain, and can be shown to have beneficial effects on human health. Consequently, plant extracts or individual bioactive compounds must be identified and developed for the food market to supplement a balanced diet.

Polyphenols are widely distributed as secondary metabolites from plants. Many researchers have found health properties in phenolic compounds, especially in flavonoids. Several studies report a relationship between higher intakes of flavonoids and lower prevalence of some types of cancer and cardiovascular diseases (Zhang and Tsao, 2016; Belščak-Cvitanović et al., 2018; Bever et al. 2021). Accumulated evidence supports the notion that diets rich in specific flavonoids may play a role not only in promoting health but also in preventing and mitigating the consequences of a range of chronic degenerative diseases. A popular term used to describe the action of phenolic compounds is antioxidant. This term refers to polyphenols used on the assumption that antioxidant activity *in vitro* translates to a similar function *in vivo* (Rossi et al., 2018).

Many authors question if the health benefits from phenolic compound come from their antioxidant activity (AA), since it is not proved that *in vitro* AA has the same impact as *in vivo* (Harnly, 2015). This does not mean that these compounds are not health-promoting, but that they have antioxident action and are also involved in broader physiological mechanisms.

Among all the varieties of native maize, those that contain the greatest amounts of bioactive compounds are those with colors, such as blue, red, or purple. Anthocyanins are the main water-soluble pigment in colored maize. In grain, the presence of anthocyanins has been reported mainly in the pericarp (Guillén Sánchez et al., 2014). Although there are several scientific researches on the biological properties of purple maize anthocyanins and their antioxidant power *in vitro*, there is a lack of *in vivo* evidence (Smorowska et al., 2021; Mazewski et al., 2017). Therefore, a comprehensive knowledge of the bioavailability and metabolism of anthocyanins is essential to understanding their health effects.

Repo-Carrasco-Valencia et al. (2019) studied the content of total phenolic compounds and the antioxidant capacity of kañiwa, quinoa, and kiwicha. The kañiwa sample had the highest value compared with the other grains

(57.87 mg gallic acid/100 g of sample). This same tendency was found in the values of the antioxidant capacity of the studied grains. Flavonoids are phenolic compounds with many health promoting properties, for example, antioxidant activity, anti-inflammatory, and anticarcinogenic properties. In kañiwa the principal flavonoids are quercetin and isorhamnetin (Repo-Carrasco-Valencia, 2011). Individual phenolic compounds present in amaranth and quinoa are 4-dihydroxybensoic acid, ferulic acid, rutin, ellagic acid, kaemferol-3-rutinoside and quercetin (Stikić et al., 2020). Berries have been considered an excellent source of flavonoids, especially quercetin and myricetin. For example, lingonberry contains 10 mg/100 g wb of quercetin and cranberry contains 10.4 and 6.9 mg/100 g wb quercetin and myricetin, respectively (Mattila et al., 2000). The levels in these flavonoid-rich berries are 5–10 times lower than those found in *Chenopodium* seed samples. When compared on a dry weight basis, the flavonoid content in berries and *Chenopodium* samples are of the same magnitude. Abderrahim et al. (2012) studied the effect of germination on total phenolic compounds, antioxidant capacity, Maillard reaction products, and oxidative stress markers in kañiwa. They observed an increase in the content of phenolic compounds and antioxidant capacity after 72 h of germination.

Kañiwa is exceptionally rich in resorcinol compounds, not very common in plants (Peñarrieta et al., 2008). Of the major cereals, resorcinols have been reported to be present in high levels in wheat, rye, and triticale and in low amounts in barley, millet, and maize. Cereal alkylresorcinols (ARs) have been reported to have anticancer and antimicrobial effects, as well as the ability to inhibit some metabolic enzymes *in vitro*. ARs have also been reported to have antioxidant activity (Ross et al., 2003).

Phytosterols are natural components of plant cell membranes and are abundant in vegetable oils, seeds, and grains. Phytosterols have various biological effects, such as anti-inflammatory, antioxidative, and anticarcinogenic activity, as well as cholesterol-lowering capacity (Abugoch James, 2009). Repo-Carrasco-Valencia et al. (2019) studied the phytosterol content of Andean grain oils and found that total phytosterol content in kañiwa was 74.74 mg/100 g dry weight. Besides campesterol, stigmasterol, and β-sitosterol, which are the most common phytosterols, ergost-7-en-3-ol, gramisterol, Δ7-sitosterol, Δ7-avenasterol, 24-methylenecycloartanol, and citrostadienol were identified.

Regarding chia, there are higher values of total polyphenols in its black seeds than in white seeds 295 and 185 mg/GAE (gallic acid equivalent), respectively. Of these polyphenols, the majority are flavonoids, most of which are anthocyanidins and proanthocyanidins (Muller Tito, 2015).

The total content of phenol in the 16 varieties of the sacha inchi almonds studied is between 64.6-80.0 mg GAE/100 g; this is higher compared to other oleaginous fruits such as almonds and hazelnuts (Chirinos et al., 2013).

Pérez-Hernández et al. (2020) reported that in *P. vulgaris* seeds, there was a wide array of polyphenols including hydroxycinnamic acids, flavonoids, condensed tannins, and anthocyanins. The major anthocyanins in black beans are delphinidin-3-O-β-D-glucoside, petunidin-O-β-D-glucoside, and malvidin-3-O-β-D-glucoside. Catechins were the predominant forms of anthocyanidins (Hall et al., 2017).

In natural tarwi grains, the total polyphenols were found in ranges between 518.9 to 533.9 mg GAE/100 g (wb). Within the polyphenol groups, tarwi contains flavonoids and phenolic acids (Sabelino Francia, 2020).

Amaranth and quinoa have a series of antinutrients such as saponins, oxalic acid, tannins, trypsin inhibitors, and phytic acid. Phytic acid is found in large quantities in amaranth and has the property of forming complexes with metal cations such as Mg^{+2}, Fe^{+2}, Fe^{+3}, Ca^{+2}, Zn^{+2}, Co^{+2}, and Cu^{+2} causing a loss of intestinal bioavailability of these components. However, some authors have also described beneficial functions of phytic acid, such as preventing chronic and coronary disease, among others (Ranilla, 2019). Saponins are characterized by the formation of foam, having detergent and surfactant properties; they largely determine the bitter taste in some seeds such as quinoa and sacha inchi. Consumption of these products without de-saponification or when consumed raw can cause disorganization in the cell membranes, and in excess its symptoms begin in the gastrointestinal tract with nausea, vomiting, and diarrhea. A method of eliminating saponins from quinoa is scarification (Mackay et al., 1995) followed by five washes; this leaves a concentration of less than 0.1% and eliminates its bitter taste (Calliope et al., 2015).

8.7   ANTI-NUTRITIONAL COMPOUNDS IN LATIN-AMERICAN GRAINS OF INTEREST

Oboh et al. (2010) found that the roasting process produced a significant decrease in phytic acid, in yellow maize (26.1%) and white maize (36.81%); it also caused a significant decrease of flavonoids in both types of maize. Grundy et al. (2020) demonstrated that fermentation, in addition to improving the nutritional value and digestibility of amaranth grain proteins, degrades phytates and oxalates. Cooking amaranth in excess of water, even without bursting the grain, as well as other thermal processes (sterilization by autoclaving and blanching) and germination reduce the content of phytic acid, oxalate, and tannins.

Legumes contain anti-nutritional factors such as trypsin inhibitors, phytic acid, tannins, and oligosaccharides that limit the use of proteins and minerals. Heat treatment significantly improves protein quality in legumes by destroying or inactivating heat-labile anti-nutritional factors. Cooking produces a significant reduction in the activity of the trypsin inhibitor, from 9.9 to 1.1 mg/g of dry matter; it also reduces tannins from 13 to 3.2 g/kg of dry matter but has little effect on phytic acid (Wang et al., 2010).

He et al. (2015) highlighted that excessive use of *P. vulgaris* beans could cause toxic and allergenic reactions. Besides the beneficial roles of lectins for the plant, they can from an agronomic viewpoint bind to the surface of epithelial cells in the digestive system due to their high affinity with carbohydrates and can result in toxic reactions with changes in intestinal permeability (Menard et al. 2010). Like other anti-nutritional factors, lectins are heat-stable proteins; they can be inactivated after special heating, such as cooking (around 100°C), with an improvement in the accessibility of proteins to enzymatic attack. Thermal inactivation of lectin from black turtle beans can be achieved by heating at 90°C for 5 min, demonstrating the importance of high temperature and short time (HTST) treatment in the packaged product. In an autoclave, it can be completely inactivated at 121°C for 5 min.

Valles Ramírez (2012) used a chemical peeling by immersion in NaOH for 3 min to separate the tegument and precooking at 100°C for 20 min to inactivate anti-nutritional factors from the sacha inchi seed. To inactivate the fixed coumarins and alkaloids in sacha inchi, a pre-toast at temperatures between 103 and 110°C was also used.

Curti et al. (2018) reported that grains of *Lupinus* species contain between 1 and 4 g/100 g dw of toxic alkaloids that must be eliminated before consumption. The traditional debittering process applied in *L. mutabilis* includes a soaking stage of seeds for 18 h, followed by a cooking step (0.5–6 h). The cooking stage is essential to inactivate the germination capacity of the seeds, their enzymes (lipase, lipoxygenase), to eliminate occurring microorganisms for food safety, to reduce the loss of proteins through their coagulation, and to facilitate the leaching of the alkaloids by increasing cell wall permeability. The process is completed with a wash for several days in which most of the alkaloids are removed. The debittering process is a necessary step required by *Lupinus* species to ensure safe human consumption.

Once the anti-nutritional compounds are eliminated, these grains are good ingredients to be included in foods since they provide beneficial bio-functional compounds and dietary fiber for the prevention of non-transmissible chronic diseases.

8.8   EFFECTS OF DIFFERENT TECHNOLOGICAL AND CULINARY PROCEDURES ON NUTRITIONAL COMPOSITION

Food composition tables (FCT) generally include the nutritional composition of foods in their natural state, even when consumed after industrial or culinary processes, heating, or storage. These treatments can affect the nutritional quality and content of nutrients and bioactive compounds of raw foods because they can cause degradation or leaching of their components after processing or treatments. Therefore, these changes must be analyzed to know the true amount of nutrients. Few studies have been carried out on this subject. This is done by calculating the nutrient retention factors (NRF%) that correspond to the percentage of preservation of nutrients, especially vitamins and minerals. To calculate NRF%, it is necessary to know the yield factor as well the nutrient content of raw and cooked or processed foods.

It is important to know how the different proteins (and their breakdown during digestion) are affected by treatments including baking, extrusion, cooking, etc. It is well-known that the physicochemical and biological characteristics, including antigenicity, of proteins, are changed after heating and also that the extent of their degradation during digestion can be either increased or lowered after heat processing (Davis, 1998). All of these modifications can affect the characteristics of the digestion products of proteins and, consequently, their effects on human physiology (Passini et al., 2001).

Yield factor (YF) is the term used to indicate the change in the food or dish weight after preparation or processing.

$$YF = \frac{weight\ of\ cooked\ or\ processed\ food}{weight\ of\ raw\ food} \qquad \text{Ec.8.1}$$

$$NRF\% = \frac{nutrient\ content\ per\ g\ of\ cooked\ food}{nutrient\ content\ per\ g\ of\ raw\ food} \times YF \times 100 \qquad \text{Ec.8.2}$$

When the NRFs are known for each processed food under standardized conditions, it is possible to apply them to other similar products processed under identical conditions. Some food composition tables (FCT) include NRF% for the most commonly consumed processed foods, but there is a lack of information on the grains discussed in this chapter.

8.9 CHANGES IN NUTRITIONAL VALUE AND MODIFICATIONS OF BIOACTIVE COMPOUNDS CONTENT THROUGH PROCESSING

Maize is prepared and consumed in various ways; it can be dried in the sun, cooked, fermented, roasted, mashed, or crushed to use in different preparations, depending on the region or ethnic group. Within these processes, toasting is a traditional practice. It is used to increase nutritional properties, improve flavor, and reduce anti-nutritional factors. Oboh et al. (2010) studied this process in two races of yellow and white maize. After roasting, they found a decrease in protein content caused by denaturation and the participation of some amino acids in the Maillard reaction; there was an 18% increase in fat content in yellow maize and 20% in white maize after roasting; there were no significant differences in ash content between raw and roasted maize. Nutrient content of both yellow and white maize is shown in Table 8.5.

Modifications in the content of nutrients in the final product are variable due to the fact that two factors cause them: (i) modification of the total weight of the final product; (ii) the loss/gain of nutrients (leaching, transformation, reaction with other components). In this case the first factor has not been considered.

Nixtamalization is nowadays international technology to process corn and produce dough for the preparation of tortillas and other foods. It was used by the Mayans and Aztecs and continues to be used today with some modifications (Cravioto et al., 1945). Briefly, maize undergoes aqueous cooking with lime, is left to rest for about 12 h and then washed to produce the nixtamal; finally it is ground to form dough (Bressani, 1990). The process produces chemical and nutritional changes; the nixtamal produced after alkaline cooking depends on

TABLE 8.5 PROXIMAL COMPOSITION OF MAIZE ACCORDING TO TREATMENT (G/100 G DRYBASE)

Maize	Treatment	Composition			
		Protein	**Crude fat**	**Ash**	**Moisture**
Yellow	Raw	$8.45^a \pm 1.5$	$6.21^b \pm 1.6$	$1.98^d \pm 1.2$	$13.67^a \pm 1.5$
	Roasted	$7.85^b \pm 1.7$	$7.34^c \pm 2.0$	$1.85^e \pm 0.3$	$10.93^b \pm 1.0$
White	Raw	$12.97^c \pm 2.1$	$5.32^d \pm 1.2$	$1.93^d \pm 0.7$	$16.92^c \pm 2.0$
	Roasted	$10.86^d \pm 1.6$	$6.39^b \pm 0.9$	$2.00^d \pm 0.1$	$14.31^f \pm 1.2$

Source: Oboh et al. (2010).

Notes: Data are presented as mean±standard deviation; n = 3. Values with the same letter per column are not significantly different.

the processing parameters used during cooking and steeping, as well as the physicochemical properties of the maize used. Nixtamalization alters the structure and solubility of maize proteins; this together with the cooking process to make tortillas reduces the solubility of albumins, globulins, and prolamines, and induces the appearance of high molecular weight protein fractions as glutelins; digestibility of the protein decreases slightly in both the nixtamal and the tortilla but improves the overall protein quality. Paredes López et al. (2009) reported that when yellow maize underwent nixtamalization, the lipids decreased, up to 3.4% in the yellow maize tortilla and 2.5% in the white maize. These losses are due to the elimination of the pericarp, aleurone, and the germ – which can be partial or total – where most of the lipids are located. The losses in vitamins that alkaline cooking and tortilla production caused are variable; in nixtamalized yellow maize losses of 15–28% of carotenes and thiamine and riboflavin were reduced by up to 60–70%. However, during this process, the soluble dietary fiber goes from 0.9% in the grain to 1.3% in the dough and 1.7% in the tortilla. The alkaline cooking process improves the bioavailability of niacin because this vitamin is present in the maize kernel attached to other components and therefore not then available to be released as nicotinic acid, an active component of niacin. During alkaline cooking, calcium content in the dough increases. The phosphorus in maize is mainly found as part of phytic acid, a chemical compound that strongly interferes with the absorption of various bivalent elements, including calcium, and whose content decreases from 1% in the maize grain to 0.4% in the tortilla, which leads to the relationship between calcium and phosphorus from 1:20 in maize grains become 1:1 in the tortilla. López-Martínez et al. (2011) found that the nixtamalization process reduced phenolic and anthocyanin compounds, and total antioxidant activities of white, blue, red, and purple maize. Also significantly reduced were aflatoxin (up to 90%) and fumonisins (up to 80%). Overall, there was no significant difference among the sensory properties prepared from nixtamal and non-nixtamal maize flour. The study findings suggest that nixtamalization is a promising and affordable processing technology for reducing mycotoxin levels in maize and enhances the nutrient profile of maize products, simultaneously increasing consumer acceptability (Bressani, 1990; Paredes López et al., 2009; Maureen et al., 2020). Additionally, the nixtamalization process significantly increases calcium availability, which represents an outstanding nutritional fact. Many alternative nixtamalization technologies have emerged, mostly based on the use of extrusion, with scientific evidence in improving the nutrimental/nutraceutical properties, reduction of contaminant effluents, processing time and water use, although none of them has been widely adopted by the tortilla industry; the reaction of consumers to the change is mainly explained by the sensorial properties provided by traditional nixtamalization processes (Escalante-Aburto et al., 2020).

Salvador-Reyes et al. (2020) highlight *chicha Morada* and *mazamorra Morada* as highly consumed products in Peru. In this regard, Bhornchai et al. (2014) reported that the boiling process to obtain these products leads to a 60% reduction in the content of anthocyanins and other phenolic compounds in purple waxy maize. Jing and Giusti (2007) maintained that the loss of anthocyanins in the cooking water of purple maize during the process of obtaining sweets and beverages was due to the denaturation of proteins at 100°C, associated with their complexation with anthocyanins and tannins. De la Parra et al. (2007) associated the decrease in bioactive compounds to the indicated treatment (total phenols, anthocyanins, ferulic acid, carotenoids) with making pasta, tortillas, and chips using white, yellow, blue, and red maize.

According to Ramos Diaz et al. (2013) the first step toward the industrialization of Andean grains was the adoption of low-cost technological processes, such as extrusion cooking, which is characterized by its ability to increase the digestibility of starch and proteins. The extrusion of starchy products causes gelatinization of starch, denaturation of proteins, and the formation of starch-lipids and protein-lipids complexes. These authors studied the effects of extrusion on lipid stability in extruded products made with mixtures of maize with quinoa, amaranth, and kañiwa flours with different humidity conditions. Depending on the type of interaction between amylose and fatty acids in solid matrices, oxidation of lipids can lead to the formation of volatile compounds during storage. A tendency to lipid oxidation was observed in extrudates containing quinoa and kañiwa exposed to 11% humidity. However, the lipid stability of extrudates containing amaranth or just pure maize did not appear to be affected by exposure to different moisture conditions and physical modification. It was also observed that extrudates with higher fiber content (lower hardness) were more prone to oxidation. This could be because plasticizers (and possibly fiber) could have limited chain-to-chain interaction, thus reducing the ability of macromolecules (amylose and amylopectin) to form a protective *layer* around lipids (entrapment). The effect of this process on micronutrients and bioactive compounds remains a subject of scientific research.

Phytate content of kañiwa was studied by Repo-Carrasco-Valencia et al. (2009) and was found to be 8.0 mg/g, which is higher than that in amaranth, 3.4–6.1 mg/g (Guzman-Maldonado and Paredes-Lopez, 1998), but lower than common cereals. Gualberto et al. (1997) found 14.2, 43.2, and 52.7 mg/g of phytate in oat, rice and wheat bran, respectively. Phytic acid has long been considered as an anti-nutrient because it chelates minerals and trace elements. However, its antioxidant potential is now recognized (Fardet et al., 2008). Repo-Carrasco-Valencia et al. (2010) studied the effect of common processing methods, such as roasting and boiling, on the content of three micronutrients (Fe, Ca, Zn) in Andean grains. There was a significant decrease in iron content during the boiling process of kañiwa. Wet processing procedures in general cause loss of dry matter and iron. Boiling reduced the content of zinc in quinoa and kañiwa, but not in amaranth. Roasting negatively affected the content of calcium in quinoa but not in kañiwa and amaranth. Boiling enhanced the iron, zinc, and calcium dialyzability in kañiwa. To increase the potential contribution of minerals from Andean crops, and prevent losses, it is important to study the effect on their nutritional composition of the different processes to which they are subjected.

Zare et al. (2019) indicated that soaking chia seeds increases the extraction capacity of ω3 and ω6 fatty acids compared to seeds without soaking.

Currently, interest in sacha inchi is growing because its oil is a new source of polyunsaturated fatty acids with a favorable ratio of ω6/ω3, and also because it contains other functional ingredients (Wang et al., 2018). The residual cake of sacha inchi seeds, after the oil extraction process by cold pressing, preserves a high percentage of proteins (56.6 g/100 g cake wb), oil (4.1 g/100 g), and moisture (2.6 g/100 g sample wb) so it would be a good source for protein extraction as a way to add value to the crop. The pressed cakes contain 4.2% sucrose, 2.2% starch, and 11.1% dietary fiber. The lipid composition shows saturated and unsaturated fatty acids, and minerals such as potassium, which was the highest element followed by phosphorus, magnesium, calcium, and, to a lesser extent, zinc, sodium, manganese, and copper; it also contains 0.51 mg GAE/g. All these components suggest that sacha inchi seed press cake could be a useful ingredient for the formulation of food for human consumption (Rawdkuen et al., 2016).

Curti et al. (2018) studied the effect of the debittering process on the fatty acid composition of two species of bitter lupine (*L. albus* and *L. mutabilis*). They found that after applying the process, there were higher changes in the fatty acid composition of *L. mutabilis* than *L. albus*; the content of linoleic, palmitic, and stearic acid increased in *L. mutabilis*, while in *L. albus* PUFA/SFA, behenic, linolenic, and arachidic acids increased. Changes in the fatty acid composition between the processed *Lupinus* species were explained by the differences in their initial composition. Carvajal-Larenas et al. (2016) found that the debittering process can induce decreasing the content of soluble carbohydrates, fiber, and fat and increasing the protein content in *L. mutabilis*.

Tannins form complexes with proteins, which leads to a decrease in their digestibility. Wang et al. (2010) reported that heat treatment significantly improves legume protein quality by destroying or inactivating heat-labile antinutritional factors such as phytic acid and tannins. The boiling method affects the chemical composition of legumes. Wang et al. (2010) evaluated the effect of boiling on the nutritional and anti-nutritional components of black turtle beans. Proteins and starch apparently increased since the loss of soluble solids during cooking and the decrease in ash due to leaching of some minerals in the cooking water modify the total weight of the cooked product. The calcium, copper, iron, and zinc in turtle bean were not significantly modified while potassium and magnesium decreased significantly, from 1415–1070 mg/100 g (db) and 200–156 mg/100 g (db) respectively, and manganese and phosphorus increased from 1.5–1.7 mg/100 g (db) and 370–443 mg/100 g (db) respectively (Wang et al., 2010).

All of the above examples indicate that it is of great importance to choose suitable processing/cooking methods to achieve maximum preservation of nutrients and bioactive compounds and also to maintain the natural/typical flavor of all these Latin-American crops after processing (Guillén-Sánchez et al., 2014).

On the other hand, Motta et al. (2016b) provided information on the effects of cooking (boiling and steaming) on NRF% in quinoa and amaranth. These processes produced a significant increase in grain moisture, probably because heating affected the water absorption capacity of the starch. These processes had effects on mineral content and NRF%. In Table 8.6 mineral content in raw and processed grains and NRF% are shown. These results indicate that, in amaranth, both types of culinary method produce similar losses, while in quinoa the steam process produces lower losses for all minerals. Different effects on minerals can be explained by their cellular location in the seed matrix.

The term folic acid or folates is applied to a family of vitamers with equivalent biological activity. It is a water-soluble vitamin whose function resides mainly in its ability to donate and capture carbon units. Folic acid is an essential nutrient for cell life; its deficiency leads to the development of numerous pathologies. Several studies reported the negative effects of cooking processes on folate content in foods, especially legumes and vegetables (Delchier et al., 2013, Stea et al., 2007). Quinoa and amaranth are good sources of folates.

Motta et al. (2017) studied the folate content and the effect of steaming, boiling, and malting methods applied to quinoa and amaranth grains. The boiled amaranth samples showed a significant decrease in total folate content, while in the malted it increased significantly with respect to the raw grains (Figure 8.2). This increase in amaranth after the malting process was also previously reported in legumes and wheat by other authors (Hefni and Witthöft, 2011; Koehler et al., 2007; Shohag et al., 2012). This is possibly because during the germination phase and before the grain drying process, folate production increases (Nelson et al., 2013).

The boiled and steamed quinoa did not show significant differences on folate content and were higher than the raw grain and lower than boiled samples (Figure 8.2). The increase in total folate content in quinoa after cooking was reported by other authors (Stea et al., 2007). As folates occur intracellularly, they are released from the matrix when the cell/seed structure is destroyed. According to Motta et al. (2017), amaranth seeds are smaller and have a higher surface area than quinoa. This may explain how boiling and steaming increases the extracted folates from amaranth more than from quinoa.

TABLE 8.6   MOISTURE AND MINERAL CONTENT IN RAW AND PROCESSED AMARANTH AND QUINOA AND NUTRIENT RETENTION FACTORS (NRF)

Cooking method	Amaranth**				
	Raw	Boiled	% NRF Boiled	Steamed	% NRF Steamed
Moisture (g/ 100 g wb)	12.2^a±0.09	73.6^b±0.84	NA	52.4^c±0.13	NA
Cu (mg/100 g-db)	0.572^a±0.02	0.652^b±0.02	98.29^a ±4.67	0.601^a±0.02	91.09^a ±1.03
Mn (mg/100 g-db)	4.42^a±0.34	4.56^a±0.06	89.30^a ±7.57	4.38^a±0.09	86.21^a ±7.01
Fe (mg/100 g-db)	7.35^a±0.46	7.42^a±0.07	87.32^a ±5.86	7.25^a±0.28	85.70^a ±6.13
Zn (mg/100 g-db)	4.55^a±0.17	4.70^a±0.12	89.05^a ±4.05	4.68^a±0.15	89.31^a ±6.16
Mg (mg/100 g-db)	328^a±9.2	307^b±4.7	80.72^a ±2.63	302^b±8.7	79.96^a ±4.60
Ca (mg/100 g-db)	200^a±7.2	207^a±1.5	89.06^a ±3.87	205^a±2.1	88.98^a ±3.94
P (mg/100 g-db)	663^a±13	655^a±8.3	85.23^a ±2.14	647^a±9.1	84.69^a ±2.77
K (mg/100 g-db)	552^a±10	538^b±2.8	84.05^a ±1.83	535^b±5.1	84.07^a ±2.37
Na (mg/100 g-db)	<LoQ	<LoQ	NA	<LoQ	NA
Cooking method	Quinoa**				
	Raw	Boiled	%NRF Boiled	Steamed	%NRF Steamed
Moisture (g/ 100 g wb)	11.7^a±0.18	66.6^b±0.18	NA	62.8^c±0.02	NA
Cu (mg/100 g-db)	0.502^a±0.005	0.461^b±0.02	81.65^a ±2.71	0.465^b±0.02	89.7^b ±3.77
Mn (mg/100 g-db)	1.89^a±0.030	1.91^a±0.16	96^a ±7.99	1.99^a±0.24	109.93^a ±10.97
Fe (mg/100 g-db)	4.29^a±0.07	4.44^a±0.29	92.98^a ±4.03	4.34^a±0.12	98.46^a ±3.24

(continued)

TABLE 8.6   (CONTINUED)

Cooking method	Raw	Boiled	% NRF Boiled	Steamed	% NRF Steamed
			Amaranth**		
Zn (mg/100 g-db)	2.97[a]±0.05	2.93[a]±0.24	84.42[a] ±7.14	2.88[a]±0.22	90.37[a] ±6.27
Mg (mg/100 g-db)	196[a]±3.1	192[a]±7.9	89.76[a] ±4.83	189[a]±8.7	96.20[a] ±3.51
Ca (mg/100 g-db)	77.6 [a]±2.1	74.2[a]±10	77.58[a] ±10.37	71.4[a]±7.4	85.06[a] ±9.09
P (mg/100 g-db)	436[a]±4.7	447[ab]±10	92.68[a] ±1.76	451[b]±10	100.04[b] ±2.88
K (mg/100 g-db)	559[b]±7.5	536[a]±7.8	87.18[a] ±1.62	520[c]±9.7	90.70[a] ±2.94
Na (mg/100 g-db)	<LoQ	<LoQ	NA	<LoQ	NA

Source: Motta et al. (2016a).

Notes: LoQ (limit of quantification) = 2.5 mg/100 g; * dry base; **mean ± standard deviation (n = 3). Different letters in each file indicate significant differences. NA: Not applicable.

Table 8.7 shows the NRF% of folates in quinoa and amaranth subjected to the three mentioned processes. It can be seen that these were dependent on both the food matrix and the method of processing (boiling, steaming, and malting). Retention values provided information on nutrient gains and losses, facilitating the selection of the most suitable cooking/processing method.

When incorporating processed grains or products derived from them into a recipe, the nutrient content must be adjusted by NRF% to create the nutritional profile of the final product.

8.10   CONCLUSIONS

The information compiled in this chapter, regarding the content of nutrients and bioactive compounds of the following grains of Latin-American origin, corn, quinoa, amaranth, chia, and beans such as black turtle bean, tarwi, and some

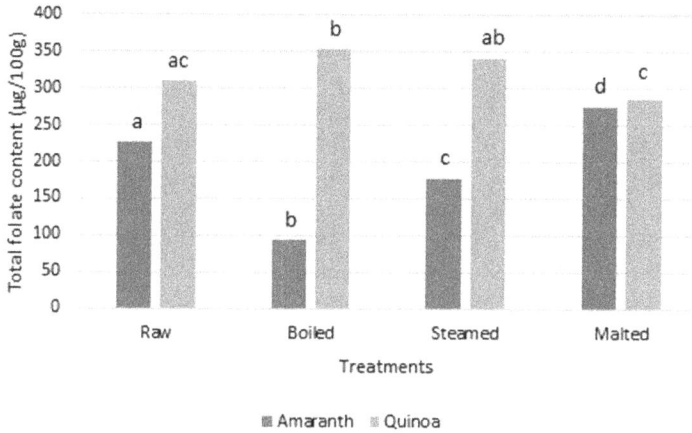

Figure 8.2   Effect of cooking methods and malting on total folate content (μg/100 g dry weight basis) in quinoa and amaranth. Different letters in the same grain indicate significant differences. **Source: Motta et al. (2017).**
Notes: Bar graph depicting effect of cooking methods and malting on total folate content (μg/100 g dry weight basis) in quinoa and amaranth. Gray bars indicate raw, boiled, steamed, and malted treatments.

TABLE 8.7   FOLATE NUTRIENT RETENTION FACTORS FOR PROCESSED QUINOA AND AMARANTH

	NRF (%)		
	Process		
Grain	Boiled	Steamed	Malted
Amaranth	41	78	121
Quinoa	118	110	92

Source: Motta et al. (2017).

lesser known examples, such as kañiwa and sacha inchi, allow us to affirm that they are excellent foods to be included in the diet. Some, such as quinoa, amaranth, and kañiwa have been shown to be mainly a source of good biological quality protein; others such as chia, sacha inchi, and turtle bean provide essential fatty acids; all contain different minerals and vitamins required for a healthy diet. Corns, mainly colored ones, contain high amounts of bioactive compounds with antioxidant activity, which can contribute to the prevention of chronic non-communicable diseases and can also be used as natural colorants.

Some of these grains contain anti-nutritional factors, which by simple methods and also by culinary processes can be eliminated; this is necessary and useful information for the consumer and for those who have never previously consumed them.

The by-products of all these crops – depending on the process they are subjected to – can potentially be raw material rich in nutritional and functional components to be incorporated into formulated foods. An example of this is the residual cakes from the extraction of chia and sacha inchi, sources of essential amino acids, unsaturated fatty acids, dietary fiber, and minerals.

The technological processes applied for the transformation of these grains and also the culinary examples produce changes in the components, nutrients, and non-nutrients, of the food which can cause losses in different proportions; to safeguard this, it is necessary that FCTs include information on the composition corrected by NRFs. When incorporating processed grains or products derived from them into a recipe, the nutrient content for calculating intakes must be adjusted to create the nutritional profile of the final product.

ACKNOWLEDGMENTS

This work was supported by grant laValSe-Food-CYTED (Ref. 119RT0567); Consejo Nacional de Investigaciones Científicas y Técnicas (CONICET) and Universidad Nacional de Jujuy, Argentina.

REFERENCES

Abderrahim, F., E. Huanatico, R. Repo-Carrasco-Valencia, S. M.Arribas, M. C. Gonzalez, and L. Condezo-Hoyos. 2012. Effect of germination on total phenolic compounds, total antioxidant capacity, Maillard reaction products and oxidative stress markers in canihua (*Chenopodium pallidicaule*). *Journal of Cereal Science* 56(2): 410–417.

Abugoch-James, L. E. 2009. Quinoa (*Chenopodium quinoa* Willd.): Composition, chemistry, nutritional, and functional properties. *Advances in Food and Nutrition Research* 58: 1–31.

Alcocer-Villacís, I. R. 2018. Determinación del contenido de azúcares reductores, totales, almidón y de los elementos minerales calcio (Ca), hierro (Fe) y magnesio (Mg) en harinas de Amaranto variedad Alegría (*Amaranthus Caudatus* L.), Chía (*Salvia Hispanica* L.), papa Puca-Shungo (*Solanum Tuberosum*), y zanahoria blanca (ArracachaXxanthorrhiza Bancroft) Bachelor's Thesis, Universidad Técnica de Ambato. Facultad de Ciencia e Ingeniería en Alimentos. Carrera de Ingeniería en Alimentos.

Arévalo, J. B., C. M. Zegarra, B. C. Amado, A. M. R. Del-Castillo, and G. Vargas-Arana 2019. Composición nutricional y capacidad antioxidante de tres especies de sacha inchi plukenetia spp. de la Amazonía Peruana. *Folia Amazónica* 28(1): 65–74.

Audu, S. S., M. O. Aremu., and L. Lajide 2013. Effects of processing on physicochemical and antinutritional properties of black turtle bean (*Phaseolus vulgaris* L.) seeds flour. *Oriental Journal of Chemistry* 29(3): 979–989.

Ayerza, R. 2010. Effects of seed color and growing locations on fatty acid content and composition of two chia *(Salvia hispanica* L.) genotypes. *Journal of the American Oil Chemists' Society* 87(10): 1161–1165.

Ayerza, R. 2013. Seed composition of two chia (*Salvia hispanica* L.) genotypes which differ in seed color. *Emirates Journal of Food and Agriculture* 25(7): 495–500.

Bañuelos-Pineda, J., C. C. Gómez-Rodiles, R. Cuéllar-José, and O. Aguirre López. 2018. *The Maize Contribution in Human Health.* Corn: Production and Human Health in Changing Climate, 29. www.intechopen.com/books/corn-production-and-human-health-in-changing-climate/the-maize-contribution-in-the-human-health

Bello-Pérez, L. A., G. A. Camelo-Mendez, E. Agama-Acevedo, and R. G. Utrilla-Coello. 2016. Aspectos nutracéuticos de los maíces pigmentados: Digestibilidad de los carbohidratos y antocianinas. *Agrociencia* 50(8): 1041–1063.

Bello-Perez, L. A., P. C. Flores-Silva, E. Agama-Acevedo, and J. Tovar. 2020. Starch digestibility: Past, present, and future. *Journal of the Science of Food and Agriculture* 100(14): 5009–5016.

Belščak-Cvitanović, A., K. Durgo, A. Huđek, V. Bačun-Družina, and D. Komes. 2018. Overview of polyphenols and their properties. In *Polyphenols: Properties, recovery, and Applications,* 3–44. Cambridge, UK: Woodhead Publishing.

Bever, A. M., A. Cassidy, E. B. Rimm, M. J. Stampfer, and D. J. Cote. 2021. A prospective study of dietary flavonoid intake and risk of glioma in US men and women. *The American Journal of Clinical Nutrition* 114(4): 1314–1327.

Bhornchai, H., S. Bhalang, T. Ratchada, M. Scott, and L. Kamol. 2014. Anthocyanin, phenolics and antioxidant activity changes in purple waxy corn as affected by traditional cooking. *Food Chemistry* 164: 510–517.

Bhosale, S. S., B. S. Agarkar, R. B. Kshirsagar, and B. M. Patil. 2021. Studies on physico-chemical properties of cereals (Rice, sorghum, finger millet, amaranth) and pulses (Green gram, black gram and chickpea). *Pharma Innovation* 10(4): 110–114.

Bravo, M., J. Reyna, and M. Huapaya, 2013. Estudio químico y nutricional de granos andinos germinados de quinua (*Chenopodium Quinoa*) y Kiwicha (*Amarantus Caudatus*). *Revista Peruana de Química e Ingeniería Química* 16(1): 54–60.

Bressani, R. 1990. Chemistry, technology, and nutritive value of maize tortillas. *Food Reviews International* 6(2): 225–264.

Bressani, R. 1991. Protein quality of high-lysine maize for humans. *Cereal Foods World* 36(9): 806–811.

Bressani, R. (2018). Composition and nutritional properties of amaranth. In *Amaranth Biology, Chemistry, and Technology,* edited by Paredes Lopez, 185–205. London: CRC Press.

Calliope, S. R., M. O. Lobo, and N. C. Sammán. 2015. Proceso de elaboración de hojuelas cocidas de quínoa (*Chenopodium quinoa* Willd). *Archivos Latinoamericanos de Nutrición* 65(4): 234–242.

Carvajal-Larenas, F. E., A. R. Linnemann, M. J. R. Nout, M. Koziol, and M. A. J. S. van Boekel. 2016. Lupinus mutabilis: Composition, uses, toxicology, and debittering. *CriticalReviews in Food Science and Nutrition* 56(9): 1454–1487.

Cázares-Sánchez, E., J. L. Chávez-Servia, Y., Salinas-Moreno, F. Castillo-González, and P. Ramírez-Vallejo. 2015. Variación en la composición del grano entre poblaciones de maíz (*Zea mays* l.) nativas de Yucatán, México. *Agrociencia* 49(1): 15–30.

Chirinos, R., G. Zuloeta, R. Pedreschi, E. Mignolet, Y. Larondelle, and D. Campos. 2013. Sacha inchi (*Plukenetia volubilis*): A seed source of polyunsaturated fatty acids, tocopherols, phytosterols, phenolic compounds and antioxidant capacity. *Food Chemistry* 141(3): 1732–1739.

Chirinos, R., O. Necochea, R. Pedreschi, and D. Campos. 2016. Sacha Inchi (*Plukenetiavolubilis* L.) shell: An alternative source of phenolic compounds and antioxidants. *International Journal of Food Science & Technology* 51: 986–993.

Chirinos R., K. Ochoa, A. Aguilar-Galvez et al. 2018. Obtaining of peptides with in vitro antioxidant and angiotensin I converting enzyme inhibitory activities from cañihua protein (*Chenopodium pallidicaule* Aellen). *Journal of Cereal Science* 83: 139–146.

Clavijo, D. B., F. V. Rodríguez, and J. E. C. Estupiñán. 2015. Utilización de *plukenetia volubilis* (sacha inchi) para mejorar los componentes nutricionales de la hamburguesa. *Enfoque UTE* 6(2): 59–76.

Coates, W. 2011. Protein content, oil content and fatty acid profiles as potential criteria to determine the origin of commercially grown chia (*Salvia hispanica* L.). *Industrial Crops and Products* 34(2): 1366–1371.

Coelho, M. S., and M. D. L. M. Salas-Mellado. 2014. Chemical characterization of chia (*Salvia hispanica* L.) for use in food products. *Journal of Food and Nutrition Research* 2(5): 263–269.

Coelho, L. M., P. M. Silva, J. T. Martins, A. C. Pinheiro, and A. A. Vicente. 2018. Emerging opportunities in exploring the nutritional/functional value of amaranth. *Food and Function* 9(11): 5499–5512.

Colín-Chávez, C., J. J. Virgen-Ortiz, L. E. Serrano-Rubio, M. A. Martínez-Téllez, and M. Astier. 2020. Comparison of nutritional properties and bioactive compounds between industrial and artisan fresh tortillas from maize landraces. *Current Research in Food Science* 3: 189–194.

Costantini, L., L. Lukšič, R. Molinari, I. Kreft, G. Bonafaccia, L. Manzi, and N. Merendino. 2014. Development of gluten-free bread using tartary buckwheat and chia flour rich in flavonoids and omega-3 fatty acids as ingredients. *Food Chemistry* 165: 232–240.

Cravioto, R. O., R. K. Anderson, E. E. Lockhart, F. de P. Miranda, and R. S. Harris.1945. Nutritive value of the Mexican tortilla. *Science* 102: 91–93.

Cuevas Espinal, M. M., and N. R. Lozano Julián 2017. Actividad antioxidante y composición de ácidos grasos, tocoferoles y tocotrienoles en tres variedades de *Chenopodium quinoa* Willdenow y elaboración de una crema dermocosmética antienvejecimiento. Thesis de grado. Lima: Universidad Nacional Mayor de San Marcos. https://cybertesis.unmsm.edu.pe/bitstream/handle/20.500.12672/7113/Cuevas_em.pdf?sequence=1.

Curti, C. A., R. N. Curti, N. Bonini, and A. N. Ramón. 2018. Changes in the fatty acid composition in bitter Lupinus species depend on the debittering process. *Food Chemistry* 263: 151–154.

Da Silva, B. P., P. C. Anunciação, J. C. da Silva Matyelka, C. M. Della Lucia, H. S. D. Martino, and H. M. Pinheiro-Sant'Ana. 2017. Chemical composition of Brazilian chia seeds grown in different places. *Food Chemistry* 221: 1709–1716.

D'Amico, S., and R. Schoenlechner. 2017. Amaranth: Its unique nutritional and health-promoting attributes. In *Gluten-Free Ancient Grains*, 131–159. Cambridge, UK: Woodhead Publishing.

Das Gupta, S., and N. Suh. 2016. Tocopherols in cancer: An update. *Molecular Nutrition and Food Research* 60: 1354–1363.

Davis, P. J., and S. C. Williams (1998). Protein modification by thermal processing. *Allergy* 53: 102–105.

De la Parra, C., S. S. O. Serna, and L. R. Hai. 2007. Effect of processing on the phytochemical profiles and antioxidant activity of corn for production of masa, tortillas, and tortilla chips. *Journal of Agriculture and Food Chemistry* 55(10): 4177–4183.

De Souza, A. H. P., A. K. Gohara, Â. C. Rodrigues, N. E. De Souza, J. V. Visentainer, and M. Matsushita. 2013. Sacha inchi as potential source of essential fatty acids and tocopherols: Multivariate study of nut and shell. *Acta Scientiarum Technology* 35(4): 757–763.

Delchier, N., C. Ringling, J. Le Grandois, D. Aoudé-Werner, R. Galland, S. Georgé, M. Rychlik, and C. M. Renard. 2013. Effects of industrial processing on folate content in green vegetables. *Food Chemistry* 139(1–4): 815–824.

Dyner, L., S. Drago, A. Pineiro, H. Sanchez, R. Gonzalez, and M. E. Valencia. 2007. Composición y aporte potencial de hierro, calcio y zinc de panes y fideos elaborados con harinas de trigo y amaranto. *Archivos Latinoamericanos de Nutrición* 57: 69–77.

Ejigui, I., L. Savoie, I. Marin, and T. Desrosiers. 2005. Antinutritional factors of red peanuts (*Arachis hypogea*) and small red kidney beans (*Phaseolus vulgaris*). *Journal of Biological Sciences* 5(5): 597–605.

Elena, M., L. Marroú, V. González, S. Elizabeth, and P. Flores 2011. Composición química de oca (*Oxalis tuberosa*), arracacha (*Arracaccia xanthorriza*) y tarwi (*Lupinus mutabilis*). Formulación de una mezcla base para productos alimenticios. *Revista Venezolana de Ciencia y Tecnología de Alimentos* 2(2): 239–252.

Escalante-Aburto, A., R. M. Mariscal-Moreno, D. Santiago-Ramos, and N. Ponce-García. 2020. An update of different nixtamalization technologies, and its effects on chemical composition and nutritional value of corn tortillas. *Food Reviews International* 36(5): 456–498.

FAO (Food and Agriculture Organization). 1985. *Joint FAO/WHO/UNU Expert Consultation on Protein-Energy Requirements*. Geneva: WHO.

FAO (Food and Agriculture Organization). 2016. Legume consumption and production has lost strength in Latin America and the Caribbean compared to more commercial crops. www.fao.org/americas/noticias/ver/es/c/455947/.

FAO (Food and Agriculture Organization) and FINUT (Fundación Iberoamericana de Nutrición). 2017. *Evaluación de la Calidad de la proteína de la dieta humana. Consulta de Expertos*. Granada: Publicado por Organización de las Naciones Unidas para la Alimentación y a Agricultura (FAO) y la Fundación Iberoamericana de Nutrición.

Fardet, A., E. Rock, and C. Remesy. 2008. Is the in vitro antioxidant potential of whole-grain cereals and cereal products well reflected *in vivo? Journal of Cereal Science* 48(2): 258–276.

Fernández Mejía, J. L., and C. L. Guivar Delgado. 2020. Formulación de harina proteica y extruida a base de harina de: arveja (*Pisum sativum*), kiwicha (*Amaranthus caudatus*) y tarwi (*Lupinus Mutabilis*). https://alicia.concytec.gob.pe/vufind/Record/UPRG_dbc98eb5d2d403153f75cbfcd8b29b7b/Details.

Filho, A. M. M., M. R. Pirozi, J. T. D. S. Borges, H. M. Pinheiro Sant'Ana, J. B. P. Chaves, and J. S. D. R. Coimbra. 2017. Quinoa: nutritional, functional, and anti-nutritional aspects. *Critical Reviews in Food Science and Nutrition* 57(8): 1618–1630.

Gade, D. W. 1970. Ethnobotany of cañihua (*Chenopodium pallidicaule*), rustic seed crop of the Altiplano. *Econonic Botany* 24(1): 55–61.

Galinat, W. C. 1971. The origin of maize. *Annual Review of Genetics* 5(1): 447–478.

Ganesan, B., C. Brothersen, and D. J. McMahon. 2014. Fortification of foods with omega-3 polyunsaturated fatty acids. *Critical Reviews in Food Science and Nutrition* 54(1): 98–114.

García, D. C., A. Mateo del Pino, and N. Pascual Soler 2015. Comidas bastardas. Gastronomía, tradición e identidad en América Latina. *Revista de Filología y Lingüística de la Universidad de Costa Rica* 41(2): 198–202.

Gargiulo, L., Å Grimberg, R. Repo-Carrasco-Valencia, A. S. Carlsson, and G. Mele. 2019. Morpho-densitometric traits for quinoa (Chenopodium quinoa Willd.) seed phenotyping by two X-ray micro-CT scanning approaches. *Journal of Cereal Science* 90: 102829.

Giménez, M. A., C. N. Segundo, M. O. Lobo, and N. Samman. 2020. Physicochemical and techno-functional characterization of native corn reintroduced in the Andean Zone of Jujuy, Argentina. *Procceding la ValSe-Food* 53(1): 7.

González, J. A., Y. Konishi, M. Bruno, M. Valoy, and F. E. Prado. 2012. Interrelationships among seed yield, total protein and amino acid composition of ten quinoa (*Chenopodium quinoa*) cultivars from two different agroecological regions. *Journal of the Science of Food and Agriculture* 92(6): 1222–1229.

Grundy, M. M., D. K. Momanyi, C. Holland, F. Kawaka, S. Tan, M. Salim, and W. O. Owino. 2020. Effects of grain source and processing methods on the nutritional profile and digestibility of grain amaranth. *Journal of Functional Foods* 72: 104065.

Gualberto, D., C. Bergman, M. Kazemzadeh, and C. Weber. 1997. Effect of extrusion processing on the soluble and insoluble fiber and phytic acid contents of cereal brands. *Plant Foods for Human Nutrition* 51: 187–198.

Guillén-Sánchez, J., S. Mori-Arismendi, and L. M. Paucar-Menacho. 2014. Características y propiedades funcionales del maíz morado (*Zea mays* L.) var. subnigroviolaceo. *Scientia Agropecuaria* 5(4): 211–217.

Gutiérrez, L. F., L. M. Rosada, and A. Jiménez. 2011. Chemical composition of Sacha Inchi (*Plukenetia volubilis* L.) seeds and characteristics of their lipid fraction. *Grasas y aceites* 62(1): 76–83.

Guzman-Maldonado, S., and O. Paredes-Lopez. 1998. Functional products of plants indigenous to Latin America. Amaranth and quinoa, common beans and botanicals. In *Functional Foods. Biochemical and Processing Aspects*, edited by G. Mazza, 293–328. Lancaster: *Technomic Publishing Company*.

Guzman-Maldonado S. H., M. G. Vazquez-Carrillo, J. A. Aguirre-Gómez, and I. Serrano-Fujarte. 2015. Contenido de ácidos grasos, compuestos fenólicos y calidad industrial de maíces nativos de Guanajuato. *Revista Fitotecnia Mexicana* 8(2): 213–222.

Hall, C., C. Hillen, and J. G. Robinson. 2017. Composition, nutritional value, and health benefits of pulses. *Cereal Chemistry* 94(1): 11–31.

Harnly, J. 2015. Antioxidant methods. *Journal of Food Composition and Analysis* 64(2): 145–146

Haros, C. M., and R. Schöenlechner (Eds.). 2017. *Pseudocereals: Chemistry and Technology.* Hoboken, NJ: John Wiley and Sons.

Hayat, I., A. Ahmad, T. Masud, A. Ahmed, and S. Bashir. (2014). Nutritional and health perspectives of beans (*Phaseolus vulgaris* L.): An overview. *Critical Reviews in Food Science and Nutrition* 54(5): 580–592.

He, S., B. K. Simpson, H. Sun, M. O. Ngadi, Y. Ma, and T. Huang. 2015. Phaseolus vulgaris lectins: A systematic review of characteristics and health implications. *Critical Reviews in Food Science and Nutrition* 58(1): 70–83.

Hefni, M., and C. M. Witthöft. 2011. Increasing the folate content in Egyptian baladi bread using germinated wheat flour. *LWT – Food Science and Technology* 44: 706–712.

Hemalatha, S., K. Platel, and K. Srinivasan. 2007. Zinc and iron contents and their bioaccessibility in cereals and pulses consumed in India. *Food Chemistry* 102: 1328–1336.

Hernández, A. D. S. C. 2018. Composición química, características de calidad y actividad antioxidante de pasta enriquecida con harina de amaranto y hoja de amaranto deshidratada. http://ri-ng.uaq.mx/handle/123456789/921.

Huaccho-Human, C. V., and M. Lope-Surichaqui. 2007. Elaboración de una mezcla alimenticia a base de tarwi (*lupinus mutabilis sweet*), Quinua (*chenopodium quinoa willd*), Maca (*lepidium peruvianum chacón*), y Lúcuma *(pouteria lucuma)* mediante extrusión. Doctoral dissertation, Universidad Nacional del Centro de Peru, Junin, Peru. http://hdl.handle.net/20.500.12894/1872.

Jing, P., and M. Giusti. 2007. Effects of extraction conditions on improving the yield and quality of an anthocyanin-rich purple corn (*Zea mays* L.) color extract. *Journal of Food Science* 72(2): 366–368.

Katunzi-Kilewela, A., L. D. Kaale,O. Kibazohi, and L. M. Rweyemamu. 2021. Nutritional, health benefits and usage of chia seeds (Salvia hispanica): A review. *African Journal of Food Science* 15(2): 48–59.

Khattab, R. Y., and M. A. Zeitoun. 2013. Quality of flaxseed oil obtained by different extraction techniques. *LWT – Food Science and Technology* 53: 338–345.

Koehler, P., G. Hartmann, H. Wieser and M. Rychlik. 2007. Changes of folates, dietary fiber, and proteins in wheat as affected by germination. *Journal of Agriculture and Food Chemistry* 55: 4678–4683.

Laurente Flores, Y. R. 2016. Obtención del concentrado proteico y determinación del perfil de aminoácidos de dos variedades de Tarwi (*Lupinus mutabilis Sweet*). Doctoral dissertation. Universidad Nacional del Altiplano, Puno, Peru. https://repositorioslatin oamericanos.uchile.cl/handle/2250/3275226.

López-Martinez, L., K. L. Parkin, and H. S. García. 2011. Phase II – inducing, polyphenols content and antioxidant capacity of corn (*Zea mays* L.) from phenotypes of white, blue, red and purple colors processed into masa and tortillas. *Plant Foods for Human Nutrition* 66(1): 41–47.

Los, F. G. B., A. A. F. Zielinski, J. P. Wojeicchowski, A. Nogueira and I. M. Demiate 2018. Beans (Phaseolus vulgaris L.): Whole seeds with complex chemical composition. *Current Opinion in Food Science* 19: 63–71.

Lv, J. S., X. Y. Liu, X. P. Zhang and L. S. Wang. 2017. Chemical composition and functional characteristics of dietary fiber-rich powder obtained from core of maize straw. *Food Chemistry* 227: 383–389.

Mackay, W., T. Davis, and D.Sankhla. (1995). Influence of scarification and temperature treatments on seed germination of Lupinus havardii. *Seed Science and Technology* 23(3): 815–821.

Mattila, P., J. Astola, and J. Kumpulainen. 2000. Determination of flavonoids in plant material by HPLC with diode-array and electro-array detections. *Journal of Agricultural and Food Chemistry* 48(12): 5834–5841.

Maureen, M., A. N. Kaaya, J. Kauffman, C. Narrod and A. Atukwase. 2020. Enhancing nutritional benefits and reducing mycotoxin contamination of maize through nixtamalization. *Journal of Biological Sciences* 20: 153–162.

Maurer, N. E., B. Hatta-Sakoda, G. Pascual-Chagman, and L. E Rodriguez-Saona 2012. Characterization and authentication of a novel vegetable source of omega-3 fatty acids, sacha inchi (*Plukenetia volubilis* L.) oil. *Food Chemistry* 134(2): 1173–1180.

Mazewski, C., K. Liang, and E. G. de Mejia. 2017. Inhibitory potential of anthocyanin-rich purple and red corn extracts on human colorectal cancer cell proliferation in vitro. *Journal of Functional Foods* 34: 254–265.

Mburu, M. W. 2016. *Properties of a Complementary Food based on Grain Amaranth (Amaranthus cruentus)*. Doctoral dissertation. Food Science and Technology, JKUAT) http://journals.jkuat.ac.ke/index.php/pgthesis_abs/article/view/408.

Menard, S., N. Cerf-Bensussan, and M. Heyman. 2010. Multiple facets of intestinal permeability and epithelial handling of dietary antigens. *Mucosal Immunology* 3(3): 247–259.

Miranda, M., A. Vega-Gálvez, I. Quispe-Fuentes, M. J. Rodríguez, H. Maureira, and E. A. Martínez. 2012. Nutritional aspects of six quinoa (*Chenopodium quinoa* Willd.) ecotypes from three geographical areas of Chile. *Chilean Journal of Agricultural Research* 72(2): 175.

Motta, C., A. C. Nascimento, M. Santos, I. Delgado, I. Coelho, A. Rego, A. S Matos, D. Torres, and I. Castanheira. 2016a. The effect of cooking methods on the mineral content of quinoa (*Chenopodium quinoa*), amaranth (*Amaranthus* sp.) and buckwheat (*Fagopyrum esculentum*). *Journal of Food Composition and Analysis* 49: 57–64.

Motta, C., M. Santos, R. Mauro, N. Samman, A. S. Matos, D. Torres, and I. Castanheira .2016b. Protein content and amino acids profile of pseudocereals. *Food Chemistry* 193: 55–61.

Motta, C., I. Delgado, A. S. Matos, G. B. Gonzales, D. Torres, M. Santos, V. Chandra-Hioe, J. Arcot, and I. Castanheira. 2017. Folates in quinoa (Chenopodium quinoa), amaranth (Amaranthus sp.) and buckwheat (Fagopyrum esculentum): Influence of cooking and malting. *Journal of Food Composition and Analysis* 64: 181–187.

Motta, C., I. Castanheira, G. B. Gonzales, I. Delgado, D. Torres, M. Santos, and A. S. Matos. 2019. Impact of cooking methods and malting on amino acids content in amaranth, buckwheat and quinoa. *Journal of Food Composition and Analysis* 76: 58–65.

Muller-Tito, K. 2015. Capacidad antioxidante y contenido de flavonoides entre las semillas de chia negra (*Salvia Nativa*) y chia blanca (*Salvia hispánica* L.). Bachelor dissertation. Universidad Nacional del Altiplano, Puno, Peru.

Muñoz, L. A., A. Cobos, O. Diaz, and J. M. Aguilera. 2013. Chia seed (*Salvia hispanica*): an ancient grain and a new functional food. *Food Reviews International* 29(4): 394–408.

Nascimento, A. C., C. Mota, I. Coelho, S. Gueifão, M. Santos, A. S. Matos, A. Gimenez, M. Lobo, N. Samman, and I. Castanheira. 2014. Characterization of nutrient profile of quinoa (*Chenopodium quinoa*), amaranth (*Amaranthus caudatus*), and purple corn (*Zea mays* L.) consumed in the north of Argentina: Proximate, minerals and trace elements. *Food Chemistry* 148: 420–426.

Navarro Cortez, R. O., C. A. Gómez-Aldapa, E. Aguilar-Palazuelos, E. Delgado-Licon, J. Castro Rosas, J. Hernández-Ávila, A. Solís-Soto, L. A. Ochoa-Martínez, and H. Medrano-Roldán. 2016. Blue corn (*Zea mays* L.) with added orange (*Citrus sinensis*) fruit bagasse: Novel ingredients for extruded snacks. *CyTA Journal of Food* 14(2): 349–358.

Nelson, K., L. Stojanovska, T. Vasiljevic, and M. Mathai. 2013. Germinated grains: A superior whole grain functional food? *Canadian Journal of Physiology and Pharmacology* 91(6): 429–441.

Nuss, E. T., and S. A. Tanumihardjo. 2010. Maize: A paramount staple crop in the context of global nutrition. *Comprehensive Reviews in Food Science and Food Safety* 9(4): 417–436.

Oboh, G., A. O. Ademiluyi, and A. A. Akindahunsi. 2010. The effect of roasting on the nutritional and antioxidant properties of yellow and white maize varieties. *International Journal of Food Science and Technology* 45(6): 1236–1242.

Pachon, H., D. A. Ortiz, C. Araujo, M. W. Blair, and J. Restrepo. 2009. Iron, zinc, and protein bioavailability proxy measures of meals prepared with nutritionally enhanced beans and maize. *Journal of Food Science* 74: 147–154.

Paredes López, O., F. Guevara Lara, and L. A. Bello Pérez. 2009. La nixtamalización y el valor nutritivo del maíz. *Ciencias* 92: 60–70. www.revistacienciasunam.com/es/component/content/article/41-revistas/revista-ciencias-92-93/205-la-nixtamalizacion-y-el-valor-nutritivo-del-maiz-05.html.

Pasini, G., B. Simonato, M. Giannattasio, A. D. Peruffo, and A. Curioni (2001). Modifications of wheat flour proteins during in vitro digestion of bread dough, crumb, and crust: An electrophoretic and immunological study. *Journal of Agricultural and Food Chemistry* 49(5): 2254–2261.

Paucar-Menacho, L. M., R. Salvador-Reyes, J. Guillén-Sánchez, J. Capa-Robles, and C. Moreno-Rojo. 2015. Estudio comparativo de las características físico-químicas del aceite de sacha inchi (*Plukenetia volubilis* L.), aceite de oliva (*Olea europaea*) y aceite crudo de pescado. *Scientia Agropecuaria* 6(4): 279–290.

Peñarrieta, J. M., J. A. Alvarado, B. Åkesson, and B. Bergenståhl. 2008. Total antioxidant capacity and content of flavonoids and other phenolic compounds in canihua (*Chenopodium pallidicaule*): an Andean pseudocereal. *Molecular Nutrition and Food Research* 52: 708–717.

Perez-Hernandez, L. M., K. Nugraheni, M. Benohoud, W. Sun, A. J. Hernández-Álvarez, M. R. A. Morgan, C. Boesch, and C. Orfila. 2020. Starch digestion enhances bioaccessibility of anti-inflammatory polyphenols from borlotti beans (Phaseolus vulgaris). *Nutrients* 12(2): 295.

Písaříková, B., S. Kráčmar, and I. Herzig. 2005. Amino acid contents and biological value of protein in various amaranth species. *Czech Journal of Animal Science* 50(4): 169–174.

Ramos Diaz, J. M., S. Kirjoranta, S. Tenitz, P. A. Penttilä, R. Serimaa, A. M. Lampi, and K. Jouppila. 2013. Use of amaranth, quinoa and kañiwa in extruded corn-based snacks. *Journal of Cereal Science* 58(1): 59–67.

Ranilla, L. G. 2019. Bioactive ingredients from corn and lactic acid bacterial biotransformation. In *Functional Foods and Biotechnology*, 19–45). Boca Raton, FL: CRC Press.

Rawdkuen, S., D. Murdayanti, S. Ketnawa, and S. Phongthai. 2016. Chemical properties and nutritional factors of pressed-cake from tea and sacha inchi seeds. *Food Bioscience* 15: 64–71.

Repo-Carrasco-Valencia, R. 2011. Andean Indigenous Food Crops: Nutritional Value and Bioactive Compounds. Doctoral dissertation on Food Chemistry, Finland: University of Turku.

Repo-Carrasco-Valencia, R., C. Espinoza, and S. E. Jacobsen. 2003. Nutritional value and use of the Andean crop's quinoa (*Chenopodium quinoa*) and kañiwa (*Chenopodium pallidicaule*). *Food Reviews International* 19: 179–189.

Repo-Carrasco-Valencia, R., A. de la Cruz, J. C. Alvarez, and H. Kallio. 2009. Chemical and functional characterization of kañiwa (*Chenopodium pallidicaule*) grain, extrudate and bran. *Plant Foods for Human Nutrition* 64: 94–101.

Repo-Carrasco-Valencia, R., C. R. Encina, C. Binaghi, M. Greco, and P. Ronayne de Ferrer. 2010. Effects of roasting and boiling of quinoa, kiwicha and kaniwa on composition and availability of minerals in vitro. *Journal of Science of Food and Agriculture* 90: 2068–2073.

Repo-Carrasco-Valencia R., S. Melgarejo-Cabello, and J. M. Pihlava. 2019. Nutritional value and bioactive compounds in quinoa (*Chenopodium quinoa* Willd.), kañiwa (*Chenopodium pallidicaule* Aellen) and kiwicha (*Amaranthus caudatus* L.). In *Quinoa: Cultivation, Nutritional Properties and Effects on Health*, edited by P. Peiretti and F. Gai, 83–113. New York: Nova Science Publishers.

Rincon-Londoño, N., L. J. Vega-Roja, M. Contreras-Padilla, A. A. Acosta-Osorio, and M. E. Rodrıguez-Garcıa. 2016. Analysis of the pasting profile in corn starch: Structural, morphological, and thermal transformations. Part I. *International Journal of Biological Macromolecules* 91: 106–114.

Ross, A.B., M. J. Shepherd, and M. Schüpphaus. 2003. Alkylresorcinols in cereals and cereal products. *Journal of Agriculture and Food Chemistry* 51: 4111–4118.

Rossi, M. C., M. N. Bassett, and N. C. Samman. 2018. Dietary nutritional profile and phenolic compounds consumption in school children of highlands of Argentine Northwest. *Food Chemistry* 238: 111–116.

Ruiz, C., C. Diaz, J. Anaya, and R. Rojas. 2013. Análisis proximal, antinutrientes, perfil de ácidos grasos y de aminoácidos de semillas y tortas de 2 especies de Sacha inchi (*Plukenetia volubilis y Plukenetia huayllabambana*). *Revista de la Sociedad Química del Perú* 79(1): 29–36.

Sabelino-Francia, Z. D. P. 2020. Modelos de calibración del contenido de proteína y fenólicos totales usando espectroscopia del infrarrojo medio en Tarwi (*Lupinus mutabilis*). Doctoral dissertation, Universidad Nacional Agraria La Molina, Lima, Peru. https://repositorio.lamolina.edu.pe/handle/20.500.12996/4428.

Salvador-Reyes, R., and M. T. Pedrosa-Silva-Clerici. 2020. Peruvian Andean maize: General characteristics, nutritional properties, bioactive compounds, and culinary uses. *Food Research International* 130: 108934.

Salvatierra-Pajuelo, Y. M., M. E. Azorza-Richarte, and L. M. Paucar-Menacho. 2019. Optimización de las características nutricionales, texturales y sensoriales de cookies enriquecidas con chía (*Salvia hispánica*) y aceite extraído de tarwi (*Lupinus mutabilis*). *Scientia Agropecuaria* 10(1): 7–17.

Sánchez Ortega, I., and E. Pérez-Urria-Carril. 2014. Maíz I (*Zea mays*). *REDUCA Biologia* 7(2): 151–171.

Sánchez-Sánchez G. L. 2012. Caracterización y cuantificación de los ácidos grasos omega 3 y omega 6 presentes en el aceite de sacha inchi (Plukenetia volubilis L). Escuela de Química. Masters thesis. Bogotá: Universidad Nacional de Colombia. https://repositorio.unal.edu.co/handle/unal/11657.

Serna-Saldivar, S. O., and C. Chuck-Hernandez. 2019. Food uses of lime-cooked corn with emphasis in tortillas and snacks. In *Corn* (3rd edn). *Chemical and Technology*, Chap 17, 469–500. https://doi.org/10.1016/B978-0-12-811971-6.00017-.

Shah, Tajamul Rouf, K. Prasad, P. Kumar, and F. Yildiz. 2016. Maize. A potential source of human nutrition and health: A review. *Cogent Food and Agriculture* 2(1): 1166995.

Shohag, M. J. I., Y. Wei, and X. Yang. 2012. Changes of folate and other potential health-promoting phytochemicals in legume seeds as affected by germination. *Journal of Agriculture and Food Chemistry* 60: 9137–9143.

Sierra-Macías, M., P. Andrés-Meza, A. Palafox-Caballero, and I. Meneses-Márquez. 2016. Diversidad genética, clasificación y distribución racial del maíz nativo en el estado de Puebla, México. *Revista de Ciencias Naturales y Agropecuarias* 3(9): 12–21.

Siyuan-Sheng, L. T., and H. Liu-Rui. 2018. Corn phytochemicals and their health benefits. *Food Science and Human Wellness* 7: 185–195.

Smorowska, A. J., A.K. Żołnierczyk, A. Nawirska-Olszańska, J. Sowiński, and A. Szumny. 2021. Nutritional properties and in vitro antidiabetic activities of blue and yellow corn extracts: A comparative study. *Journal of Food Quality* https://doi.org/10.1155/2021/8813613.

Stea, T. H., M. Johansson, M. Jägerstad, and W. Frølich. 2007. Retention of folates in cooked, stored and reheated peas, broccoli and potatoes for use in modern large-scale service systems. *Food Chemistry* 101: 1095–1107.

Stikić, R. I., D. D. Milinčić, A. Ž. Kostić, Z. B. Jovanović, U. M. Gašić, Ž. L.Tešić, N. Z. Djordjević, S. K. Savić, B. G. Czekus, and M. B. Pešić. 2020. Polyphenolic profiles, antioxidant, and in vitro anticancer activities of the seeds of Puno and Titicaca quinoa cultivars. *Cereal Chemistry* 97(3): 626–633.

Suárez, P. A., J. G. Martínez, and J. R. Hernández. 2013. Amaranto: Efectos en la nutrición y la salud. *Tlatemoani: Revista Académica de Investigación* 12: 1. https://econpapers.repec.org/article/ervtlatem/y_3a2013_3ai_3a12_3a14.htm.

Suca, G. R. 2015. Potencial del tarwi (*Lupinus mutabilis* Sweet) como futura fuente proteínica y avances de su desarrollo agroindustrial. *Revista Peruana de Química e Ingeniería Química* 18(2): 55–71.

Ullah, R., M. Nadeem, A. Khalique, M. Imran, S. Mehmood, A. Javid, and J. Hussain. 2016. Nutritional and therapeutic perspectives of Chia (*Salvia hispanica* L.): A review. *Journal of Food Science and Technology* 53(4): 1750–1758.

Valenzuela-Zamudio, F., and M. R. Segura-Campos. 2020. Amaranth, quinoa and chia bioactive peptides: A comprehensive review on three ancient grains and their potential role in management and prevention of Type 2 diabetes. *Critical Reviews in Food Science and Nutrition*: 62(10): 2707–2721.

Valles Ramírez, S. M. 2012. Obtención de Leche de Sacha Inchi (*Plukenetia Volubilis* Linneo). Tesis Ingeniería. Tarapoto: Universidad Nacional de San Martín, Facultad de Ingeniería Agroindustrial. https://repositorio.unsm.edu.pe/handle/11458/2271.

Vera, J. L. 2014. Perfil de ácidos grasos en granos tres cultivares de quinua (*Chenopodium quinoa* Willd.) sometidos a tres tipos de procesamiento. *Revista Investigaciones Altoandinas* 16(1): 13–20.

Vilcacundo, R., and B. Hernández-Ledesma. 2017. Nutritional and biological value of quinoa (Chenopodium quinoa Willd.). *Current Opinion in Food Science* 14: 1–6.

Villanueva, E., G. Rodríguez, E. Aguirre, and V. Castro. 2017. Influence of antioxidants on oxidative stability of the oil chia (*Salvia hispanica* L.) by rancimat. *Scientia Agropecuaria* 8(1): 19–27.

Wang, N., D. W. Hatcher, R. T. Tyler, R. Toews, and E. J. Gawalko. 2010. Effect of cooking on the composition of beans (*Phaseolus vulgaris* L.) and chickpeas (*Cicer arietinum* L.). *Food Research International* 43(2): 589–594.

Wang, S., F. Zhu, and Y. Kakuda. 2018. Sacha inchi (*Plukenetia volubilis* L.): Nutritional composition, biological activity, and uses. *Food Chemistry* 265: 316–328

WHO (World Health Organization) and United Nations University. 2007. *Protein and Amino Acid Requirements in Human Nutrition* (Vol. 935). Geneva: World Health Organization.

Yang, C., and Suh, N. 2013. Cancer prevention by different forms of tocopherols. *Topics in Current Chemistry* 329: 21–33.

Zare, T., T. W. Rupasinghe, B. A. Boughton, and U. Roessner. 2019. The changes in the release level of polyunsaturated fatty acids (ω-3 and ω-6) and lipids in the untreated and water-soaked chia seed. *Food Research International* 126: 108665.

Zettel, V., and B. Hitzmann. 2018. Applications of chia (*Salvia hispanica* L.) in food products. *Trends in Food Science & Technology* 80: 43–50.

Zhang, H., and R. Tsao. 2016. Dietary polyphenols, oxidative stress and antioxidant and anti-inflammatory effects. *Current Opinion in Food Science* 8: 33–42.

Chapter 9

Contributions from Latin-American Grains to Nutrition and Health

Carla Motta[1], Norma Sammán[2], and Isabel Castanheira[1]
[1]National Institute of Health Doutor
Ricardo Jorge, Lisbon, Portugal
[2]Universidad Nacional de Jujuy and Consejo
Nacional de Investigaciones Científicas y Técnicas
(CONICET), San Salvador de Jujuy, Argentina

CONTENTS

9.1 INTRODUCTION

Health effects from Andean crops have been a matter of discussion in recent years. In this chapter, we are addressing the health benefits, and evidence regarding the influence of Latin-American grains such as quinoa (*Chenopodium quinoa*), kiwicha (*Amaranthus caudatus*), kañiwa (*Chenopodium pallidicaule*), chia (*Salvia hispanica*), and legumes such as black turtle bean (*Phaseolus vulgaris*) and tarwi (*Lupinus mutabilis*) on the improvement and maintenance of different health-related aspects. The impact of these grains has shown promising results on disorders like cancer, type-2 diabetes, allergic conditions, celiac disease, and on the modulation of cardiovascular risk factors (through hypocholesterolemic properties) either in *in vitro* or *in vivo* approaches or clinical studies with biological evaluations. Some of the grains, like quinoa and amaranth, have been reported as gluten-free foods, revealing their importance to celiac disease or non-celiac gluten sensitivity patients, either in the maintenance of a gluten-free diet or in the reduction of nutritional deficiencies, respectively (Bergamo et al., 2011; Zevallos et al., 2014). Due to the high content of bioactive peptides and antioxidant molecules (such as phenolic compounds) and micronutrient availability of minerals and vitamins of these grains, different impacts on immunological and cellular activity can be observed (Ayyash et al., 2019; Lee and Joo, 2018; Tang and Tsao, 2017). Some process methods (e.g., *fermentation, germination*) have recently been reported to increase the bioaccessibility of some of these nutrients. These studies also suggest that prebiotic and probiotic effects on microbiota modulation can be associated with specific health benefits. Fast-growing *in vitro* and *in vivo* studies evidence that Andean grain consumption, like quinoa and amaranth, can modify specific intestinal bacteria (Gullón et al., 2016). Intestinal microbiota can be associated, for instance, with the decrease of obesity levels and inflammation-mediated chronic disorders, working as a modulator for several chronic disease biomarkers (Ugural and Akyol, 2020). Compared with other grains or legumes used in western diets, Latin-American grains are a good alternative for a balanced diet to prevent noncommunicable diseases, the major causes of disability, ill-health, health-related retirement, and premature death.

9.2 NONCOMMUNICABLE DISEASES: DIABETES MELLITUS, CARDIOVASCULAR RISK FACTORS AND OBESITY

9.2.1 Diabetes Mellitus

Diabetes mellitus is a chronic and metabolic disease characterized by high levels of blood glucose (or blood sugar). High glucose levels lead over time to severe health conditions: damage to the kidneys, heart, blood vessels, eyes, and nerves. Type 2 diabetes is common in adults and is characterized by insulin resistance or low insulin production. Diabetes is a global health problem, presenting high prevalence and an incidence that rises every day. In 2016, WHO reported 422 million people worldwide with diabetes, especially in low- and middle-income countries (WHO, 2021).

Several research publications are available concerning the association between dietary factors and the increase in the incidence of type 2 diabetes. Comprehensive systematic reviews and meta-analyses have been published in recent years showing evidence of association between dietary behaviors or diet quality and this pathology (Neuenschwander et al., 2019). Some studies have also demonstrated strong evidence of the correlation between higher consumption of whole grains, beans, nuts, or grains, and dietary fiber and the decrease of the incidence of type 2 diabetes (Bellou et al., 2018; Micha et al., 2017; Schwingshackl et al., 2017).

As already described throughout the book, almost all Latin-American crops, like quinoa, amaranth, kañiwa, or black turtle beans, are rich in dietary fiber and should be used as a healthy base diet to prevent and control type 2 diabetes and obesity.

Indeed, amaranth's capacity to modulate glycemic and insulinemic response was studied by Guerra-Matias and Arêas (2005), showing that the presence of dietary fiber is correlated with a decrease in glycemic response. The content of pectins, the main fiber fraction not digested and not absorbed in the small intestine, can be responsible for the decrease of rises in glucose after meals and for a satiety effect. The same conclusion was addressed in a review study conducted by Marventano et al. (2017), which resumes several clinical trials that evaluate and demonstrate the effects in healthy subjects, with high consumption of whole grain diets rich in fiber, the improvement of postprandial glucose and insulin response.

Graf et al. (2014), in their studies using mice observed that the hyperglycemic condition improved with quinoa consumption. Biologically active phytoecdysteroids, flavonoids, and proteins present at high levels in quinoa grains lead to lower fasting blood glucose in obese, hyperglycemic mice. The same conclusion was reported by the Fornasini team in two different human studies addressing lupin consumption (Baldeón et al., 2012; Fornasini et al., 2012). Purple corn extracts are characterized by a rich composition of anthocyanins and functional phenolics (Baldeón et al., 2012). Anthocyanins have been reported to exhibit antiangiogenic, antidiabetic, and anticarcinogenic properties (Burton-Freeman et al., 2019; Sancho and Pastore, 2012). In *in vitro* studies, the anthocyanins present in purple corn extracts can attenuate high glucose levels, retarding diabetes-associated renal fibrosis (Li et al., 2012). Recently, other authors in an *in silico* essay refer to the role and importance of purple maize, due to its phenolic compounds, in controlling obesity, diabetes, and anti-inflammatory action (Zhang et al., 2019; Damián-Medina et al., 2020). *In vitro* and clinical studies address the importance of essential nutrients like amino acids, phytochemicals (phenolic acids and flavonoids), fatty acids, especially the polyunsaturated fatty acids (PUFAs), carotenoids, and tocopherols, present in quinoa and amaranth as crucial in lowering the risk of type 2 diabetes by assessing antihyperglycemic and antihypertension activity (Tang and Tsao, 2017).

9.2.2 Cardiovascular Risk Factors

Together, diabetes, obesity, and cardiovascular diseases (CVD), where hypertension is included, are responsible for more deaths than any other conditions. Approximately 10 million people die each year because of hypertension (Frieden and Jaffe, 2018). Reduction of sodium from food and increasing legume consumption, like beans (Bazzano et al., 2011; Jiang et al., 2020; Mudryj et al., 2014) and diets that include Andean grains such as quinoa and amaranth can reduce the risk factors in the affected individuals (Chmelík, et al., 2019; Jiang et al., 2020; Tang and Tsao, 2017).

Quinoa and amaranth provide all the essential amino acids (Motta et al., 2019), which have been identified as modulators of several systemic pathways due to their high nutraceutical potential demonstrating benefits for human health (Orona-Tamayo et al., 2019). Grain amino acids and peptides produced by protein hydrolysis are associated with a wide range of beneficial effects such as modulation of blood pressure and lipid metabolism; with anticancer, immunomodulatory, anti-thrombotic, anti-atherosclerotic, and anti-inflammatory potential as well (Orona-Tamayo et al., 2019; Millán-Linares et al., 2014). Latin-American ancient grains such as amaranth, quinoa, maize, common bean, and chia grains have been identified as important sources of these bioactive peptides, with potential anti-inflammatory activity, by several authors (López et al., 2019; Orona-Tamayo et al., 2019; Valenzuela et al., 2020). The anti-inflammatory activity was tested in cell cultures, demonstrating an essential role in preventing chronic diseases, like diabetes and CVD associated with chronic inflammation (Millán-Linares et al., 2014). Lupins, due to their protein profile content, also contributed to the plasma low density lipoprotein (LDL) cholesterol modulation. Bähr et al. (2013) reported that lupin protein could positively impact cardiovascular risk factors, especially in higher hypercholesterolemic subjects. Besides bioactive peptides, the antioxidant activity also demonstrated by these grains provides important molecules that can modulate immune response. The role of bioactive molecules and the antioxidant activity provided by the consumption of ancient grains may have a positive impact on several functions in the chronic disease pathways with health benefits that have been identified in several studies in recent years including illness prevention (López et al., 2019; Orona-Tamayo et al., 2019).

9.2.3   Obesity

With the advance of knowledge of Andean grains' chemical composition, their use as an anti-obese food is growing at the same rate at which obesity has become a serious disease. World Health Organization (WHO, 2021) reports more than two billion overweight people in the world and several comorbidities associated with obesity such as diabetes, hypertension, cardiovascular diseases; and recently, their correlation with the increase of death risk because of SARS-CoV-2 infection. Furthermore, levels of hormones such as leptin and ghrelin, with influence on metabolic homeostasis, are increased in obese subjects. Other biochemical compounds, such as cytokines and interleukins released by adipose tissues, seem to be linked to insulin resistance and atherosclerosis. Since obesity is a disease related to genomics, diets, lifestyle, and environmental factors, interest in Andean grains as anti-obesity activity foods is growing mainly due to their chemical components such as fiber and flavonoids. An intervention study reported that patients under Salba-chia (*Salvia hispanica* L.) treatment lose more weight than do the control patients. Participants on this diet maintained glycemic control and a reduction of risk factors (Vuksan et al., 2017). Reductions in C-reactive protein (hs-CRP) and increased adiponectin concentrations were observed. At present, mechanisms underlying the observed effects remain unknown, and more studies are needed. Studies on chemical composition reveal that Salba-chia is lignin-free with low available carbohydrate fraction and a rich source of magnesium, calcium, and iron; and rich in antioxidants. Information on antinutrients is lacking. It can also be observed that not all literature studies agree.

Mithila and Khanum (2015) evaluated the effect of amaranth and quinoa in rats. In this study, food intake and satiety were evaluated by monitoring biochemical parameters. Leptin, produced by adipose cells, and ghrelin segregated by cells of the gastrointestinal tract, are regulated by each other. In this work, a decrease in ghrelin was observed, and an improvement in leptin and cholecystokinin was detected, resulting in hunger and satiety changes. A faster response to glucose blood level was observed in rats submitted to Andean grain diets. The studies also linked the richness of the amino acids profile and the fiber of Andean crops as quinoa with the observed results on reducing obesity (Mithila and Khanum, 2015).

Recent research demonstrated the anti-obesity properties of quinoa in obese mice. Animals receiving a quinoa diet were revealed to have a reduced level of plasma cholesterol, interleukin, and hepatic steatosis (Martínez-Villaluenga et al., 2020). Other research has focused on the mechanism underlying the effects of this grain on increasing satiety with a concomitant decrease in leptin (Tang and Tsao, 2017).

A clinical trial was conducted among Australian participants with different body mass indexes (25 < BMI < 40), where subjects were segregated into over-weight: obese type I; obese type II; and obese type III groups. Reduction in triglycerides in serum and metabolic syndrome was correlated with the quinoa dose. For the subjects that consumed 50 g/day of quinoa, better results were obtained, with a decrease of 70% in the prevalence of metabolic syndrome, than those who received 25 g/day (Navarro-Perez et al., 2017).

Chaiittianan and Sutthanut (2017) demonstrated *in vivo* experiments that purple corn has anti-obesity effects, with action in the adipose cycle through inhibition of adipogenesis and induction on lipolysis and apoptosis in adipocytes (Chaiittianan et al., 2017). Several works suggest that polyphenols including anthocyanins, quercetin, and phenolic acids and derivatives are the main ones responsible for these effects. However, since weak evidence has been observed in *in vitro* studies, authors recommend more *in vivo* investigations and clinical trials to support the stimulating impact of polyphenols as anti-obesity agents. The inhibition of the adipose cycle has been demonstrated with quinoa (Teng et al., 2020).

Berti et al. (2005) studied the effect on satiety of oat bread, buckwheat oat-meal pasta, and quinoa compared to their wheat counterparts; they worked with healthy young male volunteers. The results showed that the satiety efficiency indices (SEI) for alternative crop foods were higher in contrast to traditional cereal foods, suggesting that they can be exploited for their potential impact on eating behavior, especially considering that they have good nutritional value and are good sources of functional compounds (Berti et al., 2005).

In brief, it is clear that all studies enhance the need for future investigation to clarify the gaps observed in present levels of knowledge. In spite of important advances, more research is needed, either *in vivo* or in clinical trials, with quinoa and other Andean grains.

Table 9.1 shows some clinical and preclinical studies related to the nutritional effects and health benefits of Andean grains in non-communicable diseases.

9.3 CANCER

Together with cancer, chronic diseases from several etiologies increase the risk of cancer itself and death (Shu-Ju et al., 2018). Cancer is a group of diseases that can affect almost any organ or tissue of the body. It can be caused by different risk factors and constitutes the second highest cause of death in many countries. Lung, colorectum, liver, stomach, and breast cancers contribute to about 50% of cancer deaths worldwide (Sung et al., 2021).

TABLE 9.1 CLINICAL AND PRECLINICAL STUDIES ON THE NUTRITIONAL EFFECTS AND HEALTH BENEFITS REGARDING ANDEAN GRAINS' IMPACT ON NONCOMMUNICABLE DISEASES

Andean crop	Dietary Compound	Type of Trial	Main Biological Effect	Reference
Amaranth	Starch	*In vivo*; Human; female (22–34 years); BMI: 22	↓ Plasma glucose; ↑Insulin response	Guerra-Matias and Aréas (2005)
	Phytosterols, tocopherols, tocotrienols	*In vivo*, animal studies	Improve cardiovascular risk profiles by modifying cardiovascular risk factors such as cholesterol, diabetes and hypertension	Chmelík, Šnejdrlová, and Vrablík (2019)
Amaranth; Quinoa	Phytochemicals phenolics, betacyanins and lipophilic, fatty acids, tocopherols, and carotenoids	vitro enzyme assays and the *in vivo* anti-obesity effect by using obese, hyperglycemic mice model	Lowering risk of type 2 diabetes by assessing the antihyperglycemia and antihypertension	Tang and Tsao (2017)
Amaranth, Chia and Quinoa	Bioactive peptides	*In vitro* and *in vivo* in diabetic mice	Regulation of blood glucose level, increase insulin production or enhanced insulin sensitivity	Valenzuela Zamudio and Segura Campos (2020)
Beans; Whole grains	Dietary fiber	Meta-analyses of prospective studies or randomized clinical trials	Causal cardiometabolic effects, CVD and diabetes	Micha et al. (2017)

(continued)

TABLE 9.1 (CONTINUED)

Andean crop	Dietary Compound	Type of Trial	Main Biological Effect	Reference
Blue corn (*Zea mays* L.) and black bean (*Phaseolus vulgaris* L.)	Phenolic compounds; anthocyanins	In silico study	Modulate the activity of proteins involved in the main pathways of type 2 diabetes mellitus such as insulin secretion, insulin resistance and carbohydrate absorption	Damián-Medina et al. (2020)
Chia (*S. hispanica*)	Benefits of Salba-chia consumption	*In vivo* human 6-month randomized, double-blind, parallel design study with 77 eligible participants	Beneficial role of Chia ↑ weight loss ↓ obesity related risk factors, maintain glycemic control	Vuksan *et al.* (2017)
Corn; Maize; Sorghum; Whole Grains	Carbohydrates/starch	Clinical trials	Improve postprandial glucose and insulin response	Marventano et al. (2017)
Lupinus	Protein hydrolysates	*In vitro* - THP-1-derived macrophage model	↓ proinflammatory cytokines; ↑ expression of anti-inflammatory marker genes	Millán-Linares et al. (2014)
	Lupin consuption; alkaloids	*In vivo*; human; n=30, males and females	↓ Plasma glucose; ↑Insulin response	Baldeón et al. (2012)
	Lupin consuption, conglutin-γ and alkaloids	Case control study (Young healthy vs high glucose level volunteers)	Young healthy no changes; high glucose level volunteers ↓ Plasma glucose; ↑Insulin response	Fornasini et al. (2012)

Purple corn (*Z. mays* L)	Bioactive peptides	*In vivo* human, randomized, controlled crossover study in 33 hypercholesterolemic subjects	↓ plasma LDL, cholesterol significantly reduced after 4 wks	Bähr et al. (2013)
	Phenolic compounds; anthocyanins	*In vitro* - cell culture HRMCs	Retarded diabetes-associated renal fibrosis and mesangial inflammation	Li et al. (2012)
	Phenolic composition; Anthocyanin	In silico study	anti-inflammatory and anti-diabetic properties, improving insulin sensitivity in insulin-resistant adipocytes	Zhang et al. (2019)
Quinoa	Ecdysterone flavonoids	*In vivo* - diet-induced in obese, hyperglycemic mice (metabolic syndrome)	Significantly lowered fasting blood glucose	Graf et al. (2014)
	Gluten free	*In vivo*, human healthy volunteers no celiacs	Effect on appetite control	Berti et al. (2005)

Throughout the years, research works have reported the influence of physical activity and food habits on the maintenance of health and the decrease of chronic diseases and cancer (Sung et al,. 2021). Several *in vitro* studies show that grains can be used as an anti-inflammatory and antiproliferative tool for controlling the disease, especially for colorectum cancers. In particular, quinoa, amaranth, purple and red corn, and chia were associated with the decrease in oxidative stress (Gawlik-Dziki et al., 2013; Lee and Joo, 2018; Sabbione et al., 2019), cancer cell viability and the inhibition of tumor cell growth factors with an antiproliferative activity, cell necrosis, and apoptosis (Mazewski et al., 2017; Sabbione et al., 2019; Srdić et al., 2020; Vilcacundo et al., 2018). Research results have assigned these properties to those grains and plant products containing bioactive peptides and antioxidant compounds, total phenols including flavonoids, and anthocyanins, whose properties may be mainly responsible for the described benefits in cells used for experimentation. Apart from intestinal cancer cells, some research works also evidence the importance of quinoa, chia, and lupinus for treating individuals with breast cancer and in subjects with a multidrug resistance phenotype. The use of chia seed mucilage in cell culture studies gives promising results on improving the treatment in these particular individuals. The results found in cell culture studies evidence that the chia oligosaccharides from mucilage and the antioxidant from quinoa, chia, and lupinus constitute the main bioactive compounds responsible for decreasing cell proliferation (Ayyash et al., 2019; Rosas-Ramírez et al., 2017).

Purple corn, chia, and quinoa evidence a promissory impact on liver cancer. *In vivo* studies focused on transgenic animal models evidence that Andean cereals can influence carcinogenic mechanisms due to their chemical composition. Bioactive compounds, including peptides, antioxidants, and fatty acids, have a beneficial immunometabolic effect, chemopreventive potential, and promote resistance to hepatocarcinogenesis under pro-tumorigenic inflammation (Laparra and Haros, 2019; Llopis et al., 2020; Yokohira et al., 2008). The protease inhibitory activity was achieved by studying two groups of rats after receiving the extracts enriched with serine-type protease inhibitors (STPIs) from quinoa and chia grains. Only STPIs from quinoa and chia seemed to promote the production of inflammatory mediators contributing to the antitumoral macrophage phenotype activity. This functionality of STPIs can be used to control hepatocellular cancer aggressiveness (Laparra and Haros, 2019). Due to its content in anthocyanins and bioactive peptide composition, purple corn was employed as a chemopreventive agent for prostate and mammary tumors from non-transgenic rat models. Consumption of purple corn seemed to induce apoptosis in mammary tumors by suppressing the Ras protein that controls signaling pathways to malignant transformation of cells (Fukamachi et al., 2008). Purple corn also evidences an inhibitory activity on prostate carcinogenesis through the action of cyanidin-3-glucoside and pelargonidin-3-glucoside present as an active compound (Long et al., 2013).

A randomized clinical trial on 75 patients diagnosed with duodenal ulcers caused by Helicobacter pylori was conducted to evidence the antimicrobial activity and the inhibitory effect of anthocyanins, present in purple corn and amaranth oil, on a Helicobacter pylori bacterium and its toxin control. Helicobacter has been described as a risk factor to developing gastric cancer by induction of oxidative stress in gastric mucosa. For the experimentation, two groups were created: apart from the standard treatment, supplementation with amaranth oil was given in the test group. Compared with the control group, the duodenal peptic ulcer patients significantly reduced the accumulation of the oxidative stress markers in gastric mucosa and improved the histological parameters (Cherkas et al., 2018).

Table 9.2 shows the biological effect of Andean grains on tumor cell proliferation and presents the *in vitro* and *in vivo* studies that evidence the impact of several bioactive compounds on immunometabolism and as chemoprotective agents.

9.4 CELIAC DISEASE

Celiac disease (CD) is an autoimmune digestive disorder triggered by gluten ingestion in genetically predisposed individuals (Lebwohl et al., 2018). This pathology involves activation of an inflammatory process in the digestive tract, particularly the intestine, which produces hyperplasia of the crypts and villous atrophy in the small intestine (Caio et al., 2019). CD presents various clinical pictures that range from tangible symptoms such as diarrhea, weight loss, and osteoporosis to more vague symptoms such as iron and folic acid deficiency, arthralgia, fatigue, and abdominal discomfort that can manifest in childhood and adulthood (Mezquita Cerezal et al., 2011). It is considered one of the most common gluten-related disorders in western society. The highest CD prevalence is found in people with family predisposition and is associated with autoimmune diseases. CD is common worldwide, affecting 1 in 100 people in the world's population (Butterworth and Los, 2019). Destruction of enterocytes caused by the immune system with atrophic intestinal epithelium results in a decrease in the surface area for mineral absorption, which leads to vitamin and protein deficiencies, consequently to general malnutrition, triggering reduced body mass index. Although CD is a well characterized disease, it is highly underdiagnosed, despite the serious consequences of prolonged gluten ingestion, such as increased autoimmunity, refractory CD, and intestinal T-cell lymphoma (Mirijello et al., 2019). The main treatment for CD is lifetime

TABLE 9.2   ANDEAN GRAINS' BEHAVIOR IN DIFFERENT TUMOR CELLS

Andean crop	Dietary Compound	Type of Trial	Main Biological Effect	Reference
Amaranth	Amaranth Oil	*In vivo* (75 patients with Duodenal Ulcera by H. pylori)	Reduced HNE (oxidative stress marker) in gastric mucosa. Improvement of histological parameters in Duodenal Peptid Ulcera patients, reduced manifestations of chronic oxidative stress	Cherkas et al. (2018)
	Antioxidant; total phenol, and total flavonoid	*In vitro* CT-26 cell	Amaranth antioxidation activity inhibits growth of CT-26 cell Rectal cancer cells	Lee and Joo (2018)
	Bioactive peptides	*In vitro* HT-29 colon tumor cells	Amaranth peptides reveals a potential antiproliferative activity over HT-29 colon tumor cells by induction of cell necrosis and apoptosis	Sabbione et al. (2019)
Chia (*S. hispanica*)	Oligosaccharides	*In vitro* - colon (HCT-15 and HCT-116), cervix (HeLa), breast (MCF7 and MDA-MB-231) carcinoma cell lines	Chia oligosaccharides from mucilage modulate compounds of the multidrug resistance phenotype to chemotherapeutic in breast cancer cells	Rosas-Ramírez et al. (2017)
Lupinus, Quinoa	Antioxidant activitie	*In Vitro* - Caco-2 and MCF-7 cancer cell lines of intestinal and mammary origin, respectively	Antiproliferative activities Fermented Quinoa had greater proliferative inhibition against the breast cancer cell line (MCF-7) when compared with the colon cancer cell line	Ayyash et al. (2019)
Purple and red corn	Anthocyanin - Antioxidant activitie	*In vitro* human colon cancer cells, HT-29 and HCT-116	↓ Colon cancer cell viability in a dose-dependent manner. Inhibition of colon cancer cell proliferation	Mazewski et al. (2017)

Purple Corn	Antioxidant; Anthocyanin	*In vivo*; 100 male F344 rats	Liver carcinogenesis effect; chemopreventive potential against liver preneoplastic lesion development. Evidence of antioxidant power *in vivo*	Yokohira et al. (2008)
	Bioactive peptides	*In vivo* heterozygous male transgenic rats for adeno- carcinoma of prostate	Prostate cancer chemoprevention; inhibit prostate carcinogenesis by cyanidin-3-glucoside and pelarg- onidin-3-glucoside as active compounds from corn	Long et al. (2013)
	Anthocyanin - Antioxidant activities	*In vivo* female c-Ha-ras transgenic (Hras128, Tg) and non-transgenic (non-Tg) rats	Induces apoptosis in mammary tumors by decreasing ras protein levels; can be used for screening for chemopreventive agents that act via suppressing the Ras signaling pathway	Frukamachi et al. (2008)
Quinoa	*Chenopodium quinoa* Leaves Extract; phenolic content	*In vitro* bioaccessibility and bioavailability study; Rat prostate cancer AT-2 and MAT-LyLu cells (the Dunning rat model), HTB-140 and normal mouse 3T3 fibroblasts	Inhibitory effect on lipoxygenase activity; antioxidative, antiradical and reducing power. Chemopreventive and anticarcinogenic effect on oxidative stress; prevention of Reactive oxygen species related deseases	Gawlik-Dziki et al. (2013)
	Bioactive peptides	*In vitro* - Human colorectal cancer cell lines (Caco-2, HT-29, and HCT-116)	↓ oxidative stress-associated diseases, including cancer	Vilcacundo et al. (2018)

(continued)

TABLE 9.2 (CONTINUED)

Andean crop	Dietary Compound	Type of Trial	Main Biological Effect	Reference
Quinoa and Chia (*C. quinoa* and *S. hispanica*)	Fatty acids and iron	*In vivo* mice C57BL/6 hepatocarcinoma-developing mice	Beneficial immunometabolic effects of protease (serine-type) from quinoa and chia promote resistance to hepatocarcinogenesis	Llopis et al. (2020)
	Bioactive peptides	*In vitro* - Human-like macrophages Cells (HB-8902)	Immunometabolic effects of bioaccesible polypeptides; prevention of diet-associated innate immune imbalances	Srdić et al. (2020)
	Serine-type protease inhibitors (STPIs)	*In vivo* C57Bl/6 mice	STPIs from C. quinoa and S. hispanica can be used to control Hepatocarcinoma aggressiveness. S. hispanica showed positive effects, ↑ F4/80+ cells normalizing the expression (mRNA) of CD36 and the innate immune receptors.	Laparra and Haros (2019)

adherence to a gluten-free diet (GFD) from the time it is diagnosed. Gluten is a protein found naturally in cereals such as wheat, barley, rye, oats, kamut, spelt, and hybrid varieties, and products derived from these cereals such as flours and starches or semolina. In general, GFD is low in nutrients, vitamins, minerals, and dietary fiber (Saturni et al., 2010). Calcium and vitamin D, among other components of importance for normal development of the human being, generally have poor absorption, which leads to increased hypocalcemia and vitamin D deficiency (Krupa-Kozak, 2014). This suggests that more emphasis should be placed on nutritional quality in GFDs; the raw materials used to make these dietary products are mainly cereals such as corn, rice, sorghum, millet, quinoa, buckwheat, and amaranth. In general, the quality and availability of GFD products on the market have continuously improved over the past decades. However, many CD patients are still dissatisfied with these products, especially due to low palatability, texture, mouthfeel, and nutritional quality compared to their gluten-containing counterparts. At present, achieving high production in GF quality baked goods is a great challenge due to the absence of the technological properties that gluten confers on the dough (Rai et al., 2018). In this area, Jagelaviciute and Cizeikiene (2021) evaluated the potential use of *Lactobacillus* sanfranciscensis for the fermentation of chia, quinoa, and hemp flour in the production of gluten-free bread. The application of unfermented chia and hemp flour increased the firmness and ageing rate of the bread, while the use of non-traditional hemp and quinoa sourdough reduced the bread's ageing rate. In many cases, chia, hemp, and quinoa flour increased the acceptability of gluten-free corn/rice bread (Jagelaviciute and Cizeikiene, 2021).

Food scientists, nutritionists, biochemists, and the food industry actually study the technological and nutritional properties of new grains to make GF products as substitutes for wheat (Rai et al., 2018). At present, there is considerable interest in the consumption of alternative crops, such as quinoa, amaranth, and kañiwua as potential ingredients for healthy food production and special dietary use. These are the most consumed Andean grains in Latin America; they possess high nutritional value and are formidable food alternatives for celiac patients and/or those suffering from gluten-sensitivity (Jnawali et al., 2016).

Mezquita Cerezal et al. (2011) found that the mixtures of quinoa (*Chenopodium quinoa* Willd) and lupine (*Lupinus albus* L) flours, with two traditional cereals, corn (*Zea mays* L.) and rice (*Oryza sativa* L.), were suitable for the elaboration of cakes and other sweet foods as a good alternative and a supplement for the nutrition of children between 6 and 24 months who suffer from celiac disease. The gluten-free premix developed by Coronel et al. (2021) was based on buckwheat flour supplemented with chia flour obtained as a by-product of cold-pressed oil extraction, with and without the addition of xanthan gum, for the production of gluten-free bread. It had better nutritional characteristics (higher protein and crude fiber content), higher antioxidant activity, and a significantly high content of essential polyunsaturated fatty acids when compared with the control bread (commercial premix). Other work also found that the inclusion of quinoa in alternative grain-based products significantly increased the nutrient profile of GF dietary products in protein, iron, calcium, and fiber content (Lee et al., 2009).

In vitro research works using quinoa and amaranth demonstrate that both grains can be used as GF alternatives for celiac patients. The results evidenced low content of protein cross-reacting with anti-gliadin antibodies either with animals or serum immunoglobulins from celiac patients, demonstrating that amaranth and quinoa are safe (Ballabio et al., 2011; Bergamo et al., 2011; Peñas et al., 2014). Other studies also evaluate quinoa compliance with the Codex Alimentarius conditions of GF products (gluten <20 mg/Kg). Zevallos et al. (2012) evaluated 15 quinoa cultivars from different origins and found that four cultivars had quantifiable concentrations of celiac-toxic epitopes but were below the maximum allowed for GF foods. However, two cultivars stimulated T-cell lines to levels similar to gliadin and caused cytokine secretion from biopsy samples cultured at levels comparable to those of gliadin. They concluded that most quinoa is potentially suitable for CD patients but that further investigation with *in vivo* studies is required.

In other studies, the addition of quinoa to the GFD of celiac patients was well tolerated. There was a positive trend towards improvement in histological and serological parameters, particularly a mild hypocholesterolemic effect. Overall, these are the first clinical data suggesting that celiac patients can safely tolerate 50 g of quinoa daily for 6 weeks. However, more research is needed to determine the long-term effects of quinoa consumption (Zevallos et al., 2012; Zevallos et al., 2014). The evaluation of amaranth as a GF alternative for children was tested by Bavykina et al. (2017), who found that it was well tolerated, and that 89.2% of parents of the 37 evaluated children did not notice allergic or dyspeptic reactions.

Table 9.3 summarizes some in vivo and in vitro studies on the effect of Andean grains on gluten related diseases.

9.5   PREBIOTIC EFFECT AND MICROBIOTA MODULATION

There is a consensus within the literature that many health aspects are influenced by gut microbiota despite information dealing with the mechanism of action still missing. Presently diet is considered one of the most important modifiers of gut microbiota (Danneskiold-Samsøe et al., 2019; Ferrario et al., 2017). Food pattern as a direct mediator of gut microbiota offers a tremendous opportunity to prevent noncommunicable disease (Spencer et al., 2019).

TABLE 9.3 THE EFFECT OF ANDEAN GRAINS ON GLUTEN RELATED DISEASES

Andean crop	Type of Trial	Main Biological Effect	Reference
Amaranth	*In vitro* reaction with Serum from Celiac Subjects	Amaranth is safe for celiac; ↓ content of proteins cross-reacting with anti-gliadin antibodies	Ballabio et al. (2011)
	In vivo study, in 37 children from 1–17 years long term gluten free diet	Amaranth tested in the study was well tolerated, allergic and dyspeptic reactions were not noted. 89.2% of parents commented positively on the new gluten-free amaranth products	Bavykina et al. (2017)
Amaranth; Quinoa	*In vitro* – intestinal T-cell lines, cultures of duodenal explants from HLA-DQ21 CD patients and HLA-DQ8 transgenic mice for signs of activation	Amaranth and quinoa did not show any immune cross-reactivity confirming their safety in the diet of CD patients	Bergamo et al. (2011)
Quinoa	*In vitro* culture of celiac duodenal biopsy samples and exame unknown cultivars in human inume response in T cells	Quinoa cultivars do not present quantifiable amounts of celiac-toxic epitopes; Well tolerated among individuals with CD	Zevallos et al. (2012)
	19 treated celiac patients, with evaluation of diet, serology, and gastrointestinal parameters	Addition of quinoa to GFD of celiac patients, is well tolerated, improuve histological and serological parameters; mild hypocholesterolemic effect	Zevallos et al. (2014)
	In vitro reaction with serum from Celiac subjects	11 quinoa demonstrate safe for celiac; absence of gliadin proteins and no binding affinity to animal IgG or serum IgA from celiac patients	Peñas et al. (2014)

The western diet, characterized by monosaccharides (glucose, galactose, and fructose) and disaccharides (saccharose) consumption, causes microbiota and health impairment changes. Scientific evidence suggests that regular sugar and sugar substitute consumption, as well as other nocuous food ingredients, can cause dysbiosis, intestinal inflammation, and even noncommunicable diseases (Ferrario et al., 2017). The composition of gut microbiota has been associated with dysregulation of blood pressure, chronic kidney disease, cardiovascular disease, age-related health impairment, immunomodulation, allergy, or asthma (Danneskiold-Samsøe et al., 2019).

In contrast, the consumption of fiber, polyphenolic rich foods, saponins, and prebiotics and probiotics has a positive effect on gut microbiota due to their balance being restored (Del Hierro et al., 2020; Ferrario et al., 2017).

Studies on the characterization of Andean grains' nutrient profile reveal a high content of dietary fiber, polyunsaturated fatty acids, and a high-quality protein source. Furthermore, these grains contain an abundance of anti-inflammatory phytochemicals (Liu et al., 2018). These substances have shown a range of immunomodulation effects due to the enhancement of *Bifidobacterium* spp. and *Lactobacillus* spp., which positively regulate a wide range of physiological functions. Andean grains (amaranth and quinoa) present a profile of fiber polysaccharides, more similar to those found in fruits and vegetables than in cereals (Zhu, 2020). In amaranth and quinoa grains, pectins, when compared with other whole grains are quantitatively predominant, as recently demonstrated (Ciudad-Mulero et al., 2019; Martínez-Villaluenga et al., 2020). Pectin is a complex group of polysaccharides formed by D-galacturonic acid monomers, linked by α-(1–4) glycosidic bonds and branched regions primarily formed by various types of neutral monosaccharides (mainly rhamnose, xylose, mannose, and arabinose) (Ciudad-Mulero et al., 2019). Soluble dietary fiber polysaccharides promote bacteria growth, increasing the production of short-chain fatty acids (SCFAs) in the gut. They improve immunity while decreasing excessively stimulated immune responses through regulating gut microbiota (Zhu, 2020). Among carbohydrates, resistant starch cannot be digested and absorbed in the small intestine reaching the colon, and here it is slowly fermented by microorganisms to produce short-chain fatty acids (Lehmann and Robin, 2007). Andean grains present more than 20% of resistant starch, considering European Food Safety Authority (EFSA) dietary guidelines (EFSA Panel on Dietetic Products, 2011) that impose regulation of intestinal microbiota resulting in health benefits (Martínez-Villaluenga et al., 2020).

Polyphenols are the other category of bioactive compounds with a positive effect on the modulation of gut microbiota. The profile of quinoa polyphenols has been characterized by Tang et al. (2015). In quinoa grains, 23 different phenolic compounds can be identified on red, white, and dark cultivars, where vanillic and ferulic acids and their derivatives, and main flavonoids quercetin, kaempferol, and their glycosides represent the main compounds. Betanin and isobetanin are the predominant nitrogen-containing phenols of betacyanins (Tang et al., 2015).

Saponins are a class of glycosides whose aglycones can be either triterpenes or helical spirostanes. Considered an anti-nutrient, it reveals immunostimulatory, hypocholesterolemic, antitumoral, anti-inflammatory, antibacterial, antiviral, antifungal, or antiparasitic activities (He et al., 2019). Saponins consist of a hydrophobic aglycone backbone designated as sapogenin linked to the hydrophilic sugar chain. Glycosides are poorly absorbed in the gastrointestinal tract and transformed there into sapogenins through hydrolysis. Due to the chemical properties of sapogenins caused by lack of a sugar chain, these compounds exhibit high bioactive properties.

However, the contribution by the gastric passage to the digestion of saponins remains unclear, and it is suggested that, at colon level, saponins are transformed by the resident microbiota into sapogenins (Spencer et al., 2019). Recently, Del Hierro et al. (2020) studied the transformation of saponin into sapogenins using lentil and quinoa as triterpenoid and fenugreek as steroid saponins. Quinoa reveals the highest production of sapogenins, namely oleanolic acid, hederagenin, serjanic acid, and phytolaccagenic acid.

Bioactive components of Andean grains have increasingly interested the scientific community due to their interaction with human microbiota and consequent health benefits. These interactions determine the possible beneficial effects of polysaccharides, polyphenols, and saponins provided by their prebiotic actions. Therefore, interaction between bioactive components of Andean grains and human gut microbiota may impact human host health.

On the other hand, the effect of intestinal microbiota on the biotransformation of these compounds into metabolites with health impact constitutes a new topic of interest for scientific communities. Relevant reviews have been published as well as research papers focusing on *in vitro* studies (Ugural and Akyol, 2020; Valero-Cases et al., 2020). However, our knowledge of the impact of Andean grains on the human intestinal microbiota at different stages of life is still limited. Among Andean grains, quinoa and amaranth received attention. In Northern Europe, Australia, and New Zealand, interest in its cultivation has grown since the 1990s. One of the principal reasons for interest in these grains is their nutritional composition and the prebiotic effect of their components. Table 9.4 presents the most relevant studies reporting recent findings on the impact of Andean grains on the modulation of gut microbiota. The studies use grains in several ways as foodstuffs where the prebiotic effect is evaluated *in vivo* or *in vitro* in a starter culture where the microorganisms are inoculated with Andean grains to enhance the prebiotic effect.

TABLE 9.4   IMPACT OF DIETARY INTERVENTION OF ANDEAN GRAINS ON GUT MICROBIOTA APPLYING *IN VITRO* AND *IN VIVO* STUDIES

Andean crop	Dietary Compound	Type of Trial	Main Microbiota Effect	Other Biological Effects	References
Amaranth; Quinoa	Prebiotic effect	*In vitro* cultures with human faecal microbiota	↑ *Bifidobacterium* and Atopobium or Bacteroides	↑ short-chain fatty acids (SCFAs)	Gullón et al. (2016)
Amaranth; Fermented quinoa	Probiotic/prebiotic effect	*In vitro* and *in vivo* studies on fermented grains	↓ pathogenic bacterias; ↑ *Peptoclostridium*, Prevotellaceae, *Lactobacillus*, *Bifidobacterium*, *Enterococcus*, and Eubacteriaceae	↓ inflammation and colonic damage clinical symptoms and dysbiosis	Ugural and Akyol (2020)
Fermented Quinoa beverage	Prebiotic (fructo-oligosaccharide); Probiotic (*Lactobacillus* casei Lc-01); Synbiotic (fructo-oligosaccharide and L. casei Lc-01)	*In vitro* simulator of the human intestinal microbial ecosystem	↓ pathogenic bacteria *Clostridium* spp., *Bacteroides* spp., enterobacteria and *Enterococcus* spp: ↑ *Lactobacillus* spp. and *Bifidobacterium* spp	No changes in short-chain fatty acids (SCFAs);	Gullón et al. (2016)
Quinoa	Water soluble, nondigestible polysaccharides and fibers	*In vitro* human fecal microbiome fermentation and *in vivo* studies in obesity C57BL/6J mice.	Preserve microbial diversity; ↑ beneficial bacteria	↑ short-chain fatty acids (SCFAs).	Possomato-Vieira et al. (2016)

Serjanic acid; saponins	*In vitro* colonic fermentation	↑ *Bifidobacterium* spp. and *Lactobacillus* spp	Transformation of saponin-rich extracts of quinoa by human gut microbiota to sapogenins	Hierro et al. (2020)
Polysaccharide	*In vivo*, colitis induced by dextran sodium sulfate, in C57BL/6 mice	↓ abnormal expansion of phylum Proteobacteria, and ↓ overgrowth of genera *Escherichia/Shigella* and *Peptoclostridium* and family Lachnospiraceae	Quinoa modified the dysbiosis of intestinal microbiota	Liu et al. (2018)

All works reinforce the need for more research to clarify the mechanisms involved in gut microbiota modulation, both in microbiota species and in the metabolites of bioactive compounds (Zhu, 2020).

Preclinical studies have illustrated that Andean crops can modulate gut microbiota, suggesting quinoa and amaranth as possible alternatives to pharmaceutical drugs to improve dysbiosis (Ugural and Akyol, 2020).

Gullón et al. (2016) studied the prebiotic effect of quinoa and amaranth through its incubation with fecal human inocula. The authors observed the decrease of pH and production of SCFAs (acetate, propionate, and butyrate). A significant difference was observed in these parameters. The presence of propionate is very relevant. It is a glucogenerator that inhibits biosynthesis of cholesterol and fatty acids in the liver and can reduce the risk of cardiovascular diseases. Furthermore, acetate is considered a lipogenic substance. Regarding the microbiota profile, they observed changes in *Bifidobacterium* spp., *Lactobacillus-Enterococcus, Atopobium, Bacteroides-Prevotella, Clostridium coccoides*-Eubacterium rectale, *Faecalibacterium prausnitzii,* and *Roseburia intestinalis.* The authors concluded that Andean grains maintain a balance on gut microbiota enhanced dysbiosis. All these bacterial groups affect the production of SCFA and decrease pH in the gut. Furthermore, the study recommends more preclinical intervention to elucidate the role of bioactive compounds, both interaction of main nutrients as well as its bioavailability.

Recently Liu et al. (2018), in a comprehensive work, studied the effect of quinoa to modulate gut microbiota in murine colitis model rats induced by dextran sodium sulfate, a colitogen agent with anticoagulant properties. The robust model was informative in demonstrating quinoa components' effect on reducing the disease index, epithelium damage, and microbiota profile. *Escherichia/ Shigella* and *Peptoclostridium* expanded strongly due to colitis and dysbiosis. On the other hand, in animals fed with quinoa, a decrease of pathogenic family Lachnospiraceae was observed. Also, proteobacteria phylum inhibition, designated as a *microbial signature* in dysbiosis, occurred in animals submitted to a quinoa diet.

The *in vitro* assays presented by del Hierro et al. have studied the effect of human fecal microbiota on saponin extract of quinoa, lentil, and fenugreek and its biotransformation in sapogenin (del Hierro et al., 2020). The authors observed an increase in *Bifidobacterium* spp and *Lactobacillus* spp after the addition of quinoa extracts to the feces of the volunteers. This is promising since, as far as we know, it is the first study using saponin quinoa extracts to detect a modular effect in human gut microbiota. Furthermore, a decrease in *Enterococcus* spp was observed. These findings contribute to understanding the communication routes between quinoa and microbiota and their association with the impact on health of Andean grains.

A study evaluated the effect of prebiotic, probiotic, and symbiotic beverage formulations containing aqueous extracts of soy and quinoa, using a dynamic model of human gut simulator of the human intestinal microbiota ecosystem (Bianchi et al., 2014). This model allows for monitoring the effect of formulations in different colon compartments. In the work, the effect of a beverage with soy and quinoa water extracts on colon compartments – ascending, transverse, and descending – with a combination of prebiotic and probiotic agents acting synergistically was tested. The beverage conferred significant increase of *Lactobacillus* spp. and *Bifidobacterium* spp., and reduced *Clostridium* spp., *Bacteroides* spp., enterobacteria, and *Enterococcus* spp. The bacterial modulation promoted by this type of beverage also contributes to understanding how to enhance gut health by expanding the beneficial bacteria and reducing pathogenics.

9.5 CONCLUSIONS

This chapter has evaluated the research literature concerning the dietary intake of crops of Latin-American origin (corn, quinoa, amaranth, chia, black turtle bean, tarwi, kañiwa, and sacha inchi) associated with several health effects. Based on the data presented in scientific studies, especially *in vivo* or *in vitro*, that correlates health with nutritional and/or phytochemical compounds, it is possible to conclude that these crops are revealed to be influential on human health. Impact on noncommunicable diseases such as diabetes mellitus; on cardiovascular risk factors such as on the reduction of cholesterol and arterial hypertension; on obesity; on different types of cancer cells, as well as on the fields of gastrointestinal health with a relevant prebiotic effect and microbiota modulation and the reduction of the inflammatory response on celiac disease, reveals the importance of these crops in improving human health.

While Andean and Mesoamerican crops exhibit various protective effects in the context of cardiovascular diseases, many key questions remain unanswered considering the limited number of randomized clinical trials or case-control studies, and the limited sample size of some studies. These facts show that extensive prospective epidemiological studies need to be conducted. Future studies should include a well-defined population in order to address the dose-response relationship to assess more directly the association of each Andean and Mesoamerican crop with their respective health impact, especially considering increased consumer demands toward healthy foods.

ACKNOWLEDGMENTS

This work was supported by grants laValSe- Food- CYTED (Ref. 119RT0567); Consejo Nacional de Investigaciones Científicas y Técnicas (CONICET), and Universidad Nacional de Jujuy, Argentina.

REFERENCES

Ayyash, M., S.K. Johnson, S.Q. Liu, N. Mesmari, S. Dahmani, A.S. Al Dhaheri, and J. Kizhakkayil. 2019. In vitro investigation of bioactivities of solid-state fermented lupin, quinoa and wheat using lactobacillus spp. *Food Chemistry* 275: 50–58.

Bähr M, A. Fechner, J. Krämer, M. Kiehntopf, and G. Jahreis. 2013. Lupin protein positively affects plasma LDL cholesterol and LDL:HDL cholesterol ratio in hypercholesterolemic adults after four weeks of supplementation: A randomized, controlled crossover study. *Nutrition Journal* 12: 107.

Baldeón, M.E., J. Castro, E. Villacrés, L. Narváez, and M. Fornasini. 2012. Efecto hipoglicemiante de lupinus mutabilis cocinado y sus alcaloides en sujetos con diabetes tipo-2. *Nutricion hospitalaria* 27(4): 1261–1266.

Ballabio, C., F. Uberti, C. Di Lorenzo, A. Brandolini, E. Penas, and P. Restani. 2011. Biochemical and immunochemical characterization of different varieties of amaranth (amaranthus l. ssp.) as a safe ingredient for gluten-free products. *Journal of Agricultural and Food Chemistry* 59(24): 12969–12974.

Bavykina, I.A., A.A. Zvyagin, L.A. Miroshnichenko, K.Y. Gusev, and I.M. Zharkova. 2017. Efficient products from amaranth in gluten-free nutrition of children with gluten intolerance. *Voprosy Pitaniia* 86(2): 91–99.

Bazzano, L.A., A.M. Thompson, M.T. Tees, C.H. Nguyen, and D.M. Winham. 2011. Non-soy legume consumption lowers cholesterol levels: a meta-analysis of randomized controlled trials. *Nutrition, Metabolism and Cardiovascular Diseases* 21(2): 94–103.

Bellou, V., L. Belbasis, I. Tzoulaki, and E. Evangelou. 2018. Risk factors for type 2 diabetes mellitus: An exposure-wide umbrella review of meta-analyses. *PLoS ONE* 13(3): 1–27.

Bergamo, P., F. Maurano, G. Mazzarella, G. Iaquinto, I. Vocca, A.R. Rivelli, E. De Falco, C. Gianfrani, and M. Rossi. 2011. Immunological evaluation of the alcohol-soluble protein fraction from gluten-free grains in relation to celiac disease. *Molecular Nutrition and Food Research* 55(8): 1266–1270.

Berti, C., P. Riso, A. Brusamolino, and M. Porrini. 2005. Effect on appetite control of minor cereal and pseudocereal products. *British Journal of Nutrition* 94(5): 850–858. www.cambridge.org/core/product/identifier/S0007114505002564/type/journal_article.

Bianchi, F., E.A. Rossi, I.K. Sakamoto, M.A.T. Adorno, T. Van de Wiele, and K. Sivieri. 2014. Beneficial effects of fermented vegetal beverages on human gastrointestinal microbial ecosystem in a simulator. *Food Research International* 64: 43–52.

Burton-Freeman, B., M. Brzeziński, E. Park, A. Sandhu, D. Xiao, and I. Edirisinghe. 2019. A selective role of dietary anthocyanins and flavan-3-ols in reducing the risk of type 2 diabetes mellitus: a review of recent evidence. *Nutrients* 11(4): 1–16 .

Butterworth, J., and L. Los. 2019. Coeliac disease. *Medicine* 47(5): 314–319. https://linking hub.elsevier.com/retrieve/pii/S135730391930043X.

Caio, G., U. Volta, A. Sapone, D.A. Leffler, R. De Giorgio, C. Catassi, and A. Fasano. 2019. Celiac disease: A comprehensive current review. *BMC Medicine* 1(17): 1–20.

Chaiittianan, R., K. Sutthanut, and A. Rattanathongkom. 2017. Purple corn silk: A potential anti-obesity agent with inhibition on adipogenesis and induction on lipolysis and apoptosis in adipocytes. *Journal of Ethnopharmacology* 201: 9–16.

Cherkas, A., K. Zarkovic, A. Cipak Gasparovic, M. Jaganjac, L. Milkovic, O. Abrahamovych, O. Yatskevych, G. Waeg, O. Yelisyeyeva, and N. Zarkovic. 2018. Amaranth oil reduces accumulation of 4-hydroxynonenal-histidine adducts in gastric mucosa and improves heart rate variability in duodenal peptic ulcer patients undergoing helicobacter pylori eradication. *Free Radical Research* 52(2): 135–149.

Chmelík, Z., M. Šnejdrlová, and M. Vrablík. 2019. Amaranth as a potential dietary adjunct of lifestyle modification to improve cardiovascular risk profile. *Nutrition Research* 72: 36–45.

Ciudad-Mulero, M., V. Fernández-Ruiz, M.C. Matallana-González, and P. Morales. 2019. Dietary fiber sources and human benefits: The case study of cereal and pseudocereals. *Advances in Food and Nutrition Research*, 90: 83–134. DOI:Org/10.1016/bs.afnr.2019.02.002.

Coronel E.B., E.N. Guiotto, M.C. Aspiroz, M.C. Tomás, S. M Nolasco, and M. I. Capitani. 2021. Development of gluten-free premixes with buckwheat and chia flours: Application in a bread product. *Lebensmittel-Wissenschaft Technologie (LWT)* 141: 110916.

Damián-Medina, K., Y. Salinas-Moreno, D. Milenkovic, L. Figueroa-Yáñez, E. Marino-Marmolejo, I. Higuera-Ciapara, A. Vallejo-Cardona, and E. Lugo-Cervantes. 2020. In silico analysis of antidiabetic potential of phenolic compounds from blue corn (*Zea mays* l.) and black bean (*Phaseolus vulgaris* l.). *Heliyon* 6(3): 1–13.

Danneskiold-Samsøe, N.B., H. Dias de Freitas Queiroz Barros, R. Santos, J.L. Bicas, C.B.B. Cazarin, L. Madsen, K. Kristiansen, G.M. Pastore, S. Brix, and M.R. Maróstica Júnior. 2019. Interplay between food and gut microbiota in health and disease. *Food Research International* 115: 23–31. .

Del Hierro, J.N., C. Cueva, A. Tamargo, E. Núñez-Gómez, M.V. Moreno-Arribas, G. Reglero, and D. Martin. 2020. In vitro colonic fermentation of saponin-rich extracts from quinoa, lentil, and fenugreek. Effect on sapogenins yield and human gut microbiota. *Journal of Agricultural and Food Chemistry* 68 (1): 106–116.

EFSA (European Food Safety Authority). 2011. Panel on Dietetic Products, N. and A. Scientific opinion on the substantiation of health claims related to resistant starch and reduction of post-prandial glycaemic responses (ID 681), "Digestive Health Benefits" (ID 682) and "Favours a Normal Colon Metabolism" (ID 783) Pursuant to Article 13. *EFSA Journal* 9(4): 1–17.

Ferrario, C., R. Statello, L. Carnevali, L. Mancabelli, C. Milani, M. Mangifesta, and S. Duranti. 2017. How to feed the mammalian gut microbiota: Bacterial and metabolic modulation by dietary fibers. *Frontiers in Microbiology* 8: 1–11.

Fornasini, M., J. Castro, E. Villacres, L. Narvaez, M.P. Villamar, and M.E. Baldeon. 2012. Efecto hipoglicemiante de lupinus mutabilis en voluntarios sanos y sujetos con disglicemia. *Nutricion Hospitalaria* 27(2): 425–433.

Frieden, T.R., and M.G. Jaffe. 2018. Saving 100 million lives by improving global treatment of hypertension and reducing cardiovascular disease risk factors. *Journal of Clinical Hypertension* 20(2): 208–211.

Fukamachi, K., T. Imada, Y. Ohshima, J. Xu, and H. Tsuda. 2008. Purple corn color suppresses ras protein level and inhibits 7,12-dimethylbenz[a]anthracene-induced mammary carcinogenesis in the rat. *Cancer Science* 99(9): 1841–1846.

Gawlik-Dziki, U., M. Świeca, M. Sułkowski, D. Dziki, B. Baraniak, and J. Czyz. 2013. Antioxidant and anticancer activities of chenopodium quinoa leaves extracts - in vitro study. *Food and Chemical Toxicology* 57: 154–160.

Graf L.B., A. Poulev, P. Kuhn, M.H. Grace, M.A. Lila, and I.Raskin. 2014. Quinoa seeds leach phytoecdysteroids and other compounds with anti-diabetic properties. *Food Chemistry* 163: 178–185.

Guerra-Matias, A.C., and J.A.G. Arêas. 2005. Glycemic and insulinemic responses in women consuming extruded amaranth (*Amaranthus cruentus* L). *Nutrition Research* 25(9): 815–822.

Gullón, B., P. Gullón, F.K. Tavaria, and R. Yáñez. 2016. Assessment of the prebiotic effect of quinoa and amaranth in the human intestinal ecosystem. *Food & Function* 7(9): 3782–3788.

He, Y., Z. Hu, A. Li, Z. Zhu, N. Yang, Z. Ying, J. He, C. Wang, S. Yin, and S. Cheng. 2019. Recent advances in biotransformation of saponins. *Molecules* 24 (13): 1–23.

Jagelaviciute, J., and D. Cizeikiene. 2021. The influence of non-traditional sourdough made with quinoa, hemp and chia flour on the characteristics of gluten-free maize/rice bread. *LWT – Food Science and Technology* 137: 110457. https://linkinghub.elsevier.com/retrieve/pii/S0023643820314456.

Jiang, Y.T., J.Y. Zhang, Y.S. Liu, Q. Chang, Y.H. Zhao, and Q.J. Wu. 2020. Relationship between legume consumption and metabolic syndrome: A systematic review and meta-analysis of observational studies. *Nutrition, Metabolism and Cardiovascular Diseases* 30(3): 384–392.

Jnawali, P., V. Kumar, and T.B. Anwar. 2016. Celiac disease: Overview and considerations for development of gluten-free foods. *Food Science and Human Wellness* 5: 169–176.

Krupa-Kozak, U. 2014. Pathologic bone alterations in celiac disease: Etiology, epidemiology, and treatment. *Nutrition* 30(1) (January): 16–24. https://linkinghub.elsevier.com/retrieve/pii/S0899900713000289X.

Laparra, J.M., and C.M. Haros. 2019. Plant seed protease inhibitors differentially affect innate immunity in a tumor microenvironment to control hepatocarcinoma. *Food and Function* 10(7): 4210–4219.

Lebwohl, B., D.S. Sanders, and P.H.R. Green. 2018. Coeliac disease. *The Lancet* 391(10115): 70–81.

Lee, A.R., D.L. Ng, E. Dave, E.J. Ciaccio, and P.H.R. Green. 2009. The effect of substituting alternative grains in the diet on the nutritional profile of the gluten-free diet. *Journal of Human Nutrition and Dietetics* 22(4): 359–363.

Lee, H., and N. Joo. 2018. Antioxidative properties of amaranth cauline leaf and suppressive effect against ct-26 cell proliferation of the sausage containing the leaf. *Korean Journal for Food Science of Animal Resources* 38(3): 570–579 .

Lehmann, U., and F. Robin. 2007. Slowly digestible starch - its structure and health implications: A review. *Trends in Food Science and Technology* 18(7): 346–355.

Li, J., S.S. Lim, J.Y. Lee, J.K. Kim, S.W. Kang, J.L. Kim, and Y.H. Kang. 2012. Purple corn anthocyanins dampened high-glucose-induced mesangial fibrosis and inflammation: Possible renoprotective role in diabetic nephropathy. *Journal of Nutritional Biochemistry* 23(4): 320–331.

Liu, W., Y. Zhang, B. Qiu, S. Fan, H. Ding, and Z. Liu. 2018. Quinoa whole grain diet compromises the changes of gut microbiota and colonic colitis induced by dextran sulfate sodium in C57bl/6 mice. *Scientific Reports* 8 (1): 1–9.

Llopis, J.M.L., D. Brown, and B. Saiz. 2020. Chenopodium quinoa and salvia hispanica provide immunonutritional agonists to ameliorate hepatocarcinoma severity under a high-fat diet. *Nutrients* 12(7): 1–15.

Long, N., S. Suzuki, S. Sato, A. Naiki-Ito, K. Sakatani, T. Shirai, and S. Takahashi. 2013. Purple corn color inhibition of prostate carcinogenesis by targeting cell growth pathways. *Cancer Science* 104(3): 298–303.

López, D.N., M. Galante, G. Raimundo, D. Spelzini, and V. Boeris. 2019. Functional properties of amaranth, quinoa and chia proteins and the biological activities of their hydrolyzates. *Food Research International* 116: 419–429.

Martínez-Villaluenga, C., E. Peñas, and B. Hernández-Ledesma. 2020. Pseudocereal grains: Nutritional value, health benefits and current applications for the development of gluten-free foods. *Food and Chemical Toxicology* 137: 111178.

Marventano, S., C. Vetrani, M. Vitale, J. Godos, G. Riccardi, and G. Grosso. 2017. Whole grain intake and glycaemic control in healthy subjects: A systematic review and meta-analysis of randomized controlled trials. *Nutrients* 9(7): 769.

Mazewski, C., K. Liang, and E. Gonzalez de Mejia. 2017. Inhibitory potential of anthocyanin-rich purple and red corn extracts on human colorectal cancer cell proliferation *in vitro. Journal of Functional Foods* 34: 254–265.

Mezquita Cerezal, P., V. Gatica Urtuvia, V. Quintanilla Ramírez, and R. Zavala Arcos. 2011. Desarrollo de producto sobre la base de harinas de cereales y leguminosa para niños celíacos entre 6 y 24 meses; II: Propiedades de Las Mezclas. *Nutricion Hospitalaria* 26(1): 161–169.

Micha, R., M.L. Shulkin, J.L. Peñalvo, S. Khatibzadeh, G.M. Singh, M. Rao, S. Fahimi, J. Powles, and D. Mozaffarian. 2017. Etiologic effects and optimal intakes of foods and nutrients for risk of cardiovascular diseases and diabetes: Systematic reviews and meta-analyses from the nutrition and chronic diseases expert group (NutriCoDE). *PLoS ONE* 12(4): 1–25.

Millán-Linares, M. del C., B. Bermúdez, M. del M. Yust, F. Millán, and J. Pedroche. 2014. Anti-inflammatory activity of lupine (*Lupinus angustifolius* l.) protein hydrolysates in thp-1-derived macrophages. *Journal of Functional Foods* 8(1):224–233.

Mirijello, A., C. D'Angelo, S. De Cosmo, A. Gasbarrini, and G. Addolorato. 2019. Management of celiac disease in daily clinical practice: Do not forget depression! *European Journal of Internal Medicine* 62(17).

Mithila M. V., and F. Khanum. 2015. Effectual comparison of quinoa and amaranth supplemented diets in controlling appetite; A biochemical study in rats. *Journal of Food Sience and Technology* 52(10): 6735–6741.

Motta, C., I. Castanheira, G.B. Gonzales, I. Delgado, D. Torres, M. Santos, and A.S. Matos. 2019. Impact of cooking methods and malting on amino acids content in amaranth, buckwheat and quinoa. *Journal of Food Composition and Analysis* 76: 58–65.

Mudryj, A.N., N. Yu, and H.M. Aukema. 2014. Nutritional and health benefits of pulses. *Applied Physiology, Nutrition, and Metabolism* 39(11): 1197–1204.

Navarro-Perez D., J. Radcliffe, A. Tierney, and M. Jois. 2017. Quinoa seed lowers serum triglycerides in overweight and obese subjects: A dose-response randomized controlled clinical trial. *Current Developments in Nutrition* 1(9): 1–9.

Neuenschwander, M., A. Ballon, K.S. Weber, T. Norat, D. Aune, L. Schwingshackl, and S. Schlesinger. 2019. Role of diet in type 2 diabetes incidence: Umbrella review of meta-analyses of prospective observational studies. *The BMJ Clinical Research* 366: 1–18.

Orona-Tamayo, D., M.E. Valverde, and O. Paredes-López. 2019. Bioactive peptides from selected Latin-American food crops – A nutraceutical and molecular approach. *Critical Reviews in Food Science and Nutrition* 59(12): 1949–1975.

Peñas, E., F. Uberti, C. di Lorenzo, C. Ballabio, A. Brandolini, and P. Restani. 2014. Biochemical and immunochemical evidences supporting the inclusion of quinoa (*Chenopodium quinoa* Willd.) as a gluten-free ingredient. *Plant Foods for Human Nutrition* 69(4): 297–303.

Possomato-Vieira, José S. and R.A.K. Khalil. (2016). Prebiotics from acorn and sago prevent high-fat diet-induced insulin resistance via microbiome-gut-brain axis modulation. *Physiology & Behavior* 176(12): 139–148.

Rai, S., A. Kaur, and C.S. Chopra. 2018. Gluten-free products for celiac susceptible people. *Frontiers in Nutrition* 5(116): 1–23.

Rosas-Ramírez, D.G., M. Fragoso-Serrano, S. Escandón-Rivera, A.L. Vargas-Ramírez, J.P. Reyes-Grajeda, and M. Soriano-García. 2017. Resistance-modifying activity in vinblastine-resistant human breast cancer cells by oligosaccharides obtained from mucilage of chia seeds (*Salvia Hispanica*). *Phytotherapy Research* 31(6): 906–914.

Sabbione, A.C., F.O. Ogutu, A. Scilingo, M. Zhang, M.C. Añón, and T.H. Mu. 2019. Antiproliferative effect of amaranth proteins and peptides on ht-29 human colon tumor cell line. *Plant Foods for Human Nutrition* 74(1): 107–114.

Sancho, R.A.S., and G.M. Pastore. 2012. Evaluation of the effects of anthocyanins in type 2 diabetes. *Food Research International* 46(1): 378–386.

Saturni, L., G. Ferretti, and T. Bacchetti. 2010. The gluten-free diet: Safety and nutritional quality. *Nutrients* 2(1): 16–34.

Schwingshackl, L., G. Hoffmann, A.M. Lampousi, S. Knüppel, K. Iqbal, C. Schwedhelm, A. Bechthold, S. Schlesinger, and H. Boeing. 2017. Food groups and risk of type 2 diabetes mellitus: A systematic review and meta-analysis of prospective studies. *European Journal of Epidemiology* 32(5):362–375.

Shu-Ju T., W. Chih-Wei, P. Kuang-Tse, W. Yi-Cheng, and W. Chen-Te. 2018. Localized thin-section CT with radiomics feature extraction and machine learning to classify early-detected pulmonary nodules from lung cancer screening. *Physics in Medicine & Biology* 63(6): 065005.

Spencer, S.P., G.K. Fragiadakis, and J.L. Sonnenburg. 2019. Pursuing human-relevant gut microbiota-immune interactions. *Immunity* 51(2): 225–239.

Srdić, M., I. Ovčina, B. Fotschki, C.M. Haros, and J.M. Laparra Llopis. 2020. C. Quinoa and S. Hispanica L. seeds provide immunonutritional agonists to selectively polarize macrophages. *Cells* 9(3): 593.

Sung, H., J. Ferlay, R.L. Siegel, M. Laversanne, I. Soerjomataram, A. Jemal, and F. Bray. 2021. Global cancer statistics 2020: GLOBOCAN estimates of incidence and mortality worldwide for 36 cancers in 185 countries. *CA: A Cancer Journal for Clinicians* 71(3): 209–249. DOI.org/10.3322/caac.21660.

Tang, Y., and R. Tsao. 2017. Phytochemicals in quinoa and amaranth grains and their antioxidant, anti-inflammatory, and potential health beneficial effects: A review. *Molecular Nutrition and Food Research* 61(7): 1–16.

Tang, Y., X. Li, B. Zhang, P.X. Chen, R. Liu, and R. Tsao. 2015. Characterization of phenolics, betanins and antioxidant activities in seeds of three *chenopodium quinoa* willd. genotypes. *Food Chemistry* 166: 380–388.

Teng, C., Z. Shi, Y. Yao, and G. Ren. 2020. Structural characterization of quinoa polysaccharide and its inhibitory effects on 3t3-l1 adipocyte differentiation. *Foods* 9(10): 1511.

Ugural, A., and A. Akyol. 2020. Can pseudocereals modulate microbiota by functioning as probiotics or prebiotics? *Critical Reviews in Food Science and Nutrition*. https://doi.org/10.1080/10408398.2020.1846493.

Valenzuela Zamudio, F., and M.R. Segura Campos. 2020. Amaranth, quinoa and chia bioactive peptides: A comprehensive review on three ancient grains and their potential role in management and prevention of type 2 diabetes. *Critical Reviews in Food Science and Nutrition* www.tandfonline.com/doi/full/10.1080/10408398.2020.1857683.

Valero-Cases, E., D. Cerdá-Bernad, J.J. Pastor, and M.J. Frutos. 2020. Non-dairy fermented beverages as potential carriers to ensure probiotics, prebiotics, and bioactive compounds arrival to the gut and their health benefits. *Nutrients* 12(6). https://doi.org/10.3390/nu12061666.

Vilcacundo, R., B. Miralles, W. Carrillo, and B. Hernández-Ledesma. 2018. In vitro chemopreventive properties of peptides released from quinoa (*Chenopodium quinoa* Willd.) Protein under simulated gastrointestinal digestion. *Food Research International* 105: 403–411.

Vuksan, V., A.L. Jenkins, C. Brissette, L. Choleva, E. Jovanovski, A.L. Gibbs, R.P. Bazinet. 2017. Salba-Chia (*Salvia Hispanic*a L.) in the treatment of overweight and obese patients with type 2 diabetes: A double-blind randomized controlled trial. *Nutrition, Metabolism and Cardiovascular Diseases* 27(2): 138–146.

WHO (World Health Organization). 2021. *Obesity and Overweight.* www.who.int/news-room/fact-sheets/detail/obesity-and-overweight.

Yokohira, M., K. Yamakawa, K. Saoo, Y. Matsuda, K. Hosokawa, N. Hashimoto, T. Kuno, and K. Imaida. 2008. Antioxidant effects of flavonoids used as food additives (purple corn color, enzymatically modified isoquercitrin, and isoquercitrin) on liver carcinogenesis in a rat medium-term bioassay. *Journal of Food Science* 73(7): 561–568.

Zevallos, V.F., H.J. Ellis, T. Suligoj, L.I. Herencia, and P.J. Ciclitira. 2012. Variable activation of immune response by quinoa (*Chenopodium quinoa* Willd .) prolamins in celiac disease. *The American Journal of Clinical Nutrition* 96: 337–344.

Zevallos, V.F., L.I. Herencia, and P.J. Ciclitira. 2014. Gastrointestinal effects of eating quinoa (*Chenopodium quinoa* Willd.) in celiac patients. *The American Journal of Gastroenterology* 109(2): 270–278.

Zhang, Q., E. Gonzalez de Mejia, D. Luna-Vital, T. Tao, S. Chandrasekaran, L. Chatham, J. Juvik, V. Singh, and D. Kumar. 2019. Relationship of phenolic composition of selected purple maize (*Zea Mays* L.) genotypes with their anti-inflammatory, anti-adipogenic and anti-diabetic potential. *Food Chemistry* 289: 739–750.

Zhu, F. 2020. Dietary fiber polysaccharides of amaranth, buckwheat and quinoa grains: A review of chemical structure, biological functions and food uses. *Carbohydrate Polymers* 248: 116819. https://linkinghub.elsevier.com/retrieve/pii/S0144861720309929.

Chapter 10

Ingredients of High Nutritional Value Obtained from Latin-American Crops through Biotechnology

Manuel Oscar Lobo, Ana Laura Mosso, María Dolores Jiménez, and Norma Sammán
Universidad Nacional de Jujuy and Consejo Nacional de Investigaciones Científicas y Técnicas (CONICET), San Salvador de Jujuy, Argentina

CONTENTS

10.1   INTRODUCTION

Andean and Mesoamerican crops are characterized by a wide spectrum of interesting nutritional and functional compounds, such as gluten-free proteins of good biological quality, micronutrients and antioxidants. For this reason, quinoa, amaranth, chia, Andean corn, black bean, tarwi and sacha inchi, are suitable options for reinforcing diets that are deficient in proteins, essential nutrients and functional components. Furthermore, biotechnologies such as fermentation, germination and protein hydrolysis can improve the nutritional and functional properties of these crops, by implementing simple, economical and easily adjustable processes, which may lead to highly nutritional value foods and ingredients.

In this chapter, the terms seeds and grains will be used indiscriminately.

10.2 FERMENTATION

Fermentation is an economically and sustainable method of producing and preserving foods applied since ancient times. The process is determined by the metabolic action of microorganisms (bacteria, fungi or yeasts) on different substrates, in order to obtain favorable transformations. In this sense, traditionally, microorganisms naturally present in the source of origin to improve and control natural processes have been used. Moreover, in recent decades, research on microbial strains with specific characteristics has led to a wide spectrum of interesting, fermented products.

Fermentation can improve nutritional value, sensory profile, textural and rheological features, and functional properties of foods (Rizzello et al., 2017). Some examples include yeast fermentation of the sugars present in wheat flour for making bread or of malted barley sugars for making beer; also, the use of symbiotic bacterial cultures for the fermentation of milk to produce cheese, yogurt and other fermented milk derivatives and the fermentation of juices derived from different fruits and grains to produce alcoholic beverages, among others (Carrizo et al., 2020).

Advances in fermentation studies lead to the development of high-value composite products, especially with traditional crops – such as Latin-American grains – either fermented with specific lactic bacteria and yeast, or complex microbial combinations, as in spontaneous and sourdough fermentation.

10.2.1 Lactic Fermentation of Latin-American Grains

Lactic acid bacteria (LAB) have been at the core of microbial fermentation of foods and beverages as far back as 6000 BC. LAB strains form a heterogeneous group of microorganisms that reside in a wide variety of habitats and produce organic acids – mainly lactic acid – as a consequence of substrate fermentation (Carrizo et al., 2020).

Furthermore, the metabolic action of these bacteria leads to amino acids and B-group vitamins synthesis, anti-nutrients degradation, shelf life improvement, and in some cases functional features, like probiotics (Blandino et al., 2003). Also, smaller molecule spreadables such as peptides, amino acids and other nitrogenous compounds, which possess physiological functions, are formed during fermentation via enzymatic degradation.

Currently, research is focused on the advantage of fermentation to increase the bioactive and nutritional values of cereals and vegetables, including some Latin and Mesoamerican grains. Several studies have evaluated quinoa fermentation, from improving the quality of quinoa seeds to the development of fermented vegan products, such as: quinoa-based yogurt (Zannini et al., 2018), spoonable products (Väkeväinen et al., 2020), quinoa pasta (Carrizo et al., 2020) and beverages (Canaviri Paz et al., 2020; Jeske et al., 2018).

Li et al. (2018) studied the fermentation of quinoa seeds with *Lactobacillus casei* and found that proteins, free amino acids, carbohydrates and B1 and B2 vitamins were significantly higher in fermented quinoa seeds. However, fat and dietary fiber content decreased by 52.05% and 45.87%, respectively. Regarding fermented product development, Väkeväinen et al. (2020) studied the potential of two quinoa varieties – Pasankalla and Rosada de Huancayo – to obtain spreadable vegan products fermented with a potentially probiotic *Lactobacillus plantarum* strain. Products were flavored with date, wild berry and banana. The strain showed high functionality, nevertheless, cereal products as replacements for yogurt or desserts usually present sensorial difficulties. In the case of this quinoa spreadable product, consumers indicated the presence of an unpleasant aftertaste and sandy mouthfeel. Jeske et al. (2018) developed a milk substitute with quinoa fermented with *L. citreum* TR116 and *L. brevis* TR055, which in combination with exogenous amylolytic enzymes, enabled the reduction of glucose by 40%. In this case, the use of mannitol-producing LAB showed potential in a nutritious and sugar-reduced, plant-based beverage.

In addition to quinoa, some authors studied the fermentation of less well-known crops, such as kiwicha, tarwi and sacha inchi. Aguilar et al. (2019) studied the effect of adding the liquid portion of hydrolyzed kiwicha in the production of probiotic drinks with tarwi juice using *Lactobacillus paracasei*, *Bifidobacterium longum* and a culture of both microorganisms. Results showed an improvement in LAB viability, especially when the amount of hydrolyzed kiwicha was increased. Furthermore, the viability of *B. longum* was reduced with the decrease of pH, but improved in the co-culture with *L. paracasei*. Acidity levels enhance the stability of fermented drinks against pathogen microorganisms. In addition, the sensorial characteristics of the drinks were favorable.

Vanegas et al. (2018) studied the addition of sacha inchi seeds – which are rich sources of n-3 and n-6 fatty acids, tocopherols, phytosterols, phenolic compounds, protein of high nutritional value and dietary fiber – and betaglucans in yogurt with the aim of improving nutritional and functional characteristics. This incorporation did not affect fermentation kinetics but significantly increased α-linolenic (49.3%) and linoleic (32.2%) acids, whose values were about 50- and 25-fold higher than those of the control yogurt. Despite textural parameters (firmness, consistency, cohesiveness and index of viscosity) of the enriched yogurts, they were significantly lower than those of the control samples during the whole storage period; all enriched yogurts showed a sensorial acceptance higher than 70% by untrained panelists, reaching values similar to control.

Furthermore, Micanquer et al. (2020) evaluated the formulation of an unconventional fermentation substrate from agro-industrial waste of pineapple and sacha inchi to produce biomass. Nowadays, the main use of sacha inchi is to obtain oil. This process generates a sub-product of 59% protein content (Wang et al., 2018), which represents between 60 and 75% of the seed's total weight. The authors studied the use of this high protein and carbohydrate portion with pineapple peel to formulate a fermentation substrate, appropriate for the growth of *Weissella cibaria,* an LAB with potentially multiple uses in the food industry. Different homogenization and heating treatments were assessed to increase the feasibility of reducing sugars and improving the efficiency of these wastes as fermentation substrates.

Regarding chia fermentation, Dentice-Maidana et al. (2020) isolated LAB from chia sourdoughs to develop sorghum gluten-free bread. Strains – *Weissella cibaria, Lactobacillus plantarum* and *Lactobacillus fermentum* – were selected based on techno-functional and safety properties. Fermentation products present in sourdoughs after 24 h showed a marked increase in lactate, xylose, arabinose, free amino acids and hydrogen peroxide while glucose was undetectable. Bread manufactured with 30 and 40% of chia sourdoughs significantly improved specific volume and visual appearance compared to 100% sorghum bread.

10.2.2 Production of Vitamin and Antioxidant Compounds through Lactic Fermentation

Fermentation is a complex process where different metabolic products are developed – and some conditions favor the presence of specific ones of particular interest – such as B group vitamins and antioxidants. The study of strains, substrates and proper fermentation conditions may be a promising bioenrichment strategy.

For instance, Silva-Vieira et al. (2017) and Mosso et al. (2020) studied folate enrichment through lactic fermentation of different substrates with the incorporation of amaranth flour. In both cases, this Andean grain addition increased folate production by *B. longum* subsp *infantis* BB-0 in the first and *L. sakei* CRL2210 in the latest. Results demonstrated that this vitamin production is not only strain-dependent but also influenced by the addition of different substrates in the growth media.

High antioxidant activities have been found in many fermented cereals, such as wheat, rice, oat, maize and sorghum, which were the basis for investigations regarding fermentation of Latin-American seeds and grains as an effective approach for the preparation of bioactive substances. According to Rizzello et al. (2016), the antioxidant properties of quinoa flour were improved via fermentation with *Lactobacillus plantarum.* Regarding quinoa seeds, Li et al. (2018) found that fermentation with *L. casei* increased the 2.2-diphenyl-1-picrylhydrazyl (DPPH) radical scavenging activity, reducing ability and Fe^{2+}-chelating activity. Fermented quinoa seeds showed a higher total phenolic content (16.53 mg gallic acid equivalent (GAE)/g extract, dry weight) than unfermented seeds (13.85 mg GAE/g). Fermentation increased free phenolics and decreased bound phenolics.

Antioxidant activity related to amino acid residues was studied by Montemurro et al. (2019) who fermented quinoa flour with autochthonous *L. plantarum* T0A10. Five peptides, released during fermentation and having from 5–9 amino acid residues, were purified and identified as responsible for the increase in antioxidant activity. Such sequences, encrypted in native quinoa proteins and released through the proteolytic activity of the LAB strain, showed antioxidant activity on human keratinocytes NCTC 2544 artificially subjected to oxidative stress.

10.2.3 Reducing Antinutritional Factors through Fermentation

Mineral absorption in humans is reduced in the presence of anti-nutrients, such as alkaloids, in tarwi and phytic acid in quinoa, kañihua and amaranth.

Phytic acid binds to positively charged divalent cations such as iron, zinc, calcium and proteins, forming phytate complexes that are stable at intestinal pH (6–7), thus inhibiting the absorption of minerals in the small intestine (Castro Alba et al., 2019). Phytase, an enzyme present in plants and microorganisms, including certain strains of LAB and yeasts, decreases phytate content and improves mineral bioavailability (Carrizo et al., 2020).

For instance, Castro Alba et al. (2019) studied the phytic acid degradation during spontaneous fermentation, specifically using *Lactobacillus plantarum* 299v in quinoa, kañiwua and amaranth grains and their flours. Higher phytic acid degradation was found during the fermentation of flours (64–93% from original content) than with grains (12–51%). The addition of *Lactobacillus plantarum* 299v increased the concentration of lactic acid, with a concomitant pH decrease, which provided the conditions necessary for the activation of endogenous and microbial phytase.

Tarwi consumption is limited due to the presence of bitter alkaloids and other anti-nutritional factors, such as phytic acid, tannins, nitrates and trypsin inhibitors that have undesirable physiological effects. Nowadays, debittering is performed through a conventional aqueous thermal treatment. Villacres et al. (2020) assessed the efficacy of fermentation with *Rhizopus oligosporus* to reduce these compounds. The fungal treatment decreased the following antinutrients from debittered grain by the following percentages: nitrates (94.6%), tannins (82.1%), alkaloids (94.0%), urease activity (93.7%), phytic acid (70.0%) and trypsin inhibitors (76.7%), and increased total carotenoids (165.4%), phenolic compounds (1055.5%) and antioxidant capacity (1515.6%).

10.2.4 Traditional Beverages through Maize Fermentation

Chicha is a traditional beverage prepared in several Andean countries from ancient times – the first reports of *chicha* production date back to 200 BC, before the establishment of the Incas in the region. This type of clear, yellow and frothy beer is prepared mainly from malted yellow maize grains; nevertheless, other traditional maize strains are usually used, according to the region/country, e.g. white, black, chulpi and morocho maize (Barbosa Piló et al., 2018).

Grains are first germinated (left in water for a day) and then put onto straw mats or plastic tarps under the sun for two days until they are completely dried. After boiling, the mixture is strained and then placed in a container to ferment. Usually, after two days of spontaneous fermentation by indigenous microorganisms, the beverage is ready for consumption (Vallejo et al., 2013). Though fermentation is mainly alcoholic, by *S. cerevisiae* and other yeasts, LAB and other bacteria are also involved and play a key role in the flavor development of *chicha*. This rich biodiversity has been studied by Bassi et al. (2020) and Grijalva-Vallejos et al. (2020) as a resource of biodiversity with potential technological and health-beneficial properties.

Pozol is one of the most ancient Mesoamerican fermented products still consumed in south Mexico and in several places of Central America. This traditional beverage is used for alimentary and ceremonial purposes in many ethnic groups of southeastern Mexico. Typical elaboration starts with little nixtamalized maize balls wrapped in banana leaves and left to ferment with endogenous microbiota for a certain time, which can be as short as a few hours or as long as a month. After natural fermentation, the maize balls are mixed with water and consumed as a refreshing beverage (Pérez-Armendáriz et al., 2020).

10.2.5 Impact of Fermentation on Matrix Structural Changes

In recent decades, research on gluten-free products has increased. Dentice-Maidana et al. (2020) used sorghum to develop gluten-free bread augmented with different percentages of chia sourdoughs (0–40%), and studied techno-logical and sensorial features. Based on techno-functional and safety properties, *Weissella cibaria* CH28 together with *Lactobacillus plantarum* FUA3171 and *Lactobacillus fermentum* FUA3165, were used as inoculants to ferment chia sourdoughs used for bread making. Sensory evaluation showed that breads with 40% of fermented chia were the most accepted by panelists who were also able to discern bread inoculated with fermented and unfermented chia dough.

Another trend in the food market is development of lactose-free products that mimic milk, cheese and yogurt characteristics. In this sense, fermentation also plays a key role. Zannini et al. (2018) developed a novel beverage fermented with *Weissella cibaria* MG1 based on aqueous extracts of wholemeal quinoa flour to simulate yogurt. This strain is characterized by exo-polysaccharides (EPS) production at the end of the fermentation, which improves water holding capacity and viscosity. Microstructure observation indicated that the network structures of EPS-protein improve the texture of fermented quinoa milk. Overall, *Weissella cibaria* MG1 showed satisfactory technological properties and great potential for further possible application in the development of high viscosity fermented grains.

10.3   GERMINATION

10.3.1   Germination Conditions

Germination is another traditional and economical, biotechnological process, which can easily be performed with a wide variety of crops such as quinoa, amaranth, kañiwa, chia, black beans, corn, tarwi and sacha inchi, among others. Under controlled conditions of temperature, humidity, light and time, germination improves protein digestibility, increases soluble sugars, free fatty acids and antioxidants, and decreases anti-nutritional compounds.

Different antifungal procedures are recommended before seed germination. In most cases, prior to germination, seeds are soaked to induce water imbibition. Germination occurs in nature when seeds find proper conditions and start the first steps of plant development; in order to mimic that process, several parameters must be studied specifically for each seed. Some germination conditions are described in Table 10.1. After germination, seeds are dried to stop the enzymatic activity (in a vacuum oven, forced air circulation oven or by lyophilization), usually milled, and finally stored until analysis.

10.3.2   Changes in Nutritional Profile during Germination

10.3.2.1   Protein Content, Amino Acids Profile and Protein Digestibility

Germination increased the total protein content of amaranth, quinoa (Padmashree et al., 2018), chia (Gómez-Favela et al., 2017; Pająk et al., 2019) and sacha inchi (Chandrasekaran and Liu, 2014), possibly due to protein synthesis activation and/or dry matter reduction, particularly through the oxidation of carbohydrates and the loss of carbon dioxide, water and small amounts of ethanol during respiration.

Changes in protein profiles have been found during germination. For instance, after 18 h of amaranth germination, most of the essential amino acids increased,

TABLE 10.1 GERMINATION CONDITIONS

Crop	Antifungal Treatment	Germination Conditions	Observations	Germinative capacity	Reference
Amaranth	NaClO (20% v/v) 20 min	Soaking: 6 h 30°C – Agitation 18 and 24 h	Plastic boxes with absorbent paper	For 18 h: 52% (MSL[a]: 0.40 cm) For 24 h: 73% (MSL[a]: 0.55 cm)	Guardianelli et al. (2019)
Quinoa	-	Soaking: 6 h (RSW[b] 1:5) 22–14°C – RH[c] 90% – Dark 24 h	Plastic trays covered with wet filter papers	90% (MSL[a]: 1.0–1.5 cm)	Jimenez et al. (2019)
Purple corn	NaClO (0.1% v/v) RSW[b] 1:5 30 min	Soaking: 24 h at room temperature (RSW[b] 1:5) 12–28°C (optimal 26°C) – RH[c] 90% – Dark 12–72 h (optimal 63 h)	Trays on wet filter paper	-	Paucar-Menacho et al. (2017)
Black turtle bean	-	Soaking: 24 h (RSW[b] 1:3) with aeration 20°C – RH[c] 92% - Dark 5 days	-	90% (MSL[a]: 2.1 cm)	Guajardo-Flores et al. (2012)
	-	25°C – Dark 6 days	Into a semi-automatic germination machine	75% (MSL[a]: 18.5 cm)	Xue et al. (2016)
Chia	Etanol(96%) 1 min	22±2°C – 12/12 h day/night 7 days	Sterile stackable trays	-	Pajak et al. (2019)
Kañiwa	NaClO (0.3% v/v) 15 min	Hydrated: 14 h at 20°C and RH[c] 45% 20°C – Dark 48, 72 or 96 h	-	-	Abderrahim et al. (2012)

(continued)

TABLE 10.1   (CONTINUED)

Crop	Antifungal Treatment	Germination Conditions	Observations	Germinative capacity	Reference
Sacha inchi	-	18°C Dark or 26 °C Light 3, 6, 10, 20 and 30 days	On sterile filter paper in Petri dish	90%	Chandrasekaran and Liu (2014)
Tarwi	-	Soaking: 24 h at 16°C with stirring 16°C and 20°C – RH[c] 45% and 50% 2, 3 and 4 days	-	-	Villacrés et al. (2015)

Notes: [a]MSL: Mean sprout length; [b]RSW: Ratio seeds:water; [c]RH: Relative humidity.

except for histidine and tyrosine levels that were maintained and sulfur amino acids (methionine and cysteine) which significantly decreased. In addition, after 24 h of germination, the content of proline significantly increased, possibly due to having an adaptive role in plant stress tolerance (Guardianelli et al., 2019).

Regarding the germination of black bean, amaranth and quinoa, a reduction in high-molecular-weight proteins and an increase in polypeptides was observed, because part of the energy used for germination and plant development comes from the mobilization of storage proteins that were hydrolyzed by proteolytic enzymes and converted into soluble peptides and amino acids (Xue et al., 2016; Guardianelli et al., 2019; Jimenez et al., 2019; Piñuel et al., 2019).

Protein digestibility of quinoa and amaranth significantly improved with germination (24% and 20%, respectively) (Chaparro et al., 2010; Jimenez et al., 2019), possibly due to protein hydrolysis. Villacrés et al. (2015) observed an increase in protein digestibility reaching values between 85 and 87% during germination of tarwi; and Gómez-Favela et al. (2017) also observed a significant increase in protein digestibility (5%) with chia germination.

10.3.2.2 Lipid Content and Fatty Acids' Profile

During germination of white quinoa, the lipid content was kept the same (Padmashree et al., 2018); however, a decrease in the amount of lipids was observed after germination of red quinoa, amaranth (Jimenez et al., 2019), chia (Pająk et al., 2019) and sacha inchi (Chandrasekaran and Liu, 2015). A decrease in lipid content may be due to catabolic activity since, during germination, glucose molecules can be obtained from fatty acids of lipids in the glyoxylate cycle.

Regarding amaranth germination, Guardianelli et al. (2019) observed a decrease in the content of palmitic acid, stearic acid and oleic acid; while linoleic and α-linolenic acids increased, probably due to the conversion of oleic acid by desaturase enzymes. On the other hand, Jiménez et al. (2020a) found that saturated fatty acids decreased with both quinoa and amaranth germination, caused by the reduction in palmitic and behenic acids, which was possibly due to the lipolytic activity and the decomposition of triglycerides and polar lipids into simpler compounds.

Thin layer chromatography (TLC) of quinoa and amaranth grains showed a decrease in triglycerides and an increase in free fatty acids, phospholipids, diacylglycerols and sterols during germination (Jimenez et al., 2020a); the pathway could be related to the breakdown of reserve triglycerides into simpler compounds through hydrolytic degradation by the action of lipases (Jan et al., 2018). During sacha inchi germination, a rapid decline in the triacylglycerol and diacylglycerol content was observed after the early stages (3–10 d after imbibition) followed by a steady breakdown during the following stages of germination. Also, a rapid increase in free fatty acids content between 10 and 20 days of sacha inchi germination was observed, after which, free fatty acids decreased during the late stage of seed germination, which could indicate the possible bio-synthesis of fatty acids during the later stages of germination (Chandrasekaran and Liu, 2015).

10.3.2.3 Carbohydrates Profile

The carbohydrate content decreased after 48 h of germination of both white and red quinoa varieties (62–52% and 58–50%, respectively), possibly because carbohydrates were utilized as an energy source for the growth of the embryo during germination (Padmashree et al., 2018). The starch amount significantly decreased, due to its degradation into low-molecular-weight compounds such as glucose and fructose to provide energy for cell division. Villacrés et al. (2015) observed a reduction of starch from 1.6 to 1.1% after 4 d of tarwi germination. Jiménez et al. (2019) found a reduction in digestible starch of 52% and 64% during the germination of amaranth and quinoa, respectively; however, resistant starch content was kept constant, since it cannot be hydrolyzed by endogenous enzymes.

Besides, changes in starch conformation occur with germination. Apparently, amylose content incremented with germination between 13 and 65% in quinoa and over 100% in amaranth grains, possibly due to the hydrolysis of primary amylopectin by the action of the amylolytic enzymes α-amylase and β-amylase, which could release the linear branches of glucan chains and dextrin that react with iodine such as amylose (Jimenez et al., 2019).

The effect on dietary fiber – total, soluble and insoluble – depends on ger-mination conditions and the nature of the crop, due to the different structures and compositions of cell walls (Omary et al., 2012). Repo-Carrasco-Valencia and Serna (2011) and Jimenez et al. (2019) observed that total, soluble and insol-uble fiber content did not change with germination of quinoa and amaranth. Nevertheless, Padmashree et al. (2018), Dueñas et al. (2016) and Gómez-Favela et al. (2017) observed an increase in both insoluble and total fiber with the ger-mination of quinoa, black bean and chia respectively, which may be related to the hemicellulose, cellulose and pectic polysaccharides produced in advanced stages of germination.

In the case of chia germination, soluble dietary fiber decreased 13.5%, pos-sibly due to the loss of mucilage during the hydration of the seeds. However, other authors discerned an increase in soluble fiber and a decrease in insoluble fiber with corn germination (Gong et al., 2018).

10.3.3 Antioxidant Content and Antioxidant Capacity

During germination, the bonds of phenolic compounds with non-starch polysaccharides localized in the cell walls are degraded by enzymatic action, mainly esterase, which produces a consequent increase in free phenolic acids (ferulic, p-coumaric and synapic acids). On the other hand, secondary metabolism (specifically the phenylpropanoid metabolic pathway) could be initiated by the activation of phenylalanine lyase (a key enzyme in phenolic biosynthesis), causing the production of phenolic compounds (such as p-coumaric, caffeic and ferulic acids) from aromatic compounds such as amino acids (phenylalanine, tyrosine and tryptophan) (Gómez-Favela et al., 2017; Pajak et al., 2019). An increase in total phenolic and flavonoid is correlated with enhanced antioxidant activity during germination.

The effect of light in germination parameters was studied for some crops. For instance, in amaranth, the light had no effect on gallic acid content but increased rutin concentration; whereas, when germination was performed in the dark, a higher content of isovitexin and vitexin was recorded (Nemzer et al., 2019).

An increase in phenolic compounds and γ-aminobutyric acid (GABA) was observed during chia germination under optimal conditions (Gómez-Favela et al., 2017). Nevertheless, in purple corn, a reduction in phenolic compounds, anthocyanins and flavonoids content was found when performed in optimal germination conditions to produce GABA (26°C, 63 h). This phenomenon may be attributed to polyphenol-oxidase activation – an enzyme responsible for phenolic compound degradation during germination – or to the leaching of these compounds during the soaking period. GABA increase may be related to the activation of enzymes involved in its production and glutamate synthesis, primarily by the decarboxylation of L-glutamic acid and catalyzed by glutamate decarboxylase during seed germination (Paucar-Menacho et al., 2017).

Tocopherols are lipophilic antioxidants that are also modified during germination. Carciochi et al. (2016) observed that the total tocopherol content (TTC) of quinoa significantly increased after 72 h of germination; however, Jiménez et al. (2020a) found a significant decrease in TTC after 24 h-quinoa germination, with a decrease in β and γ-tocopherols, despite the increase in α-tocopherol. On the contrary, after 48 h-amaranth germination, TTC was raised due to the significant increase of α- and β and γ-tocopherols, despite the reduction in δ-tocopherol.

Changes in TTC are related to the activation of metabolic pathways during germination, such as cytosolic shikimate and plastid methylerythritol phosphate. Besides, biosynthesis and the increase or decrease in TTC depends on stress conditions. Lushchak and Semchuk (2012) explained that in dicots (like soybean, quinoa and amaranth) α-tocopherol levels initially increased during germination and then decreased.

10.3.4   Antinutrients and Mineral Bioavailability

An increase in mineral bioavailability is often observed during germination, due to the hydrolysis of phytic acid by the increment in phytate enzyme activity (possibly myoinositol hexakisphosphate phosphohydrolase), which consequently leads to the release of minerals associated with phytic acid, mainly divalent minerals such as calcium, magnesium, iron, copper, manganese and zinc (Świeca, 2016). An increase in the bioavailability of minerals was observed with the germination of chia and tarwi (Pająk et al., 2019; Villacrés et al., 2015). Padmashree et al. (2018) also observed a reduction of phytic acid after 48 h of both white and red quinoa germination (52.3–42.7 mg/100 g and 63.7–44.9 mg/100 g, respectively), where phytic acid might have been broken down for the utilization of phosphorus during germination. Saponin and tannin contents decreased in red quinoa germination (0.6–0.4 g/100 g and 3.4–2.9 g/100 g, respectively), while no significant variation was found after white quinoa germination.

Decrease in saponins may be due to leaching during washing and soaking before germination. During the three days of black turtle bean germination, the total saponin amount in the seed increased due to the higher concentration detected in sprouts and cotyledons, which represents 90.8% of the total seed weight. This increase is possibly due to the synthesis and activation of different enzyme systems with germination, which enhances the production of these secondary metabolites and the weakening of the seed structure, which facilitated solvent extraction procedures. However, saponins of the seed coats decreased, possibly due to leaching (Guajardo-Flores et al., 2012).

10.3.5   Technological, Thermal and Sensory Changes during Germination

Changes occurring during germination cause modifications in technological, rheological, textural, thermal and sensory properties which may be considered, particularly, when sprouts are used as ingredients in food formulation.

10.3.5.1   Rheological and Textural Characteristics

Several authors have studied the incorporation of germinated flours in bread, purees and pasta, among other bakery products. For instance, in bread making, the replacement of native quinoa flour with germinated quinoa flour (ratio 80:20 wheat to quinoa) produced an increase in dough water absorption and in the softening degree. Nevertheless, stability decreased, suggesting a weakening of the gluten network. Additionally, there was improved dough development and gas production during leavening, which produced a high specific volume and low crumb firmness, probably due to increased sugar content (maltose, sucrose and glucose) (Suárez-Estrella et al., 2020).

The germinated quinoa flours showed lower pasting properties than those made with ungerminated grains. Viscosity decreased after 48 h of quinoa germination, from 177 rapid-visco analyser units (RVU to 42 RVU). In addition, setback viscosity also decreased with germination; thus, starch molecules have a lower ability to disperse in hot paste and reassociate during cooling (Padmashree et al., 2018).

Ujiroghene et al. (2019) made a functional yogurt from sprouted and non-sprouted quinoa milk. The use of sprouted quinoa milk increased the total content of phenols and flavonoids, and the antioxidant capacity. Furthermore, the water-holding capacity of "germinated quinoa yogurt" was significantly higher than the non-germinated one, which may be attributed to the production of compounds such as soluble sugars with good water-holding capacity.

Jiménez et al. (2019) formulated baby purees with quinoa and amaranth flours and studied the changes produced by their replacement with sprouted grain flours. Purees made with germinated grain flours showed a significant decrease in the consistency coefficient, and in the storage (G') and in the loss (G") modules in both stress and frequency sweeps. In addition, the flow curves (shear stress versus shear rate) showed downward displacement, indicating a weaker food matrix and greater fluidity with the use of germinated grains flours, which may be related to the enzymatic hydrolysis of starch and proteins (Cornejo et al., 2019). Also, purees elaborated with germinated grain flours had significantly less hardness and adhesiveness than those made with non-germinated flours, due to the lower number of polysaccharide chain networks within the food matrix caused by lower starch content.

10.3.5.2 Thermal Behaviors

The thermal properties of germinated and non-germinated amaranth flours were determined by a differential scanning calorimetry (Guardianelli et al., 2019). The gelatinization peak temperature increased with germination time (18 and 24 h) because protein fractions were changed by germination. Besides, significantly lower changes of enthalpy (ΔH) values were obtained in the germinated grain flours (6.7 J/g and 4.7 J/g for 18 and 25 h of germination, respectively) as opposed to the non-germinated grain flours (9.6 J/g), which could be attributed to lower starch content and changes in protein structure due to enzymatic hydrolysis during germination. Jiménez et al. (2019) agreed with the reduction of ΔH with germination of quinoa (24 h) and amaranth (48 h), but no changes in gelatinization temperatures (initial, peak or final) were found.

10.3.5.3 Sensory Characteristics

In some foods, the use of sprouted grains showed no significant changes in sensory evaluation; for instance, in the use of sprouted quinoa liquid to obtain a functional yogurt-like product, germination had little or no importance in the overall organoleptic acceptability of the final product (Ujiroghene et al., 2019). Nevertheless, replacement of raw quinoa flour with germinated quinoa flour in pasta formulation significantly changed the taste (Demir and Bilgicli, 2020).

Moreover, bread with sprouted quinoa flour was characterized by decreased bitterness assessed by electronic tongue in comparison to bread with unsprouted quinoa flour (Suárez-Estrella et al., 2020).

In addition, the use of sprouted quinoa and amaranth flours for the elaboration of infant purees had a negative impact on sensory analysis, since the attributes most frequently used to describe purees made with sprouted grain flours were: 'more liquid', 'slightly acidic', 'with a bitter taste', 'intense taste with an aftertaste'. Therefore, it is necessary to continue studying ways to mask unpleasant flavors in using some sprouts as food ingredients (Jimenez et al., 2020b).

10.4   BIOACTIVE PEPTIDE PRODUCTION

According Morales et al. (2020), the Food and Agriculture Organization (FAO) has classified plant proteins as highly important for the human diet, highlighting their natural abundance and nutritional value. For these reasons, their biological properties are currently being studied, especially as a source of biologically active peptides when included in the diet. These peptides – amino acid sequences with no activity in the precursor protein – may display physiological functions in the body after their release by gastrointestinal digestion or *in vitro* digestion by chemical or enzymatic hydrolysis, fermentation and/or germination (Orona Tamayo et al., 2019; Apostolopoulos et al., 2021).

10.4.1   Enzymatic Hydrolysis

Enzymatic hydrolysis has prevailed over chemical hydrolysis since the hydrolysates and peptides led by the first technique show higher activity and absorption, in addition to the easy regulation of the bioprocess, low cost and simple implementation. The bioprocess can be easily implemented; furthermore, peptides with better absorption and activity can be produced. Peptide length, amino acid constitution, presence of polar and ionizable groups and hydrophobicity play a key role in the resulting functional and bioactive properties.

The enzyme catalase is the most used in protein hydrolysis studies of Latin-American crops, usually used at high temperatures (45°C) and alkaline pH (8.5–10). In addition, other enzymes have been studied, such as trypsin, α-chymotrypsin, protease, bromelain, papain, pepsin, pancreatin, flavourzyme, neutrase, in *in vitro* gastrointestinal systems and even enzymatic extracts from the stomach and intestines of different animals.

Studies on hydrolysates and peptides obtained by enzymatic hydrolysis of proteins from Latin-American crops are mainly focused on antioxidant properties (through antioxidant assays of ABTS [2,2'-azino-bis(3-ethylbenzothiazoline-6-sulfonic acid)], DPPH and metal chelating capacity-ferrozine), antihypertensive properties (by angiotensin-I converting enzyme – ACE-inhibitory activity) and antimicrobial properties (well diffusion method).

Otherwise, there is little but promising information on the inhibitor activity of adipogenesis, antihaemolytic activity in red blood cells, antihypertensive activity in spontaneously hypertensive rats and renin inhibition activity, inhibitory activity of colon cancer cell viability and antidiabetic and hypocholesterolemic cholesterolemic activity, as well as immunomodulatory activity (Morales et al., 2020; Moughan et al., 2014; Daliri et al., 2017).

Several enzymes and enzyme systems are used for protein hydrolysis of different quinoa cultivars, where incubation intervals and the enzyme to substrate ratio varies considerably. For instance, chymotrypsin showed higher efficiency than bromelian and protease enzymes, reaching 85% degree of hydrolysis in 6 h, for an enzyme to substrate ratio of 1:100. Glutelins were the most susceptible proteins to enzymatic attack and globulins showed the highest resistance (Mudgil et al., 2019). *In vitro* simulation of gastrointestinal digestion was also assessed, where pepsin and pancreatin were studied successively and separately; moreover, the influence of bile salts in the process was determined. The most common enzyme to substrate ratio was 1:100, reaching over 40% of the degree of hydrolysis in 2 h. Peptide fractions were separated by size (>/<5 kilodalton (kDalton), even a separation was performed by reversed-phase-high performance liquid chromatography (RP-HPLC), identifying peptides with encrypted structures in the 11 S globulin protein (Vilcacundo et al., 2018a; Shi et al., 2019). Papain was also used, in natural and refined extract, verifying a high degree of hydrolysis on quinoa protein concentrates, using an enzyme to substrate ratio of 2:100. In this case, the peptides fraction of less than 5 KDalton represented 50% of the hydrolysate, after 180 min of incubation at 50°C (Nongonierma et al., 2015).

Furthermore, the enzymatic hydrolysis of amaranth proteins was extensively studied, finding that in all cases the process was slower than quinoa proteins' hydrolysis. Mudgil et al. (2019) determined that the hydrolysis degree of amaranth proteins with chymotrypsin was around 40% after 6 h. Ayala-Niño et al. (2019) obtained a similar hydrolysis degree in *Amaranth hypochondriacus* spp proteins using alcalase (pH 10) and flavorzyme (pH 7), during 6 h, achieving a greater efficiency with the use of both enzymes sequentially.

The protein isolates of black beans were studied by hydrolysis with a simulated gastrointestinal system (pepsin and pancreatin) and with alcalase (ratio 1:100) for the production of peptides with different functional properties, especially hypoglycemic. The hydrolysates were separated into nine fractions with molecular weights between 400 and 1200 Dalton. Peptides with the highest concentration were identified for their subsequent synthesis and study (Mojica et al., 2017a, 2017b).

Aguilar-Toalá et al. (2020) hydrolyzed chia proteins from defatted flours from which their characteristic mucilage was removed. Chia seeds were hydrolyzed using single (alcalase) or sequential (alcalase + flavorzyme) enzymatic processes at 2%, with the conventional water bath- and microwave-assisted process. For the antimicrobial activity tests, they used the whole hydrolysate and the peptide fractions of 3–10 and <3 kDalton. On other hand, Martínez-Leo and Segura-Campos (2020) hydrolyzed chia protein using a sequential pepsin (pH2)-pancreatin (pH7.5) system at 37°C. The protein hydrolysate was fractionated by an ultrafiltration process to three peptide fractions < 1 kDalton, 1–3 kDalton and 3–5 kDalton in order to study pro-inflammatory modulation of HMC3 microglial cells.

10.4.2   Physiological Functions

10.4.2.1   Antioxidant Activity

The antioxidant activity of quinoa protein hydrolysates produced with chymotrypsin, protease and papain was studied. Generally, this activity increased with the hydrolysis degree – the smallest molecular size fractions (less than 5 kDalton) – being the most efficient. Oxygen radical absorption capacity (ORAC) was measured with ABTS and DPPH tests (Mudgil et al., 2019, Nongonierma et al., 2015). ORAC of quinoa protein hydrolysates peptides obtained from simulated *in vitro* digestion was measured by the fluorescein test. Results show that successive digestion with pepsin and pancreatin is necessary to give the hydrolysates a high antioxidant capacity, since gastric digestion alone did not increase the antioxidant property in the same proportion as gastroduodenal (Vilcacundo et al., 2018a).

Mudgil et al. (2019) determined that the antioxidant properties of amaranth protein hydrolysates were significantly higher than those determined in native proteins and even quinoa hydrolysates, especially when hydrolysis was performed with chymotrypsin. This capacity, determined by ABTS and DPPH tests, increased with hydrolysis degree so that the enzymes with greater efficiency for hydrolysis (chymotrypsin and protease) generated hydrolysates with greater antioxidant capacity. On the other hand, Ayala-Niño et al. (2019) measured the antiradical (ABTS and DPPH) and Fe-chelating (FRAP) activity of *Amaranth hypochondriacus* protein hydrolysates produced with alcalase and flavorzyme, finding an activity increase between 500 and 1000% compared to native proteins.

The peptides obtained by *in vitro* gastrointestinal hydrolysis of black bean proteins showed excellent antioxidant properties. The complete hydrolysates were twice as effective as pure peptides in scavenging radicals, indicating that the property was based on a joint action of various peptides or proteins contained in the hydrolysates (Mojica et al., 2017b).

The chia peptide fraction of 1–3 kDalton presented a protective effect after the induction of tert-Butyl hydroperoxide (TBHP) damage. HMC3 cell mortality was significantly reduced. If peptides were used at concentrations of 25, 50 and 100 µg/mL, a protective effect of 62.3%, 73.4% and 79.0% respectively was shown. In addition, the generation of reactive oxygen species (ROS) such as H_2O_2, was reduced by 50% (Martínez-Leo and Segura-Campos, 2020).

10.4.2.2   Cytotoxicity

Regarding cytotoxicity, several studies were performed using cells in order to determine the functional activity of hydrolysates prepared with quinoa, amaranth and other Latin-American crop proteins. In general, protein concentrates and hydrolyzates, as well as the different peptide fractions, did not present cytotoxic activity against different cells (Caco-2 and pre-adipocyte cells, HMC3 cells) up to concentrations closer to 3000 µg of hydrolyzate/mL for quinoa and amaranth hydrolysates; and up to 400 µg/mL for chia hydrolysates. This determination is essential to carrying out studies of different activities of *ex vivo* protein hydrolysates.

10.4.2.3   Anti-Inflammatory Activity

The most widely used *in vitro* tests to determine the anti-inflammatory activity of protein hydrolysates from Latin-American cultures refer to inhibition of the enzyme renin and ACE, which regulate blood pressure. The combined hydrolysis with alcalase and flavourzyme on quinoa proteins caused an ACE inhibition of up to 60% (Ayala-Niño et al., 2019). On the other hand, there are studies referring to inhibition of the nitrous oxide production of phagocytic cells, caused by quinoa protein hydrolysates produced with papain, pepsin and pancreatin enzymes. Although they have a positive effect, no significant differences were determined with respect to the native protein (Shi et al., 2019).

The anti-inflammatory activity of hydrolyzed amaranth proteins and synthetic peptides encrypted in amaranth proteins were studied, in order to assess the inhibition of renin and ACE. Amaranth protein concentrates and hydrolysates were tested together with seven synthetic peptides that regulated the kinetics of the enzyme, demonstrating that amaranth proteins, hydrolysates and different encoded sequences were able to inhibit both of the enzymes studied. In addition, a heptapeptide (FNLPILR) was identified as that with the highest efficiency; also, probable sites of the renin enzyme in which the peptides can associate to achieve their inhibition were determined (Nardo et al., 2020; Suárez-Estrella et al., 2020). A decrease of up to 40 mmHg in systolic blood pressure was also verified in hypertensive rats (renin-angiotensin system) through treatments with concentrates, hydrolysates and synthetic peptides encrypted in the *Amaranth hypochondriacus* proteins and even by the administration of biscuits supplemented with amaranth hydrolysates. The antihypertensive capacity of these foods showed a decrease comparable to that achieved with drugs recommended for this purpose. The authors determined that the bioactive amaranth peptides also behaved as vasorelaxant agents on the vascular system (Suárez-Estrella et al., 2020; Ontiveros et al., 2020).

In vitro anti-inflammatory activity of black bean protein hydrolysates was also verified, inhibiting the action of dipeptidyl peptidase IV (DPPIV) and ACE enzymes with IC 50 of 0.14 and 0.29 mg/mL (Mojica et al., 2017b).

Martínez-Leo and Segura-Campos (2020) verified *in vitro* the anti-inflammatory action of different fractions of peptides, separated from chia protein hydrolysates, especially fraction 1-3 kDalton. The hydrolysates (100 µg/mL) decreased the concentration of nitrous oxide and cytokines in lipopolysaccharide induced HMC3 cells by up to 50%.

10.4.2.4 Antihemolytic Activity

The antihemolytic activity of quinoa hydrolysates increased with the degree of hydrolysis and depended on the type of enzyme used. Quinoa hydrolysates produced with protease increased this property five-fold compared with protection of the native protein on human erythrocytes.

In amaranth, enzymatic hydrolysis enhanced the antihemolytic capacity of proteins, especially when hydrolysis was carried out with protease. In addition, a pronounced increase in capacity was concomitant with hydrolysis time, that is, with a decrease in molecular size (Shi et al., 2019).

10.4.2.5 Antimicrobial Activity

Even though quinoa proteins lack antimicrobial activity, it was demonstrated that their hydrolysates express this activity after a certain degree of hydrolysis. Mudgil et al. (2019) determined antimicrobial activity of peptides produced by the protease enzyme, contrasted with *S. aureus*, *S. typhimurium*, *E. coli* and *E. aerogenes*. Amaranth protein concentrates and hydrolysates showed the same behavior: native proteins lacked activity, nevertheless, the hydrolysates presented antimicrobial activity, especially those produced with the enzyme chymotrypsin and with a high degree of hydrolysis (Mudgil et al., 2019). Contrary to what Mudgil et al. (2019) found, there is experimental evidence that antimicrobial activity increases with the degree of hydrolysis, reaching a maximum in which peptides have the correct molecular size and amino acid composition to be able to interact with the phospholipids of the cell membrane, increasing their thickness and creating pores that cause cell disruption (Mirzaei et al., 2016), a situation that would be lost with the increasing degree of hydrolysis.

Chia seed protein hydrolysates reported antimicrobial activity against *Escherichia coli*, *Salmonella enterica* and *Listeria monocytogenes*. Aguilar Toalá et al. (2020) observed that the <3 kDalton peptide fraction had greater antimicrobial activity than chia seed hydrolysate and the 3–10 kDalton fraction; also, peptides obtained by microwave-assisted enzymatic hydrolysis had higher efficiency than those obtained by the traditional method. Peptide fraction <3 kDalton caused modifications in the lag phase, maximum growth and growth rate of the growth curves and promoted multiple indentations (transmembrane tunnels), membrane wrinkling, and deformation in the integrity of the bacterial cell membranes, against Gram-positive and Gram-negative strains. Besides, different peptide sequences were determined in the <3 kDa fraction by liquid chromatography mass spectrometry (LC-MS/MS), some of them with potential antimicrobial scoring.

10.4.2.6 Antiadipogenic Activity

The adipogenesis inhibitory activity of quinoa and amaranth hydrolysates was investigated. On insulin-induced adipocytes, the hydrolysates showed positive and significantly higher activity (IC50 786.58 µg/mL) than the proteins without hydrolyzing up to a concentration of 1,600 µg/mL, especially in the hydrolysates obtained with pepsin and with a higher degree of hydrolysis (Shi et al., 2019).

10.4.2.7 Antidiabetic Activity

The antidiabetic action of quinoa hydrolysates obtained with papain and those processed by gastroduodenal digestions were assessed, measuring the inhibition of the metabolic enzyme dipeptidyl peptidase IV (DPPIV), responsible for the incretin hormones degradation to decrease insulin secretion in pancreatic beta-cells. The hydrolysates showed twice the inhibition of enzyme DPPIV with respect to protein concentrates (Nongonierma et al., 2015; Vilcacundo et al., 2018a).

Black bean hydrolysates were able to reduce glucose uptake *in vitro* using a monolayer of Caco-2 cells. These cells decreased glucose uptake by up to 20% after being in contact with hydrolysate solutions for 30 min, and up to 60% when using pure peptides (Mojica et al., 2017a). This property was also determined in live tests, achieving a reduction of up to 20% of the postprandial glucose present in the serum of animals treated with these hydrolysates. Furthermore, the hydrolysates showed a 50% *in vitro* inhibition of alpha-glucosidase enzyme. For this reason, consumption of black bean protein hydrolysates may be recommended for the treatment of type 2 diabetes (Mojica et al., 2017b).

10.4.2.8 Antithrombotic Activity

The antithrombotic activity of *Amaranth hypochondriacus* protein peptides was tested by the inhibition of thrombin enzymes *in vitro*. These hydrolysates, produced with alcalase and flavourzyme, reached an inhibition of 90%, while the amaranth proteins reached 10% inhibition (Ayala-Niño et al., 2019).

10.4.2.9 Anticancer Activity

Recently Vilcacundo et al. (2018a, 2018b) reported that quinoa and amaranth protein hydrolysates (gastroduodenal hydrolysis) showed antiproliferative effects of colon cancer cells. Cytotoxic activity against human colorectal cancer Caco-2 cells and colon cancer HT-29 and HCT-116 cells was studied. The hydrolysates and especially the high molecular weight peptide fractions (fraction >5 kDalton) showed a growth inhibition (cytotoxic activity on cancer cells) of up to 80%. Significant differences were observed between gastric and gastroduodenal digestions indicating the importance of pancreatin in the generation of bioactive peptides.

10.4.2.10 Immunomodulatory Activity

The immunomodulatory properties of quinoa protein hydrolysates and 1–3 kDalton fractions were evaluated in BALB/c mice. The *in vitro* cytokine profile and *ex vivo* macrophage activation were also evaluated. The quinoa hydrolysates showed no cytotoxic effect on peritoneal macrophages at 32.5–1000 µg/mL and induced peritoneal and spleen macrophage activation through the production of cytokines INFγ and TNFα. Increased INFγ may serve as a stimulant, enhancing phagocytic activity in peritoneal and spleen macrophages which is beneficial for the immune surveillance system against pathogens. Macrophage activation was down-regulated by the co-production of IL-10, which guarantees a balanced activation-regulation response for tissue homeostasis. Hence, hydrolysates proved to be safe for oral administration and enhanced the phagocytic activity of peritoneal and spleen macrophages by stimulating the innate immune system (Rueda, 2020).

10.4.3 In Silico Approach and Molecular Docking

For the determination of bioactivity in peptides, computer tools are used more and more frequently. There are databases such as PepBank, BIOPEP, RCSB PDB, UniProtKB, in which the sequence of proteins, bioactive peptides, allergenic proteins and sensory peptides are stored. With all this information, researchers can compare the amino acid sequence of peptides found in different hydrolysis assays with those of biofunctional peptides already identified. Therefore, a peptide with biofunctional potential can be predicted by an in silico approach. Consequently, it is not necessary to isolate or concentrate peptides from hydrolysates to study their functional properties, since knowing their composition they can be synthesized and tested on peptides of 95% purity. Furthermore, availability of the protein and peptide sequence, together with knowledge of the specificity of the enzymes commonly used for protein hydrolysis, allows for knowledge of the sequence of peptides encrypted in the proteins that could be released in potential hydrolysis. In addition, the inhibitory or enhancing action of these peptides on different enzymes could also be tested by a molecular approach using a different software (SYBYL-X 2.1.1- Tripos a Certara Company, Maestro 9.1 software – Schrödinger Software Suite) in which the degree to which the inhibitors bind to the enzyme can be identified, only with molecular models (Zheng et al., 2019; Mojica et al., 2017a; Guo et al., 2020).

10.5 CONCLUSIONS

Fermentation, germination and bioactive peptides production are suitable bioprocesses to apply on crops such as quinoa, amaranth, kañiwa, maize, black turttle bean, tarwi, chia and sacha inchi, prior to their incorporation into food formulation, since the bio- and techno-functional changes produced are promising. In particular, fermentation and germination are low-cost and eco-friendly methods, which may improve the nutritional, technological and sensorial properties of products. Although they require specific controls; they are easy-to-make processes, which can be implemented by small enterprises.

Fermentation is a traditional biotechnology that has been applied in the manufacture and preservation of foods since ancient times. In order to obtain favorable transformation of raw materials, a proper selection of microorganisms is fundamental. In this way, fermentation can improve the nutritional value, sensory profile, textural and rheological features and functional properties of foods.

Germination is considered a straightforward and economic bioprocess for improving the nutritional value of the seeds and, in some cases, improve the nutraceutical level. Germination – by the action of proteolytic, lipolytic and amylolytic enzymes, and activation of secondary metabolic ways – increases protein bioavailability, the content of free amino acids and free fatty acids; it also improves the availability of minerals and vitamins, and increases phytochemical content and, in some cases, decreases anti-nutritional factors; unfortunately, germination may, eventually, negatively affect some key components of the substrate.

Quinoa and amaranth are the grains most studied for production of bioactive peptides; however, currently, there are many scientific publications on other crops such as chia, tarwi, kañiwa, different races of Andean maize and sacha inchi, among others.

The latest results obtained indicate that the human gastrointestinal system would produce peptides with functional properties from the consumption of these traditional foods, so the reevaluation of their production and consumption would notably improve the intake of nutrients and especially functional components in the region.

On the other hand, the generation of bioactive peptides by enzymatic hydrolysis has great potential for industrial development since it involves simple and cheap processes. With current technology, it is possible to achieve good regulation of the degree of hydrolysis and therefore of molecular size, as well as the composition of peptides to optimize their functional properties. Generally, the best yields were observed in the 1–3 and 3–10 kDalton peptide fractions. A set of tests *in vivo, ex vivo, in vitro* and even by computer, is currently available, able to identify peptides and peptide fractions with functional action and quantifying antioxidant, antihypertensive, antimicrobial, antiadipogenesis, antihaemolytic, antidiabetic, anticancer, hypocholesterolemic and immunomodulatory activities, among others.

The nutritional benefits obtained by germinating most crops, and the fermentation and protein hydrolysis of their flour and/or concentrates make these processes advisable in and applicable to the food industry. So, new products with high nutritional and functional value can be offered to consumers, but care must be taken so that technological changes are appropriate for food products and do not affect sensory properties of them and they are acceptable to consumers.

ACKNOWLEDGMENTS

This work was supported by grant laValSe-Food-CYTED (Ref. 119RT0567); Consejo Nacional de Investigaciones Científicas y Técnicas (CONICET) and Universidad Nacional de Jujuy, Argentina.

REFERENCES

Abderrahim, F., E. Huanatico, R. Repo-Carrasco-Valencia and S.M. Arribas. 2012. Effect of germination on total phenolic compounds, total antioxidant capacity, Maillard reaction products and oxidative stress markers in canihua (*Chenopodium pallidicaule*). *Journal of Cereal Science* 56(2):410–417.

Aguilar, E., and E. Flores. 2019. Assessment of the use of the hydrolyzed liquid fraction of the kiwicha grain in the fermentation process of probiotic drinks from tarwi juice: microbiological, chemical and sensorial analysis. *Food Science and Technology* 39(3): 592–598.

Aguilar-Toalá, J.E., A.J. Deering, and A.M. Liceaga. 2020. New insights into the antimicrobial properties of hydrolysates and peptide fractions derived from chia seed (*Salvia hispanica* L.). *Probiotics and Antimicrobial Proteins* 12: 1571–1581.

Apostolopoulos, V., J. Bojarska, T.-T. Chai, S. Elnagdy, K. Kaczmarek, J. Matsoukas, R. New, K. Parang, O. Paredes-López, O.H. Parhiz, C.O. Perera, M. Pickholz, M. Remko, M. Saviano, M. Skwarczynski, Y. Tang, W.M. Wolf, T. Yoshiya, J. Zabrocki, J. Zielenkiewicz, M. Alkhazindar, M. Barriga, K. Kelaidonis, E.M. Sarasia., and I.Toth. 2021. A global review on short peptides: Frontiers and perspectives. *Molecules* 26(2): 1–45.

Ayala-Niño, A., G.M. Rodríguez-Serrano, L.G. González-Olivares, E. Contreras-López, P. Regal-López, and A. Cepeda-Saez. 2019. Sequence identification of bioactive peptides from amaranth seed proteins (*Amaranthus hypochondriacus* spp.) *Molecules* 24(17): 3033.

Barbosa-Piló, F., E. Carvajal-Barriga, C. Guamán-Burneo, P. Portero-Barahona, A. Morato Dias, L. Daher de Freitas, F. Oliveira-Gomes, and C. Rosa. 2018. Saccharomyces cerevisiae populations and other yeasts associated with indigenous beers (*chicha*) of Ecuador. *Brazilian Journal of Microbiology* 49: 808–815.

Bassi, D., L. Orrù, J. Cabanillas-Vasquez, P. Cocconcelli, and C. Fontana. 2020. Peruvian chicha: A focus on the microbial populations of this ancient maize-based fermented beverage. *Microorganisms* 8(1): 93.

Blandino A, M.E. Al-Aseeri, S. Pandiella, D. Cantero, and C. Webb. 2003. Cereal-based fermented foods and beverages. *Food Research International* 36: 527–543.

Canaviri-Paz, P., R. Janny, and A. Hakansson. 2020. Safeguarding of quinoa beverage production by fermentation with Lactobacillus plantarum DSM 9843. *International Journal of Food Microbiology* 324: 108630.

Carciochi, R.A., L. Galván-D'Alessandro, P. Vandendriessche, and S. Chollet. 2016. Effect of germination and fermentation process on the antioxidant compounds of quinoa seeds. *Plant Foods for Human Nutrition* 71(4): 361–367.

Carrizo, S., A. de Moreno de LeBlanc, J. LeBlanc, and G. Rollán. 2020. Quinoa pasta fermented with lactic acid bacteria prevents nutritional deficiencies in mice. *Food Research International* 127: 108735.

Castro-Alba, V., C. Lazarte, D. Perez-Rea, N. Carlsson, A. Almgren, B. Bergenståhla, and Y. Granfeld. 2019. Fermentation of pseudocereals quinoa, canihua, and amaranth to improve mineral accessibility through degradation of phytate. *Journal of the Science of Food and Agriculture* 99: 5239–5248.

Chandrasekaran, U., and A. Liu. 2014. Stage-specific metabolization of triacylglycerols during seed germination of Sacha Inchi (*Plukenetia volubilis* L.). *Journal of the Science of Food and Agriculture* 95(8): 1764–1766.

Chaparro, D.C., P. Pismag, E. Correa, V. Quila, and C.A. Caicedo. 2010. Effect of germination on the protein content and digestibility in amaranth, quinoa, soybean and grandul seeds. *Biotecnología en el Sector Agropecuario y Agroindustrial* 8(1): 35–42.

Cornejo, F., G. Novillo, E. Villacrés, and C.M. Rosell. 2019. Evaluation of the physicochemical and nutritional changes in two amaranth species (*Amaranthus quitensis* and *Amaranthus caudatus*) after germination. *Food Research International* 121: 933–939.

Daliri, E., D.H. Oh, and B.H. Lee. 2017. Bioactive peptides. *Foods* 6(5): 32.

Demir B., and N. Bilgiçli. 2020. Changes in chemical and anti-nutritional properties of pasta enriched with raw and germinated quinoa (*Chenopodium quinoa* Willd.) flours. *Journal of Food Science and Technology* 57: 3884–3892.

Dentice-Maidana, S., S. Finch, M. Garro, G. Savoy, M. Gänzle, and G. Vignolo. 2020. Development of gluten-free breads started with chia and flaxseed sourdoughs fermented by selected lactic acid bacteria. *LWT – Food Science and Technology* 125: 109189.

Dueñas, M., T. Sarmento, Y. Aguilera, V. Benitez, E. Mollá, R.M. Esteban, and M.A. Martín-Cabrejas. 2016. Impact of cooking and germination on phenolic composition and dietary fibre fractions in dark beans (*Phaseolus vulgaris* L.) and lentils (*Lens culinaris* L.). *LWT – Food Science and Technology* 66: 72–78.

Gómez-Favela, M.A., R. Gutiérrez-Dorado, E.O. Cuevas-Rodríguez, V.A. Canizalez-Román, C.R. León-Sicairos, J. Milán-Carrillo, and C. Reyes-Moreno. 2017. Improvement of chia seeds with antioxidant activity, GABA, essential amino acids, and dietary fiber by controlled germination bioprocess. *Plant Foods for Human Nutrition* 72(4): 345–352.

Gong, K., L. Chen, X. Li, L. Sun, and K. Liu. 2018. Effects of germination combined with extrusion on the nutritional composition, functional properties and polyphenol profile and related in vitro hypoglycemic effect of whole grain corn. *Journal of Cereal Science* 83: 1–8.

Grijalva-Vallejos, N., A. Aranda, and E. Matallana. 2020. Evaluation of yeasts from Ecuadorian chicha by their performance as starters for alcoholic fermentations in the food industry. *International Journal of Food Microbiology* 317: 108462.

Guajardo-Flores, D., M. García-Patiño, D. Serna-Guerrero, J.A. Gutiérrez-Uribe, and S.O Serna-Saldívar. 2012. Characterization and quantification of saponins and flavonoids in sprouts, seed coats and cotyledons of germinated black beans. *Food Chemistry* 134(3): 1312–1319.

Guardianelli, L.M., M.V. Salinas, and M.C. Puppo. 2019. Chemical and thermal properties of flours from germinated amaranth seeds. *Journal of Food Measurement and Characterization* 13: 1078–1088.

Guo, H., A. Richel, Y. Hao, X. Fan, N. Everaert, X. Yang, and G. Ren. 2020. Novel dipeptidyl peptidase-IV and angiotensin-I-converting enzyme inhibitory peptides released from quinoa protein by in silico proteolysis. *Food Science and Nutrition* 8: 1415–1422.

Jan, R., D.C. Saxena, and S. Singh. 2018. Comparative study of raw and germinated Chenopodium (*Chenopodium album*) flour on the basis of thermal, rheological minerals, fatty acid profile and phytocomponents. *Food Chemistry* 269(15): 173–180.

Jeske, S., E. Zannini, K. Lynch, A. Coffey, and E. Arendt. 2018. Polyol-producing lactic acid bacteria isolated from sourdough and their application to reduce sugar in a quinoa-based milk substitute. *International Journal of Food Microbiology* 286: 31–36.

Jiménez, D., M. Lobo, B. Irigaray, M.A. Grompone, and N. Sammán. 2020a. Oxidative stability of baby dehydrated purees formulated with different oils and germinated grain flours of quinoa and amaranth. *LWT – Food Science and Technology* 127: 109229.

Jimenez, D., M. Miraballes, A. Gámbaro, M. Lobo, and N. Samman. 2020b. Baby purees elaborated with andean crops. Influence of germination and oils in physico-chemical and sensory characteristics. *LWT – Food Science and Technology* 124: 108901.

Jimenez, M.D., M. Lobo, and N. Sammán. 2019. 12th IFDC 2017 Special Issue Influence of germination of quinoa (*Chenopodium quinoa*) and amaranth (*Amaranthus*) grains on nutritional and techno-functional properties of their flours. *Journal of Food Composition and Analysis* 84: 103290.

Li, S., C. Chen, Y. Ji, J. Lin, X. Chen, and B. Qi. 2018. Improvement of nutritional value, bio-activity and volatile constituents of quinoa seeds by fermentation with *Lactobacillus casei*. *Journal of Cereal Science* 84: 83–89.

Lushchak, V.I., and N.M. Semchuk. 2012. Tocopherol biosynthesis: Chemistry, regulation and effects of environmental factors. *Acta Physiologiae Plantarum* 34: 1607–1628.

Martínez-Leo, E.E., and M.R. Segura-Campos. 2020. Neuroprotective effect from *Salvia hispanica* peptide fractions on pro-inflammatory modulation of HMC3 microglial cells. *Journal of Food Biochemistry* 44(6): e13207.

Micanquer, A., M. Cortes, and L. Serna-Cock. 2020. Formulation of a fermentation sub-strate from pineapple and sacha inchi wastes to grow Weissella cibaria. *Heliyon* 6: e03790.

Mirzaei, M., M. Aminlari, and E. Hosseini. 2016. Antioxidant, ACE-inhibitory and anti-microbial activities of Kluyveromyces marxianus protein hydrolysates and their peptide fractions. *Functional Foods in Health and Disease* 6(7): 425–439.

Mojica, L., E. Gonzalez de Mejia, M.A. Granados-Silvestre, and M. Menjivar. 2017a. Evaluation of the hypoglycemic potential of a black bean hydrolyzed protein iso-late and its pure peptides using in silico, in vitro and in vivo approaches. *Journal of Functional Foods* 31: 274–286.

Mojica, L., D.A. Luna-Vitala, and E. González de Mejía. 2017b. Characterization of peptides from common bean protein isolates and their potential to inhibit markers of type-2 diabetes, hypertension and oxidative stress. *Journal of the Science of Food and Agriculture* 97: 2401–2410.

Montemurro, M., E. Pontonio, and C. Rizzello. 2019. Quinoa flour as an ingredient to enhance the nutritional and functional features of cereal-based foods. In *Flour and Breads and their Fortification in Health and Disease Prevention*, edited by V. Preedy and R. Watson, 453–464. Cambridge, MA: Academic Press.

Morales, D., M. Miguel, and M. Garcés-Rimón. 2020. Pseudocereals: A novel source of bio-logically active peptides. *Critical Reviews in Food Science and Nutrition* ISSN: 1040–8398 (Print) 1549–7852 (Online) Journal homepage: www.tandfonline.com/loi/bfsn20.

Mosso, A. L., J. LeBlanc, C. Motta, I. Castanheira, P. Ribotta, and N. Sammán. 2020. Effect of fermentation in nutritional, textural and sensorial parameters of vegan-spread products using a probiotic folate-producing *Lactobacillus sakei* strain. *LWT – Food Science and Technology* 127: 109339.

Moughan, P.J., S.M. Rutherfurd, C.A. Montoya, and L.A. Dave. 2014. Food-derived bioactive peptides – A new paradigm. *Nutrition Research Reviews* 27(1): 16–20.

Mudgil, P., L.S. Omar, H. Kamal, B.P. Kilari, and S. Maqsood. 2019. Multi-functional bio-active properties of intact and enzymatically hydrolysed quinoa and amaranth proteins. *LWT – Food Science and Technology* 110: 207–213.

Nardo, A.E., M.C. Añón, and A.V. Quiroga. 2020. Identification of renin inhibitor peptides from amaranth proteins by docking protocols. *Journal of Functional Foods* 64: 103683.

Nemzer, B., Y. Lin, and D. Huang. 2019. Chapter 3: Antioxidants in Sprouts of Grains. Sprouted Grains – Nutritional Value, Production, and Applications. Cambridge and Washington, DC: Woodhead Publishing and AACC International Press.

Nongonierma, A.B., S.L. Maux, C. Dubrulle, C. Barre, and R.J. FitzGerald. 2015. Quinoa (*Chenopodium quinoa* Willd.) protein hydrolysates with in vitro dipeptidyl peptidase IV (DPP-IV) inhibitory and antioxidant properties. *Journal of Cereal Science* 65: 112–118.

Omary, M.B., C. Fong, J. Rothschild, and P. Finney. 2012. Effects of germination on the nutritional profile of gluten-free cereals and pseudocereals: A review. *Cereal Chemistry* 89(1): 1–14.

Ontiveros, N., V. López-Teros, M.J. Vergara-Jiménez, A.R. Islas-Rubio, F.I. Cárdenas-Torres, E.O. Cuevas-Rodríguez, C. Reyes-Moreno, D.M. Granda-Restrepo, S. Lopera-Cardona, G.I. Ramírez-Torres, and F. Cabrera-Chávez. 2020. Amaranth-hydrolyzate enriched cookies reduce the systolic blood pressure in spontaneously hypertensive rats. *Journal of Functional Foods* 64: 103613.

Orona Tamayo, D., M.E.Valverde, and O. Paredes López. 2019. Bioactive peptides from selected Latin American food crops - A nutraceutical and molecular approach. *Critical Reviews in Food Science and Nutrician* 59(12): 1949–1975.

Padmashree, A., N. Negi, S. Handu, M. Khan, A. Semwal, and G. Sharma. 2018. Effect of germination on nutritional, antinutritional and rheological characteristics of quinoa (*Chenopodium quinoa*). *Defence Life Science Journal* 4(1): 55–60.

Pająk, P., R. Socha, J. Broniek, K. Królikowska, and T. Fortuna. 2019. Antioxidant properties, phenolic and mineral composition of germinated chia, golden flax, evening primrose, phacelia and fenugreek. *Food Chemistry* 275: 69–76.

Paucar-Menacho, L.M., C. Martínez-Villaluenga, M. Dueñas, J. Frias, and E. Peñas. 2017. Optimization of germination time and temperature to maximize the content of bioactive compounds and the antioxidant activity of purple corn (*Zea mays* L.) by response surface methodology. *LWT – Food Science and Technology* 76: 236–244.

Pérez-Armendáriz, B., and G. Cardoso-Ugarte. 2020. Traditional fermented beverages in Mexico: Biotechnological, nutritional, and functional approaches. *Food Research International* 136: 109307.

Piñuel, L., P. Boeri, F. Zubillaga, D.A. Barrio, J. Torreta, A. Cruz, G. Vásquez, A. Pinto, and W. Carrillo. 2019. Production of white, red and black quinoa (*Chenopodium quinoa* Willd Var. Real) protein isolates and its hydrolysates in germinated and non-germinated quinoa samples and antioxidant activity evaluation. *Plants* 8(8): 257.

Repo-Carrasco-Valencia, R., and A.L. Serna. 2011. Quinoa (*Chenopodium quinoa* Willd.) as a source of dietary fiber and other functional components. *Ciência e Tecnologia de Alimentos* 31(1): 225–230.

Rizzello, C.., A. Lorusso, M. Montemurro, and M. Gobbetti. 2016. Use of sourdough made with quinoa (*Chenopodium quinoa*) flour and autochthonous selected lactic acid bacteria for enhancing the nutritional, textural and sensory features of white bread. *Food Microbiology* 56: 1–13.

Rizzello, C., A. Lorusso, V. Russo, D. Pinto, B. Marzani, and M. Gobbetti. 2017. Improving the antioxidant properties of quinoa flour through fermentation with selected autochthonous lactic acid bacteria. *International Journal of Food Microbiology* 241: 252–261.

Rueda, J. 2020. *Actividad inmunomoduladora de hidrolizados proteicos de quínoa y su aplicación a alimentos funcionales.* Doctoral thesis. Faculty of Engineering of the National University of Jujuy. Argentina.

Shi, Z., Y. Hao, C. Teng, Y. Yao, and G. Ren. 2019. Functional properties and adipogenesis inhibitory activity of protein hydrolysates from quinoa (*Chenopodium quinoa* Willd.) *Food Science and Nutrition* 7: 2103–2112.

Silva-Vieira, A.D., R. Bedani, M.A.C. Albuquerque, V. Biscola, and S.M. Isay-Saad. 2017. The impact of fruit and soybean by-products and amaranth on the growth of probiotic and starter microorganisms. *Food Research International* 97: 356–363.

Suárez-Estrella, D., G. Cardone, S. Buratti, M.A. Pagani, and A. Marti. 2020. Sprouting as a pre-processing for producing quinoa-enriched bread. *Journal of Cereal Science* 96: 103111.

Świeca, M. 2016. Hydrogen peroxide treatment and the phenylpropanoid pathway precursors feeding improve phenolics and antioxidant capacity of quinoa sprouts via an induction of L-Tyrosine and L-Phenylalanine Ammonia-Lyases activities. *Journal of Chemistry* 2016(4): 1–7.

Ujiroghene, O.J., L. Liu, S. Zhang, J. Lu, C. Zhang, J.L.X. Pang, and M. Zhang. 2019. Antioxidant capacity of germinated quinoa-based yoghurt and concomitant effect of sprouting on its functional properties. *LWT – Food Science and Technology* 116: 108592.

Väkeväinen, K., F. Ludena-Urquizo, E. Korkala, A. Lapveteläinen, S. Peräniemi, A. von Wright, and C. Plumed-Ferrer. 2020. Potential of quinoa in the development of fermented spoonable vegan products. *LWT – Food Science and Technology* 120: 108912.

Vallejo, J., P. Miranda, J. Flores Félix, F. Sánchez-Juanes, J. Ageitos, J. González-Buitrago, E. Velázquez, and T. Villa. 2013. Atypical yeasts identified as *Saccharomyces cerevisiae* by MALDI-TOF MS and gene sequencing are the main responsible of fermentation of chicha, a traditional beverage from Peru. *Systematic and Applied Microbiology* 36(8): 560–564.

Vanegas, A., and L. Gutiérrez. 2018. Physicochemical and sensory properties of yogurts containing sacha inchi (*Plukenetia volubilis* L.) seeds and β-glucans from *Ganoderma lucidum*. *Journal of Dairy Science* 101: 1020–1033.

Vilcacundo, R., B. Miralles, W. Carrillo, and B. Hernandez-Ledesma. 2018a. In vitro chemopreventive properties of peptides released from quinoa (*Chenopodium quinoa* Willd.) protein under simulated gastrointestinal digestion. *Food Research International* 105: 403–411.

Vilcacundo, R., C. Martínez-Villalueng, B. Miralles, and B. Hernández-Ledesma. 2018b. Release of multifunctional peptides from kiwicha (*Amaranthus caudatus*) protein under in vitro gastrointestinal digestion. *Journal of Science of Food and Agricultural* 99: 1225–1232.

Villacrés, E., V. Allauca, E. Peralta, G. Insuasti, G., J. Álvarez, and M.B. Quelal. 2015. Germination, an effective process to increase the nutritional value and reduce non-nutritive factors of Lupine Grain (*Lupinus mutabilis* Sweet). *International Journal of Food Science and Nutrition Engineering* 5(4): 163–168.

Villacrés, E., M.B. Quelai, E. Fernández, G. Garcia, G. Cueva, and C. Rosell. 2020. Impact of debittering and fermentation processes on the antinutritional and antioxidant compounds in *Lupinus mutabilis* sweet. *LWT – Food Science and Technology* 131: 109745.

Wang, S., F. Zhu, and Y. Kakuda. 2018. Sacha inchi (*Plukenetia volubilis* L.): nutritional composition, biological activity, and uses. *Food Chemistry* 265: 316–328.

Xue, Z., C. Wang, L. Zhai, W. Yu, H. Chang, X. Kou, and F., Zhou. 2016. Bioactive compounds and antioxidant activity of mung bean (*Vigna radiata* L.), soybean (*Glycine max* L.) and black bean (*Phaseolus vulgaris* L.) during the germination process. *Czech Academy of Agricultural Sciences* 34(1): 68–78.

Zannini, E., S. Jeske, K. Lynch and E. Arendt. 2018. Development of novel quinoa-based yoghurt fermented with dextran producer Weissella cibaria MG1. *International Journal of Food Microbiology* 268: 19–26.

Zheng, Y., X. Wang, Y. Zhuang, Y. Li, H. Tian, P. Shi and G. Li 1. 2019. Isolation of novel ACE-inhibitory and antioxidant peptides from quinoa bran albumin assisted with an in silico approach: Characterization, in vivo antihypertension, and molecular docking. *Molecules* 24: 4562.

Chapter 11

Current Position of Legislation on Latin-American Grains and Its Regional Socioeconomic Impact

María Dolores Jiménez[1], Ana Laura Mosso[1],
Claudia M. Haros[2], and Norma Sammán[1]
[1]National University of Jujuy and National Council
for Scientific and Technical Research (CONICET),
San Salvador de Jujuy, Argentina
[2]Institute of Agrochemistry and Food Technology (IATA).
Spanish Council for Scientific Research (CSIC), Valencia, Spain

CONTENTS

DOI: 10.1201/9781003088424-11

11.1   AGRO-ECONOMIC CONTEXT IN LATIN AND MESOAMERICA

The Latin and Mesoamerican region is now experiencing a spectacular expansion of its agricultural frontier. As a result, social, economic and ecological transformation has occurred.

In the Central Andes of South America – one of the eight most important genocentres of plant species in the world – about 8,000 years ago local populations domesticated plants such as quinoa, amaranth, kiwicha and tarwi, among others. These species reached our times due to the dedication and perseverance of native communities, who preserved them as part of their cultural heritage (Tapia, 1993). On the other hand, Mesoamerica, which encompasses the lands from Mexico to Costa Rica, is also considered one of the most important domestication centers of plant species in the world. Maize, tomato, cocoa, chia, chili and common beans, among hundreds of others, are crops that originated in this region (Pickersgill, 2007).

Traditionally, peasant communities maintained their own agricultural tradition, taking care of the biodiversity of crops, ancestral technologies and the proper conservation of environments for their production. In this sense, several governmental research organizations, international research centers and non-governmental organizations (NGOs) have collected, registered and characterized the main species, and to a lesser extent those of restricted areas and usage, despite their being of significant value for local economies and nutrition. For instance, in Peru and Bolivia alone, estimates indicate that there are more than 8,000 peasant communities – heirs of the pre-Hispanic Ayllus – that cultivate ancestral varieties despite technical assistance programs that seek to impose cultivation of the so-called *improved varieties*. There are crops which may potentially be more productive, but in many cases they are not appropriate for these agricultural systems or food sovereignty (Izquierdo and Roca, 1998).

Outside South and Mesoamerica, there are many underexploited crops that are rapidly disappearing because of social unrest and environmental damage, while others – like quinoa and amaranth – have crossed frontiers and are nowadays found in Europe and the United States.

In recent decades, scientists have highlighted the benefits and new uses of these crops to boost their global demand and generate economic opportunities in regions where family incomes are far less than the minimum required for food security.

The region is now one of the poorest in the world because of a decline in farming, high rates of population growth, migration and misuse of natural resources, despite once being the home of some of the most advanced cultures at a global level (Izquierdo and Roca, 1998). According to a Food and Agriculture Organization report, FAO (2020), the socioeconomic situation is critical and hunger affects 42.5 million people in Latin America and the Caribbean (LAC), 6.5% of the regional population. The increase in hunger is closely related to the general economic slowdown in the region. The fall of commodity prices since 2011 has led to a deterioration in the public finances of many countries dependent on the export of commodities.

In this sense, the member states of FAO in LAC agreed at the thirty-sixth regional conference on three major priorities to guide the actions of the organization during the 2020–2021 biennium: (a) sustainable food systems to provide healthy diets for all; (b) *hand-in-hand initiatives* to achieve prosperous and inclusive rural societies (c) sustainable and resilient agriculture. Furthermore, the United Nations (UN) declared the period 2019–2028 as the Decade of Family Farming to bring a new perspective on what it means to be a family farmer in a rapidly changing environment, highlighting its role in eradicating hunger and building the future of food. Family farming provides 70% of the food in the world -reaching 97% in countries such as Peru – and represents the basis of the population's food security.

11.2 PRODUCTION SYSTEMS AND FARMERS' SITUATIONS

According to MINAGRI (2018), in Bolivia and Peru, most agricultural producers belong to family farming, about 40% are under the poverty line, 17% have at least one unsatisfied basic need, 64% only reached first grade in elementary school and 48% must work outside their farm since if they were dedicated only to their plot, they would not have enough income to survive. This situation is closely related to the fact that small-scale producers have limited investment capacity, low bargaining power and inadequate management of production with regard to the demands of the market; and limited commercial knowledge and low market opportunities. In addition, the scarce productive infrastructure and road development, low coverage of technical assistance and training, poor agricultural information services and few incentives for associativity aggravate the situation.

In Mexico, small-scale producers form a very important agricultural subsector. For instance, some crops produced in Puebla – such as maize, chia, and black common bean – constitute a basic resource for the subsistence of rural populations, generating jobs for communities and securing food for neighboring cities. About 74% of rural economic units are smallholdings of fewer than 5 hectares cultivated for subsistence or self-consumption (Muñoz et al., 2019).

The practice of growing *milpa* (maize combined with common beans) is the foundation of food security in many Guatemalan rural communities. Although most peasants are aware of the potential to increase their returns from cash crops or alternative economic activities, 99% of the households surveyed maintained that the practice was important to their family food security. In this sense, the contribution of *milpa* to food security represents much more than the nutrients that it generates. It also guarantees that the basic sustenance needs of the family are met (Altieri and Toledo, 2011).

Across all Latin and Mesoamerica, small-scale farmers regularly integrate rural trading systems, to sell or exchange their surplus production to acquire other goods they do not produce. In these traditional systems, marketing channels are not formally established and producers continually deal with several problems, such as intermediaries, inadequate infrastructure for distributing agricultural products and unequal terms between all the economic agents (producers, collectors and merchants) who participate in the local, national and international markets. The consequences of this informal trade are reflected in the lack of official data on the volume of commercialization or on exchange prices and profit margins in the value chains, and represent a disadvantage mainly for primary producers; additionally it also carries phytosanitary risks and difficulties with the maintenance of quality standards necessary for international trade (Ofstehage, 2011).

11.3   DIET SHIFTS AND FOOD SYSTEM TRANSFORMATION

The transition from traditional value chains to modern food systems began in the 1960s when governments set up grain and processed products parastatals in wholesale, processing and retail. In the 1990s a phase of liberalization with privatization and globalization of the food system started, which initiated rapid investment by foreign firms. The improved infrastructure helped the proliferation of small and medium local enterprises (Pengue, 2004).

Five meta-conditions have encouraged and facilitated food system changes and diet shifts. These conditions are mutually dependent: income growth, policy liberalization, infrastructure improvement, urbanization and the increase of rural nonfarm employment. Regarding the last meta-condition, nowadays in Latin America 25% of the population lives in rural areas, and the majority of the rural population lives near a city. Barbier and Hochard (2014) showed that less than 10% of the rural population lives far from a town or a city, which implies proximity to processed food stockists and large companies' van networks.

The shifts in dietary patterns in Latin and Mesoamerican communities are the result of decades of changes in land tenure and local economies: liberalization, rural-urban migration and food aid programs, among other factors, have affected the consumption of traditional foods. Over time, peasants have transitioned from a diet based on what they produced on their land to one that incorporates external foods from the market. Of course, diet is influenced by socioeconomic status, however, the consumption of traditional crops is influenced by history, migration, globalization and changing traditions. Dietary habits are also shaped by desire for convenience and less labor-intensive foods, changing tastes, and the availability of other foods (from imported products to a greater diversity of cheaper local goods available for purchase). Once staples for the region, legumes such as tarwi have declined as a critical part of the diet over the past decade, representing less than 5 or 10% of the daily energy intake (Popkin and Reardon, 2018).

In particular, in the Andes, diets of peasant families are also being influenced by recent local and international efforts to promote quinoa products that are changing the value and meaning of the ancestral crop in these communities. Quinoa cultivated by Aymara farmers in Puno, Peru, has been traditionally destined for home consumption, but peasants have been experiencing a shift in their dietary habits in recent years (Bedoya-Perales, 2018). Typical dishes containing quinoa are still prepared (especially after harvest and for special occasions), but the convenience and availability of other foods like rice, wheat flour and pasta often make them more attractive, especially for younger generations.

11.4 ACCESS TO LAND AND GENDER EQUITY

In the last decades, the lands of small peasants have decreased significantly. This situation, according to an FAO report (2017), especially affects women, who only own 8% of the land in Guatemala and 31% in Peru, properties that are usually smaller and lower in quality. About 23% of Latin-American lands are managed or owned by indigenous people (Bose, 2017). From this situation, also in close relation with traditional agroecology rooted in Andean culture, new small farmer associations have emerged such as AOPEB (*Asociación de Organizaciones de Productores Ecológicos de Bolivia*), founded in 1991 and composed of 75 organizations and about 70,000 families. In Peru, ANPE (*Asociación de Productores Ecológicos*) has 12,000 members from 22 different regions of the country.

The revolution at the beginning of the 20th century in Mexico generated the first agrarian reform on the continent, leaving a great part of the land, forests and native germplasm in the hands of indigenous communities. Today, the so-called social property includes more than 100 million hectares. In addition, in

Bolivia, a regulation of agro-silvo-pastoral production was promoted, in order to protect indigenous rights on lands.

Gender equality is a pending debt and a continuous challenge in this region. According to Katz (2002) a very small percentage of women (between 4 and 15%) in Chile, Colombia, Costa Rica, El Salvador, Honduras, Mexico, Nicaragua and Peru have benefited from land distribution programs. An FAO report (2017) indicated that the *Gini* coefficient, which measures gender inequality, applied to the distribution of land in the region as a whole, reaching 0.79 in LAC, far surpassing Europe (0.57), Africa (0.56) and Asia (0.55) (Bose, 2017).

11.5   IMPACT OF GLOBALIZATION ON ECONOMIC GROWTH AND THE MODERNIZATION OF PRODUCTION CHAINS: THE QUINOA BOOM

Scientists and development experts have especially taken an interest in the potential of quinoa as a crop to feed the world, and have even promoted it as a *superfood* throughout the western world (Bazile et al., 2015). Beyond its nutritional profile, quinoa's adaptability to different climates and growing conditions makes it an important crop in respect of climate change. The UN declared 2013 as the *International Year of Quinoa*. This promoted cultivation areas mainly in Bolivia, Peru and Ecuador, as well as increasing export volumes, which changed the use of quinoa from a domestic food supply to being introduced to the global market.

While quinoa prices rose, the majority of Andean producers depleted their own consumption and sold their products, thus acquiring cheaper foods with a lower nutritional value, such as noodles, wheat and rice (Kerssen, 2015).

The term *quinoa boom* refers to the globalization impact of this highly nutritional grain that is expensively sold overseas to consumers interested in their health. This phenomenon resulted in socioeconomic and environmental impacts on indigenous and local communities. Initially, traditional cultivation was conducted through manual labor, which was carefully carried out along with nature in a sustainable way. When production of quinoa intensified, mechanical facilities were introduced with concomitant land degradation (Bedoya-Perales, 2018).

After 2013, commercial quinoa production started in Italy, India and China, and intensified in the USA and Canada, where investment by market-oriented farmers with access to capital and technology increased. Moreover, quinoa is cultivated in low-altitude regions with conditions amenable to capital-intensive agricultural production. A similar situation occurred in the Andes when farmers with no cultural link to quinoa but with capital to invest and productive farms entered the market. On the other hand, Andean small farmers who had previously monopolized quinoa production were then unable to compete in an even more crowded market. Nevertheless, even with quinoa's global popularity on the rise, production overshot demand, and as a result, prices fell, and the economy of many small farmers was crushed. The *quinoa boom* also created new challenges in terms of environmental degradation, for example, manifesting as pesticide contamination or drastic land-use change from extensively used pastures to intensively used agricultural land (Tschopp et al., 2018).

Equitable access to land is a human rights issue, and improving gender and social inclusion in the land administration effectively enhances the sustainable management of resources. Robinson et al. (2014) reported a particular case of the collective land tenure system of indigenous women after the *quinoa boom*: by Quechua women closer to Titicaca Lake, in Bolivia, who have an ancestral tradition of land management including access to communal and family lands which was affected after the *quinoa boom*. Traditionally, small-scale community-based production was mainly used for household consumption. Since 2013, the diversity of quinoa cultivation has been declining because the market demanded white and red varieties. According to Bose (2017), Quechua women, who trad-itionally bartered kañiwa and quinoa in the local market, started losing their roles both as collective entrepreneurs and as individuals failing to barter quinoa for household items. Furthermore, the *quinoa boom* has resulted in some forms of land dispute, particularly communal land, which historically had been col-lectively used for sowing. This was a practice of convenience to manage crops rotationally and especially helped women support each other at times of vulner-ability, due to drought or extreme frost. Nevertheless, after 2013, several changes occurred in the way land was traditionally managed: the collective land has been reclaimed and sold away as individual land to outsiders, and large-scale white and red quinoa cultivation has been introduced. Women continued supporting the farming of quinoa crops, but decision making and marketing have been taken over by men, mainly because men are the formal landowners, along with the fact that most women only speak Quechuan and traders speak Spanish. Price vola-tility of quinoa in the global market and individualization of land – mostly by men´s interest in land tenure – were the main causes of vulnerability. Quechuan women's collective activities of protecting quinoa and kañiwa biodiversity have been challenged due to an exclusively export-driven orientation of mechaniza-tion of agricultural production.

11.6 NAGOYA PROTOCOL

The Nagoya Protocol was adopted on October 29, 2010 in Nagoya, Japan, and entered into force on October 12, 2014. Its objective is the fair and equitable sharing of benefits arising from the utilization of genetic resources, thereby contributing to the conservation and sustainable use of biodiversity. The utilization of genetic resources is understood as research and development (R&D) into the genetic and biochemical composition of plant seeds, animals or microorganisms. Access to genetic resources and the fair and equitable sharing of benefits arising from their utilization to the *Convention on Biological Diversity* (CBD, 2011) is an international agreement, which aims at sharing the benefits arising from the utilization of genetic resources in a fair and equitable way. This protocol helps to create greater legal certainty and transparency for

TABLE 11.1   STATUS OF RATIFICATION, ACCEPTANCE, APPROVAL OR ACCESSION OF NAGOYA PROTOCOL IN IBERO-AMERICAN COUNTRIES

Country Name	Signed	Ratification		Party
Argentina	2011-11-15	2016-12-09	rtf	2017-03-09
Bolivia (Plurinational State of)	--	2016-10-06	acs	2017-01-04
Brazil	2011-02-02	2021-03-04	rtf	2021-06-02
Cuba	--	2015-09-17	acs	2015-12-16
Dominican Republic	2011-09-20	2014-11-13	rtf	2015-02-11
Ecuador	2011-04-01	2017-09-20	rtf	2017-12-19
Guatemala	2011-05-11	2014-06-18	rtf	2014-10-12
Honduras	2012-02-01	2013-08-12	rtf	2014-10-12
Mexico	2011-02-24	2012-05-16	rtf	2014-10-12
Nicaragua	--	2020-06-12	acs	2020-09-10
Panama	2011-05-03	2012-12-12	rtf	2014-10-12
Peru	2011-05-04	2014-07-08	rtf	2014-10-12
Portugal	2011-09-20	2017-04-11	apv	2017-07-10
Spain	2011-07-21	2014-06-03	rtf	2014-10-12
Uruguay	2011-07-19	2014-07-14	rtf	2014-10-12
Venezuela (Bolivarian Republic of)	--	2018-10-10	acs	2019-01-08
Colombia	2011-02-02	--	--	--
Costa Rica	2011-07-06	--	--	--
El Salvador	2012-02-01	--	--	--

Source: CBD, 2020.

Notes: rtf: Ratification; acs: Accession; acp: Acceptance; apv: Approval; scs = Succession.

both providers and users of these resources. These establish more predictable conditions for accessing genetic resources, ensuring benefit sharing when these leave their providing country. The Nagoya Protocol also protects traditional knowledge associated with the genetic resources covered by the CBD, and the benefits arising from their utilization. Seed and plant collecting is still governed by the laws of countries where the plants originate. However, the Nagoya Protocol will ensure that anyone wishing to develop or research these plants can access the relevant contracts to perform their labor. Not all countries have signed the agreement. In this sense, as of December 2020, 128 parties have been ratified, which includes 127 United Nations member states and the European Union. Table 11.1 provides information on dates of signing and the status of ratification, acceptance, approval or accession of the Nagoya Protocol in Ibero-American countries (CBD, 2020).

11.7   FOOD LEGISLATION: THE GLOBAL SITUATION

Legislation on crops and food products is necessary to ensuring the achievement of nutritional quality and safety. Additionally, national and international regulations and standards govern local and international trade, and guarantee consumer protection (FAO/WTO, 2018; OPS, 2020).

FAO and the World Trade Organization (WTO) provide governments with the means to establish a framework that internationally facilitates agreements on rules-based food trade. Through the joint FAO/WHO *Codex Alimentarius* Commission, governments develop science-based food standards (FAO/WTO, 2018).

The *Codex Alimentarius* is the highest international body of food standards where all 188 UN members and associated members have negotiated science-based recommendations in all areas related to food safety and quality (*Codex Alimentarius*, 2020). The WTO considers the *Codex Alimentarius* the only basis for harmonizing food safety measures in the context of international food trade.

Furthermore, most countries – or economic blocs, such as the EU or the Southern Common Market (MERCOSUR), among others – have established specific regulations based on *Codex Alimentarius* in accordance with the particular situation of each country or bloc's situation. In general, recommendations related to hygiene during production, processing and food handling are included in sanitary codes developed by health ministries. However, food standard codes are generally formulated by independent government agencies specially designated for this labor.

11.8   CROPS' LEGAL FRAMEWORK IN LATIN AND MESOAMERICAN COUNTRIES

The development of novel foods with Latin and Mesoamerican seeds and grains has recently achieved extraordinary levels. However, the lack of specific food regulations for these crops hinders their production and commercialization. Nevertheless, the regulatory framework for regional crops and their derivatives has been improved in recent decades. An important advance in international legislation on Andean crops was the approval of the first international standard on quinoa in 2019 by the *Codex Alimentarius* Commission (CXS 333–2019), which was a decisive step toward promoting its production and consumption throughout the world.

Regarding Latin-American countries, several differences among laws and regulations were found and are summarized in Table 11.2.

The *Código Alimentario Argentino* (CAA) [Argentinian Food Code] is an open-access technical regulation under constant update, which regulates all foods, condiments, food additives and beverages, or their raw materials, that are prepared, preserved and transported, as well as any person, commercial firm or establishment that produces them. Therefore, it establishes the hygienic-sanitary, bromatological, quality and commercial identification standards. CAA's main objectives are to protect public health and to set legal frameworks for commercial transactions (CAA, 2020). In recent years, CAA has incorporated some regional crops such as amaranth, black turtle bean, chia, quinoa and chia and quinoa flours.

Bolivia and Peru – and especially since the *quinoa boom* – have been highlighted to have paid more attention due to the legislative framework achieved regarding Andean crops. Bolivia has developed food standards through the *Instituto Boliviano de Normalización y Calidad* (IBNORCA) [Bolivian Institute of Standardization and Quality] – a private non-profit association. All participants in the food supply chain, such as farmers, manufacturers or retailers, can benefit from the guidelines and practices set out in the technical standards, which include each link in the production chain, from food collection harvesting to product packaging (IBNORCA, 2020). In Peru, the *Normas Técnicas Peruanas* (NTP) [Peruvian Technical Standards], of the *Instituto Nacional de Calidad* (INACAL) [National Institute of Quality], are documents that establish test methods, sampling, packaging and labeling of foods, as well as the quality specifications of products, processes and services (INACAL, 2020).

In Chile, *Biblioteca del Congreso Nacional de Chile* (BCN) [Library of the National Congress of Chile] (BCN, 2020) and *Reglamento Sanitario de los Alimentos del Ministerio de Salud* [Food Sanitary Regulations of the Ministry of Health] (MINSAL, 2019) establishes the sanitary conditions that food must meet during production, import, preparation, packaging, storage, distribution

TABLE 11.2   LEGISLATION OF LATIN AND MESOAMERICAN CROPS IN LATIN-AMERICAN COUNTRIES

Country	Crops/Products Included	Legislation	Description	Reference
Codex Alimentarius (International Food Standards)	Quinua	CXS 333-2019	This standard defines quinoa (grain obtained from *Chenopodium quinoa* Willd.) and processed quinoa (quinoa grains that have been subjected to cleaning and removal of the pericarp containing saponin). It defines the requirements for human consumption and includes the classification by color and size.	*Codex Alimentarius*, 2020
	Standard for maize (corn)	CXS 153-1985 Adopted in 1985. Revised in 1995. Amended in 2019.	This standard applies to maize (corn) for human consumption.	
	Whole maize (corn) meal	CXS 154-1985 Adopted in 1985. Revised in 1995. Amended in 2019.	This standard applies to whole maize (corn) meal for direct human consumption prepared from kernels of common maize.	
Argentina	Amaranth. Leafy vegetables	Chapter XI Vegetable foods. Article 822 and Article 853	Amaranthus leaves for industrial use that includes heat treatment and/or extrusion, having discarded the juices produced in the process.	CAA, 2020

(continued)

TABLE 11.2 (CONTINUED)

Country	Crops/Products Included	Legislation	Description	Reference
	Black turtle bean	Chapter XI Vegetable foods. Article 877 and Article 885	This classification includes fresh legumes of recent harvest and for immediate consumption, and dried. The name bean is understood as the fresh or dried seed of the *Phaseolus vulgaris* L. species, which includes the black bean.	
	Chia seeds	Chapter XI Vegetable foods. Article 917 and Article 918	Denomination, requirements and classification of seeds for human consumption.	
	Quinoa or Quinua seeds	Chapter XI Vegetable foods. Article 917		
	Amaranth flour	Chapter IX Farinaceous foods – Cereals, flours and derivatives. Article 660.	Denomination and requirements of seeds and seed flours for human consumption.	
	Quinoa or Quinua seeds	Chapter IX Farinaceous foods – Cereals, flours and derivatives. Article 682.		
	Quinoa or Quinua flour	Chapter IX Farinaceous foods – Cereals, flours and derivatives. Article 682 bis.		

				IBNORCA, 2020
	Chia flour	Chapter XIX Flours, concentrates, isolates and derivatives Protein. Article 1407 bis.	Denomination and requirements of types of chia flour for human consumption.	
	Chia oil	Chapter VII Fatty foods, food oils. Article 527 bis.	Denomination and requirements of chia oil for industrial use.	
Bolivia	Amaranth grains	Grain foods NB 336003:2005	Definitions of amaranth grains.	
	Amaranth grains	Grain foods NB 336004:2006	Classification and characteristics for amaranth to establish its class and grade for marketing.	
	Cañahua grains	Grain foods NB 336001:2004	Definition of cañahua grains.	
	Cañahua (or kañiwua) Grains	Grain foods NB 336002:2005	Characteristics to establish the class and grade of cañahua grains for commercialization.	
	Chia seed	Oil seed standards NB 313025:2014	Classification and requirements for chia for human consumption.	
	Corn	Cereals NB 312008:2003	Classification and requirements of corn in grain for its commercialization and industrialization.	
	Quinoa grains	Cereal standard NB 312032:2006	Energy determination of quinoa in grain.	

(continued)

TABLE 11.2   (CONTINUED)

Country	Crops/Products Included	Legislation	Description	Reference
	Raw corn flour	Standard Flours and derivatives NB 583:1990	Characteristics and specifications of raw corn flour for human consumption.	
	Tarwi (chocho). Bitter grain	Legumes NB/NA 0094:2011	Quality requirements and test methods for tarwi grain (chocho).	
	Tarwi (chocho). Debittered grain	Leguminous standard NB/NA 0097:2011	Tarwi grain for marketing.	
	Tarwi cookie	Flour and its derivatives NB 39027:2009	Standards for the development, characteristics and requirements of tarwi products.	
	Tarwi queque	Flour and derivatives NB 39026:2009		
Chile	Chia seed oil	Exempt Resolution 214	Authorizes the obtaining of edible oil from chia seed.	BCN, 2020
	Quinoa seeds	Exempt Resolution 9179	Establishes a specific standard for the certification of quinoa seeds.	
	Sacha Inchi oil	Exempt Resolution 528	Authorizes the obtaining of edible oil from *Plukenetia volubilis* seeds.	
Colombia	Beans	NTC 871:2005	Requirements for beans, including black bean, for human consumption.	INCONTEC, 2020
	Corn	NTC 2227:1986	Determination of moisture content in corn grains, on milled grains and on whole grains.	

	Corn tortilla	NTC 6173:2016	Quality requirements that must be fulfilled and the test methods to which the corn tortilla obtained from nixtamalized corn for human consumption must be submitted.
	Quinoa flakes	NTC 6071:2014	Requirements for quinoa flakes for human consumption.
	Quinoa flour	NTC 6069:2014	Requirements with which quinoa flour must comply for human consumption.
Ecuador	Amaranth grains	NTE INEN 2646:2012	Quality requirements of amaranth grain for marketing and test methods for the evaluation and verification of these requirements.
	Corn flour precooked without germ	NTE INEN 1737:2016	Requirements for precooked cornmeal without germ.
	Corn grains	NTE INEN 187:2013	Requirements for grain corn, of any variety, intended for human consumption, zootechnical food and industrial use.
	Quinoa grains	NTE INEN 1671:2013	Determination of infestation and impurities levels.
	Quinoa grains	NTE INEN 1672:2013	Determination of saponins content of quinoa grains by the foam method (routine method).

INEN, 2020

(continued)

TABLE 11.2   (Continued)

Country	Crops/Products Included	Legislation	Description	Reference
	Quinoa grains	NTE INEN 1673:2013	Requirements for quinoa intended for human consumption. It does not apply to quinoa destined for seed.	
	Maize (milled grains and on whole grains)	NTE INEN-ISO 6540 (identical translation to the international standard ISO 6540:1980)	Maize. Determination of moisture content (on milled grains and on whole grains).	
	Sacha inchi oil	NTE INEN 2688:2014	Requirements that sacha inchi oil must meet for human and/or industrial consumption.	
Peru	Amaranth grains	NTP 205.054:2020 NTP 205.055:2017	Good manufacturing practices for processing plants and test methods.	INACAL, 2020
	Amaranth expanded	NTP 011.461:2017	Good manufacturing practices for processing plants and test methods.	
	Cañihua grains	NTP 011.452:2019	Requirements for obtaining products.	
	Quinua and cañihua grains	NTP 011.453:2014		
	Black turtle bean	NTP 205.015:2016	Requirements for dry bean grain, including black bean, for human consumption.	

			IMPO, 2020
	Cañihua flakes	NTP 011.456:2015	Requirements for obtaining products.
	Cañihua flour	NTP 011.454:2015	Quality and safety requirements of oil for direct human consumption.
	Cañihua. Toasted flour	NTP 011.455:2015	
	Sacha inchi oil	NTP 151.400:2019	
	Tarwi or chocho. Debittered grain	NTP 205.090:2018	Requirements for debittering tarwi for direct human consumption or as raw material for the food industry (fresh or dehydrated grain), and the test method to determine alkaloids.
Uruguay	Amaranth. Leafy vegetables	Decreto N° 14/013	Maximum limits of inorganic contaminants in food.
	Chia seeds	Decreto N° 80/019	Incorporation of chia in the list of raw materials of vegetable origin used to produce vegetable oils and fats. Fatty acid composition for chia oil.
	Corn	Decreto N° 80/019	Incorporation of seed germ of corn in the list of raw materials of vegetable origin used for the production of vegetable oils and fats. Fatty acid composition for corn oil.
	Corn on the cob	Norm no number	Determination of phytosanitary requirements for the introduction to the country of corn on the cob (*zea mays*).
	Quinoa flour	Decreto N° 315/994	Description and specific provisions for flour.

and sale. This regulation includes the obtention and commercialization of chia, quinoa, and sacha inchi oils. In addition, Resolution 5482 Exemption modifies Resolution 2677, of 1999, which establishes import regulations for grains and other products, intended for consumption and industrialization, including amaranth imported from Mexico (BCN, 2020).

Normas Técnicas Colombianas (NTC) [Colombian Technical Standards] of the *Instituto Colombiano de Normas Técnicas y Certificación* [INCONTEC] (Colombian Institute of Technical Standards and Certification) – a private multinational organization – offers free previews of the technical standards, but full access must be paid for (INCONTEC, 2020). In Ecuador, *Normas Técnicas Ecuatorianas* (NTE) [Ecuadorian Technical Standards] of the *Servicio Ecuatoriano de Normalización* (INEN) [Ecuadorian Standardization Service] is mandatory and with open access. Nevertheless, in some cases, standards are adopted from other standardization bodies which are not available for free download due to copyright (INEN, 2020). In recent years, in both Colombia and Ecuador, some Latin and Mesoamerican crops (such as amaranth, beans, corn and quinoa) and their by-products (such as amaranth and quinoa flours, precooked corn flour, quinoa flakes and sacha inchi oil) were included in the regulations.

Uruguay has an open-access *Banco Electrónico de Datos Jurídicos Normativos* [Electronic Bank of Legal Normative Data] of the *Dirección Nacional de Impresiones y Publicaciones Oficiales* (IMPO) [National Directorate of Prints and Official Publications], which includes the national regulations and legal notices (IMPO, 2020). The databank contains some requirements and specifications for Latin and Mesoamerican crops and their products such as amaranth leafy vegetables, chia seeds, corn and quinoa flour.

In Brasil, *Agência Nacional de Vigilância Sanitária* (ANVISA) [National Agency of Health Surveillance] is a regulatory body linked to the Ministry of Health, whose aim is to promote public health protection through sanitary control of the production and commercialization of products and services subject to sanitary regulation, including manufacturing, facilities and processes (ANVISA, 2020). In Paraguay, *Instituto Nacional de Alimentación y Nutrición* (INAN) [National Institute of Food and Nutrition] is technically responsible for the implementation and development of the National Food and Nutrition Plan that integrates human, physical and administrative resources, in the areas of standardization, laboratory control, medical care, food education and control of food and nutrition outlets (INAN, 2020).

Paraguay also has the *Instituto Nacional de Tecnología, Normalización y Metrología* (INTN) [National Institute of Technology, Standardization and Metrology] with a catalog of standards available on the website (INTN, 2020).

Venezuela harmonized the national standards with *Codex Alimentarius* standards. In addition, the *Comisión Venezolana de Normas Industriales*

(COVENIN) made its transition to *Fondo para la Normalización y Certificación de la Calidad* (FONDONORMA) which was created with the aim of supporting the programs established by the Ministry of Public Works in the matter of Standardization and Quality Certification. It was not possible to access the online catalog of FONDONORMA (2020); and legislation of Andean products was not found in the COVENIN database (COVENIN, 2020). However, even though the database of regulations in these countries is quite complete, no regulations were found for Andean crops or products. Possibly, their elaboration, control and commercialization are legislated by non-public regulations or are guided by the *Codex Alimentarius.*

In Mexico, the *Sistema Integral de Normas y Evaluación de la Conformidad* (SINEC) [Integral System of Standards and Conformity Assessment] contains the *Normas Mexicanas* (NMX) [Mexican Standards], where test methods for some Latin and Mesoamerican products are included; for example black turtle bean (NMX-FF-038-SCFI-2013); chia seed oil (NMX-F-592-SCFI-2017); nixtamalized corn flours (NMX-F-046-SCFI-2018); and popped amaranth grain (NMX-FF-116-SCFI-2010) (Secretaría de Economía de Mexico, 2020).

In Panamá, *Autoridad Panameña de Seguridad de Alimentos* [Panamanian Food Safety Authority] and *Normas Nacionales para la Importación de Alimentos* ([National Standards for Food Import] establishes an extension to present free sale certificates and analyses of raw materials, ingredients and industrialized or processed foods intended for human consumption (AUPSA, 2021). For example, the standards establish requirements for the importation of amaranth from the USA; chia from Argentina, Bolivia, Ecuador, Mexico, Nicaragua, Paraguay and Peru; and quinoa from Bolivia, Colombia, Spain, the USA, Italy, Mexico, Paraguay and Peru.

The *Ministerio de Salud de Costa Rica* [Ministry of Health of Costa Rica] authorized Decree 27980, which sets the maximum level of aflatoxins in corn and beans, among others (Ministerio de Salud de Costa Rica, 2021).

On the other hand, the *Cámara Guatemalteca de Alimentos y Bebidas* (CGAB) [Guatemalan Chamber of Food and Beverages] was started by the action of visionary and innovative companies to promote and enhance the union of the country's food and beverage industry; but nothing about Andean crops was found in the database (CGAB, 2021). However, through Agreement 214–2002 the National Committee of the *Codex Alimentarius* of Guatemala was created, which promotes the harmonization of national regulations on food safety and international trade with the standards, guidelines and recommendations defined and established by the *Codex Alimentarius*. Likewise, Honduras and Nicaragua adopted the *Codex Alimentarius* in March 1992 and through Decree 99–2002, respectively.

El Salvador and Belize have the *Organismo Salvadoreño de Reglamentación Técnica* (OSARTEC) [Salvadoran Agency for Technical Regulation] and Belize

Bureau of Standards as responsible for food legislation, respectively (OSARTEC, 2021; BBS, 2021) but nothing was found about Latin and Mesoamerican crops.

11.9 LATIN AND MESOAMERICAN CROPS' LEGAL FRAMEWORK IN IMPORTING COUNTRIES

The European Union (EU) is Latin America's top investor and some of their agreements include the Global Agreement with Mexico, Association Agreements with Chile and the Central America countries, and Trade Agreement with Peru and Colombia. Additionally, the EU maintains structured dialogues and regular meetings with Latin-American economic blocs (Andean Community, Mercosur, Sistema de Integración Centroamericano [SICA]), which reflects the European desire to support regional integration (EU, 2016).

The European Food Safely Authority (EFSA) was established as a source of scientific advice and communication on risks associated with the food chain. EFSA guidance documents are regularly updated, and the EFSA's scientific advice helps to protect consumers, animals and the environment from food-related risks (EFSA, 2020). The regulations and guidance documents in the nutrition area include the regulatory framework on novel foods, in which some Andean products are included, such as chia seeds, chia oil and sacha inchi oil.

Chenopodium quinoa and varieties of amaranth are mentioned in the EU Novel Food Catalog (2021). Furthermore, Regulation EU 2015/2283 of the European Parliament and of the Council on novel foods has authorized the commercialization of chia oil, establishing the labeling denomination of its by-products (such as fats and oils, puree and food supplements) (EU, 2017). In addition, the resolution has set the labeling of chia-contained products, such as prepackaged chia seeds, bread, baked products, breakfast cereals, seed mixes, fruit juices, and fruit/vegetable blend beverages, fruit spreads and yogurt. In addition, the EU Regulation 2020/500, in 2020, authorized the placing on the market of partially defatted chia seed powders as novel foods under Regulation EU 2015/2283 of the European Parliament and of the Council and amending EU 2017/2470 (European Union, 2020). On the other hand, in the list of novel foods of the EU 2017/2470, sacha inchi oil was authorized to be placed on the market.

Food safety control in relation to quinoa as a food and food ingredient was discussed and evaluated through an examination of the current statutory provisions at UK and EU levels. Currently, neither the EU nor the UK requires any specific legislation for quinoa if it complies with the provisions of the Food Safety Act 1990 as amended and Regulation (EC) 178/2002 on the general principles and requirements of food law and procedures in matters of food safety (Ojinnaka, 2016).

In the USA, the *Guidance Industry 2018 – Application of the Foreign Supplier Verification Program Regulation to Importers of Grain Raw Agricultural Commodities* includes imported corn, amaranth and quinoa grains (Food and Drug Administration (FDA), 2020).

The *National Standard of China* includes the following paid-access legislation: Milled quinoa (China Food Industry Standards LS/T 32-45-2015) and quality grading of the seeds (*Amaranthus hypochondriacus* L.) (China National Standards GB/T 26615–2011) (China National Standards, 2020).

The *Food Safety and Standards Authority of India* (FSSAI) include definitions and requirements for amaranth, chia seeds, and quinoa (FSSAI, 2020). On the other hand, annexure of regulation 1-1764/FSSAI/Imports/2018(Part-1) authorizes the importation of non-transgenic food crops, including bean and maize. Besides, a list of approved products/ingredients under the Food Safety and Standards (Approval for Non-Specified Food and Food Ingredients Regulations, 2017) mentions Inca Inchi (*Plukenetia Volubilis*) (23/Std/PA/FSSAI/2018) of the Company Maxcure Nutravedics, a chia protein (26/Std/PA/FSSAI/2019) and chia fiber powder (27/Std/PA/FSSAI/2019), both Company Amway India Enterprises.

Food Standards Australia and New Zealand (FSANZ) has established that amaranth seeds meet the primary food authorization criteria and do not require further evaluation, since no safety problems were identified after a preliminary risk profiling exercise (Consultation Paper Proposed Future Regulation of Nutritive Substances and Novel Foods in FSANZ, 2020). On the other hand, quinoa and chia are mentioned in Schedule 20 (Proposal M1013 Schedule 20) which includes maximum residue limits for agricultural and veterinary chemicals.

CONCLUSIONS

It is known that responsible trade promotes the development of regions and favors socio-economic growth. In particular, Latin and Mesoamerica are experiencing a spectacular expansion of its agricultural frontier. As a result, social, economic and ecological transformations are occurring.

Because of this, it is necessary to find appropriate strategies that accompany the development of the region in the framework of the modernization of productive chains that the world is experiencing. Therefore, the experience of the quinoa boom provides valuable information. In addition, the strategies must consider the agricultural tradition of communities and what is appropriate for local agricultural systems and food sovereignty.

Legislation is essential for proper product commercialization and, particularly in regard to food, results vital to guarantee safety. Furthermore, treaties are necessary to ensure the sustainable use of genetic diversity and to protect resources; many Ibero-American countries adhered to the Nagoya Protocol, for example, in pursuit of this objective.

This chapter presents the compilation of the legislation regarding Latin American crops in different countries of Latin America and other regions of the world. In recent years, advances in legislation allowed their commercialization in international markets. Nevertheless, there is still a long way to go both for the harmonization of existing laws and for the issuance of laws that accompany the development of new foods made with Latin and Mesoamerican seeds and grains.

ACKNOWLEDGMENTS

This work was supported by grant laValSe-Food-CYTED (Ref. 119RT0567); Consejo Nacional de Investigaciones Científicas y Técnicas (CONICET) and Universidad Nacional de Jujuy (Argentina) and Food4ImNut Food4ImNut PID2019-107650RB-C21 funded by MCIN/AEI/10.13039/501100011033, Spain.

REFERENCES

Altieri, M., and V. M. Toledo. 2011. The agroecological revolution in Latin America: Rescuing nature, ensuring food sovereignty and empowering peasants. *Journal of Peasant Studies* 38(3): 587–612.

ANVISA (Agência Nacional de Vigilância Sanitária). 2020. www.gov.br/anvisa/pt-br/setorregulado/regularizacao/alimentos (accessed December 19, 2020).

AUPSA (Autoridad Panameña de Seguridad de Alimentos). 2021. www.aupsa.gob.pa/ (accessed January 15, 2021).

Barbier, E., and J. Hochard. 2014. *Poverty and the Spatial Distribution of Rural Population.* World Bank Policy Research Working Paper 7101. Washington, DC: World Bank.

Bazile, D., D. Bertero, and C. Nieto. 2015. *State of the Art Report on Quinoa around the World in 2013.* Santiago de Chile: Montpelliere: Rome: FAO/CIRAD. www.fao.org/583/a-i4042e.pdf.

BBS (Belize Bureau of Standards). 2021. https://bbs.gov.bz/standards-catalogue/ (accessed January 15, 2021).

BCN (Biblioteca del Congreso Nacional de Chile). 2020. www.bcn.cl/leychile/ (accessed December 19, 2020).

Bedoya-Perales, N., G. Pumi, M. Talamini, and A. Padula.2018. The quinoa boom in Peru: Will land competition threaten sustainability in one of the cradles of agriculture? *Land Use Policy* 79: 475–480.

Bose, P. 2017. Land tenure and forest rights of rural and indigenous women in Latin America: Empirical evidence. *Women's Studies International Forum* 65: 1–8.

CAA (Código Alimentario Argentino). 2020. Administración Nacional de Medicamentos, Alimentos y Tecnología Médica (ANMAT) – Ministerio de Salud. www.argentina. gob.ar/anmat/codigoalimentario (accessed December 19, 2020).

CBD (Convention on Biological Diversity). 2011. Nagoya Protocol on access to genetic resources and the fair and equitable sharing of benefits arising from their utilization to the convention on biological diversity, Secretariat of the Convention on Biological Diversity, Montreal, Convention on Biological Diversity United Nations, ISBN: 92-9225-306-9, Quebec, Canada. www.cbd.int/abs/.

CBD (Convention on Biological Diversity). 2020. Parties to the Nagoya Protocol. Convention on Biological Diversity. www.cbd.int/abs/nagoya-protocol/signator ies/, accessed December 10, 2020.

CGAB (Cámara Guatemalteca de Alimentos y Bebidas). 2021. http://cgab.org.gt/ (accessed January 15, 2021).

China National Standards. 2020. www.gbstandards.org/ (accessed December 19, 2020).

Codex Alimentarius. 2020. www.fao.org/fao-who-codexalimentarius/es/ (accessed December 19, 2020).

COVENIN (Comisión Venezolana de Normas Industriales). 2020. www.sencamer.gob.ve/ sencamer/action/normas-find (accessed December 19, 2020).

EFSA (European Food Safety Authority). 2020. www.efsa.europa.eu/ (accessed December 19, 2020).

EU (European Union). 2016. EU trade relations with Latin America: Results and challenges in implementing the EU-Colombia/Peru Trade Agreement. Doi: 10.2861/056827.

EU (European Union). *Official Journal of the European Union*. 2017. https://eur-lex.europa. eu/legal-content/EN/TXT/?uri=celex%3A32017R2470.

EU (European Union). 2020. European Parliament and of the Council and amending Commission Implementing Regulation EU 2017/2470. https://ec.europa.eu/food/ safety/novel_food/authorisations/union-list-novel-foods_en (accessed December 19, 2020).

EU (European Union. 2021. *Novel Food Catalogue*. https://ec.europa.eu/food/safety/nov el_food/catalogue/search/public/index.cfm (accessed June 11, 2021).

FAO (Food and Agriculture Organization). 2020. *The State of Food Security and Nutrition in the World*. www.fao.org/3/ca9692en/online/ca9692en.html.

FAO/WTO (Food and Agriculture Organization)/(World Trade Organization). 2018. Commerce and Standards Food. www.fao.org/3/i7407es/I7407ES.pdf.

FDA (Food and Drug Administration). 2020. www.fda.gov/food (accessed December 19, 2020).

FSANZ (Food Standards Australia New Zealand). 2020. www.foodstandards.gov.au/ Pages/default.aspx (accessed December 19, 2020).

FSSAI (Food Safety and Standards Authority of India). 2011. Fondo para la Normalización y Certificación de la Calidad (FONDONORMA). www.fssai.gov.in/upload/uplo adfiles/files/Compendium_Food_Additives_Regulations_08_09_2020-compres sed.pdf.

FSSAI (Food Safety and Standards Authority of India. 2020. www.fssai.gov.in/cms/about-fssai.php (accessed December 19, 2020).

IBNORCA (Instituto Boliviano de Normalización y Calidad). 2020. www.ibnorca.org/es (accessed December 19, 2020).

IMPO (Dirección Nacional de Impresiones y Publicaciones Oficiales). 2020. Banco Electrónico de Datos Jurídicos Normativos [Electronic Bank of Normative Legal Data]. www.impo.com.uy/ (accessed December 19, 2020).

INACAL (Instituto Nacional de Calidad). 2020. Normas Técnicas Peruanas (NTP) [Peruvian Technical Standards]. www.inacal.gob.pe/principal/categoria/ntp (accessed December 19, 2020).

INAN (Instituto Nacional de Alimentación y Nutrición). 2020. www.inan.gov.py/site/ (accessed December 19, 2020).

INCONTEC (Instituto Colombiano de Normas Técnicas y Certificación) [Colombian Institute of Technical Standards and Certification]. 2020. Normas Técnicas Colombianas (NTC) [Colombian Technical Standards]. https://tienda.icontec.org/ sectores/tecnologia-de-alimentos.html (accessed December 19, 2020).

INEN (Servicio Ecuatoriano de Normalización). 2020. Normas Técnicas Ecuatorianas (NTE). http://apps.normalizacion.gob.ec/descarga/ (accessed December 19, 2020).

INTN (Instituto Nacional de Tecnología, Normalización y Metrología) [National Institute of Technology, Standardization and Metrology]. 2020. http://normas.intn.gov.py/ (accessed December 19, 2020).

Izquierdo, J., and W. M. Roca. 1998. Under-utilized Andean food crops: Status and prospects of plant biotechnology for the conservation and sustainable agricultural use of genetic resources. *ISHS Acta Horticulture* 457: *Symposium on Plant Biotechnology as a Tool for the Exploitation of Mountain Lands.*

Katz, E. 2002. *La "feminización" de la economía rural en América Latina: evidencia, causas y consecuencias.* Rome: United Nations Food and Agriculture Organization (FAO)

Kerssen, T. 2015. Food Sovereignty: Convergence and contradictions condition and challenges. Food sovereignty and the quinoa boom: Challenges to sustainable re-peasantisation in the southern Altiplano of Bolivia. *Third World Quarterly* 36(3).

MINAGRI (Ministerio de agricultura y riego). 2018. Viceministerio de políticas agrarias. Dirección general de políticas agrarias. 2018. Manejo de granos andinos https:// webcache.googleusercontent.com/search?q=cache:ITAXFyRXGa0J:https://www. minagri.gob.pe/portal/analisis-economico/analisis-2019%3Fdownload%3D14 580:manejo-agronomico-de-granos-andinos%26start%3D20+&cd=2&hl=es-419&ct=clnk&gl=ar.

MINSAL (Ministerio de Salud de la República de Chile). 2019. Reglamento sanitario de los alimentos – DTO. N° 977/96. Publicado en el Diario Oficial de 13.05.97. https:// dipol.minsal.cl/wp-content/uploads/2020/11/RSA-DECRETO_977-96-actualiz ado-a-noviembre-2019.pdf,

Ministerio de Salud de Costa Rica. 2021. www.ministeriodesalud.go.cr/ (accessed January 15, 2021).

Muñoz, M. T., F. I. Ocampo, and I. F. Parra. 2019. Socioeconomic characterization of the family production unit and the importance of the cultivation of chia (*Salvia hispanica* L.) in the municipalities of Atzitzihuacán and Tochimilco, Puebla, Mexico. *Acta Universitaria* 29: e2494.

Ofstehage, A. 2011. Nusta Juira's gift of quinoa: Peasants, trademarks, and intermediaries in the transformation of a Bolivian commodity economy. *Anthropology of Work Review* 32(2): 103–114.

Ojinnaka, D. 2016. Legislative control of quinoa in the United Kingdom and European Union. *Madridge Journal of Food Technology* 1(1): 53–57. doi: 10.18689/mjft.2016-108.

OPS (Organización Panamericana de la Salud). 2020. *Development of Food Legislation – Guidelines for the Development of Legislative and Executive Regulations in Food Control Systems.* www.paho.org/hq/index.php?option=com_content&view=article&id=10708:2015-desarrollo-de-la-legislacion-alimentos&Itemid=41373&lang=es.

OSARTEC (Organismo Salvadoreño de Reglamentación Técnica). 2021. www.osartec.gob.sv/ (accessed January 15, 2021).

Pengue, W. 2004. https://grain.org/es/article/entries/413-a-short-history-of-farming-in-latin-america

Pickersgill, B. 2007. Domestication of plants in the Americas: Insights from mendelian and molecular genetics. *Annals of Botany* 100(5): 925–940.

Popkin, O., and T. Reardon. 2018. Obesity and food system transformation in Latin America. *Obesity Reviews* 19(8): 1028–1064.

Robinson, B., M. Holland and L. Naughton-Treves. 2014. Does secure land tenure save forests? A review of the relationship between land tenure and tropical deforestation. *Global Environmental Change* 29: 281–293.

Secretaría de Economía de Mexico. 2020. www.2006-2012.economia.gob.mx/comunidad-negocios/normalizacion/catalogo-mexicano-de-normas (accessed December 19, 2020).

Tapia, M. 1993. *Semillas andinas, el banco de oro.* Lima: CONCYTEC.

Tschopp, M., S. Bieri, and S. Rist. 2018. Quinoa and production rules: How are cooperatives contributing to governance of natural resources? *International Journal of the Commons* 12(1): 402–427.

Index

A

Abiotic stress, 66
  on crop yield, 58
  quinoa, 61
Above sea level (a.s.l.), 5
Acid-soluble whey fraction, 237
Acrylamide, 185
Agência Nacional de Vigilância Sanitária
    (ANVISA), 418
Agricultural production systems
  diet shifts, 404–405
  farmers' situations, 403–404
  food system transformation, 404–405
Agro-biodiversity, 3
  importance of, 2–3
  neglected and underutilized species
    (NUS), 3
Agro-economic context, Latin and
    Mesoamerican region, 402–403
Albumins, 164, 323
Alkaline cooking, 322
Alkaloid extraction, Andean lupine
    (*Lupinus mutabilis* Sweet), 236
Alkaloids, 17
Amaranth (*Amaranthus cruentus* L./*A.*
    hypochondriacus L.), 22, 27, 55,
    63–65, 98–100, 120, 260, 304
  baking properties of, 169
  commercial gluten free foods, 171
  cultivation, new regions of, 99–100
  present/future uses, 98–99
  protein digestibility of, 381
Amaranthaceae species, 11
Amaranth flour
  bread, characteristics of, 262–263
  fiber extractions, 221
  milling, 219–220

oil extraction, 220–221
in pasta processing, 99
protein isolation, 221–222
proteins, techno-functional properties
    of, 222–223
starches, 143, 223
  granules of, 143
  modification, 224
  physico-chemical/thermal/
    rheological properties of, 223
techno-functional properties of,
    222–223
Amaranth foods, 271
Amaranth grains, 305
  lipids of, 309
  thin layer chromatography (TLC), 381
Amaranth lipids, 220
Amaranth native starch, 224
Amaranth oil, 220
Amaranth protein, 389, 390
  Amaranth hypochondriacus protein
    hydrolysates, 388
  Amaranth hypochondriacus protein
    peptides, 391
  Amaranth hypochondriacus proteins,
    390
  emulsion formation/stabilization of,
    222
Amaranth seed, 98, 222
Amaranthus caudatus, 65, 219
Amaranthus tricolor, 63
Amaranth wet grinding, 220
Amino acids profile, 141
  anti-nutritional compounds, 319–320
  bioactive compounds, 317–319
    modifications of, 322–328
  cooking methods/malting, 329
  folate nutrient retention factors, 329

For Product Safety Concerns and Information please contact our EU
representative GPSR@taylorandfrancis.com
Taylor & Francis Verlag GmbH, Kaufingerstraße 24, 80331 München, Germany

www.ingramcontent.com/pod-product-compliance
Lightning Source LLC
Chambersburg PA
CBHW060743220326
41598CB00022B/2307

9 781032 367392